PHYSICAL AND DYNAMICAL
METEOROLOGY

PHYSICAL & DYNAMICAL METEOROLOGY

By

DAVID BRUNT, M.A., Sc.D., F.R.S.

*Professor of Meteorology in the University of London, late
Superintendent of the Army Meteorological Services,
Air Ministry, London*

CAMBRIDGE
AT THE UNIVERSITY PRESS
1952

CAMBRIDGE UNIVERSITY PRESS
Cambridge, New York, Melbourne, Madrid, Cape Town,
Singapore, São Paulo, Delhi, Tokyo, Mexico City

Cambridge University Press
The Edinburgh Building, Cambridge CB2 8RU, UK

Published in the United States of America by Cambridge University Press, New York

www.cambridge.org
Information on this title: www.cambridge.org/9781107601437

© Cambridge University Press 1952

First edition 1934
Second edition 1939
Reprinted 1941, 1952
First paperback edition 2011

A catalogue record for this publication is available from the British Library

ISBN 978-1-107-60143-7 Paperback

PREFACE TO FIRST EDITION

My aim in writing this volume has been to provide a textbook of physical meteorology suitable for postgraduate students, and so to meet a need which has been felt in English-speaking countries for many years. At the same time I hope that such a book may be useful to those who are engaged in the profession of meteorology.

I have endeavoured to give an account of theoretical meteorology, especially of the physical aspects of the subject, which shall represent the present state of our knowledge as completely as possible. Where theory has so far failed to explain the observed phenomena, as for example in dealing with the general circulation of the atmosphere and the travelling cyclone and anticyclone, the treatment is limited to a description of the observed phenomena, combined with such theoretical notes as may help to an understanding of the phenomena.

My grateful thanks are due to Dr G. C. Simpson, Mr C. K. M. Douglas, Mr H. W. L. Absalom, and Mr O. G. Sutton of the Meteorological Office, London, and to Professor Sir Gilbert Walker, for helpful criticisms of the book in the typescript stage. I am especially indebted to Mr Douglas, who not only read the whole of the book in typescript, but also placed at my disposal his encyclopaedic knowledge of synoptic charts; and to Mr Sutton for considerable assistance in the preparation of the chapters on turbulence. I am deeply indebted to Mr C. E. Britton for assistance in proof-reading, and for the preparation of the index.

To Sir Napier Shaw and the Cambridge University Press I am indebted for permission to reproduce, in figures 1 to 9, 25, 27, 29, 69, 71, and 103 to 111,* diagrams from the *Manual of Meteorology*.

To Sir Napier Shaw and Messrs Macmillan and Co. I am indebted for permission to reproduce in figure 67 † a diagram from the *Dictionary of Applied Physics*.

To the Royal Meteorological Society I am indebted for permission to reproduce, and for the loan of blocks for, figures 18, 24, 30, 31,

---

* In the present edition figures 25 and 69 are omitted; the remainder are numbered 1 to 9, 30, 32, 77, and 110 to 118 respectively.

† Now figure 71.

35, 36, 60, 80, and 90 to 98; * also for permission to reproduce from a paper in the *Quarterly Journal* portions of sections 79 and 80 †below.

To H.M. Stationery Office I am indebted for permission to reproduce as figures 82 to 84 ‡ diagrams from M.O. Geophysical Memoirs, No. 50, by J. Bjerknes.

To the Director of the Meteorological Office I am indebted for permission to copy portions of autographic records shown in figure 78.§

It is difficult to express at all adequately my debt to the officials of the Cambridge University Press for their unvarying courtesy and helpfulness at all stages of the printing.

DAVID BRUNT

20 *August* 1934

* Now figures 18, 24, 33, 35, 39, 40, 62, 73, 96 to 104 respectively.
† Now sections 89 and 90.
‡ Now figures 88 to 90.
§ Now figure 84.

PREFACE TO THE SECOND EDITION

In the Preface to the first edition, the aim of this book was stated to be to provide a textbook of physical meteorology suitable for postgraduate students. The gratifying reception accorded to it shows that it has to some extent filled a gap in meteorological literature. The changes made in the second edition have been dictated not so much by advances made in the subject during the last five years as by the experience gained in using this book as a textbook in classes at the Imperial College of Science and Technology. A few portions of the original text, which now appear to have little bearing on the modern development of meteorology, have been omitted. Additions have been made in the present chapters III, IV, VI, XVI, XVII and XIX; and chapter XII, dealing with Turbulence in the Atmosphere, has been completely re-arranged. Chapters IX and XI of the first edition have been combined to form the present chapter IX.

A few reviewers of the first edition complained of the number of equations it contained. I make no apology for having perhaps added to the number of equations, as I take the view that meteorology should aim at being a metric science wherever possible, and that no physical theory can be regarded as wholly satisfactory which cannot be expressed in mathematical form.

This book makes no claim to be a textbook of weather forecasting. The subject of weather forecasting would alone fill a book of the size of the present volume, and even then it is doubtful whether the art of forecasting would be learned by its perusal. While forecasting requires a background of physical theory, it is an art which is acquired by practice rather than by the study of pure theory. It is hoped that the necessary background of theory is provided in this volume.

While I hope that the new edition will prove an improvement on the first, I cannot hope that it has entirely escaped errors and uncertainties. To reduce the voluminous literature of meteorology to what might be called a standard textbook form is no inconsiderable task.

For information as to errors in the first edition I am indebted to my own students at the Imperial College of Science and Technology;

to Mr J. Durward, Director of the Meteorological Service of Iraq; and to Professor H. Arakawa, of Tokyo. In the revision of chapter XII, on Turbulence, I have again received valuable assistance from Mr O. G. Sutton. I have also to acknowledge my indebtedness to the Royal Meteorological Society for permission to reproduce from the *Quarterly Journal* the diagrams shown in figs. 26, 27, 28 and 76. My debt to the officials of the Cambridge University Press for their unfailing courtesy and assistance at all stages of the production of this book cannot be adequately expressed.

DAVID BRUNT

20 January 1939

Advantage has been taken of the reprinting of the second edition, to correct some minor errors.

10 March 1941

TABLE OF CONTENTS

LIST OF ILLUSTRATIONS *page* xvii

SOME USEFUL CONSTANTS AND UNITS xxii

INTRODUCTION xxiii

CHAPTER I

THE FACTS WHICH CALL FOR EXPLANATION. A SKETCH OF THE SURFACE DISTRIBUTION OF THE METEOROLOGICAL ELEMENTS OVER THE GLOBE

Section

1	Temperature	1
2	Pressure	6
3	Wind	12
4	Rainfall	13
5	The general circulation and the local circulations	16
6	Surfaces of separation between winds of different origin	16
7	The observed distribution of temperature in the vertical	17
8	Correlation between different variables in the free air	21
9	The diurnal variation of meteorological factors	22

CHAPTER II

SOME STATICAL AND THERMAL RELATIONSHIPS

10	The earth	28
11	Geopotential	29
12	Composition of the atmosphere	30
13	The fundamental gas equation	30
14	The density of water-vapour and of damp air; virtual temperature	31
15	Vapour-pressure and its control by temperature	31
16	Pressure and its variation with height	33
17	Barometric altimetry	35
18	The specific heat of air	36
19	The adiabatic equation	38
20	Potential temperature	38
21	The adiabatic lapse-rate; vertical stability for dry air	39
22	Stability in a moist atmosphere (unsaturated)	41
23	The effect of vertical motion on the lapse-rate	44
24	The dry adiabatic lapse-rate in geodynamic units	45
25	The variation of density with height; the auto-convection gradient	45
26	The effect of mixing two masses of damp air	47

CHAPTER III

WATER-VAPOUR IN THE ATMOSPHERE. THE ASCENT OF DAMP AIR

Section

27 The definition of saturation *page* 49
28 Condensation on nuclei 51
29 Supercooling of water in the atmosphere 54
30 Condensation at temperatures below the freezing-point 54
31 Rain, snow and hail 55
32 Statement of the general meteorological problem 56
33 Stage (*a*). Unsaturated air 59
34 Stage (*b*). Saturated air above 0° C, or the rain stage 59
35 Stage (*c*). The freezing or hail stage 60
36 Stage (*d*). Condensation below 0° C. The snow stage 60
37 The Neuhoff diagram 61
38 The effect of the loss of precipitated water and ice 63
39 The lapse-rate of ascending damp air 65
40 Stability of saturated air 66

CHAPTER IV

THERMODYNAMICS OF THE ATMOSPHERE

41 The concept of entropy 69
42 The effect of conduction or mixing on the total entropy 74
43 Formulae for entropy 75
44 Efficiency of a heat engine 75
45 Entropy-temperature diagrams or T-ϕ diagrams 76
46 The tephigram 77
47 The use of the Neuhoff diagram or the tephigram to evaluate the energy liberated during the ascent of air 79
48 Latent heat of evaporation 82
49 The entropy of saturated air 83
50 The entropy of a mixture of damp air, water and ice 85
51 The thermodynamics of the wet- and dry-bulb hygrometer 85
52 Some thermodynamical propositions relating to the wet-bulb temperature 87
53 Variation of dry- and wet-bulb temperatures, and of dew-point in air ascending adiabatically 89
54 The wet-bulb potential temperature 90
55 Practical use of wet-bulb potential temperature 92
56 Equivalent temperature and equivalent potential temperature 94
57 Alternative definition of equivalent potential temperature 96
58 Potential instability and the variation of wet-bulb potential temperature or of equivalent potential temperature with height 98

Section

59 The Rossby diagram *page* 98
60 Latent instability 99
61 The general conditions of stability in the atmosphere 100
62 The Clausius-Clapeyron equation 101
63 The functional relation of saturation vapour-pressure to temperature 103

CHAPTER V
RADIATION

64 Radiation of light and heat 105
65 Kirchhoff's law 105
66 Black-body radiation 106
67 Grey radiation 106
68 Planck's law 106
69 Stefan's law 108
70 Range of wave-lengths in radiation from bodies at different temperatures 109
71 The spectral distribution of solar radiation 109
72 The solar constant 111
73 The variation of insolation with season and latitude 112
74 The earth's albedo 112
75 The coefficient of absorption 113
76 Absorption of long-wave radiation in the atmosphere 114
77 The absorption spectrum of water-vapour 114
78 The effect of air pressure, partial vapour pressure, and temperature, on the water-vapour spectrum 118
79 The absorption spectrum of liquid water 120
80 Diffuse reflexion and scattering 122

CHAPTER VI
RADIATION IN THE TROPOSPHERE

81 General survey of radiation phenomena in the atmosphere 124
82 Water-vapour as a controlling agent in atmospheric absorption and radiation 126
83 The heat balance of the atmosphere 126
84 The equations of radiative transfer of heat 128
85 An equation for radiative transfer 129
86 The diurnal variation of temperature as affected by radiation 133
87 Limitations of the discussion of §§ 85 and 86 135
88 The interchange of long-wave radiation between the atmosphere and the earth's surface. Nocturnal radiation with clear skies 136
89 Nocturnal radiation: conditions within the surface layers of the ground 138
90 Nocturnal radiation with cloudy skies 142
91 Conditions which favour the nocturnal cooling of the ground 145

CHAPTER VII

Radiative equilibrium and the stratosphere

Section

92 Statement of the problem   *page* 147
93 The difficulties in the way of mathematical treatment   149
94 Emden's solution for grey radiation   149
95 The effect of absorption of ultra-violet radiation by ozone on the temperature distribution in the stratosphere   152
96 The heat balance of the atmosphere   154
97 The vertical transport of heat in the atmosphere   158

CHAPTER VIII

The general equations of motion

98 The general equations of motion in spherical polar co-ordinates   160
99 Expression of the general equations of motion in Cartesian co-ordinates   163
100 The equations in hydrodynamical form   165
101 The special case of horizontal motion   166
102 The deviating force of the earth's rotation   166
103 The order of magnitude of the terms in the equations of motion   168
104 The equation of continuity   168
105 The forces X, Y, Z 169
106 Barotropic and baroclinic fluids   171
107 Circulation and vorticity   172
108 Vortex lines, filaments and sheets   175
109 The generation of circulation in the atmosphere   176
110 Velocity potential   183
111 The stresses in a viscous fluid   184

CHAPTER IX

Motion under balanced forces: the gradient wind

112 Conditions for steady motion   187
113 The gradient wind equation   188
114 Comparison of the gradient wind with observed winds   190
115 The solution of the gradient wind equation   191
116 Direct derivation of the gradient wind equation   193
117 The effect of changing pressure distribution   194
118 An alternative form of the equations of motion in two dimensions   196
119 The variation of pressure gradient and geostrophic wind with height   196
120 Some applications of the above formulae   198

CHAPTER X
SURFACES OF DISCONTINUITY

Section
121 The slope of a surface of discontinuity with steady motion *page* 203
122 Discontinuity of velocity alone; density uniform 204
123 Discontinuity of density alone 205
124 Approximate form of equation (6) 205
125 General nature of the results derived above 206
126 Extension to motion in small circles, with a surface of discontinuity 207
127 The form of the isobaric surfaces 208
128 The general equations for the slope of surfaces of discontinuity 208

CHAPTER XI
THE GENERAL ASPECTS OF TURBULENCE

129 Stream-line and turbulent flow 212
130 The Reynolds stresses 213
131 Dynamical similitude; the Reynolds number 214
132 The partition of eddy energy 216
133 The nature of eddies 219
134 The nature of convection 221
135 Stability of motion 222

CHAPTER XII
TURBULENCE IN THE ATMOSPHERE: THE EDDIES AS DIFFUSING AGENCIES

136 Diffusion by eddies and molecular diffusion 224
137 The vertical transfer of heat by turbulence 224
138 Richardson's treatment of diffusion by eddies 226
139 Application of Taylor's results to the atmosphere 227
140 The effect of turbulence on the diurnal variation of temperature 229
141 Taylor's discussion of the eddy transfer of momentum contrasted
    with that of Schmidt and Prandtl 230
142 Extension to three dimensions 234
143 Comparison of the momentum-transport and the vorticity-trans-
    port theories 235
144 Eddy diffusivity and *Austausch* for different properties 236
145 Stability and the criterion of turbulence 237
146 The maintenance of super-adiabatic lapse-rates in the atmosphere 243
147 Prandtl's theory of the Mischungsweg 244
148 Development of Prandtl's theory for the surface layers 247
149 Variation of wind with height in the surface layers; Rossby and
    Montgomery's extension to stable atmospheres 249

Section

150 The variation of wind with height   *page* 251
151 The internal friction due to turbulence   255
152 The height to which the effect of surface turbulence extends   256
153 Total flow along and across the isobars   256
154 The boundary layer   257
155 Skin friction at the ground   259
156 The diurnal variation of wind at different levels   261
157 Diffusion by continuous movement   263
158 The form of R_ξ 265
159 Sutton's extension of Taylor's theory   266
160 The variation of wind with height on Sutton's theory   267
161 Recent developments in the study of turbulence   269
162 The derivation of the hygrometer equation   269
163 Evaporation from the surface of water   270

CHAPTER XIII
The classification of winds

164 The terms in the equations of motion   276
165 The application of the classification   278
166 Geostrophic winds   279
167 Antitriptic winds   280

CHAPTER XIV
The transformations of energy in the atmosphere

168 Incoming radiant energy   282
169 The classification of energy   282
170 Transformations of energy in the atmosphere   283
171 The kinetic energy of the atmosphere, and its dissipation by turbulence   285
172 Comparison of the eddy dissipation of kinetic energy with the radiation coming in from the sun   286
173 The development of circulation between sources of heat and cold   287
174 Equations of energy   288
175 Energy liberated when vertical interchange of masses occurs   291
176 Single layer in unstable equilibrium   293
177 Effect of the presence of water-vapour   295
178 Maintenance of a difference of pressure by addition of heat   296
179 A cycle of changes in the atmosphere   297

CHAPTER XV
The growth of cyclic circulations and of pressure inequalities in the atmosphere

180 The growth of cyclic circulations in the atmosphere   299
181 The formation of revolving fluid in the atmosphere   299

Section
182 The tropical cyclone and the tornado as revolving fluid   *page* 303
183 The relation of wind and isobars in a moving cyclone   306
184 The genesis of pressure inequalities   308
185 The effect on pressure changes of the presence of a surface of
       discontinuity, or of a zone of transition   309
186 Gradient winds in a stationary pressure distribution   310

CHAPTER XVI

THE IDEA OF AIR MASSES

187 The life-history of surface air currents   313
188 The classification of air masses   315

CHAPTER XVII

THE POLAR FRONT AND ITS RELATION TO THE
DEVELOPMENT OF CYCLONES

189 The depression as a wave disturbance in a surface of discontinuity 318
190 The polar front methods of analysis of charts   319
191 The formation of secondaries   334
192 Families of depressions   335
193 The regeneration of depressions   335
194 Some general aspects of polar front depressions   338
195 The drawing of fronts   340
196 Sharp and diffuse fronts   342
197 Upper air conditions above depressions   348
198 Rainfall in depressions   352
199 The vertical structure and extent of depressions   356
200 A brief review of the wave theory of the formation of extra-
       tropical cyclones   358
201 The causes of formation of depressions   360
202 The structure of depressions at high level   365
203 The energy of depressions   368
204 Kobayasi's theory of the formation of fronts in a vortex as a result
       of horizontal temperature gradients   369
205 Exner's barrier theory of depressions   370

CHAPTER XVIII

ANTICYCLONES

206 Types of anticyclones   373
207 The cold anticyclone   374
208 The warm anticyclone   377
209 Some observational data   381
210 Subsidence and divergence in anticyclones   384

CHAPTER XIX

THE GENERAL CIRCULATION OF THE ATMOSPHERE

Section
211 The surface conditions over the earth   *page* 387
212 The circulation in the upper air   388
213 The observed distribution of winds in the upper air   394
214 The problem to be solved   402
215 Some theoretical aspects of the general circulation   403
216 Jeffrey's theory of the monsoons and similar winds   409
217 The present position regarding the theory of the general circulation   413
218 Epilogue   417

APPENDIX   419
INDEX OF NAMES   423
INDEX OF SUBJECTS   425

LIST OF ILLUSTRATIONS

FIG. 1. Normal temperature at sea level, Southern hemisphere, January. (Shaw, *Manual of Meteorology*, **2**, fig. 20) . *page* 2

2. Normal temperature at sea level, Southern hemisphere, July. (Shaw, *Manual of Meteorology*, **2**, fig. 32) . . 3

3. Normal temperature at sea level, Northern hemisphere, January. (Shaw, *Manual of Meteorology*, **2**, fig. 19) . 4

4. Normal temperature at sea level, Northern hemisphere, July. (Shaw, *Manual of Meteorology*, **2**, fig. 31) . . 5

5. Normal pressure at sea level, Southern hemisphere, January. (Shaw, *Manual of Meteorology*, **2**, fig. 133) . 8

6. Normal pressure at sea level, Southern hemisphere, July. (Shaw, *Manual of Meteorology*, **2**, fig. 145) . . 9

7. Normal pressure at sea level, Northern hemisphere, January. (Shaw, *Manual of Meteorology*, **2**, fig. 132) . 10

8. Normal pressure at sea level, Northern hemisphere, July. (Shaw, *Manual of Meteorology*, **2**, fig. 144) . . 11

9. Annual variation of the mass of air in the Northern hemisphere. (Shaw, *Manual of Meteorology*, **4**, fig. 78) 12

10. Winds of the globe, January 14

11. Winds of the globe, July 15

12. Distribution of temperature in the atmosphere. (Ramanathan) 18

13. Diurnal range of temperature at Parc St Maur and at the top of the Eiffel Tower 23

14. Diurnal variation of pressure at Batavia, in the Arctic, at Aberdeen and on Ben Nevis 25

15. Diurnal variation of wind at Parc St Maur and at the top of the Eiffel Tower 25

16. Saturation vapour-pressure as a function of temperature . 32

17. The Neuhoff diagram 61

18. Isopleths of the saturated adiabatic lapse-rate. (Brunt, *Q.J. Roy. Met. Soc.* **59**) 66

19. Indicator diagram 69

20. Cycle bounded by isothermals and adiabatics . . . 70

21. T-ϕ diagram 76

22. Tephigram 77

xvii

FIG. 23. Tephigram, schematic . . . . *page* 78

24. Energy in the Neuhoff diagram and the tephigram. (Brunt, *Q.J. Roy. Met. Soc.* **57**) 80

25. Normand's temperature-height diagram 90

26. Distribution of wet-bulb potential temperature in an air-mass on two successive days. (Hewson, *Q.J. Roy. Met. Soc.* **62**) 93

27. Estimating subsidence by use of wet-bulb potential temperature. (Hewson, *Q.J. Roy. Met. Soc.* **62**) . . 93

28. Estimating ascent of air at a warm front. (Hewson, *Q.J. Roy. Met. Soc.* **63**) 94

29. Tephigram showing latent instability 99

30. Cycle for the Clausius-Clapeyron equation. (Shaw, *Manual of Meteorology*, **3**, fig. 91) 101

31. Theoretical curve of distribution of black-body radiation . 108

32. Observed intensity distribution of solar radiation. (Shaw, *Manual of Meteorology*, **3**, fig. 73) 110

33. The absorption spectrum of water-vapour. (Brunt, *Q.J. Roy. Met. Soc.* **58**) 115

34. Absorption spectrum of water-vapour according to Weber and Randall's measurements 117

35. The absorption spectrum of liquid water and the reflecting power of liquid water. (Brunt, *Q.J. Roy. Met. Soc.* **58**) 121

36. Diagram of radiative transfer 129

37. Diagram of transfer of *W*-radiation 130

38. Diurnal variation of temperature on clear nights . . 142

39. Solar and terrestrial radiation. (Simpson, *Memoirs R. Met. Soc.* **3**, No. 21) 155

40. The heat balance of the atmosphere. (Simpson, *Memoirs R. Met. Soc.* **3**, No. 21) 156

41. Polar co-ordinates 160

42. The deviating force in two dimensions 166

43. The effect of pressure gradient 169

44. Co-ordinate axes in any direction 171

45. Circulation and vorticity 173

46. Vorticity in Cartesian co-ordinates 173

47. Irrotational symmetrical motion 174

48. Circulation and vorticity in any circuit 174

49. Circulation in the earth's atmosphere 177

FIG. 50. Line and surface integrals for $-\int\frac{dp}{\rho}$ . . . *page* 178

51. The geometrical relations between the gradients of p, ρ, θ and T 180

52. The direction of apparent spin 181

53. Land and sea breezes 181

54. Motion in a small circle 188

55. Centrifugal force on an element of air 188

56. Direct derivation of the gradient wind equation . . 193

57. Surface of discontinuity 203

58. Velocities at a surface of discontinuity 210

59. Motion at a front 211

60. Mean and eddy winds 216

61. Circulation in a Bénard cell 220

62. Durst's application of Richardson's criterion. (Durst, *Q.J. Roy. Met. Soc.* **59**) 240

63. Mallock's representation of the eddy 241

64. Rosenhead's representation of the development of the eddy 242

65. The variation of wind with height; the equiangular spiral 254

66. The height at which the geostrophic wind is attained . 254

67. Margules' diagram showing warm and cold layers inverted 292

68. An unstable layer inverted 294

69. Margules' diagram for a cycle in a vertical plane . . 297

70. The same cycle represented in a tephigram . . . 297

71. An atmospheric cycle shown in the tephigram. (Shaw, *Dictionary of Applied Physics*, **3**, fig. 17, p. 81) . . 298

72. The development of revolving fluid in the atmosphere . 300

73. Revolving fluid in any cyclone. (Douglas, *Q.J. Roy. Met. Soc.* **55**) 306

74. Changes of pressure at a moving sharp surface of discontinuity 309

75. Effect of changes of curvature of isobars on pressure changes 311

76. Regions of convergence and divergence resulting from isobaric curvature 312

77. Shaw's representation of the air currents in a cyclone. (Shaw, *Manual of Meteorology*, **2**, fig. 211) . . . 314

78 *a–d*. The development of a depression at a polar front between opposing currents 319

78 *e*. The typical polar front depression 320

Fig. 79. The development of a back-bent occlusion . . . *page* 321

80. Types of occlusion 322

81. The seclusion of depressions 323

82. The development of a depression at a boundary between two westerly currents 323

83. The depression of October 22, 1932 326

84. Autographic records at Holyhead, October 21–22, 1932 . 327

85. Upper air observations at Duxford on October 22, 1932 . 328

86, 87. The depression of January 23, 1926. Charts for 7 h. and 13 h. 329

88–90. Autographic records for January 22–23, 1926, at Valentia, Holyhead and Eskdalemuir. (J. Bjerknes, *Geophysical Memoir*, **50**) 330, 331

91. An occluded depression, November 15, 1933 . . . 332

92. Upper air observations on November 15, 1933 . . 334

93. Charts for October 8–9, 1932, showing the amalgamation of a depression and a secondary, and the rotation of two centres counter-clockwise 336

94. The effects of subsidence; upper air conditions behind a line squall 344

95. Motion at cirrus levels above depressions . . . 349

96. A depression with a marked equatorial sector, April 1, 1909. (Douglas, *Q.J. Roy. Met. Soc.* **50**) 351

97. Upper air observations on April 1, 1909. (Douglas, *Q.J. Roy. Met. Soc.* **50**) 352

98. Rainfall at a front. (Brunt and Douglas, *Memoirs R. Met. Soc.* **3**, No. 22) 353

99. Form of the cold front and its effect on rainfall. (Brunt and Douglas, *Memoirs R. Met. Soc.* **3**, No. 22) . . . 354

100. Isallobars in a deepening depression. (Brunt and Douglas, *Memoirs R. Met. Soc.* **3**, No. 22) 355

101, 102. The depression of November 17, 1910. (Douglas, *Q.J. Roy. Met. Soc.* **50**) 357

103. Formation of depressions over the Western Atlantic. (Kaye and Durst, *Q.J. Roy. Met. Soc.* **58**) 362

104. Schematic pressure field of the upper perturbations after J. Bjerknes 366

105. Kobayasi's diagram to illustrate the formation of fronts. (Kobayasi, *Q.J. Roy. Met. Soc.* **49**) 370

106. Exner's barrier scheme of formation of depressions . . 371

FIG. 107. The anticyclone of March 5, 1931 . . . . *page* 375

108. Upper air observations, March 5, 1931 376

109. The distribution of pressure at different heights . . 379

110. Normal pressure at 2 km, Northern hemisphere, July. (Shaw, *Manual of Meteorology*, **2**, fig. 170). . . 390

111. Normal pressure at 4 km, Northern hemisphere, July. (Shaw, *Manual of Meteorology*, **2**, fig. 172). . . 391

112. Normal pressure at 8 km, Northern hemisphere, July. (Shaw, *Manual of Meteorology*, **2**, fig. 167). . . 392

113. Normal pressure at 2 km, Northern hemisphere, January. (Shaw, *Manual of Meteorology*, **2**, fig. 169). . . 393

114. Normal pressure at 4 km, Northern hemisphere, January. (Shaw, *Manual of Meteorology*, **2**, fig. 164). . . 394

115. Normal pressure at 4 km, Southern hemisphere, July. (Shaw, *Manual of Meteorology*, **2**, fig. 166). . . 395

116. Normal pressure at 4 km, Southern hemisphere, January. (Shaw, *Manual of Meteorology*, **2**, fig. 165). . . 396

117. Lines of flow of cirrus, December–February (van Bemmelen). (Shaw, *Manual of Meteorology*, **2**, fig. 173) . 397

118. Lines of flow of cirrus, June–August (van Bemmelen). (Shaw, *Manual of Meteorology*, **2**, fig. 174) . . 397

119. The distribution of potential temperature in the Northern hemisphere in summer and winter 415

SOME USEFUL CONSTANTS AND UNITS

For dry air the constant R in the equation $p = R\rho T$ is $2 \cdot 8703 \times 10^3$, when p is in millibars and ρ in gm cm^{-3}; when c.g.s. units are used, and p is in dynes cm^{-2}, R for dry air is $2 \cdot 8703 \times 10^6$.

For water-vapour the corresponding values of R are $4 \cdot 62 \times 10^3$ and $4 \cdot 62 \times 10^6$.

The universal gas-constant is $83 \cdot 15 \times 10^6$ ergs.

The density of dry air at $0°$ C and 1000 mb is $1 \cdot 27617 \times 10^{-3}$ gm cm^{-3}.

Specific heats of dry air: $c_p = 0 \cdot 2396$; $c_v = 0 \cdot 1707$; $\gamma = c_p/c_v = 1 \cdot 403$.

Mean molecular weight of atmospheric air $= 28 \cdot 97$.

Latent heat of evaporation of water $= 594 \cdot 9 - 0 \cdot 51 t$, where t is the temperature in $°$ C (see Table VII, p. 421).

Latent heat of fusion of ice $= 79 \cdot 7$ calories.

Specific heat of ice $= 0 \cdot 5$.

Number of gas molecules per cm^3 at $0°$ C and 1000 mb $= 2 \cdot 66 \times 10^{19}$.

Stefan's constant $\sigma = 1 \cdot 38 \times 10^{-12}$ g-cal cm^{-2} sec^{-1}
$\qquad\qquad = 0 \cdot 826 \times 10^{-10}$ g-cal cm^{-2} min^{-1}
$\qquad\qquad = 5 \cdot 77 \times 10^{-5}$ ergs cm^{-2} sec^{-1}
$\qquad\qquad = 5 \cdot 77 \times 10^{-12}$ watt cm^{-2}.

1 gramme-calorie $= 4 \cdot 186 \times 10^7$ ergs $= 4 \cdot 184$ joules.

1 watt $= 1$ joule/sec $= 10^7$ ergs sec^{-1} $= 14 \cdot 32$ cal min^{-1}.

1 kilowatt $= 10^{10}$ ergs sec^{-1} $= 1\frac{1}{3}$ horse power (approx).

1 kilowatt-hour $= 3 \cdot 6 \times 10^{13}$ ergs.

1 kilowatt per (dekametre)^2 $= 1 \cdot 43 \times 10^{-2}$ g-cal cm^{-2} min^{-1}.

1 foot $= 0 \cdot 3048$ m; 1 m $= 3 \cdot 281$ feet.

The equatorial semi-axis of the earth $= 6378 \cdot 2$ km.

The polar semi-axis of the earth $= 6356 \cdot 5$ km.

The area of the earth's surface is approximately $5 \cdot 1 \times 10^8$ km^2.

Angular velocity of the earth's rotation $= 7 \cdot 29 \times 10^{-5}$ sec^{-1}.

The difference between geocentric and geographical longitude is $700'' \sin 2\phi$.

$A =$ reciprocal of the mechanical equivalent of heat $= 2 \cdot 392 \times 10^{-8}$.

INTRODUCTION

METEOROLOGY, the science of things in the atmosphere, is concerned with the measurement and co-ordination of pressure, temperature, density and humidity of the air, and of the motion of air relative to the earth. It seeks to explain the motions observed in terms of the changes of pressure, temperature and humidity, brought about directly or indirectly by the effects of incoming solar radiation. Meteorology in its widest sense includes the study of such phenomena as atmospheric electricity and terrestrial magnetism, but we shall not discuss these subjects in this book.

Instruments have been devised for measuring the factors mentioned, as well as some others such as incoming solar radiation, duration of bright sunlight, rate of fall of rain, etc. We may say of most of these instruments that in principle they are relatively simple, though in practice there may be considerable difficulties in using them in such a way as to give reliable results. It is almost invariably a tedious and difficult matter to obtain accurate meteorological observations. The physicist in the laboratory, wishing to measure the temperature of a fluid, merely plunges a thermometer in the fluid and reads the scale. But the meteorologist who wishes to measure the temperature of air out of doors has to take precautions to shade his thermometer from the direct rays of the sun and radiation from surrounding objects, and at the same time to provide for adequate ventilation. His problem is therefore much more complicated than that of the physicist. He may devise an electrically recording instrument which will measure the temperature of the air to 0·01° C, and then find that the record indicates a constantly fluctuating temperature. The measurement of air temperature to such a high degree of apparent accuracy is probably a waste of time, as the air does not know its own temperature to that degree of accuracy. A current of air, as we know it in the natural state in the atmosphere, is far from being homogeneous. The most casual glance at the record of wind direction and velocity provided by an anemometer shows that at a given point the motion is never steady, but the fact that the temperature is equally unsteady at a given point of observation is frequently overlooked.

The most serious of all difficulties of meteorology are probably those which arise from the fact that the air contains widely varying quantities of water-vapour. This ever-changing content is a potent source of supply of thermal energy to the atmosphere, but is a continual stumbling-block to the student of physical meteorology on account of the difficulty of applying physical reasoning to a medium of variable constitution.

A further difficulty arises from the paucity of observations over certain parts of the globe, and in certain types of weather. Relatively few observations have been made of wind, temperature, and humidity in the free atmosphere over the oceans, or in the polar regions, and even over the continents direct observation of upper air conditions in cyclonic depressions is practically impossible.

In the present work it is not proposed to discuss in detail the difficulties of observation, or the details of the methods which have been devised for obtaining observations of different meteorological factors. The reader who desires such information should consult the *Meteorological Observer's Handbook* (H.M.S.O.), the *Dictionary of Applied Physics*, **3**, and other works referred to in the subsequent pages.

Great as are the difficulties of accurate observation in meteorology, the difficulties of theoretical discussion are even greater. It is not possible to isolate a portion of the atmosphere, and to discuss the physical processes which take place in that portion, on account of the translation of pressure systems and of the variation of wind with height, both of which factors make it impossible to specify with certainty what is taking place in a given mass of air. It is customary to speak of the atmosphere as a heat engine, working between a source of heat at the equator and a source of cold at the poles. The most formidable obstacle to progress is ignorance of the laws of transfer of heat within this so-called engine.

CHAPTER I

THE FACTS WHICH CALL FOR EXPLANATION. A SKETCH OF THE SURFACE DISTRIBUTION OF THE METEOROLOGICAL ELEMENTS OVER THE GLOBE

§ 1. *Temperature*

THE distribution of temperature over the globe is here shown in four charts, for January and July, for the Northern and Southern hemispheres, the isotherms being drawn for intervals of 10° F (figs. 1–4). Looking first at the January (summer) chart of the Southern hemisphere, we are impressed by the fact that this is almost entirely an ocean hemisphere, having the Antarctic continent centrally situated. The isotherms run almost symmetrically over the oceans, with sudden swerves over the coasts of the continents. The land is everywhere warmer than the oceans in the corresponding latitudes. The July (winter) chart for the Southern hemisphere shows the same general symmetry of the isotherms over the oceans, but temperatures over the continents are now lower than the temperatures over the oceans in corresponding latitudes.

Coming to the January (winter) chart of the Northern hemisphere, we are impressed by the much greater extent and less symmetrical distribution of the land masses. The isotherms are by no means as symmetrical as those in the Southern hemisphere. The lowest temperatures are those recorded in North-east Siberia, and there is another region of intense cold centred over Greenland. The region immediately surrounding the North Pole is left free of isotherms, since the available observations are too few to yield reliable mean values. It is seen that the temperatures over the continents are lower than those over the oceans in corresponding latitudes, while in middle latitudes where westerly winds predominate (*vide* p. 12) the temperature is higher over the western coasts of the continents than over the eastern coasts. In the July (summer) chart for the Northern hemisphere the run of the isotherms is somewhat irregular. There is a long belt of high temperature (above 90° F) running across North Africa and South-west Asia, to the north of India, and a centre of high temperature over the South-west of the United States. The centre of extreme cold in North-east Siberia has now disappeared, and this region is warmer than any other region in the same latitude except the extreme North-west of Canada. The land is warmer than the sea in most latitudes, the difference being especially well marked over the North Pacific Ocean and the adjacent land masses. Greenland still interposes a tongue of cold into the now relatively warm area of the extreme north of the Atlantic Ocean.

A comparison of figs. 1 and 3, and of figs. 2 and 4 indicates that in January the highest surface temperatures observed in either hemisphere form a belt in latitude 30° S approximately, while in July the belt of highest surface temperatures is in latitude 35° N.

B D M

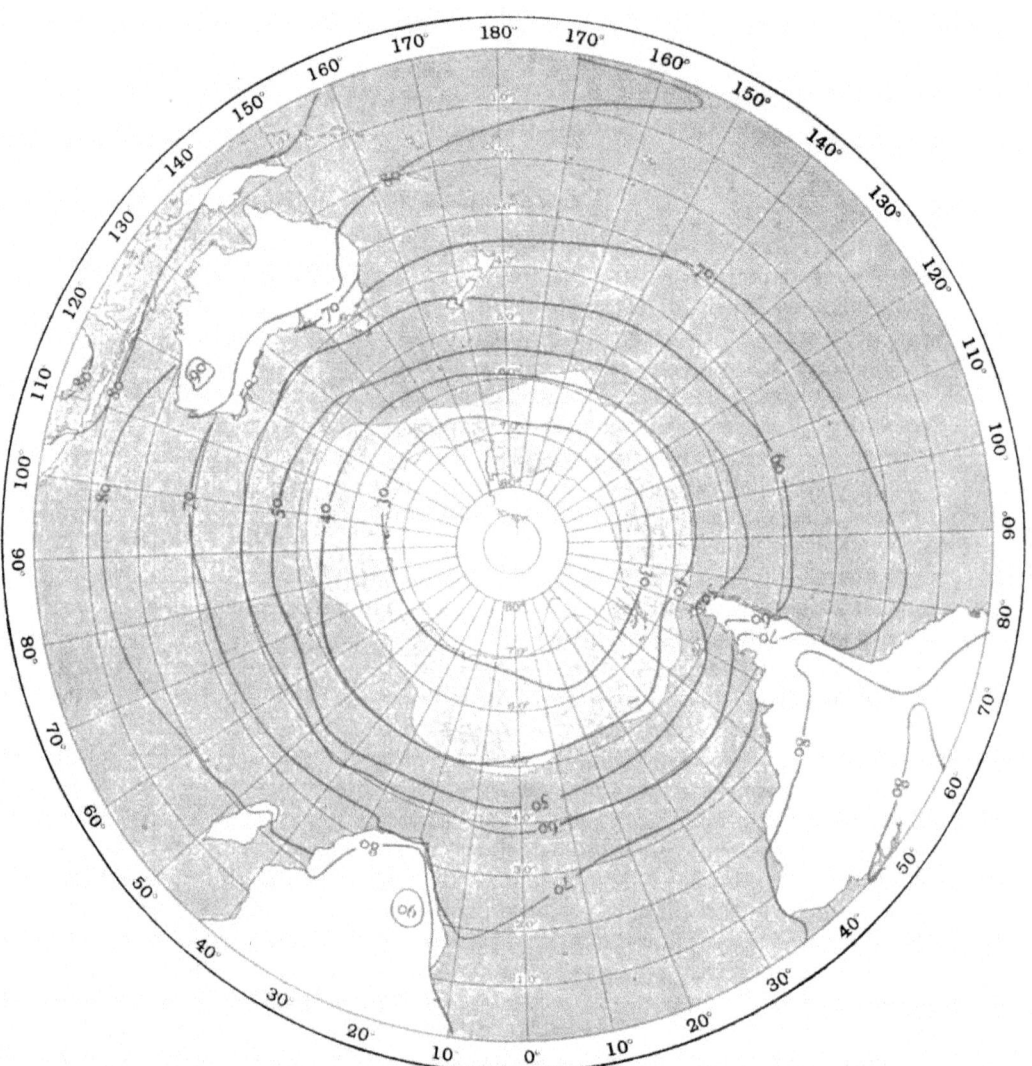

Fig. 1. Normal temperature at sea level, Southern hemisphere, January.

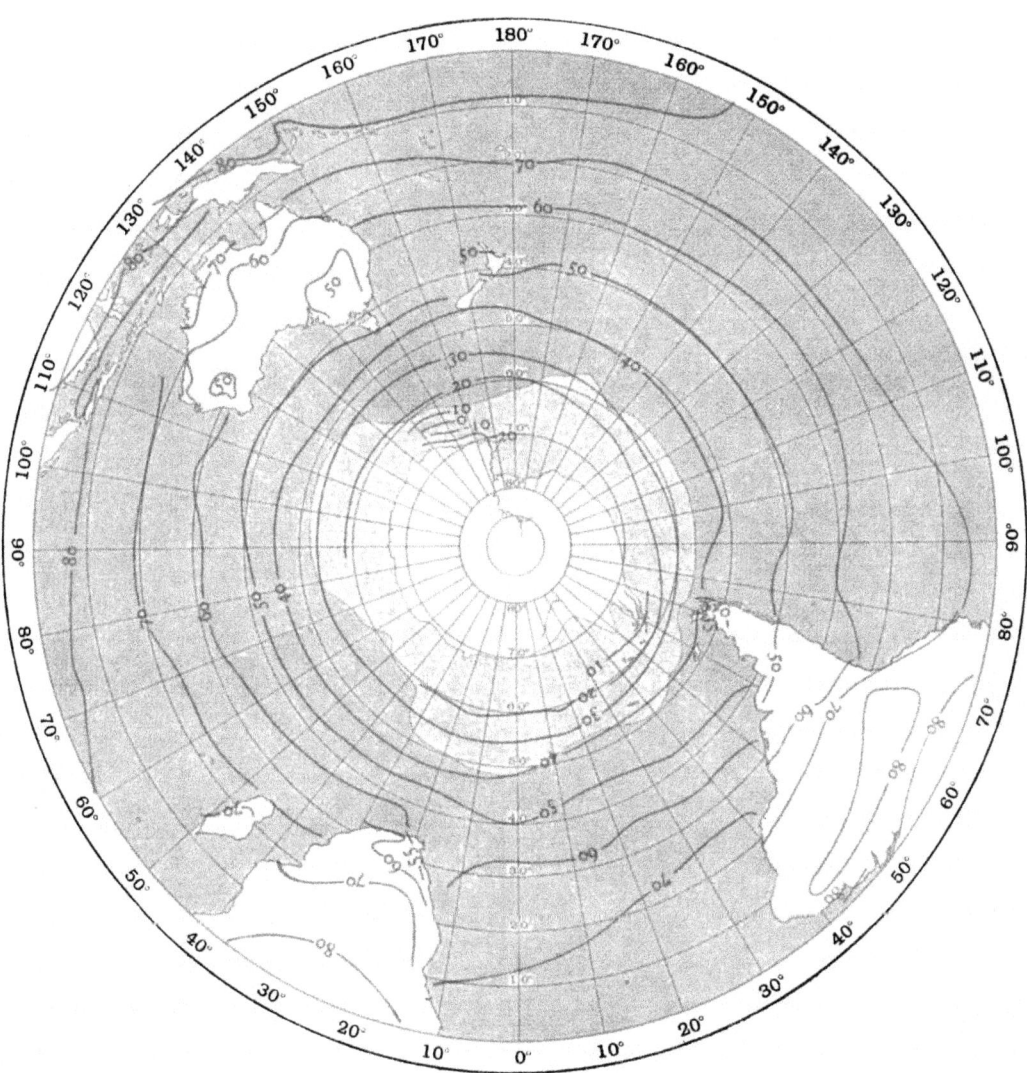

Fig. 2. Normal temperature at sea level, Southern hemisphere, July.

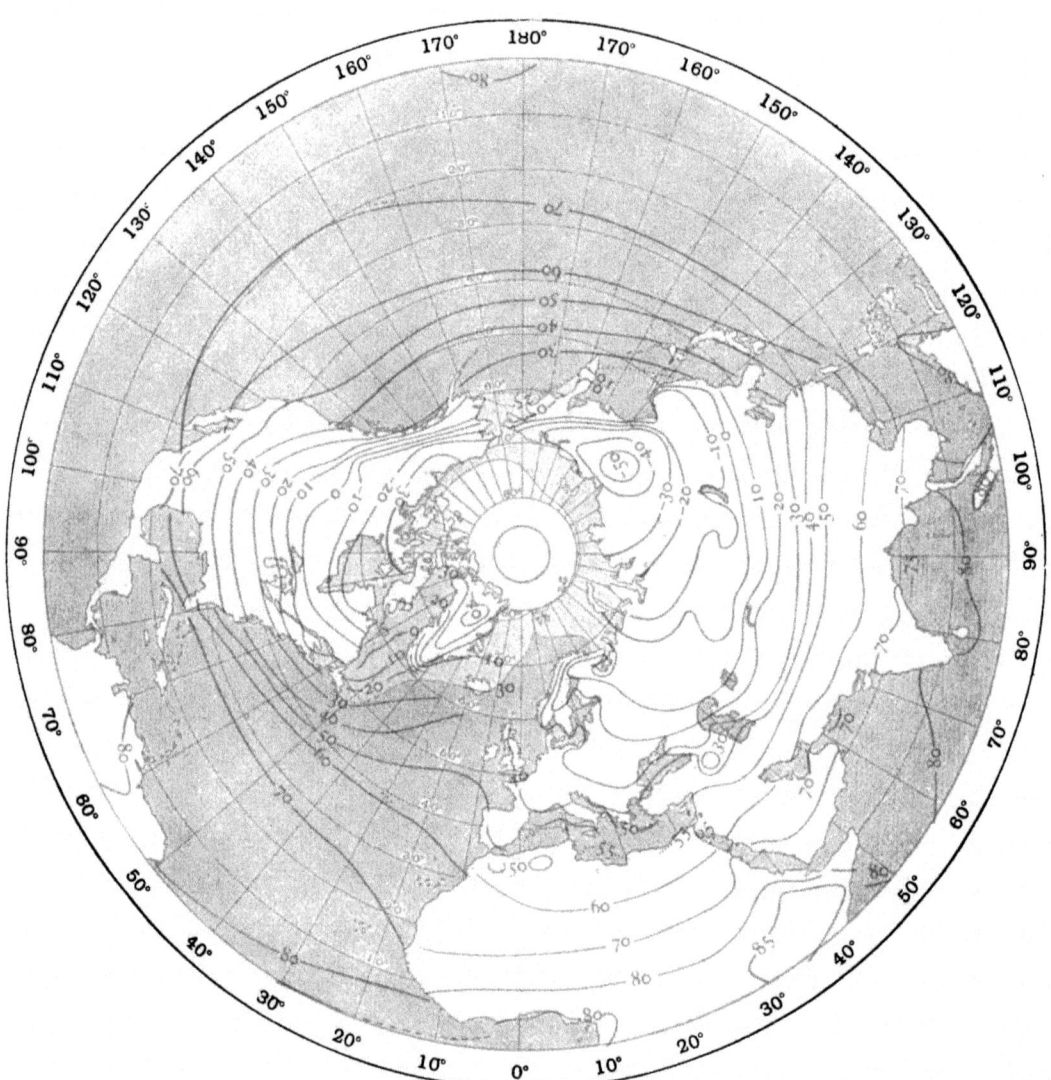

Fig. 3. Normal temperature at sea level, Northern hemisphere, January.

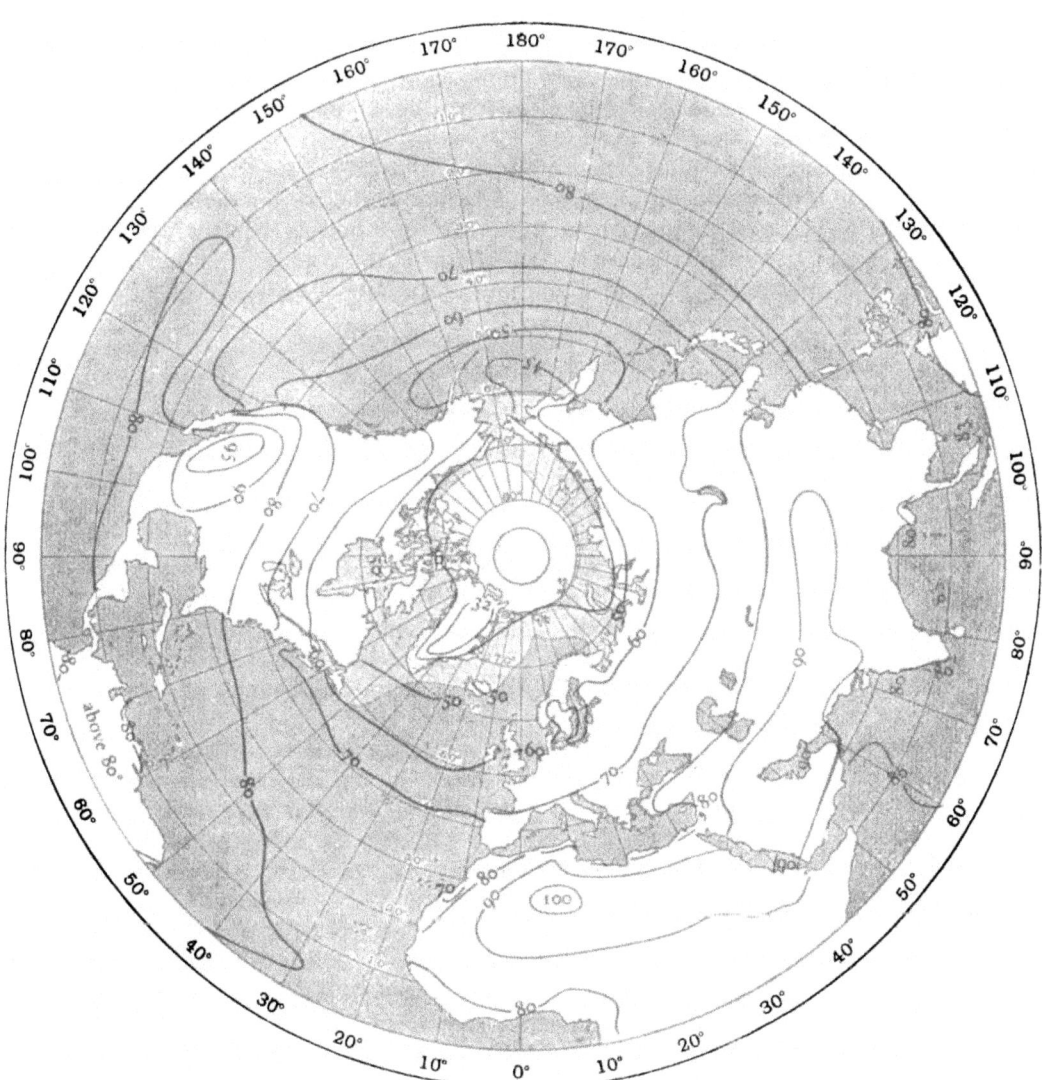

Fig. 4. Normal temperature at sea level, Northern hemisphere, July.

One further point of interest arises from the comparison of the January and July charts. The temperatures at the equator show only slight variations in the course of the year, but in say latitude 60° in the Northern hemisphere the annual variation may be anything between 10° and 50° according to the longitude. Thus the average rate of decrease of temperature with increase of latitude is much greater in winter than in summer, so that any phenomenon which depends on the variation of temperature in the horizontal will be more intense in winter than in summer.

We have recounted above the main facts which are to be gathered from an inspection of the temperature charts for January and July for the two hemispheres. The other months are in the main intermediate between the two extreme months. Similar charts for all months of the year are given in Shaw, *Manual of Meteorology*, **2**, and should be consulted for further details.

§ 2. *Pressure*

The mean distribution of pressure for January and July for the Northern and Southern hemispheres is represented in figs. 5–8. The chart for the Southern hemisphere in January shows a belt of low pressure in latitude 60–70°, north of which pressure increases up to a belt of high pressure centred about latitude 30°, having three centres of highest pressure over the oceans, with relatively lower pressure over the land masses of South America, South Africa and Australia, corresponding to three shallow centres of low pressure over the equator in the same longitudes. The July chart for the Southern hemisphere indicates the same general features, but the sub-tropical belt of high pressure is now farther north, and is more intense and more continuous around the globe.

For the Northern hemisphere the January chart is far more complicated than either chart for the Southern hemisphere. In the latter the sub-tropical belt of high pressure appears as the most striking and persistent feature, but in the January chart for the Northern hemisphere the sub-tropical belt of high pressure is only traceable over North Africa, the North Atlantic Ocean and the eastern half of the North Pacific Ocean. The dominating features of this chart are the region of high pressure centred over Siberia, and the regions of low pressure centred south of Greenland and over the Aleutian Islands in the North Pacific, respectively. The July chart shows well-marked anti-cyclones centred over the North Atlantic and North Pacific Oceans in latitude 35°, and a centre of low pressure to the north of India.

The outstanding feature shown by the pressure charts is the winter development of high pressure over the continents and low pressure over the oceans, and the summer development of low pressure over the continents and high pressure over the oceans. In winter the continents are loaded with air at the expense of the oceans, and in summer the converse occurs. These phenomena are more marked in the Northern than in the Southern hemisphere on account of the greater extent of the land masses in the Northern hemisphere, and it will be seen later that these phenomena are produced by the annual variation of temperature over the continents.

For the distribution of pressure in the other months of the year the reader is referred to Shaw, *Manual of Meteorology*, 2.

If we examine the Northern hemisphere charts of monthly mean pressures month by month, we find that the areas on these charts which show changes of 5 mb or more from one month to the next are definitely localised in certain restricted regions. The outstanding region is the Asiatic continent. In September a long tongue of pressure over 1015 mb but under 1020 mb extends from the Azores anticyclone across Europe and Asia nearly to the Pacific coast, while the depression over India is shallow with an extensive region of pressure just below 1005 mb. By October the latter region has disappeared and the pressure over Central Asia surpasses 1025 mb in the central region of the anticyclone. During the next three months the anticylone grows steadily, reaching its maximum in January, when the pressure at the centre surpasses 1035 mb. During this period the Icelandic and Aleutian lows both develop steadily, reaching their maximum intensity at about the same time, while the sub-tropical high-pressure belt over the North Atlantic shows little marked modification, beyond an initial weakening, and the sub-tropical belt of high pressure in the North Pacific becomes weaker in November and December and then becomes rather more accentuated.

The growth of the Siberian anticyclone in the interval from September to October is remarkably rapid, pressure rising by an amount varying from 5 to 10 mb over more than one half of the area of Asia. During the same period a fall of about the same magnitude occurs in the region of the Aleutian low, but only over a very much smaller area. The figures suggest that the Siberian anticyclone is fed in part by the hollowing of the Aleutian low, but mainly by the denudation of a very wide area of its share of air. This area probably extends into the Southern hemisphere.

By March the Siberian anticyclone has again diminished in intensity, so that the central pressure is slightly in excess of 1025 mb, while the Icelandic and Aleutian lows have filled up slightly. Between March and April there is a marked diminution in the intensity of all three features, and the diminution is continued in May, when a shallow depression appears over India. In June the depression has extended over nearly all Asia, and the anticyclone has disappeared. The depression maintains its intensity during July and August, during which months the Aleutian low ceases to have a separate existence on the monthly charts.

Since the pressure at the earth's surface measures the total mass of air in the column which extends from the ground to the top of the atmosphere, we could, by integration over the whole hemisphere, find the mass of air above a hemisphere of the earth's surface. This computation would be incorrect, if carried out with the pressure reduced to mean sea level, in that it would, in effect, replace the land above sea level by air, the error being greatest in winter, on account of the greater density of the air at low levels at that time. But when we take the differences from month to month, the figures can be accepted as giving an idea of the order of magnitude of the changes which take place from month to month in the amount of air in a hemisphere. Fig. 9, which reproduces the result of a computation carried out by Sir Napier Shaw, shows that the

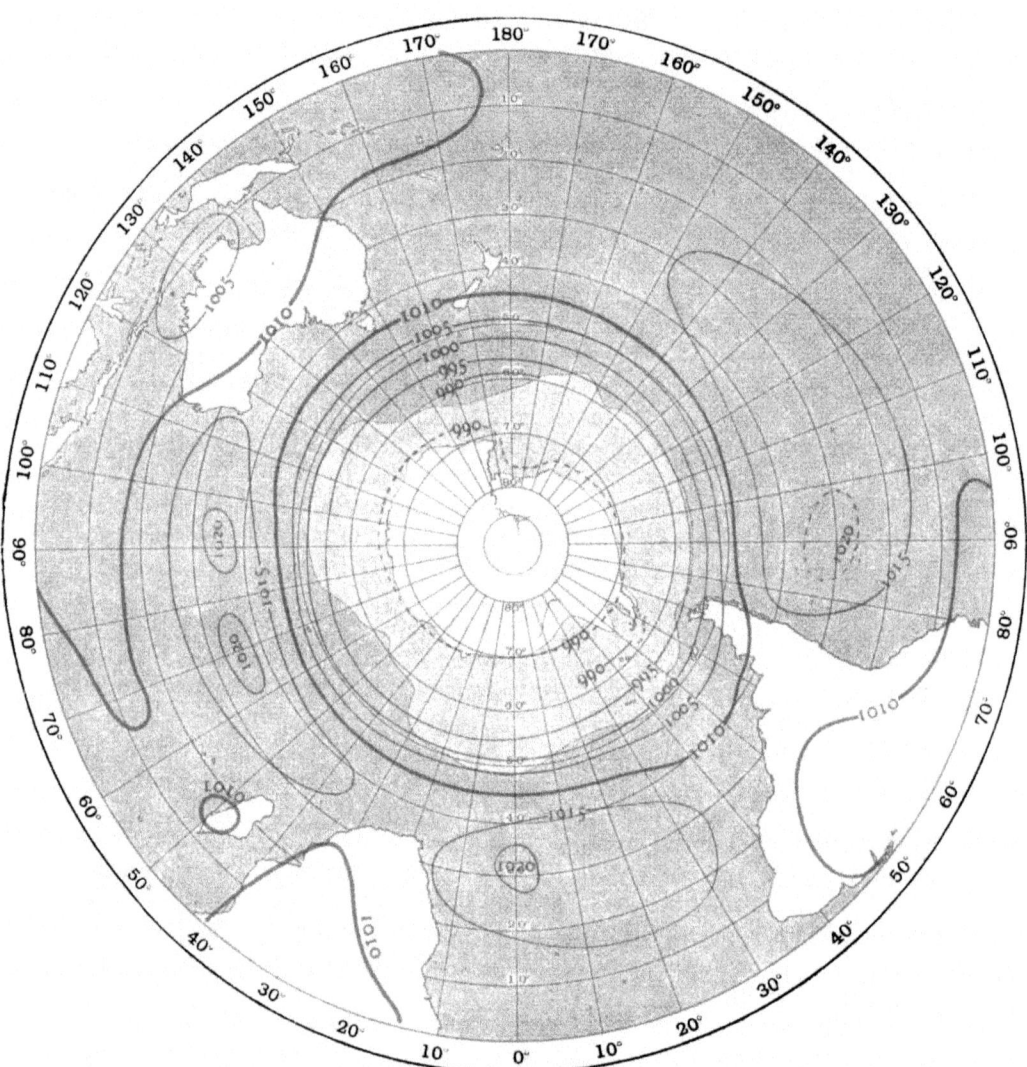

Fig. 5. Normal pressure at sea level, Southern hemisphere, January.

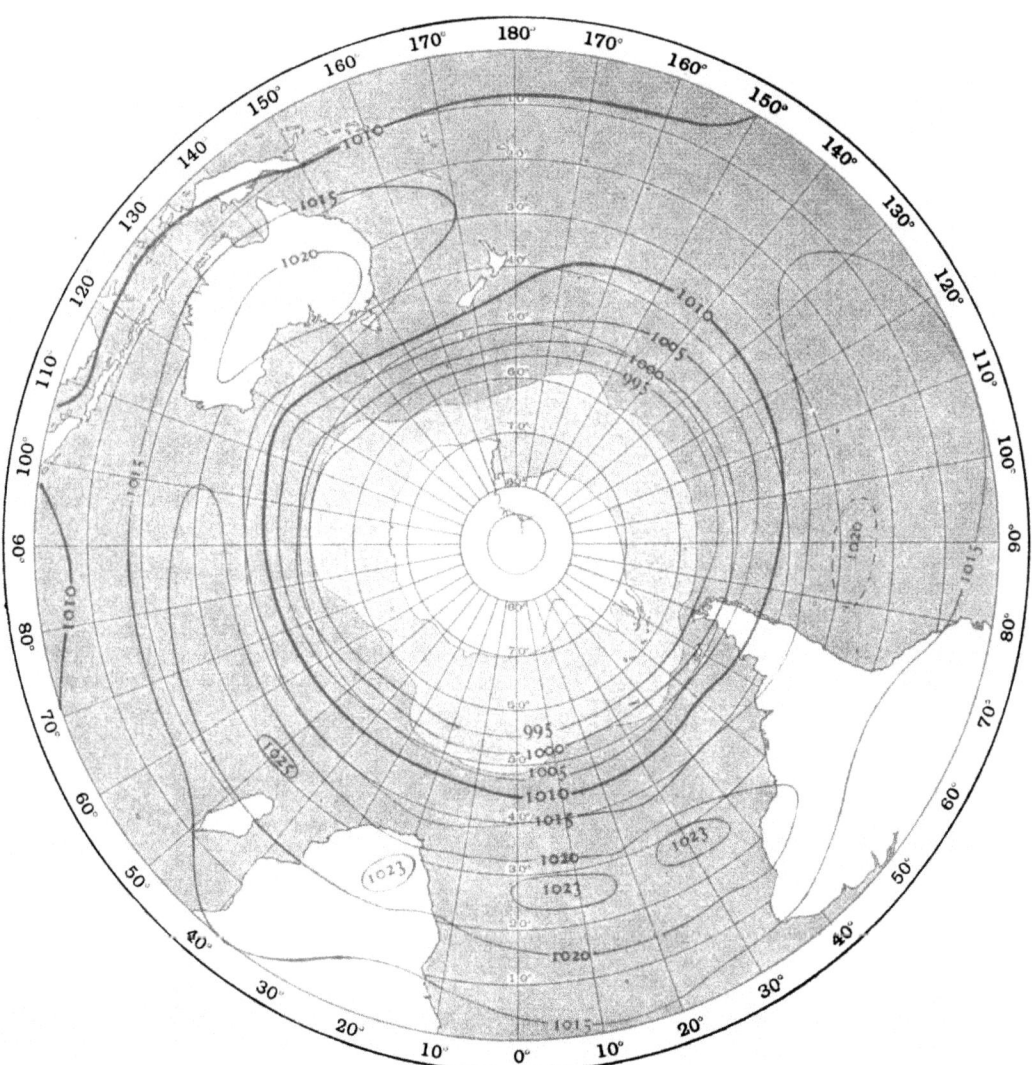

Fig. 6. Normal pressure at sea level, Southern hemisphere, July.

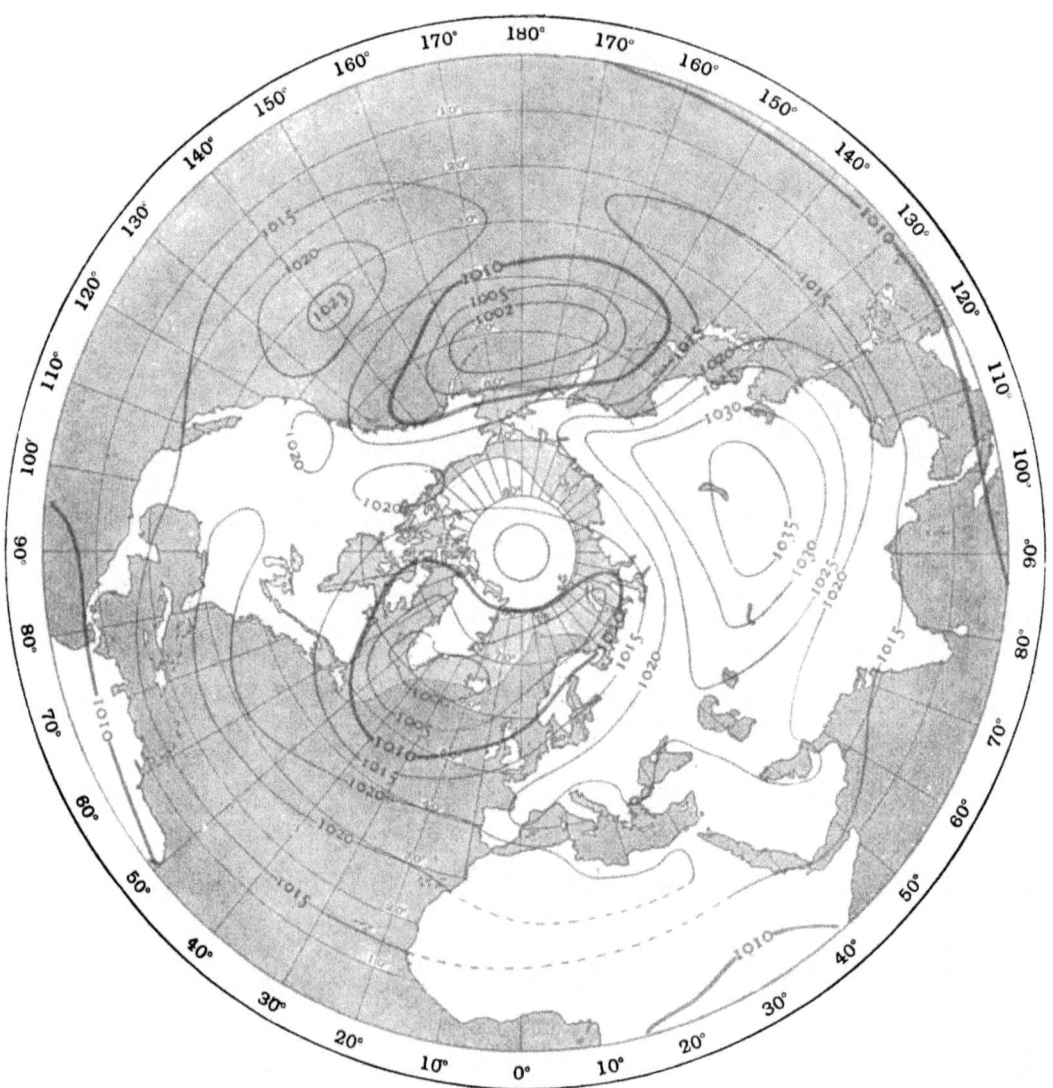

Fig. 7. Normal pressure at sea level, Northern hemisphere, January.

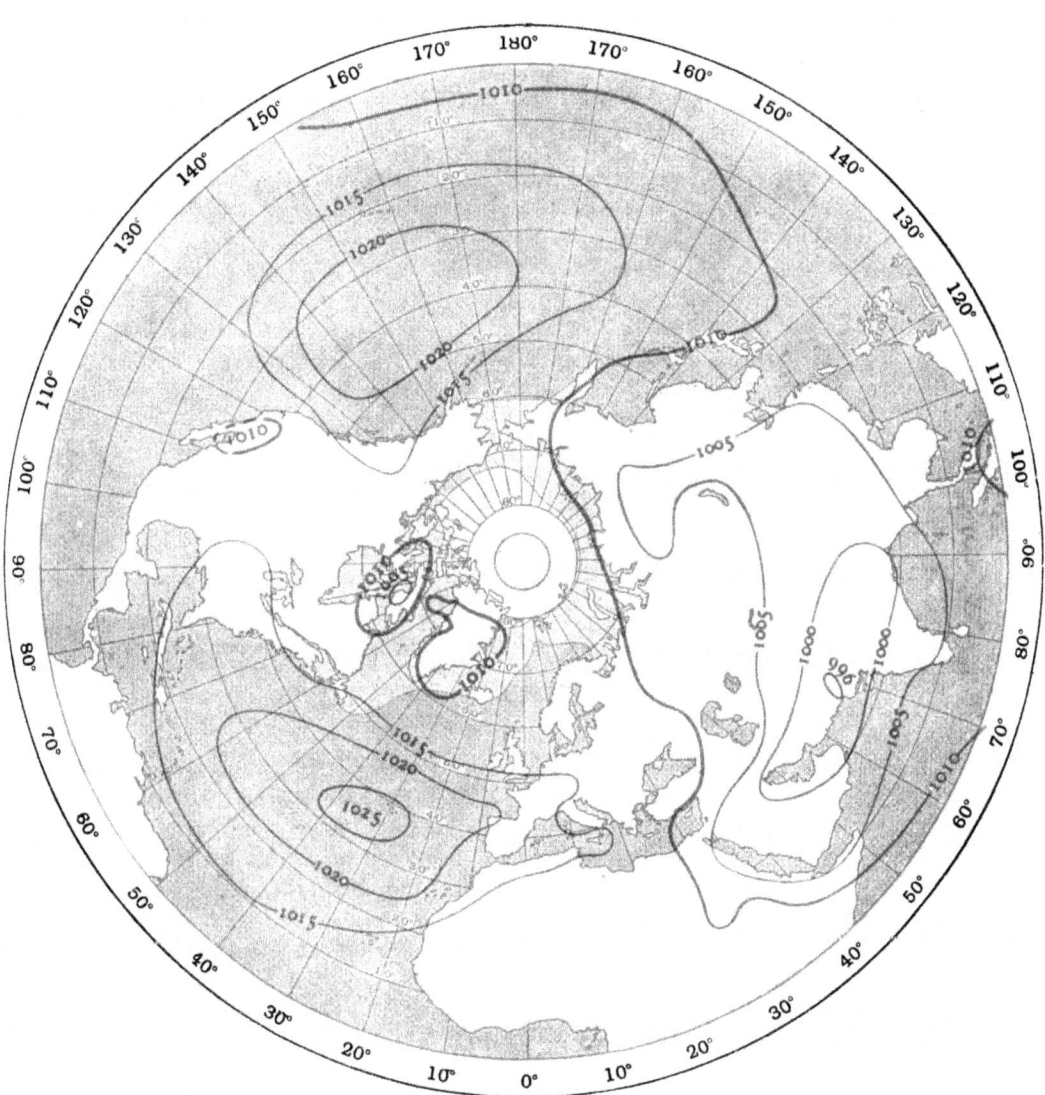

Fig. 8. Normal pressure at sea level, Northern hemisphere, July.

amount of air in the Northern hemisphere has a maximum at mid-summer, and a minimum at mid-winter, the total range of variation being about ten billion tons. (The total amount of air over a hemisphere is about 2700 billion tons.) Shaw made a correction for temperature in the figures which he computed, and the corrected figures are shown in fig. 9.

It is worthy of note that the time of most rapid flow of air from the Southern to the Northern hemisphere is in September to November.

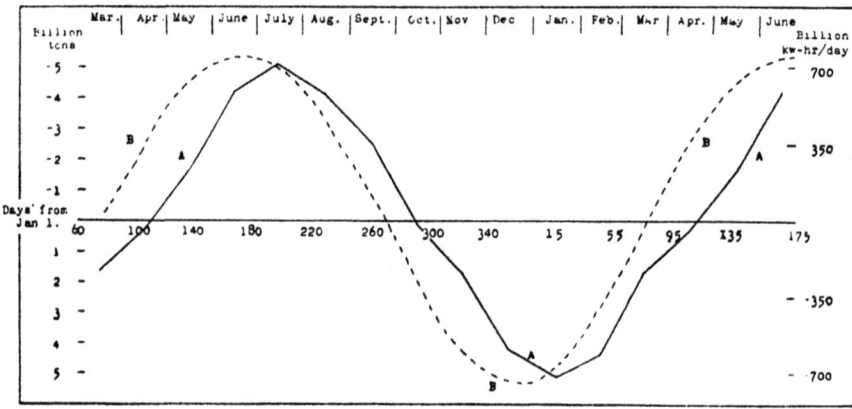

Fig. 9. Annual variation of the mass of air in the Northern hemisphere, shown reversed in curve *AA*, and the corresponding incoming solar energy, shown in curve *BB*.

§ 3. *Wind*

The distribution of winds over the globe can be most simply described by means of the distribution of pressure. There is a relation between the wind and the pressure distribution known as Buys Ballot's law, which may be stated as follows: "In the Northern hemisphere an observer who stands with his back to the wind will have lower pressure to his left than to his right; in the Southern hemisphere the contrary holds." In terms of the isobars on the chart this is equivalent to saying that the wind blows round the isobars keeping low pressure to the left and high pressure to the right in the Northern hemisphere, and the reverse in the Southern hemisphere. In actual practice it is found that the surface wind, while blowing in the general sense thus indicated, blows slightly across the isobars into low pressure, at an angle of 20–30°.

By the use of Buys Ballot's law we can readily interpret the charts of pressure distribution in terms of the prevailing winds. Take first the charts for the Southern hemisphere. The region south of the sub-tropical high-pressure belt is a region of winds blowing from a westerly direction. North of the sub-tropical high-pressure belt the winds blow from a generally easterly direction, but from a direction somewhat south of east. These winds are the South-east trade winds. The equator is marked by a shallow belt of low pressure, known as the Doldrums, a region of calms and light variable winds. This region is slightly south of the equator in the Southern summer, and north of the equator in the Northern summer, its mean position being north of the equator. Conditions in the Northern hemisphere vary widely from winter to summer. In winter there is a clockwise circulation of winds round the Siberian anticyclone, giving over India and China the winds known as the North-east monsoon; while over the Atlantic north of the sub-tropical anticyclone there is a region of prevailing westerly winds, where the average conditions are disturbed by the depressions of middle latitudes. With the approach of summer the Siberian anticyclone disappears and its place as the chief controller of conditions over Asia is taken by a depression centred to the north of India. From the centre of this depression there is a continuous increase of pressure southward to the centre of the sub-tropical high-pressure belt of the South Indian Ocean, and over the whole of the Indian Ocean blows a broad current of wind which starts in the sub-tropical belt of the Southern hemisphere as a south-easterly wind, blows across the equator, and finally appears as a south-westerly wind blowing into the southern edge of the depression over India, where it is known as the South-west monsoon. This air current passes over some thousands of miles of ocean, and so reaches India as a warm and very moist wind. When it reaches India the configuration of the land forces it to rise over the coastal ridges of mountains, so giving rise to copious rainfall. The Asiatic depression, with the associated monsoon wind, develops in June and persists until late September, when the depression fills up, the winds die away, and the rain ceases. The economic importance to India of the monsoon rainfall tends to focus attention on the south-westerly winds which blow over India in summer, but it should be noted that the influence of the summer Asiatic depression extends to China, where the southerly wind which blows around the eastern edge of the depression is known as the Southerly monsoon of the China seas, and to the Mediterranean, where the northerly wind which blows around the western edge of the depression is known as the Etesian wind.

Figs. 10 and 11 indicate the prevailing winds over the globe for January and July. These charts should be examined in conjunction with the charts showing the pressure distribution reproduced in figs. 5–8.

§ 4. *Rainfall*

Rainfall is so extremely variable with place that it is difficult to treat it in a simple manner. It is only over a very restricted portion of the globe that the rainfall of a single year will resemble at all closely the "normal" rainfall. The outstanding case in which this occurs is India, where the maximum rainfall of the year always occurs in June to September, in association with the South-west monsoon winds. Within the tropics rain usually has a well-marked maximum, and in some places two maxima; in other words there will be one or two rainy seasons each year. Thus Batavia in Java has its rainiest month in February; Colombo has two rainy months, May and October; Singapore has most rain during November, December and January; Bombay has its maximum rainfall in July, but Madras in November.

In the British Isles the averages taken over many years point to a definite maximum of rainfall in a particular month of the year. This month is October

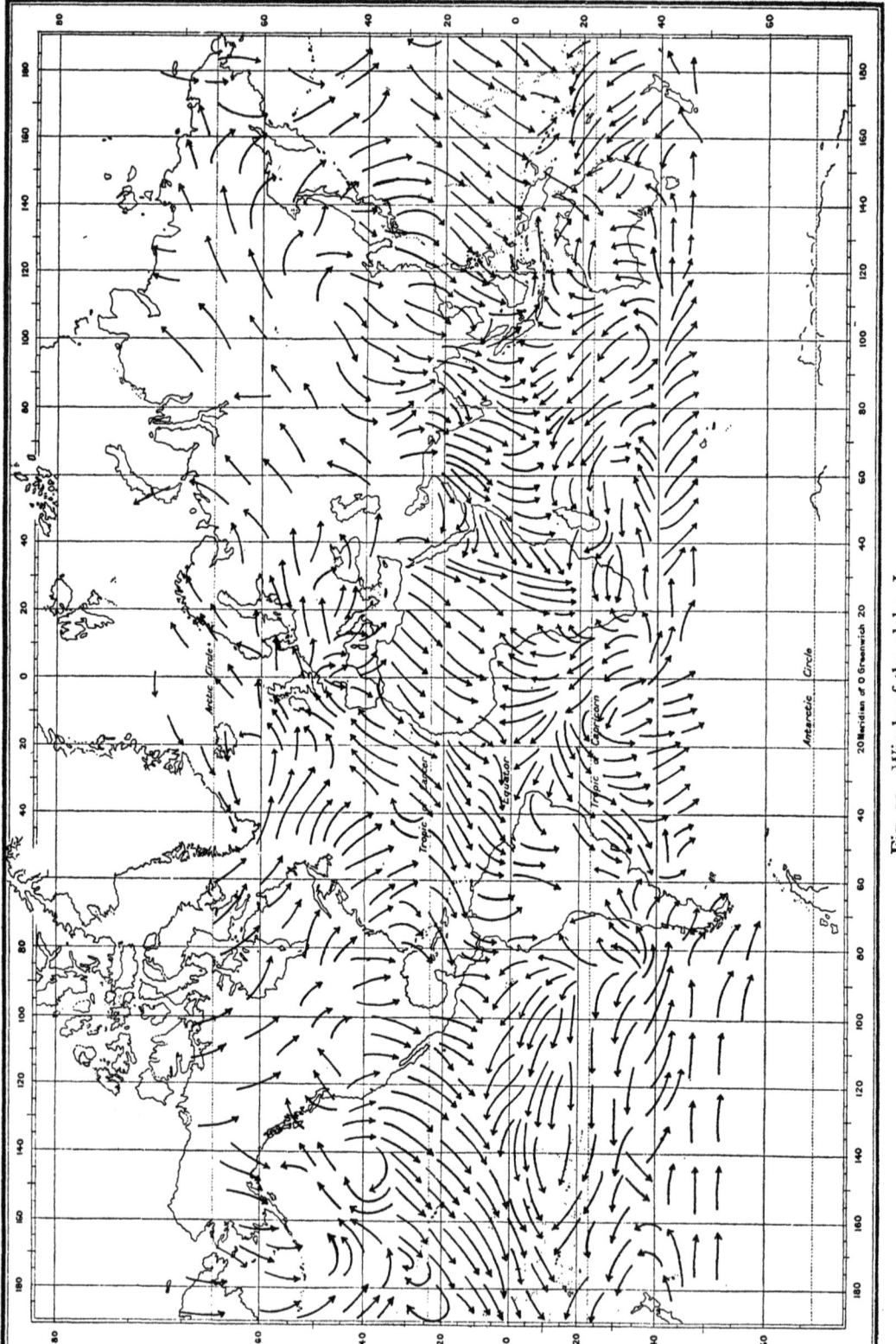

Fig. 10. Winds of the globe, January.

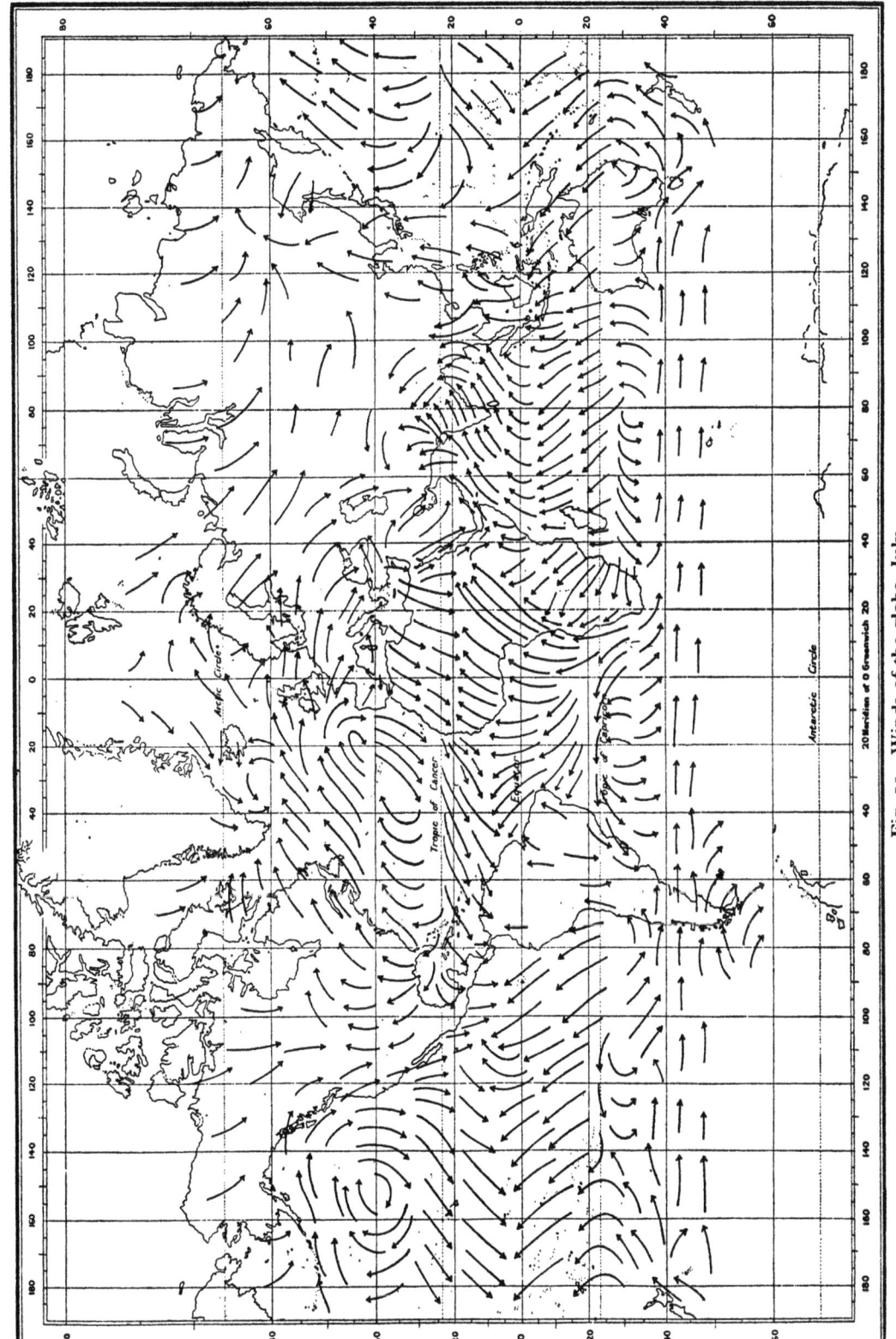

Fig. 11. Winds of the globe, July.

at Greenwich, December at Aberdeen, and July at Edinburgh, but it would be extremely hazardous to apply this result to forecast the rain of any particular year.

The Russian Meteorological Atlas gives two very interesting charts, one showing for the whole of Europe the month of maximum rainfall, and another showing for the same area the month of minimum rainfall. There is a clearly-marked tendency for the month of maximum rainfall to occur in May or June at stations in South-east Europe, but as we go farther westward over the continent the time of maximum rainfall becomes steadily later, and falls in October to December over a large part of the British Isles. The details of the annual variation of rainfall for individual stations can be most readily studied by means of tables of rainfall such as are to be found in Kendrew's *The Climate of the Continents* (Oxford, 3rd edn. 1937) or other textbooks of climatology.

§ 5. *The general circulation and the local circulations*

In the preceding paragraphs we have given a brief outline of the average distribution of some meteorological elements as represented by monthly means over a large number of years, and have given a very brief account of the distribution of winds over the globe in the months of January and July. It is scarcely necessary to emphasise the fact that on any one day in either of these months the actual situation, as represented by the distribution of pressure, temperature and wind, may vary considerably from the mean values shown in figs. 1–11. But some of the main features of these charts remain constant from day to day, and so it is found more convenient to adopt the mean situation for the month, or in other words, the *general circulation* for the month, as a standard of reference, and to regard the situations which occur from day to day as deviations from the general circulation. The most important of these deviations are the travelling depressions and anticyclones of middle latitudes, and the tropical cyclones of low latitudes, and to these we may give the name of *local circulations*.

It is clear that the general circulation is not independent of the local circulations. It is made up of the mean conditions over a long period, and so implicitly contains the integrated effect of a large number of local circulations which in part, but only in part, cancel one another. The mean circulation is in fact a climatological normal in precisely the same sense as the mean temperature at a given station.

We shall return to the problems concerned with tropical cyclones and the depressions and anticyclones of middle latitudes in later chapters. In Chapter XIX will be found a fuller description of the winds of the globe, taking account of the fact that the atmosphere is three-dimensional, and that the nature of the wind-circulation is different at different heights in the atmosphere.

§ 6. *Surfaces of separation between winds of different origin*

In the charts of winds given in fig. 10 and 11 it can readily be seen that currents of air originating in polar and equatorial regions flow along the surface of the earth. It is possible here and there to draw lines separating the currents of different origin. This was done by T. Bergeron (*Meteorologische Zeitschrift*, July 1930) for a part of the earth. Some of these lines of separation could be readily drawn in figs. 10 and 11, but charts on such a small scale could only show these lines incompletely. For the moment we are only concerned to show that the system of prevailing winds brings into juxtaposition masses of air of widely different temperatures, and that these masses are separated by lines which appear as discontinuities on a chart of this scale. We shall find later that such lines of separation, when they occur in association with the depressions of middle latitudes, are of great importance in determining the growth of depressions and the distribution of cloud and rainfall.

§ 7. *The observed distribution of temperature in the vertical*

The manner in which temperature varies with height is of fundamental importance in determining the processes of weather. On the average, temperature decreases with height at the rate of approximately 6° C per kilometre, or about 3° F per 1000 feet, from the ground up to a considerable height, but eventually a limit is reached at which the normal decrease with height ceases, and above this limit the temperature remains constant, or even increases slightly at first. The atmosphere is thus divided into two thermally distinct regions, the lower, known as the *troposphere*, in which the temperature decreases steadily with height, and an upper region, the *stratosphere*, in which the temperature remains constant or increases slightly with height. The surface of separation is known as the *tropopause*, and is at a height which varies from about 18 km at the equator, to about 11 km over Southern Europe, and to about 6 km at the poles. The height at the poles cannot be specified with accuracy, mainly on account of the paucity of observations in polar regions. The rate of diminution of temperature with height, known as the *lapse-rate*, does not diminish steadily to zero at the tropopause, but retains approximately its normal value right up to the limit of the tropopause, and then suddenly changes to zero or even changes sign.

The temperature within the stratosphere is usually reputed to increase steadily from the equator towards the poles, the generally accepted view being as represented in fig. 12, which is due to Ramanathan.* According to this diagram the coldest air in the whole atmosphere forms a limited ring round the equator in the lower stratosphere. The increase of temperature with height in the stratosphere shown in this diagram is definitely established by observations in low latitudes. The details of the diagram are less certain in high latitudes, though it probably represents the actual situation with reasonable accuracy in summer. It is probable, however, that in the polar regions in winter, in the absence of any incoming solar radiation, the distribution of temperature varies widely from that shown in fig. 12, and that lower temperatures occur there than over the equator, possibly with temperature decreasing with height in the stratosphere.

* As modified by Samuels, *Monthly Weather Review*, Sept. 1929.

L. H. G. Dines* has also shown that over the British Isles temperature increases with height in the lower stratosphere to a height of 3 km above the tropopause, followed by a slow decrease of temperature with height in the next 5 km. Thus while fig. 12 may be regarded as a useful first sketch of the temperature distribution in the atmosphere, it may be expected that careful study of the details of observations made in different regions will show that the actual temperature distribution is more complicated than is suggested by fig. 12, and possibly deviates widely from symmetry about the earth's axis.

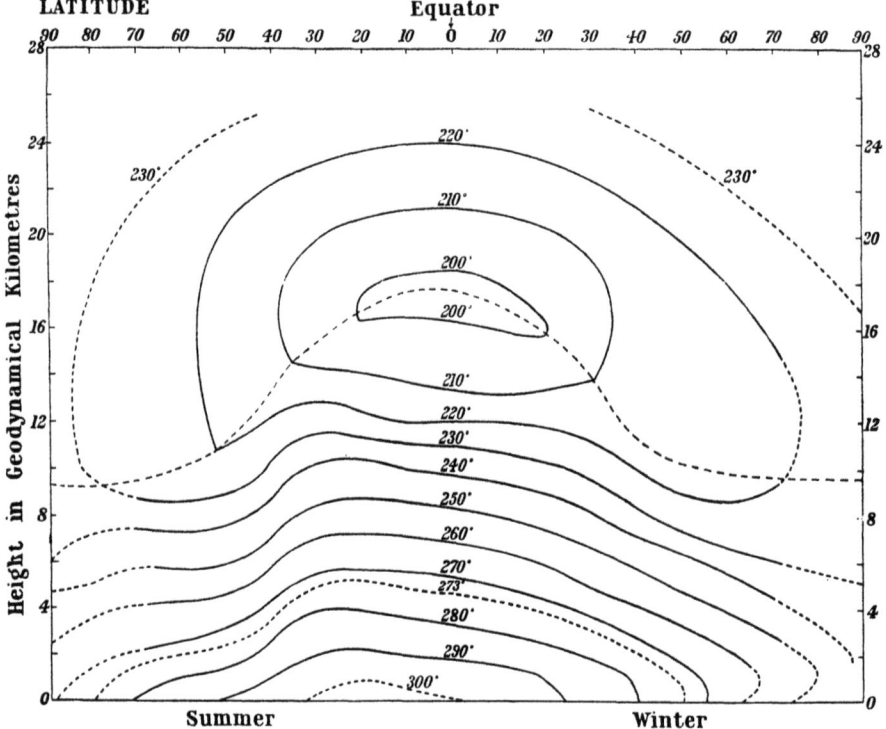

Fig. 12. Distribution of temperature in the atmosphere.

Over North-west Europe the stratosphere is on the average lower and warmer than the normal over low surface pressures, and higher and colder than the normal over high surface pressure. The correlation of the surface pressure with the height and temperature of the tropopause is not very high. It is far from being clear that the correlation is similar over other regions of the globe. The seasonal variation of the temperature of the stratosphere varies considerably with place. Over Europe the stratosphere is higher and warmer in summer than in winter, but over Canada and India it is higher and colder in summer than in winter.

* *Mem. R. Met. Soc.* **2**, No. 18, 1928.

Table 1* gives on the absolute scale the monthly mean temperatures up to 14 km for each month of the year for England. Fitting a sine-curve to the data for each height gives the amplitude, in degrees C, given in the last column but one, and the date of minimum given in the last column, measured in days from Jan. 1st. The amplitude diminishes with height up to 3 km, and then increases, attaining its highest value at about 6 km, beyond which it steadily diminishes until at 11 km the annual range is less than half the annual range at the surface. The date of occurrence of minimum temperature is sensibly constant in the first 6 km, is then steadily retarded up to 10 or 11 km, and at still greater heights becomes earlier with increasing height. It is not claimed that these figures are strictly representative, as they are based upon means of a small number of observations, but the general nature of the changes with height of both the annual range and of the time of maximum agrees with results derived for other places.

Table 1

Monthly mean temperatures for England on the absolute scale

Height in kilometres	Jan.	Feb.	Mar.	Apr.	May	June	July	Aug.	Sept.	Oct.	Nov.	Dec.	Amp. °C	Date min.
						200°+								
14	16	17	19	21	22	23	22	21	19	17	16	15	2·8	27
13	16	17	19	21	22	23	23	21	19	18	17	16	2·5	14
12	17	18	19	20	21	22	22	21	20	19	18	17	2·2	29
11	17	17	17	19	20	21	22	22	21	20	19	18	2·9	54
10	20	20	20	22	24	25	26	26	26	24	23	21	4·1	56
9	24	23	24	26	29	32	34	33	33	31	28	25	5·2	44
8	30	29	30	32	36	38	41	41	41	38	35	32	6·0	35
7	37	36	37	39	42	45	47	48	47	45	41	38	6·3	30
6	43	43	44	46	49	52	55	55	54	51	49	45	6·6	31
5	50	49	50	52	56	59	61	62	61	58	55	52	6·6	36
4	57	56	57	59	62	65	67	68	67	64	61	58	6·2	35
3	63	62	63	65	68	71	73	74	73	70	67	64	5·2	37
2	67	66	67	70	73	76	78	79	78	75	72	69	5·4	37
1	71	71	73	76	79	82	83	83	81	79	75	72	6·1	35
0	76	76	77	82	85	88	89	89	86	83	80	77	6·2	38

Table 1 gives rather a wrong idea of what occurs at the tropopause. On account of the varying height of the tropopause the effect of taking the mean of a number of observations at fixed heights is to mask the sudden change at the tropopause, and to replace it by a gradual transition. Further, the table is based on selected observations mainly restricted to fair-weather conditions.

Individual observations show considerable irregularities near the ground, where the lapse-rate in the lowest layers may be several times the "normal" value on a sunny afternoon, and where at night the normal decrease of temperature may be replaced by a condition in which temperature steadily increases from the ground upward, giving what is known as an *inversion* in the lower layers, above which there is a return to the more normal decrease of temperature with height. At cloud levels and at surfaces of separation of warm and cold currents irregularities in the form of inversions may occur within a narrow range of height. The variations of the lapse-rate at the ground and at

* From W. H. Dines, *M.O. Geophysical Memoir*, No. 13.

cloud levels are discussed later in Chapter VI, in connection with the effects of radiation and turbulence, which afford a rational explanation of the observed phenomena. Excluding these we are left with four outstanding problems in connection with the distribution of temperature in the free air:

1. The approximate constancy of the mean lapse-rate at all heights in the troposphere and in all latitudes.

2. The sudden nature of the change at the tropopause.

3. The approximate constancy of temperature at all heights within the lower stratosphere.

4. The decrease of temperature from pole to equator within the stratosphere.

These problems are discussed in later chapters, but it cannot be said that an adequate explanation has yet been given of any of them.

The temperature at the base of the stratosphere varies from about 196° A over the equator to about 222° A (in summer) and 216° A (in winter) over England. Within the limits of height investigated instrumentally (by meteorographs) the variations of temperature with height in the stratosphere have been found to be relatively small except in the tropics. Above these limits there is a further increase of temperature, and at heights of about 50 km the temperature is again as high as at the surface. This result was deduced by Lindemann and Dobson* from observations of meteors, and is to be explained as an effect of absorption of incoming ultra-violet light by ozone. Whipple† has shown that the travel of sound from explosions bears out the supposition that the temperature at 50 km is at least 300° A.

For discussions of data of observation up to and within the stratosphere the reader is referred to the following papers, which do not, however, exhaust the available material:

GOLD, International Kite and Balloon Ascents, *M.O. Geophysical Memoir*, No. 5. (European data.)
GOLD and HARWOOD, *Report to British Association*, Winnipeg, 1909.
RYKATCHEW, *Met. Zeit.* 28, pp. 1–16. (Pavlovsk.)
WAGNER, *ibid.* pp. 261–5. (Pavia.)
PEPPLER, *Beitr. Phys. fr. Atmos.* 4. (Pavlovsk and Kutschino.)
J. PATTERSON, *Upper Air Observations in Canada*, Ottawa, 1915.
Monthly Weather Review, Jan. 1918. (Fort Omaha, Indianapolis, Huron, Avalon.)
Travaux Scientifiques de l'Obs. de Met. de Trappes, 4, 1909, Paris. (North Atlantic.)
VAN BEMMELEN, K. *Mag. en Met. Obs. te Batavia*, 1916, p. 27. (Batavia.)
Annals Astr. Obs. Harvard College, 68, pt 1, p. 68. (St Louis.)
HILDEBRANDSSON, *Geografiska Annaler*, 1920, p. 110. (North Atlantic.)
BERSON, *Ergebnisse K. Preuss. Aer. Obs. Lindenberg*, 1910, p. 82. (East African Observations made by Berson in 1908.)
HARWOOD, *Mem. Ind. Meteor. Dept.* 24, pt 6, 1924. (India.)

---

* Lindemann and Dobson, *Proc. Roy. Soc.* A, 102, 1922, pp. 411–36. See also later papers by Dobson, and by Dobson and others, *Proc. Roy. Soc.* A, 110, 114, 120, 122 and 129. Also, Report of Conference on Ozone, Oxford, Sept. 9 to 11, 1936. *Q.J. Roy. Met. Soc.* 62, 1936, Appendix.
† *Q.J. Roy. Met. Soc.* 57, 1931, p. 331, and 58, 1932, p. 471.

J. REGER, *Die Arbeiten des Preuss. Aeronaut. Obs. Lindenberg*, Band XIII, 1919. (Lindenberg sounding balloon ascents, 1906–16.)

W. H. DINES, "The Characteristics of the Free Atmosphere," *Geophysical Memoir*, No. 13, M.O. Publication No. 220c, London, 1919.

For further data, the *Yearbooks* of the different official meteorological services should be consulted.

§ 8. *Correlation between different variables in the free air*

A large amount of work has been done by W. H. Dines* in correlating the variations of pressure and temperature at different heights in the free air, and of the temperature and height of the stratosphere, using European data. The main results of his researches can be readily summarised in the form of a table, in which the following symbols are used:

P_0 the barometric pressure at M.S.L.
P_9 the barometric pressure at 9 km.
T_0 the temperature at the surface.
T_n the temperature at n km.
T_m the mean temperature from 1 km to 9 km.
T_{0-4} the mean temperature from the surface to 4 km.
V the total water-vapour content of the atmosphere.
H_c the height of the tropopause.
T_c the temperature at the tropopause.

	P_0	P_9	T_m	H_c	T_c	V	T_0	T_{0-4}	T_4	T_8
P_0	—	0·68	0·47	0·68	−0·52	0·08	0·16	0·34	—	—
P_9	0·68	—	0·95	0·84	−0·47	—	0·28	—	0·82	—
T_m	0·47	0·95	—	0·79	−0·37	—	—	—	—	—
H_c	0·68	0·84	0·79	—	−0·68	0·39	0·30	0·66	0·64	0·74
T_c	−0·52	−0·47	−0·37	−0·68	—	—	—	—	—	—
V	0·08	—	—	0·39	—	—	—	0·73	—	—
T_0	0·16	0·28	—	0·30	—	—	—	—	—	—
T_{0-4}	0·34	—	—	0·66	—	0·73	—	—	—	—
T_4	—	0·82	—	0·64	—	—	—	—	—	—
T_8	—	—	—	0·74	—	—	—	—	—	—

A number of interesting conclusions can be drawn from the above table. High pressure at 9 km is accompanied by

(*a*) high pressure at the ground,
(*b*) warm air from 0 to 9 km,
(*c*) high tropopause, and
(*d*) low temperature at the tropopause.

This is effectively equivalent to saying that the stratosphere is high and cold over high surface pressures, and low and warm over low surface pressures. Again high mean temperature in the lowest 9 km is associated with high pressure and high and cold stratosphere. Indeed, the highest coefficient in the table above is that of 0·95 between T_m and P_9. High values of V, the total water-vapour content of the atmosphere, are associated with high stratosphere and high mean temperature in the lowest 4 km. There is practically no correlation between pressure and temperature at the surface.

* W. H. Dines, *M.O. Geophysical Memoirs*, Nos. 2 and 13.

In the same paper Dines gives the correlation coefficients between temperature and pressure at the same heights. The coefficients for heights 0–13 km were as follows:

0	1	2	3	4	5	6	7	8	9	10	11	12	13
0·11	0·42	0·66	0·77	0·84	0·85	0·86	0·86	0·86	0·71	0·32	−0·19	−0·36	−0·28

Thus the correlation between temperature and pressure at the ground is negligible, but increases to a maximum of 0·86 at levels of 6–8 km, then diminishes and becomes negative in the stratosphere.

Schedler[*] also computed a number of correlation coefficients similar to those given above, but in the majority of cases his values are definitely lower than those given by Dines.

Dines also gave a table of the standard deviations of temperature, pressure and density at different heights from the ground up to 13 km. The standard deviations of temperature were least in the summer quarter and greatest in the autumn quarter. They increased from the surface up to a level between 5 km and 7 km, and then decreased, but again increased at 12 and 13 km. The standard deviations of pressure were least in the summer quarter and greatest in the winter quarter. They showed no marked change from the surface up to a level of about 10 km, after which they definitely decreased. The standard deviations, expressed as percentages for each level, increased from the ground upward, to three times the surface value at 13 km, but expressed in gm/m³ they showed a decrease from the ground up to about 6 km, and an increase beyond this level. Mahalanobis[†] has shown that higher correlations are found when heights of 3 to 5 km are taken instead of the 9 km adopted by Dines.

§ 9. *The diurnal variation of meteorological factors*

(*a*) TEMPERATURE

The nature of the diurnal variation at the ground is sufficiently well known to require little discussion at the present stage. Surface temperature shows a clearly defined maximum at about two hours after noon, and a minimum shortly before sunrise. The amplitude of the diurnal variation differs for different localities, and for different times of the year, being much greater on the average in summer than in winter in England. Shaw gives in the *Manual of Meteorology*, **2**, pp. 84, 85, a chart for each hemisphere showing the variation of diurnal range of temperature over the globe. Over the centre of each continent there is an extensive region within which the mean diurnal range for the year is 17° C (30° F) or more. Over the open oceans the diurnal range is only a small fraction of this, and amounts to not more than one or two degrees F. But the annual mean values may give a false idea of the extent of the diurnal variation at different times of the year. The following small table

[*] Schedler, *Beitr. Phys. fr. Atmos.* **7**, p. 88.
[†] *Memoirs Indian Met. Dept.* **24**, pt 1, 1920.

gives the mean daily range for a few typical locations during four selected months of the year (in ° C).

	Jan.	April	July	Oct.
Arctic (1894–6)	0·7	3·3	0·7	0·6
Aberdeen	1·5	3·9	4·1	3·1
Calcutta	9·4	8·7	3·2	4·8
Batavia	4·7	5·9	6·8	6·9
Cape of Good Hope	5·8	6·1	5·1	5·6
Cape Evans (77° S, 167° E)	2·8	0·5	0·7	1·6
Paris (Parc St Maur)	3·3	8·6	9·1	6·5
Eiffel Tower	1·1	4·6	5·3	2·8

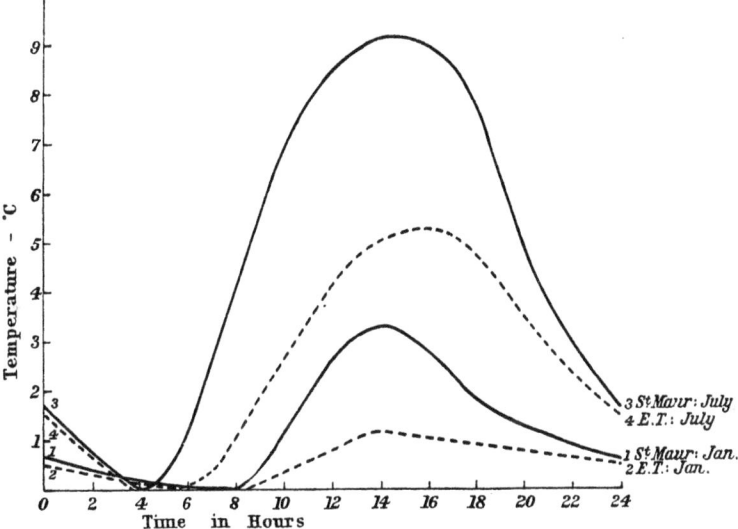

Fig. 13. Diurnal range of temperature at Parc St Maur and at the top of the Eiffel Tower.

The actual diurnal range observed within any selected interval of 24 hours will depend largely on the weather at the time. A clear sunny day followed by a clear calm night will yield a range of temperature far in excess of the average, while a cloudy day followed by a cloudy night may give no appreciable variation of temperature during the whole interval.

In the free air the daily range of temperature diminishes with increasing height. Fig. 13 gives a comparison of the curves of variation for Parc St Maur and the top of the Eiffel Tower for January and July. At greater heights the diurnal variation falls off rapidly, but it is not possible to say with certainty how rapidly. Observations made at Lindenberg have been analysed in considerable detail, but it is by no means certain that the observations are not to some extent vitiated by the effects of the sun shining on the instrument. In the lowest layers it has been shown conclusively that with increasing height the amplitude diminishes, and the time of maximum is retarded. This is in agreement with the changes up to the top of the Eiffel Tower, as shown in fig. 13. But the Lindenberg observations appear to indicate that in summer at heights of two to four kilometres the maximum occurs at mid-day, while in winter the maximum temperature at these heights is at about 8 h. The diurnal range as given by Hergesell is:

	Summer	Winter
At the ground	4·0° C	1·8° C
At 4 km	0·56	0·46

The reader is also referred to a paper by J. Durward* on the "Diurnal Variation as affected by Wind-Velocity and Cloudiness".

(b) Diurnal variation of pressure

The diurnal curve of pressure has been the subject of very comprehensive study by Hann,† Angot,‡ and Simpson.§ In the tropics the characteristic variation takes the form of a double wave whose maxima occur at 10 h and 22 h local time, and this variation shows up clearly in a barograph trace on nearly all individual days. The same features are shown at places in the temperate zone, except that there the individual days do not, except in quiet weather, show the characteristic variation, on account of the disturbance due to travelling depressions and anticyclones; and averages over a year or more must be taken in order to show the characteristic form of the curve. This semi-diurnal pressure variation can be vaguely described as a kind of wave passing round the earth two hours in advance of the sun.

In the polar regions the diurnal variation of pressure shows a combination of a single wave with a double wave whose times of maximum occur at the same time at all places on a circle of latitude, though this time varies with latitude. Simpson (loc. cit.) showed that the times of maxima at stations north of latitude 70° N were everywhere between 10 h 30 m and 13 h 30 m G.M.T.

In fig. 14 are shown for the month of July the diurnal variations of pressure (drawn from data given in the Manual of Meteorology, 2) for Batavia, the Arctic, Aberdeen and on Ben Nevis (1343 metres above M.S.L.). The time scale represents Greenwich mean time for Aberdeen and Ben Nevis, and local time for the other two stations.

(c) The diurnal variation of wind

The general nature of the diurnal variation of the surface wind over the land is represented by a maximum in the afternoon, and a minimum in the early morning. At heights of 300 metres above the ground the maximum wind occurs about midnight, and the minimum at about 10 h. Fig. 15 gives the curves of variation for the base and top of the Eiffel Tower.

The nature of the transition from one curve to the other is of some interest. At intermediate heights two maxima appear, one about midday, and one about midnight. With increasing height the day maximum diminishes while the night maximum increases, and finally the night maximum only remains. The

---

* *Professional Notes*, M.O., No. 30.
† Hann, *Wiener Sitzber. Ak. Wiss.* **127**, 1918, p. 263, and **128**, 1919, p. 379.
‡ Angot, *Ann. Bur. Cent. Météor.* Paris, 1887, Mémoires pt 1, and 1906, pt 1.
§ *Q.J. Roy. Met. Soc.* **44**, 1918, p. 1.

height at which the two maxima are of equal intensity will depend on the strength of the surface wind. Some actual observations bearing on this have

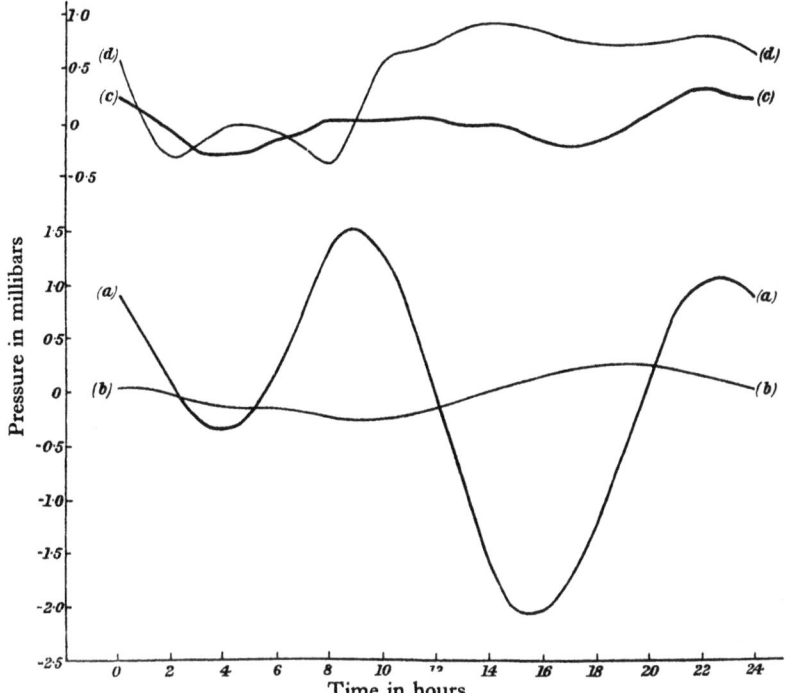

Fig. 14. Diurnal variation of pressure at Batavia, in the Arctic, at Aberdeen and on Ben Nevis. (*a*) Batavia; (*b*) Arctic; (*c*) Aberdeen; (*d*) Ben Nevis.

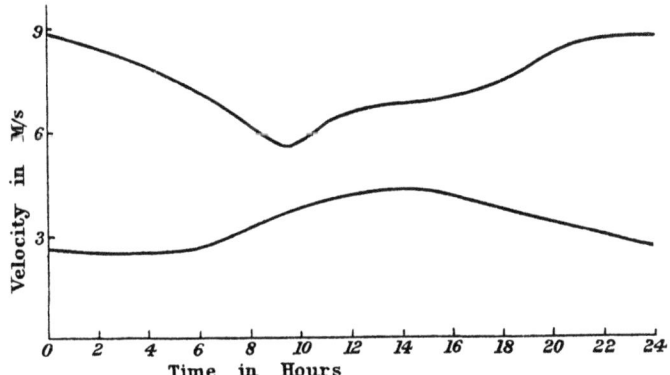

Fig. 15. Diurnal variation of wind at Parc St Maur and at the top of the Eiffel Tower.

been published by Hellmann,* who set up three anemometers at heights of 2, 16 and 32 metres above the ground in a flat meadow at Nauen. The results

* *Met. Zeit.* Jan. 1915.

of one year's observations are given in the paper referred to. Hellmann found that when strong winds were blowing all three anemometers showed a pronounced maximum in the early afternoon, and a minimum during the night. When the winds were light the anemometer at 2 metres showed the usual day maximum and night minimum; the anemometer at 16 metres showed two equal maxima at midday and midnight, with minima in the morning and afternoon; and the anemometer at 32 metres showed the same distribution of maxima and minima, the midnight maximum being the greater. At 16 metres the two maxima were just about equal in winter, but in summer there was a tendency for the day maximum to exceed the night maximum. Thus with light winds the height at which the night maximum has become equal to the day maximum is 16 metres in winter, and is between 16 and 32 metres in summer. With strong winds this height is greater than 32 metres at all seasons.

These results were confirmed by an examination of the records of wind at Potsdam, obtained from an anemometer placed 41 metres above the ground. With strong winds the maximum always occurred in the middle of the day, both in winter and in summer. With light winds, in winter the maximum is in the middle of the day, and in summer there is a weak maximum in the middle of the day, and a stronger maximum in the middle of the night.

Heywood's analysis of the observations made at Leafield* indicates that in winter the winds at 95 metres show a maximum during the day for light (<6 m/s) winds and moderate (6–9 m/s) winds, but a maximum at midday and another at midnight with strong (>9 m/s) winds. In summer light winds showed a day minimum, moderate winds a day and a night maximum, and strong winds only a day maximum.

Various writers have investigated the diurnal variation of wind at greater heights by the analysis of pilot balloon ascents. Durward† used 1736 ascents (followed by a single theodolite) made by the Meteorological Service with the British armies in France between March 1917 and September 1918. The results derived for a height of 1000 feet agreed well with those derived from the observations made at the top of the Eiffel Tower, yielding a maximum about 10 p.m., and a minimum between 9 a.m. and 10 a.m. At 2000 and 3000 feet there were two maxima, one in the morning and one about 9 p.m. Of the 1736 ascents discussed by Durward, 1101 reached 6000 feet. When all these were taken together there was no appreciable diurnal variation at 4000 and 6000 feet, but winds in different quadrants differed considerably from one another. Thus the westerly winds showed a minimum between 9 h and 10 h, while the southerly winds showed a maximum between those hours. Easterly winds showed a diurnal variation similar to the surface winds, while northerly winds showed no strongly marked feature, giving a weak minimum at night.

Tetens‡ analysed the upper wind observations at Lindenberg. In summer the winds at 4 km show a maximum in the early morning (2–4 h) and a minimum at about 14 h. At 3 km the main maximum occurs at about 6 h, and the minimum at about 14 h, but there is a secondary maximum about 16 h. At still lower levels there are complicated systems of maxima and minima, but at

---

* Q.J. Roy. Met. Soc. **57**, 1931, p. 434.   † Professional Notes, M.O., No. 15.
‡ Lindenberg Arbeiten, **14**, 1922.

the ground the maximum occurs soon after noon. In winter the main maximum at 4 km and 3 km occurs from 16 h to 18 h, and the main minimum at 8 h.

At places near the sea coast there is usually a pronounced tendency for a wind to blow from the sea to the land in the morning, and from the land to the sea in the late evening, giving the phenomena known as the land and sea breezes. These are considered more fully later in Chapter XIII.

Over the sea the diurnal variation of the wind is much less marked than over the land. Gallé* states that during the months May to October the maximum velocity of the South-east trade winds of the Indian Ocean occurs during the night hours. No detailed observations at one place are available, and it is not possible to describe with certainty the diurnal variation of winds over the sea.

(d) Diurnal variation of cloudiness

The diurnal variation of cloud amount depends to some extent on the wind direction, and to a very considerable extent on the type of cloud. At Kew Observatory, Richmond, and at certain stations in the Rhine Valley, Brunt† showed that there is a marked tendency for cloud amount to diminish in the evening. A similar investigation by Dines and Mulholland‡ for Valentia Observatory, South-west Ireland, showed a general tendency for the cloudiness to be greatest at 7 h, and to diminish steadily during the day, but the number of clear skies did not increase, though the number of overcast skies diminished. At Batavia,§ 6° S of the equator, cloudiness attains its maximum during the evening or early night, and its minimum about 4 a.m. local time. At Helwan‖ cloudiness attains its maximum during the afternoon and its minimum during the evening during the months October to May.

(e) Diurnal variation of vapour-pressure
and relative humidity

At inland stations vapour-pressure shows no very marked diurnal variation. At Kew, for example, during a month selected at random (July 1928), the mean vapour-pressure showed a maximum of 14·3 mb at about 18 h, and a minimum of 12·6 mb at about noon. On account of the small variation of vapour-pressure relative humidity shows a variation in the opposite sense to temperature.

---

* *K. Ned. Met. Inst.* No. 102.   † *Professional Notes*, M.O., Nos. 1 and 14.
‡ *Ibid.* No. 36.   § *Observations Batavia*, **38**, 1915.
‖ *Survey Dept. Cairo, Meteorological Report*, 1910.

CHAPTER II

SOME STATICAL AND THERMAL RELATIONSHIPS

§ 10. *The earth*

THE earth is a spheroid whose equatorial radius is 6378·2 km (3963 miles), and whose polar radius is 6356·5 km (3950 miles), so that it only differs slightly from a sphere. In what follows we shall in general neglect the deviation from the spherical form, except where this is explicitly stated.

The earth rotates on its axis once in 24 sidereal hours, and the rate of angular rotation, usually denoted by ω, is $7\cdot29 \times 10^{-5}$ radians per second.

Horizontal distances on the earth's surface will be given in kilometres, miles, nautical miles, or any other unit which may be convenient. The relation between these units is added here for convenience:

$$1 \text{ km} = 0\cdot62137 \text{ mile},$$

$$1 \text{ mile} = 1\cdot6093 \text{ km}.$$

The nautical mile, as used in hydrographical surveying, is identical with the geographical mile, and is defined as the length of a meridian arc of one minute of latitude. Its length varies with latitude on account of the spheroidal shape of the earth. If we denote the latitude by ϕ, the length of the geographical mile is

$$(6076\cdot8 - 31\cdot1 \cos 2\phi) \text{ feet}$$

or

$$(1852\cdot2 - 9\cdot5 \cos 2\phi) \text{ metres}.$$

Thus its length is 6045·7 feet (1842·7 metres) at the equator, and 6107·9 feet (1861·7 metres) at the poles.

The acceleration of gravity (plus the centrifugal acceleration due to the earth's rotation, see § 98 below), usually denoted by g, varies with latitude and with height above mean sea level. To a high degree of approximation the value of g at mean sea level in latitude ϕ may be represented by the formula

$$980\cdot617 \left(1 - 0\cdot00259 \cos 2\phi\right) \text{ cm/sec}^2.$$

At a height z above mean sea level the value of g in the free air is equal to the mean sea level value multiplied by $E^2/(E+z)^2$, or to a sufficient degree of approximation by $(1 - 2z/E)$, E being the radius of the earth.

Thus the value of g at height z in latitude ϕ is

$$g = 980\cdot617 \left(1 - 0\cdot00259 \cos 2\phi\right) \left(1 - 2z/E\right)$$

$$= 980\cdot617 \left(1 - 0\cdot00259 \cos 2\phi\right) \left(1 - 3\cdot14 \times 10^{-7}z\right), \text{ where } z \text{ is in metres},$$

$$= 980\cdot617 \left(1 - 0\cdot00259 \cos 2\phi\right) \left(1 - 9\cdot57 \times 10^{-8}z\right), \text{ where } z \text{ is in feet}.$$

If the isostatic compensation for elevation were perfect, the above formulae

28

would also hold at places on mountains. If there were no such compensation, the value of g would be given by

$$g = 980 \cdot 617 \, (1 - 0 \cdot 00259 \cos 2\phi) \, (1 - 5z/4E)$$

$$= 980 \cdot 617 \, (1 - 0 \cdot 00259 \cos 2\phi) \, (1 - 1 \cdot 96 \times 10^{-7} z), \text{ where } z \text{ is in metres,}$$

$$= 980 \cdot 617 \, (1 - 0 \cdot 00259 \cos 2\phi) \, (1 - 5 \cdot 97 \times 10^{-8} z), \text{ where } z \text{ is in feet.}$$

It is probable that the isostatic compensation is only partial, so that the variation of g with height should be represented by a term intermediate between $(1 - 2z/E)$ and $(1 - 5z/4E)$. The International Meteorological Tables give the latter form, which should therefore be used in reductions of pressure to mean sea level.

§ 11. *Geopotential*

The geopotential at a point at a height z above the surface of the earth is the potential energy of unit mass placed at that point. It is equal to the work done in lifting unit mass from the surface (mean sea level) up to that point, and is therefore equal to $\int_0^z g \, dz$. If we may neglect the variation of g with height the geopotential is equal to gz, where g now represents the value appropriate to the latitude.

In the c.g.s. system the unit of geopotential, which is equal to the potential of unit mass raised through unit distance in a field of force of unit strength, is $1 \text{ cm} \times 1 \text{ cm/sec}^2$. V. Bjerknes has advocated the use of a unit 10^5 times the c.g.s. unit; and has suggested for this unit the name "dynamic metre". The name "leo" has also been suggested for this unit, and would be definitely preferable, as it avoids the confusion which is inevitable if we use metres and dynamic metres side by side. But all modern writers use the name "dynamic metre". The dynamic metre has dimensions $L^2 T^{-2}$, the dimensions of (velocity)^2.

The vertical height interval separating two surfaces whose geopotentials differ by one dynamic metre is $10^5/g$ cm, and is thus approximately $10^5/981$ cm, or $1 \cdot 02$ metres.

If g_s is the surface value of g, then at height z in the free air

$$g = g_s \frac{E^2}{(E+z)^2} = g_s \, (1 + z/E)^{-2},$$

$$\int_0^z g \, dz = g_s z / (1 + z/E).$$

The error involved in using the surface value of g is equivalent to neglecting z/E by comparison with unity. Since E is over 6000 km, even at a height of 20 km the error is only of the order of $1/300$.

The variation of g with latitude is represented by the factor $(1 - 0 \cdot 00259 \cos 2\phi)$. The maximum error introduced by the adoption of a mean value for g in all latitudes is only $0 \cdot 26$ per cent, or approximately $1/400$. At first sight it may appear that such a small error is negligible, but since the pressure equivalent of a column of mercury of given length is proportional to g, an error of 1/400th in the value of g at the surface will lead to an error of about $2\frac{1}{2}$ mb in the estimate of pressure, an amount which is by no means negligible.

Table I, p. 419, gives the conversion from metres to dynamic metres, for different latitudes and for different heights.

§ 12. *Composition of the atmosphere*

The atmosphere is a mixture of gases, of which oxygen and nitrogen account for about 99 per cent, other constituents being argon, carbon dioxide, hydrogen, neon, helium, krypton and xenon, together with varying quantities of water-vapour. Apart from the water-vapour the other constituents are present in such unvarying proportions, except for local pollution due to factory chimneys and similar sources, that we may treat dry air as a uniform mixture, and leave entirely out of consideration all question of its constitution. It is only when we come to the nature of the atmosphere at very high levels which are inaccessible to direct observation that we need consider seriously the precise constitution of the air.

But while we may assume the homogeneity of the air as regards its constitution, we may not assume homogeneity of temperature or content of water-vapour. In respect to these two factors we shall find that the atmosphere shows remarkable variation from point to point, so that it is not a simple matter to decide the limits within which accuracy of observation is possible or desirable.

§ 13. *The fundamental gas equation*

If we denote pressure, absolute temperature, density and specific volume by p, T, ρ and v respectively, Boyle's law states that for a perfect gas

$$p = R\rho T \qquad \qquad \text{......(1)}$$

or $\qquad \qquad \qquad \qquad pv = RT \qquad \qquad \text{......(2),}$

where R is the gas constant appropriate to that gas. If the pressure is expressed in millibars and ρ in kilogrammes per cubic metre, or v in cubic metres per kilogramme, the value of R for dry air is 2·8703. If p is in millibars and ρ in grammes per cubic centimetre, or v in cubic centimetres per gramme, the value of R for dry air is $2 \cdot 8703 \times 10^3$. If p is measured in dynes per square centimetre and ρ in grammes per cubic centimetre, or v in cubic centimetres per gramme, the value of R is $2 \cdot 8703 \times 10^6$.

Van der Waals gave a modified form

$$(p + a/v^2)\,(v - b) = RT \qquad \qquad \text{......(3),}$$

which holds more closely than form (2). Over the range of temperatures with which we are normally concerned in meteorology the error involved in using equation (2) instead of equation (3) is negligible, and we shall use Boyle's law in the forms (1) or (2) above without further modification.

§ 14. *The density of water-vapour and of damp air; virtual temperature*

The specific gravity of water-vapour as compared with that of dry air at the same temperature and pressure is 0·6221, which may be taken as 5/8 (0·625) for all practical purposes. If the pressure of water-vapour be e and its absolute temperature be T, the density is

$$\frac{5}{8}\frac{e}{RT} \qquad\qquad ......(4),$$

where R is the constant for dry air.

If now p is the total pressure of damp air, in which the partial pressure of the water-vapour is e, the partial pressure of the dry air alone is $p-e$. The density of the mixture is therefore

$$\frac{p-e}{RT}+\frac{5}{8}\frac{e}{RT}$$

or

$$\frac{p}{RT}\left(1-\frac{3e}{8p}\right) \qquad\qquad ......(5)$$

or p/RT', where T' is the virtual temperature defined by

$$T'=T/(1-\tfrac{3}{8}e/p) \qquad\qquad ......(6).$$

Thus the virtual temperature of the damp air is the temperature at which dry air of the same pressure would have the same density as the damp air.

To a high degree of approximation, in dealing with damp air, we may use equations (1) or (2) in the form

$$p=R'\rho T \quad\text{or}\quad pv=R'T,$$

where $R'=R/(1-\tfrac{3}{8}e/p)=R(1+\tfrac{3}{8}e/p)$ approximately.

But this equation must only be used at temperatures which are sufficiently high to permit of the air remaining always unsaturated.

§ 15. *Vapour-pressure and its control by temperature*

In a closed space which contains a supply of water the amount of water-vapour is independent of the presence of other gases. In other words, the pressure of the water-vapour is independent of the pressure of the dry air. This law is known as Dalton's law. For a given temperature there is a maximum limit to the vapour-pressure, attained when the air is "saturated". The saturation vapour-pressure increases with temperature, and an equation connecting these two variables can be derived by the use of the second law of thermodynamics. This equation is derived in § 63 below.

Fig. 16 shows the variation of vapour-pressure with temperature. This diagram represents the saturation vapour-pressure over water for temperatures above freezing point, and over ice for temperatures below freezing point. The vapour-pressure is slightly greater over water than over ice at temperatures below freezing point. For example, at 250° A (−23° C) the pressure

over ice is 0·77 mb, that over water 1·77 mb. According to the Smithsonian
Meteorological Tables the vapour-pressure over ice falls to 0·021 at $-55°$ C,
to 0·0054 mb at $-65°$ C, and to 0·0026 mb at $-70°$ C. The saturation vapour-
pressure is greater when the water is present in the form of drops. A table
of saturation vapour-pressures at different temperatures is given on p. 420.

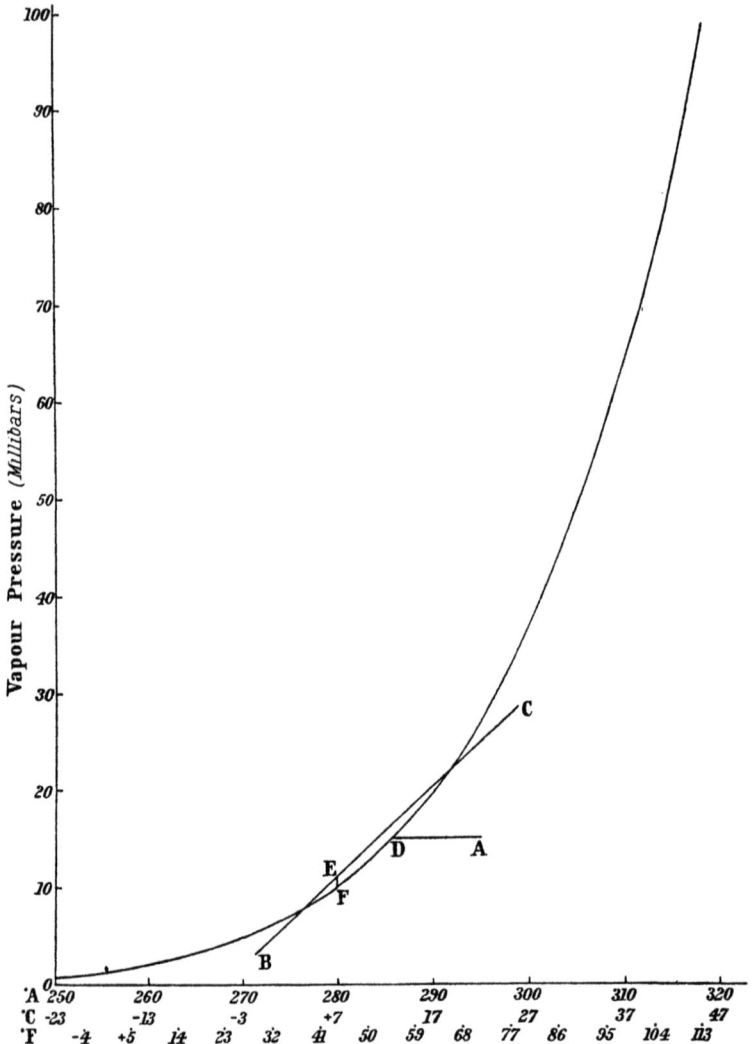

Fig. 16. Saturation vapour-pressure as a function of temperature.

The figures quoted above refer to saturated air. When air is unsaturated,
the vapour-pressure is less than the saturation value, and the ratio of the actual
vapour-pressure to the saturation value gives the relative humidity, which is
most conveniently expressed as a percentage figure. Another way of specifying

the state of air as regards content of water-vapour is by means of the "humidity mixing ratio", usually denoted by x, which expresses the relative proportions by weight of dry air and water-vapour in a given volume of the damp air. This is equivalent to saying that damp air is made up of x grammes of water-vapour to every gramme of dry air. Then

$$x = 0.622 \frac{e}{p-e} = 0.622 \frac{e}{p} \text{ approximately} \qquad \ldots\ldots(7).$$

The value of x remains unchanged for a given mass of air, so long as there is no evaporation or condensation, or mixing with other air of different humidity. So long as x remains constant the relative humidity will rise or fall as the temperature falls or rises, since the vapour-pressure remains constant while the saturation vapour-pressure varies in the same sense as the temperature. The humidity mixing ratio x is so useful in the discussion of the thermodynamics of moist air that it merits some further consideration. If we call the specific gravity of water-vapour ϵ $(= 0.622)$, then

$$\frac{x}{\epsilon} = \frac{e}{p-e}, \quad 1 + \frac{x}{\epsilon} = \frac{p}{p-e}.$$

Let v be the specific volume of the dry air alone, then

$$(p-e)\,v = RT,$$
$$ev = xRT/\epsilon,$$
$$pv = (1 + x/\epsilon)\,RT \qquad \ldots\ldots(8).$$

The density of the dry air alone is $1/v$, and the density of the mixture of dry air and water-vapour is $(1+x)/v$. If ρ be the density of the mixture, then

$$\rho = (1+x)/v = \frac{p}{RT}\frac{1+x}{1+x/\epsilon} = \frac{p}{RT'}, \qquad \ldots\ldots(9).$$

(See equation (6).) Hence

$$T' = T/(1 - \tfrac{3}{8}e/p) = T\frac{1+x/\epsilon}{1+x} \qquad \ldots\ldots(10).$$

If we prefer to define the density as $p/R'T$, then

$$R' = R/(1 - \tfrac{3}{8}e/p) = R\frac{1+x/\epsilon}{1+x} \qquad \ldots\ldots(11).$$

The value of x for a given value of e and p is readily computed. In fig. 17, p. 61, will be found a series of lines which show the values of x for saturation at different temperatures and pressures. The numbers there given are the numbers of grammes of water-vapour per kilogramme of dry air, and these numbers must be divided by 1000 in order to obtain x; e.g. for 10 grammes read 0.01.

§ 16. Pressure and its variation with height

In practice the meteorologist measures pressure by means of a mercury barometer, and deduces the pressure from the height of a column of mercury. The pressure is in fact equal to the weight of a column of mercury of unit cross-section and of the height of the observed column. To reduce the pressure to a form comparable with a fixed standard, allowance is made for the variation of gravity with latitude, and for the expansion or contraction of the mercury and the scale, by methods which are described in the *Meteorological Observer's Handbook*, or any descriptive textbook.

The pressure with which we are concerned is, however, not a length. Its dimensions are those of force/area, MLT^{-2}/L^2, or $ML^{-1}T^{-2}$. It may be measured in terms of any convenient unit of length, such as the inch or millimetre, or, preferably, the scale may be graduated so as to give pressure in terms of a unit such as the millibar, which is defined as a pressure of 1000 dynes per square centimetre.

By definition the pressure at any point is the weight of a vertical column of air of unit cross-section, whose base is centred at the given point, and which extends to the top of the atmosphere. Let p, ρ and T be the pressure, density and absolute temperature at height z in the atmosphere, and let $p+dp$ be the pressure at height $z+dz$. Then by definition, dp is the weight of a disc of air of unit cross-section and of height dz. Hence

$$dp = -g\rho\,dz \quad \text{or} \quad \frac{\partial p}{\partial z} = -g\rho \qquad \ldots\ldots(12).$$

This is the fundamental statical equation of meteorology, and will be continually required in any physical discussion. It could be integrated if ρ could be expressed as a function of z. This is not possible, and it is convenient to transform the equation by the use of equation (1) above,

$$p = R\rho T.$$

Substituting for ρ in equation (12) we find

$$\frac{1}{p}\frac{\partial p}{\partial z} = -\frac{g}{RT} \qquad \ldots\ldots(13).$$

This equation is readily integrable when T is constant, or can be represented as a function of z. The variation of g with height can be neglected in this connection. If T is constant, equation (13) yields on integration

$$\log_e p_0 - \log_e p = gz/RT \qquad \ldots\ldots(14),$$

p_0 being the pressure at $z=0$. This may be written

$$z = \frac{RT}{g}\,(\log_e p_0 - \log_e p)$$

$$= \frac{RT}{g\log_{10}e}\,(\log p_0 - \log p) \qquad \ldots\ldots(15).$$

If height is measured in feet,

$$z = 221\cdot1\,T\,(\log p_0 - \log p) \qquad \ldots\ldots(16).$$

If height is measured in metres,

$$z = 67\cdot4\,T\,(\log p_0 - \log p) \qquad \ldots\ldots(17).$$

Equation (14) or any of the three corresponding forms (15), (16) or (17) will give the relation of height and pressure in an atmosphere which has a uniform temperature T. In equation (14) the value of R is $2\cdot8703 \times 10^6$.

In an atmosphere in which the temperature decreases uniformly with height at a rate β, the temperature at height z may be represented by the expression $T_0 - \beta z$. Substituting this for T in equation (13) we find

$$\frac{1}{p}\frac{\partial p}{\partial z} = -\frac{g}{R(T_0 - \beta z)} \qquad \ldots\ldots(18).$$

On integration this yields

$$\log p = \frac{g}{R\beta}\log(T_0 - \beta z) + \text{const.}$$

or

$$\log \frac{p}{p_0} = \frac{g}{R\beta}\log \frac{T_0 - \beta z}{T_0},$$

or

$$\frac{T_0 - \beta z}{T_0} = \left(\frac{p}{p_0}\right)^{\frac{R\beta}{g}} = \left(\frac{p}{p_0}\right)^{0\cdot293\beta} \qquad \ldots\ldots(19),$$

if β is measured in degrees C per 100 metres.

§ 17. *Barometric altimetry*

(a) Isothermal atmosphere

Equations (16) or (17) above may be used to determine the height z at which a particular pressure p is observed, assuming some standard value of the temperature T. The commonest type of altimeter is an aneroid barometer, with a scale graduated in feet (or metres) instead of in millibars, the graduation being based on the assumption that the mean temperature is 50° F, or 10° C. The zero of the height scale is adjustable, and allowance is made for variations in ground pressure by rotating the height scale until it reads zero at ground level. This method is strictly correct, since according to equation (14), which may be written

$$z = \frac{RT}{g}\log_e \frac{p_0}{p},$$

the height interval from pressure p_0 to pressure p is always the same for the same value of T, whether the ground pressure is p_0 or not.

The most important correction which has to be made to this type of altimeter is that for the variation of the mean temperature from the standard value of 50° F or 10° C. The atmosphere is never strictly isothermal through any considerable range of height, and it is therefore not strictly permissible to use equation (14). It is, however, frequently used, giving T the mean value of the temperature as observed from the ground up to the level z. Then z is proportional to the mean absolute temperature T. The rule for correction for variations from the standard temperature 10° C is readily seen to be: "For every 1° C increase of temperature above the standard value, increase the height as estimated by the altimeter by 1/283 of this value; for decrease of temperature below the standard value subtract the correction from the estimated height."

Allowance can be made for the humidity of the atmosphere by taking the virtual temperature instead of the ordinary temperature.

The use of equation (14) with the mean observed temperature for T is not strictly permissible, since the atmosphere is never even approximately iso-thermal except through a very restricted range of height, but its use can be justified as theoretically sound if we use for T the harmonic mean of the observed temperatures, taken at equal intervals of height.

(b) I.C.A.N. ATMOSPHERE

The International Commission for Air Navigation has put forward a specification of a standard atmosphere with a view to uniformity in the estimation of aircraft performances. Up to 11 km this standard atmosphere is specified by equation (19) above, with $p_0 = 1013 \cdot 2$ mb, $T_0 = 288°$ A, and the lapse-rate equal to $6 \cdot 5°$ C per kilometre of height. Equation (19) then becomes

$$\frac{p}{p_0} = \left(\frac{288 - 0 \cdot 0065 z}{288}\right)^{5 \cdot 256} \qquad \ldots\ldots(20),$$

where z is now measured in metres.

Above 11 km the temperature is assumed constant at $-56 \cdot 5°$ C. For heights above 11 km we therefore use equation (14) which now takes the form

$$\log \frac{p_{11,000}}{p} = \frac{z - 11,000}{14,600} \qquad \ldots\ldots(21).$$

If the altimeter height is z' and the true height z, at a level where the pressure is p and the temperature T, and if p_0 is the sea-level pressure, a correction can be deduced to z', assuming the lapse-rate to have the standard value $6 \cdot 5°$ C per kilometre. Equation (20) becomes

$$\frac{p_0}{p} = \left(\frac{T + 0 \cdot 0065 z}{T}\right)^{5 \cdot 256} = \left(\frac{288}{288 - 0 \cdot 0065 z'}\right)^{5 \cdot 256}$$

By simple algebra we find

$$z - z' = \frac{z'\,(T - 288 + 0 \cdot 0065 z')}{288 - 0 \cdot 0065 z'},$$

or

$$z - z' = z'\left(\frac{T}{288 - 0 \cdot 0065 z'} - 1\right) \qquad \ldots\ldots(22).$$

§ 18. *The specific heat of air*

The specific heat of a substance is defined as the thermal capacity per unit mass, or, in other words, the ratio of the quantity of heat required to raise the temperature of unit mass of it by a given amount to the quantity of heat required to raise the temperature of unit mass of water by the same amount. In practice the unit of heat adopted, the gramme-calorie, is the amount of heat required to raise the temperature of 1 gramme of water by 1° C. Thus, if the specific heat of a substance is c, the amount of heat required to raise the temperature of 1 gramme of it by t degrees is ct calories.

The specific heat thus defined only has a definite meaning when the conditions under which the heating takes place are defined with precision. If the substance in question is a gas, and it is allowed to expand when heated, work is done against the external pressure, and some of the heat supplied is used in doing this work. The specific heat is thus greater when expansion is possible than it is when the volume is kept constant. In connection with gas it is customary to speak of two specific heats, that at constant volume, and that at constant pressure. These are denoted usually by the symbols c_v and c_p respectively. The ratio of these two quantities is also a physical constant of much importance; c_p/c_v is usually denoted by the symbol γ. It is possible to determine γ independently of the determination of c_p and c_v, since γ enters into a number of physical relations. For example, the velocity of sound in a gas whose temperature is T is $\sqrt{\gamma RT}$, where γ and R are the appropriate constants for that gas.

The specific heats of dry air have been determined experimentally by a great number of physicists. A full table of such determinations will be found in Partington and Shilling's book on the *Specific Heats of Gases*. In the same work will be found a table of the values of the specific heats which the authors regard as the most reliable. The values of the constants for dry air given in these tables are

$$c_p = 0.2396, \quad c_v = 0.1707, \quad \gamma = 1.403.$$

The units in which c_p and c_v are given are gramme-calories. In the present work we shall adhere to these values.

The specific heats of a gas are related by a simple equation which we shall now derive. Suppose a quantity of heat dQ is supplied to unit mass of air, whose initial absolute temperature, pressure and specific volume are T, p and v respectively. Some of the heat is used in raising the temperature by an amount dT, while the remainder is used in doing work against the external pressure during the process of expansion. If the specific volume is increased by an amount dv, the work done in expanding against the pressure is $p\,dv$. This is equivalent to an amount of heat $Ap\,dv$, where A is the reciprocal of the mechanical equivalent of heat. It follows that

$$dQ = c_v\,dT + Ap\,dv \qquad \ldots\ldots(23).$$

This equation states that

Heat added = Increase in internal energy
+ work done against external pressure.

The internal energy of unit mass of air is taken to be $c_v T$, whether the volume remains constant or not, during the series of changes of state under consideration. This question will be found discussed in detail in any textbook of heat, e.g. Preston's *Heat*, Chapter VIII, p. 662.

By Boyle's law $\quad pv = RT \quad$ and $\quad p\,dv + v\,dp = R\,dT$,

hence $\qquad\qquad dQ = (c_v + AR)\,dT - Av\,dp \qquad \ldots\ldots(24).$

If the expansion takes place under the condition that the pressure remains constant,

$$dQ = c_p\,dT = (c_v + AR)\,dT \qquad \ldots\ldots(25).$$

Hence
$$c_p = c_v + AR \quad \text{or} \quad c_p - c_v = AR \qquad \ldots\ldots(26).$$

Equation (24) may therefore be written
$$dQ = c_p\, dT - Av\, dp \qquad \ldots\ldots(27).$$

§ 19. *The adiabatic equation*

If a mass of air expands or contracts without addition or subtraction of heat, the quantity $dQ = 0$. Equation (27) then becomes
$$c_p\, dT - Av\, dp = 0,$$
$$c_p\, dT - \frac{ART}{p}\, dp = 0,$$
$$c_p\, dT/T - (c_p - c_v)\, dp/p = 0 \qquad \ldots\ldots(28).$$

If we write $c_p/c_v = \gamma$, equation (28) becomes
$$\gamma\, dT/T - (\gamma - 1)\, dp/p = 0 \qquad \ldots\ldots(29),$$
which yields on integration
$$T^\gamma/p^{\gamma-1} = \text{constant} \qquad \ldots\ldots(30).$$

This equation may be transformed into another form by substitution for T
$$\frac{(pv)^\gamma}{p^{\gamma-1}} = \text{constant},$$
or
$$pv^\gamma = \text{constant} \qquad \ldots\ldots(31).$$

Note the values
$$\frac{c_p}{AR} = \frac{c_p}{c_p - c_v} = \frac{\gamma}{\gamma - 1} = 3\cdot49.$$

§ 20. *Potential temperature*

The potential temperature of dry air is defined as the temperature it would attain if brought adiabatically to a standard pressure (which may conveniently be taken as 1000 mb). Let the standard pressure be p_0, and let the temperature and pressure of a mass of air be initially T and p respectively. When the air is brought to a pressure p_0 without addition or subtraction of heat, let it take up a temperature θ; i.e. let θ be the potential temperature of the air. Then by equation (30) above
$$\frac{T^\gamma}{p^{\gamma-1}} = \frac{\theta^\gamma}{p_0^{\gamma-1}}, \quad \text{or} \quad \theta = T\left(\frac{p_0}{p}\right)^{\frac{\gamma-1}{\gamma}} \qquad \ldots\ldots(32)$$

or
$$\log\theta - \log T = \frac{\gamma - 1}{\gamma}(\log p_0 - \log p)$$
$$= 0\cdot288\,(\log p_0 - \log p) \qquad \ldots\ldots(33).$$

This equation may be used to evaluate θ. Alternatively we may use equation (32) in the form
$$\theta = T\left(\frac{p_0}{p}\right)^{\frac{\gamma-1}{\gamma}} = T\left(\frac{p_0}{p}\right)^{0\cdot288} \qquad \ldots\ldots(34).$$

A table of values of $\left(\dfrac{p_0}{p}\right)^{0.288}$ for $p_0 = 1000$, for a range of values of p from 10 to 1090, is shown in Table II at the end of this volume (p. 419). By the use of this table θ is readily computed on a slide-rule. The potential temperature can also be read directly from a tephigram (see p. 77), with sufficient accuracy for all practical purposes.

§ 21. *The adiabatic lapse-rate; vertical stability for dry air*

The variation of pressure in the vertical direction in the atmosphere, when there is no acceleration in the vertical, is given by the relation

$$\partial p / \partial z = -g\rho$$

already deduced in § 16. The effect of motion on the truth of this relation will be considered later in Chapter VIII, in connection with the derivation of the equations of motion, but we shall find that the statical relation above is true to a very high degree of approximation.

To determine whether the atmosphere is stable or not, consider what happens to any small element of air which is displaced through a small vertical distance dz from its initial height z. If in its new position it is subjected to forces which tend to restore it to its original level, the atmosphere is said to be in stable equilibrium; if in its new position it is subjected to forces which tend to displace it still further from its original position, the atmosphere is said to be in unstable or labile equilibrium; while if the element of air in its new position is subject to no forces tending either to restore it to its original position, or to displace it still further from its original position, the equilibrium is said to be neutral.

Since a moving element of air will at all stages of its ascent take up automatically the pressure of its immediate surroundings, the relative densities of the moving element and of its environment will be determined by the absolute temperatures.

Let p, ρ and T be the pressure, density and absolute temperature at a point at height z in the atmosphere, and let p, ρ' and T' be the corresponding variables for a small element of air which is moving vertically upward, at the time when it is at the level z. In moving from a height z to height $z + dz$, let the element of air have its pressure changed to $p + dp$, and its temperature to $T' + dT'$; at this height $z + dz$ the environment will have the same pressure $p + dp$, but its temperature will be $T + dT$. If the moving mass of air neither receives nor loses heat during its displacement, it follows from equation (29) above that

$$\frac{dT'}{T'} = \frac{\gamma - 1}{\gamma} \frac{dp}{p},$$

and $\quad \dfrac{\partial T'}{\partial z} = \dfrac{\gamma - 1}{\gamma} \dfrac{T'}{p} \dfrac{\partial p}{\partial z} = -\dfrac{\gamma - 1}{\gamma} g\rho \dfrac{T'}{R\rho T} = -\dfrac{g}{R} \dfrac{\gamma - 1}{\gamma} \dfrac{T'}{T} = -\dfrac{gA}{c_p} \dfrac{T'}{T}$ ......(35).

Thus the temperature of a mass of air moving in the vertical decreases with height at a rate equal to $\dfrac{gA}{c_p} \dfrac{T'}{T}$ per unit of height. If the temperature T' of the moving air does not differ appreciably from the temperature T of the environment, the rate of decrease of temperature with height is gA/c_p. If, for example, we desire to consider the changes of temperature in an element of air displaced vertically from its original position as part of the normal environment, we have to take $T' = T$, and the rate of change of temperature with height is then given accurately by gA/c_p. We shall represent the dry adiabatic lapse-rate by the symbol Γ. Substituting for g, A and c_p, we find that

$$\Gamma = \frac{gA}{c_p} = 0\cdot0000986 \qquad \ldots\ldots(36).$$

This is equal to the rate of decrease of temperature per unit of height; i.e. per centimetre. It is more convenient to express the result as $0\cdot986°$ C per 100 metres or to a sufficient degree of approximation as $1°$ C per 100 metres. This is known as the *dry adiabatic lapse-rate.* It is the rate of decrease of temperature with height of a mass of air moving through an environment whose temperature does not differ by a finite amount from its own. The qualification as to the equality of the temperatures of moving air and environment is of importance; when there is a finite difference of temperature between the moving air and its environment, the factor T'/T must be taken into account in the rate of decrease of temperature with height.

In a dry atmosphere whose lapse-rate had the value Γ at all heights, any element of mass displaced in the vertical would have the temperature of its environment at all stages of its ascent. In any atmosphere whose lapse-rate differed from the dry abiabatic rate, any element displaced from its original level would initially have its temperature diminished at a rate Γ and would deviate from the temperature of its environment. When the displacement had extended through a finite range of height, it would be necessary to take into account the factor T'/T in the original equation for the diminution of temperature with height. (See also § 134 below.)

If the temperature in the atmosphere be T at height z, and $T + dT$ at height $z + dz$, and a small element of air be displaced from height z to height $z + dz$, its temperature in its displaced position will be $T + dT'$, where $dT' = -\Gamma dz$. In the displaced position this air will be heavier or lighter than its immediate environment according as its temperature is lower or higher than that of its environment, i.e. according as

$$dT > \quad \text{or} \quad < dT',$$

i.e. as

$$-\frac{\partial T}{\partial z} < \quad \text{or} \quad > \Gamma \qquad \ldots\ldots(37).$$

If it is heavier than its new environment, it will tend to sink back to its original level, so that it is initially stable. If it is lighter than its new environment, it will tend to move further from its original position, and the original condition is then unstable. If it has the same density as the new environment, it will stay in its displaced position, and the atmosphere is then in "neutral" equilibrium. The argument needs a slight re-arrangement if the displacement is downward, but leads to the same result.

Thus a dry atmosphere is in stable, unstable, or neutral equilibrium, according as the lapse-rate is less than, greater than, or equal to the dry adiabatic lapse-rate.

From the definition of potential temperature it is seen that the above condition for stability can be stated as follows: The atmosphere is in stable, unstable, or neutral equilibrium according as the potential temperature increases, decreases, or is constant with increasing height. This result can be readily derived algebraically as follows: Differentiating equation (34) with respect to z, we find

$$\frac{1}{\theta}\frac{\partial\theta}{\partial z}=\frac{1}{T}\frac{\partial T}{\partial z}-\frac{\gamma-1}{\gamma}\frac{1}{p}\frac{\partial p}{\partial z}$$

$$=\frac{1}{T}\left(\frac{\partial T}{\partial z}+\frac{\gamma-1}{\gamma}\frac{g}{R}\right).$$

But since $\gamma=c_p/c_v$, and $c_p-c_v=AR$,

$$\frac{\gamma-1}{\gamma}\frac{g}{R}=\frac{g}{R}\frac{c_p-c_v}{c_p}=\frac{gA}{c_p}=\Gamma \qquad\qquad ......(38).$$

Hence

$$\frac{1}{\theta}\frac{\partial\theta}{\partial z}=\frac{1}{T}\left(\frac{\partial T}{\partial z}+\Gamma\right) \qquad\qquad ......(39).$$

Combining the condition for stability given in (37) above with the relationship shown in equation (39), we find the condition for stability may be stated in the form: The atmosphere will be in stable, unstable, or neutral equilibrium according as

$$\partial\theta/\partial z>, \quad <, \quad \text{or} \quad =0 \qquad\qquad ......(40),$$

i.e. according as the potential temperature increases, decreases, or remains constant with height.

In deriving equation (35) above it was assumed that a mass of air of temperature T' moved through an environment whose temperature was T, and it was found that the rate of change of temperature of the moving air with height was proportional to T'/T. This formula will apply to the discussion of the penetration of a limited mass of air through its environment. It is not here suggested that convection in the atmosphere is to be regarded as simple penetration of this kind. The general question of convection is discussed later in § 134, p. 221, and the aim of the present discussion is to establish the results shown in (37) and (30) above, that the limiting condition for stability of dry air is the dry adiabatic lapse-rate, whose magnitude is Γ or gA/c_p, which to a high degree of approximation is equal to $1°$ C per 100 metres.

§ 22. *Stability in a moist atmosphere (unsaturated)*

If the atmosphere contains water-vapour, the above argument requires very careful reconsideration. If the composition of the damp air is uniform, the proportion of water-vapour being uniform throughout, then it is only necessary to define the specific heats c_p and c_v as the constants appropriate to the damp air. Let x be the humidity mixing ratio, and let c_p' be the specific heat of water-vapour. Then the specific heat at constant pressure of the damp air will be

$$c_p'' = \frac{c_p + xc_p'}{1 + x} \quad \text{or} \quad c_p - \frac{x(c_p - c_p')}{1 + x}.$$

For values of x which can occur in the atmosphere the difference between this expression and c_p is small, and the adiabatic lapse-rate which fixes the limit for stability will be for all practical purposes the same as for a dry atmosphere, $1°$ C per 100 metres.

We have seen that the gas constant for damp air is R', or $R/(1 - \frac{3}{8}e/p)$, where R is the gas constant for dry air. Then the density of the damp air is $p/R'T$. So long as there is neither evaporation nor condensation the value of R' remains unchanged. For under these conditions the humidity mixing ratio x remains constant, and since

$$x = 0.622\,\frac{e}{p - e} = 0.622\,\frac{e/p}{1 - e/p},$$

it follows that this involves that e/p also remains constant, and so R' is constant.

Now suppose two masses of different water-vapour content are separated by a horizontal surface of separation, the different variables being distinguished by the use of the suffixes 1 and 2 for the lower and upper respectively. At the boundary the upper mass must have the lower density, if there is to be equilibrium. Hence

$$R_1'T_1 < R_2'T_2.$$

If the lower has the greater water-vapour content, or the higher value of R', then it follows that $T_1 < T_2$, or the upper mass, has the higher temperature. This is in accord with the well-known fact that an inversion is frequently found at the upper boundary of a cloud. It appears from the analysis above that the inversion is a statical necessity, if stability is to exist. A cloud layer which had a layer of dry air above it could not remain in equilibrium unless the dry air were warmer than the cloud layer immediately below it, and a mass of rising air which is damper than its environment must come to rest with warmer air above it. If the rising air is nearly saturated, while the general environment has relative humidity 60 per cent, the inversion required is about $1°$ F, if the temperature is about $50°$ F. This is the usual order of magnitude of the observed inversions at the upper surfaces of clouds.

If the upper mass is damper than the lower mass, or $R_1' < R_2'$, it is possible, though not necessary, for T_2 to be less than T_1, i.e. for the damp mass to be colder than the dry air beneath it. This is in keeping with the observed fact that floating clouds are often slightly colder than the surrounding air at the same level, though observed differences appear to be slightly greater than can be explained by the above analysis.

The special case discussed above emphasises the need for careful consideration of the distribution of water-vapour in any discussion of stability. In the example considered the condition for stability is not a lapse but an inversion. The question is, however, amenable to discussion in quite general terms.

Let the air be initially in equilibrium in the vertical, and let R be the gas-constant, e the vapour-pressure, p the total pressure, and T the absolute temperature at height z. Let a small element of mass originally at height z be displaced to a height $z + dz$, where its pressure is p', no heat being communicated to or extracted from it in the process. The argument of § 21 can be repeated here with R' substituted for R, and the specific heat at constant pressure of the damp air substituted for c_p. This specific heat we have already seen to be

$$\frac{c_p + x c_p'}{1 + x} = c_p'' \text{ say.}$$

Then the temperature of the element of air which we are considering, after its displacement from height z to $z + dz$, will be

$$T' = T - \frac{gA}{c_p''} dz \qquad \dots\dots(41).$$

Its density at its new level will be

$$\frac{p'}{R'T'} = \frac{p'}{R'T}\left(1 + \frac{gA}{Tc_p''} dz\right) \qquad \dots\dots(42)$$

if we neglect dz^2 and higher powers. The density of its environment will be

$$\frac{p'}{R'T}\left(1 - \frac{1}{R'}\frac{\partial R'}{\partial z} dz - \frac{1}{T}\frac{\partial T}{\partial z} dz\right) \qquad \dots\dots(43).$$

Note that since R' is a function of height its variation with height must be considered. If the atmosphere is stable, the density given by (42) must be greater than that given by (43), for dz positive. Also

$$\frac{gA}{c_p''T} > -\frac{1}{R'}\frac{\partial R'}{\partial z} - \frac{1}{T}\frac{\partial T}{\partial z},$$

or

$$-\frac{\partial T}{\partial z} < \frac{gA}{c_p''} + \frac{T}{R'}\frac{\partial R'}{\partial z} \qquad \dots\dots(44).$$

If R' is constant at all heights, the second term on the right-hand side of (44) disappears, and the limiting lapse-rate is

$$gA/c_p'' \quad \text{or} \quad gA\frac{1 + x}{c_p + x c_p'}.$$

Since

$$x = 0 \cdot 622 \frac{e}{p - e},$$

it follows that the value of x will seldom exceed about $0 \cdot 04$ and the value of c_p'' may be assumed to be c_p without serious error.

Returning to (44) above, we may now assume that the first term on the right-hand side is in practice equal to the dry adiabatic lapse-rate. We can find the order of magnitude of the term $\frac{1}{R'}\frac{\partial R'}{\partial z}$ as follows: To a high degree of approximation

$$R' = R(1 + \tfrac{3}{8}e/p).$$

The larger scale variations of R' with height will be due to large variations of e with height. Again

$$\frac{\partial R'}{\partial z} = \tfrac{3}{8}R\frac{\partial}{\partial z}\left(\frac{e}{p}\right) = \frac{3}{8}\frac{R}{p}\frac{\partial e}{\partial z} \text{ approximately,}$$

and to a high degree of approximation

$$\frac{1}{R'}\frac{\partial R'}{\partial z}=\frac{3}{8}\frac{1}{p}\frac{\partial e}{\partial z}.$$

Inequality (44) may thus be written

$$-\frac{\partial T}{\partial z}<0\text{\textperiodcentered}0001+\frac{3}{8}\frac{T}{p}\frac{\partial e}{\partial z} \qquad \ldots\ldots(45).$$

We shall assume $T/p = 300/1000 = 0\text{\textperiodcentered}3$ as sufficiently near for the purpose of evaluating the order of magnitude of the terms. The two terms on the right-hand side of (44) or (45) will then be equal when

$$\frac{\partial e}{\partial z}=0\text{\textperiodcentered}0001\times\frac{8}{0\text{\textperiodcentered}9}=0\text{\textperiodcentered}001\text{ mb/cm}=1\text{ mb/10 metres}.$$

It is evident that the effect of a decrease of e with height, corresponding to dry air above damp air, makes the limiting condition for stability a lapse-rate less than the dry adiabatic, and in special cases even an inversion. When e increases with height, the limiting condition for stability is a lapse-rate greater than the dry adiabatic.

If T_v is the virtual temperature, then by its definition

$$RT_v = R'T$$

and the limiting condition for stability is given by

$$-\frac{1}{T_v}\frac{\partial T_v}{\partial z}=-\frac{1}{T}\frac{\partial T}{\partial z}-\frac{1}{R'}\frac{\partial R'}{\partial z}<\frac{gA}{c_p''T} \qquad \ldots\ldots(46).$$

If we may assume that c_p and c_p'' defined above are equal, this condition is that the lapse-rate of the virtual temperature shall equal the dry adiabatic lapse-rate multiplied by T_v/T. Since this factor is in practice very nearly unity, condition (46) effectively means that the condition for stability is that the lapse-rate of virtual temperature shall be equal to the dry adiabatic lapse-rate.

§ 23. *The effect of vertical motion on the lapse-rate*

Let a thin layer of air be raised or lowered adiabatically, from a level where the pressure is p to a level where the pressure is p'. Then since the amount of air in the layer is unchanged, and the potential temperature of the air is also unchanged, $\partial\theta/\partial p$ remains unaltered by the transfer. For if $\Delta\theta$ and Δp be the differences of potential temperature and pressure at the top and bottom of the layer, neither of these quantities is altered by the motion, and thus $\partial\theta/\partial p$, which is the limiting value of $\Delta\theta/\Delta p$ when the layer is made indefinitely shallow, is also unaltered. Hence $\frac{1}{\theta}\frac{\partial\theta}{\partial p}$ is also unaltered by the motion.

By equation (39) above

$$\frac{1}{\theta}\frac{\partial\theta}{\partial z}=\frac{1}{T}\left(\frac{\partial T}{\partial z}+\Gamma\right).$$

Then $$\frac{1}{\theta}\frac{\partial\theta}{\partial p}=\frac{1}{\theta}\frac{\partial\theta}{\partial z}\cdot\frac{\partial z}{\partial p}=-\frac{1}{g\rho}\frac{1}{\theta}\frac{\partial\theta}{\partial z}=-\frac{R}{gp}\left(\frac{\partial T}{\partial z}+\Gamma\right).$$

Hence $\dfrac{1}{p}\left(\dfrac{\partial T}{\partial z}+\Gamma\right)$ remains unchanged by adiabatic motion. If the values of z, p, T in the new position be indicated by accented letters,

$$\frac{1}{p'}\left(\frac{\partial T'}{\partial z'}+\Gamma\right)=\frac{1}{p}\left(\frac{\partial T}{\partial z}+\Gamma\right) \qquad \dots\dots(47).$$

In other words the difference between the lapse-rate and the dry adiabatic is proportional to the pressure, and the ascent of a layer will cause the lapse-rate to approach more closely to the dry adiabatic, while descent will cause the lapse-rate to deviate still further from the dry adiabatic.

The discussion above can be immediately extended to the case when the column of air changes its horizontal extent, by spreading outward or by convergence. The only change in the argument consists in assuming that the horizontal area is initially S, and changes to S'. For a given mass of air it will now be $S\Delta p$ which will remain constant, instead of Δp. The result can be written down as follows, without further discussion:

$$\frac{1}{S'p'}\left(\frac{\partial T'}{\partial z'}+\Gamma\right)=\frac{1}{Sp}\left(\frac{\partial T}{\partial z}+\Gamma\right) \qquad \dots\dots(48).$$

§ 24. The dry adiabatic lapse-rate in geodynamic units

The value of the adiabatic lapse-rate is proportional to g, and therefore differs slightly from place to place. Let position relative to the earth's surface be specified by means of the geopotential V, measured in dynamic metres. Then since the separation of two surfaces of unit difference of geopotential in dynamic metres is $10^5/g$ cm, it follows that

$$dV = dz \times g/10^5.$$

The adiabatic lapse-rate is therefore given by

$$-\frac{\partial T}{\partial V}=-\frac{\partial T}{\partial z}\times\frac{10^5}{g}=\frac{gA}{c_p}\frac{10^5}{g}=\frac{10^5 A}{c_p}=\frac{2\cdot392\times10^{-3}}{0\cdot2396}=0\cdot998\times10^{-2},$$

or $$-\frac{\partial T}{\partial V}=1^{\circ}\text{ C per 100 dynamic metres} \qquad \dots\dots(49),$$

with an accuracy of $0\cdot17$ per cent. This value is not dependent on the local value of gravity.

§ 25. The variation of density with height; the auto-convection gradient

It has been shown that the condition that the atmosphere should be in stable equilibrium is that the lapse-rate should be less than the dry adiabatic lapse-rate gA/c_p, or approximately 1° C per 100 metres. In an incompressible fluid the condition for stability is that the density shall not increase with height, and it is important to realise that in a compressible atmosphere the condition for stability is not merely that the density shall decrease with height.

The general equation for the variation of density with height is readily derived by the use of the statical equation

$$\frac{\partial p}{\partial z} = -g\rho \qquad\qquad \ldots\ldots(50)$$

and the gas-equation $\qquad\qquad p = R\rho T \qquad\qquad \ldots\ldots(51).$

Differentiating the second of these equations, we find

$$\frac{1}{p}\frac{\partial p}{\partial z} = \frac{1}{\rho}\frac{\partial \rho}{\partial z} + \frac{1}{T}\frac{\partial T}{\partial z} = -\frac{g\rho}{R\rho T} = -\frac{g}{RT}.$$

Hence $\qquad\qquad\qquad \dfrac{1}{\rho}\dfrac{\partial \rho}{\partial z} = -\dfrac{1}{T}\left(\dfrac{\partial T}{\partial z} + \dfrac{g}{R}\right) \qquad\qquad \ldots\ldots(52).$

The condition that the density shall not change with height is

$$-\frac{\partial T}{\partial z} = \frac{g}{R} = \frac{981}{2\cdot87 \times 10^6} = 3\cdot42 \times 10^{-4}\ {}^\circ\mathrm{C/cm}$$

$$= 3\cdot42^\circ\ \mathrm{C\ per\ 100\ metres.}$$

The density will increase or decrease with height according as the lapse-rate is greater or less than g/R. The limiting lapse-rate which gives density not changing with height is g/R, whereas the dry adiabatic lapse-rate is $\dfrac{g}{R}\dfrac{\gamma-1}{\gamma}$. Thus the limit we are now considering is $\gamma/(\gamma-1)$ or 3·49 times the dry adiabatic lapse-rate. The name auto-convection gradient has been suggested for this limit, because it was thought that convection must set in automatically when density increases with height. There appears to be no justification for such a supposition, since in a compressible atmosphere ordinary statical theory shows that the limiting lapse-rate for stability is the dry adiabatic lapse-rate.

"Auto-convection" is an unhappy name, and the idea has no place in accepted meteorological theory. The question of convection requires careful consideration, and some further discussion of this topic will be found later in § 134. In the idealised statical theory unstable equilibrium of any magnitude can exist, in the absence of some small initial disturbance. But in the atmosphere there is never perfect equilibrium, and the occurrence of lapse-rates much in excess of the dry adiabatic lapse-rate is to be accounted for by other reasons than the absence of small disturbances.

In some ways it would be convenient, in discussing stability, to use a conception of "potential density", which would denote the density which the air under consideration would take if brought adiabatically to some standard pressure. The equation connecting p and ρ in adiabatic changes is

$$\frac{p}{\rho^\gamma} = \text{constant} \qquad\qquad \ldots\ldots(53).$$

Hence, if ρ_i denote the potential density corresponding to a standard pressure p_0,

$$\frac{p}{\rho^\gamma} = \frac{p_0}{\rho_i{}^\gamma} \qquad\qquad \ldots\ldots(54).$$

It is readily shown that if θ be the potential temperature,

$$\frac{1}{\rho_i}\frac{\partial \rho_i}{\partial z} = -\frac{1}{\theta}\frac{\partial \theta}{\partial z} = -\frac{1}{T}\left(\frac{\partial T}{\partial z}+\Gamma\right) \qquad \ldots\ldots(55),$$

so that the condition for stability is that the potential density shall diminish with increasing height.

Density increasing with height in the layers near the ground is not uncommon on sunny days, and is the cause of most types of mirage.

§ 26. *The effect of mixing two masses of damp air*

Taylor has given an interesting application of fig. 16 above, following the lines of the earlier treatment of von Bezold. These two writers give a theorem which states that if the state of a mass of damp air as to temperature and vapour-pressure be represented by points in the diagram of fig. 16, then any mixture of the two masses will be represented by a point on the straight line joining the points which represent the two components. The theorem is not mathematically exact, and we give below a slight modification of Taylor's proof, in order to show to what extent the result can be in error.

Let m_1, T_1, e_1 and y_1 represent the total mass, the absolute temperature, the vapour-pressure and the mass of water-vapour per unit mass of moist air, in the first component of the mixture; and let similar variables with the subscript 2 represent the second mass. Let T and y represent the temperature and mass of water-vapour per unit mass of damp air in the final mixture.

The total mass of the mixture is $m_1 + m_2$, and the total mass of water-vapour in the mixture is $m_1 y_1 + m_2 y_2$. Hence the value of y is

$$\frac{m_1 y_1 + m_2 y_2}{m_1 + m_2},$$

and since the total pressure remains constant throughout the mixing, the final vapour-pressure is given by

$$e = \frac{m_1 e_1 + m_2 e_2}{m_1 + m_2} \qquad \ldots\ldots(56).$$

Let T be the temperature of the mixture, and let $T_1 > T_2$. The mass m_1 is reduced from a temperature T_1 to a temperature T. It gives up an amount of heat

$$c_p m_1 (1 - y_1)(T_1 - T) + c_p' m_1 y_1 (T_1 - T),$$

where c_p and c_p' are the specific heats at constant pressure of dry air and water-vapour respectively. The second mass gains an amount of heat

$$c_p m_2 (1 - y_2)(T - T_2) + c_p' m_2 y_2 (T - T_2).$$

These two amounts should balance exactly, since the changes take place at constant pressure. Hence

$$T = \frac{c_p (m_1 T_1 + m_2 T_2) + (c_p' - c_p)(m_1 y_1 T_1 + m_2 y_2 T_2)}{c_p (m_1 + m_2) + (c_p' - c_p)(m_1 y_1 + m_2 y_2)} \qquad \ldots\ldots(57).$$

The second terms in both numerator and denominator are readily shown to be small by comparison with the first terms, so that to a high degree of approximation

$$T_m = \frac{m_1 T_1 + m_2 T_2}{m_1 + m_2} \qquad \ldots\ldots(58).$$

It follows that, if in fig. 16 B and C represent the physical states of masses m_1 and m_2, the state of the mixture of the two masses will be represented by the point E, where

$$BE/EC = m_2/m_1$$

Should the point E fall above the saturation curve in fig. 16, the mixture will be supersaturated, by an amount measured by EF. As soon as condensation begins, some latent heat will be liberated, and the temperature will rise. Thus when all superfluous water vapour is precipitated the mixture will be at a higher temperature than that corresponding to the point E, its state being represented by a point G to the right of F. It was shown by Brunt[*] that the line EG has a slope downward to the right almost exactly equal to that of a line drawn in fig. 16 to join the point marked 20 on the vertical scale to the point marked 280 on the horizontal scale. The fraction of the apparently superfluous water vapour measured by EF which will actually be condensed will decrease rapidly with increase of temperature of the point E.

The above discussion is similar to that given by von Bezold and Taylor, except that it does not neglect the effect of the presence of water-vapour on the density and specific heat of damp air.

The practical value of the above theorem is considerable. It is readily seen from the diagram that by mixing masses of widely different temperatures, neither of which need be near saturation, it is possible to obtain a mixture which will be saturated or supersaturated. In the latter case the superfluous water-vapour will be condensed, and may produce fog or rain. If the vapour-pressure computed by the use of the diagram exceeds the saturation value for the assigned temperature, as say at the point E, condensation must occur, and the vapour-pressure must come down to the value corresponding to the point G. The formation of fog at a surface of discontinuity in the atmosphere is to be explained, at least in part, by the direct effect of the mixing of masses of air of different temperatures and humidities.

The diagram of fig. 16 can also be used to evaluate the fall of temperature which is necessary in order to produce saturation, given any initial conditions. If, for example, the air is initially represented by the point A, as the air is cooled the point moves horizontally to the left, and when it reaches the point D, the air is saturated. D is thus the dew-point, and any further cooling leads to condensation, or at least to supersaturation, which in the atmosphere leads automatically in most cases to the condensation of water-vapour in the form of water drops.

[*] *Q.J. Roy. Met. Soc.* **61**, 1935, p. 213.

CHAPTER III

WATER-VAPOUR IN THE ATMOSPHERE.
THE ASCENT OF DAMP AIR

§ 27. *The definition of saturation*

In accordance with normal practice, air is defined as saturated when the vapour it contains is in equilibrium with a plane surface of pure water at the same temperature, as many molecules passing in unit time from the liquid to the gaseous phase as from the gaseous to the liquid phase. The partial pressure which the vapour then exerts is a function of the temperature alone, being unaffected by the presence of the dry air. It follows that the density of saturated water-vapour is also a function of the temperature alone. The saturation vapour pressure is shown for different temperatures in fig. 16, p. 32, and in Table V, p. 420.

Starting from the definition of saturation given in the last paragraph, we may speak of air as *supersaturated* if it contains more water-vapour than is sufficient to saturate it. An effort to make air supersaturated may or may not lead to condensation. If saturated air is cooled, either in a closed vessel at constant volume, or in the free air at constant total pressure, it will either become super-saturated, or will condense its excess water-vapour into the liquid state.

When damp air is cooled adiabatically by ascent into regions of lower pressure, it approaches saturation, and if the ascent is continued to sufficient heights, some of the vapour will condense into liquid water, supersaturation to any marked degree being seldom produced in this way. That damp air requires adiabatic *expansion* and not *compression* in order to produce saturation is due to the specific heat of the saturated vapour being negative. It can readily be shown* that any vapour of which the latent heat is large will act in the same way as water-vapour, while vapours whose latent heat is small, as for example pentane, whose latent heat is 75, require adiabatic *compression* in order to produce condensation. It has been surmised that the clouds in the atmospheres of some of the outer planets are of liquid ammonia, and it is not without interest that the latent heat of ammonia is large, and therefore the specific heat of the saturated vapour negative, so that the clouds on these planets are formed in ascending currents.

The saturation vapour pressure defined above is the maximum pressure which water-vapour can exert when in contact with a plane surface of pure water. In the free atmosphere, the liquid water present is in the form of spherical drops, and these drops are not usually of pure water. There are three aspects of condensation on water-drops to be considered.

---

* See Brunt, "The possibility of condensation by the descent of air", *Q.J. Roy. Met. Soc.* **60**, 1934, p. 279.

(a) It was shown by Kelvin* that the equilibrium vapour pressure over a drop of pure water of radius r, which is here represented by e_{rs}, is greater than the saturation vapour-pressure, e_s, defined above, the relationship between the two being given by the equation†

$$\frac{\rho'R'T}{M}\log\frac{e_{rs}}{e_s}=\frac{2\mathbf{T}}{r} \qquad\qquad ......(1)$$

where R' is the gas constant for the vapour, ρ' the density of the liquid, T the absolute temperature, $\mathbf{T}$ the surface tension of the liquid, and $M=0.4343$. Leaving aside any doubt as to the validity of adopting in this equation the surface tension as measured for a plane surface, for a water drop at $10°$ C, we may write this in the form

$$\log\frac{e_{rs}}{e_s}=\frac{0.5\times10^{-7}}{r} \qquad\qquad ......(2)$$

where r is in centimetres.

If $r=10^{-5}$ cm $=0.1\mu$, $e_{rs}=1.012e_s$.

If $r=10^{-6}$ cm $=0.01\mu$, $e_{rs}=1.127e_s$.

If $r=10^{-7}$ cm $=0.001\mu$, $e_{rs}=3.10e_s$.

(b) If the water contains a salt in solution, the equilibrium vapour pressure above a plane surface is reduced by a factor $(1-kc)$, where c is the molecular concentration of the solution, and k is a constant depending on the nature of the solute. Whether this applies to a small drop has not been clearly established. Some idea of the possible magnitude of the effect of salts in solution may be gathered from the following facts. The vapour pressure of a saturated solution of sodium chloride at $10°$ C is about 22 per cent below that of pure water, and that of a saturated solution of ammonium nitrate is 29 per cent below that of pure water. Thus condensation could begin on large drops of saturated solution of sodium chloride if the relative humidity of the atmosphere exceeded 78 per cent, and on large drops of saturated solution of ammonium nitrate if the relative humidity exceeded 71 per cent. The restriction to large drops is made here in order to avoid considering for the moment the effect of curvature shown in equation (1) above.

(c) J. J. Thomson‡ showed that if a water drop has a charge E, the equilibrium vapour pressure over the drop is reduced, equation (1) above taking the form

$$\frac{\rho'R'T}{M}\log\frac{e_{rs}}{e_s}=\frac{2\mathbf{T}}{r}-\frac{E^2}{8\pi r^4} \qquad\qquad ......(3).$$

The effect of a charge, consisting of a few unit electronic charges, on a drop of radius 10^{-6} cm, would be negligible. For a drop of this size the second term on the right-hand side of equation (3) is equal to the first only when E consists of about 130 electronic charges. Multiple electronic charges are of frequent occurrence on fog-droplets and on rain drops, but the charge on a nucleus of condensation (see § 28) seldom exceeds one electronic charge. Thus

---

* *Proc. Roy. Soc. Edin.* Feb. 1870.
† All logarithms are given to base 10 in this and the next chapter.
‡ J. J. Thomson and G. P. Thomson, *Conduction of electricity through gases*, 2nd ed.

for all practical purposes the effect of electronic charges on the early stages of condensation phenomena may be neglected.

It is seen from equation (1) that over a small drop of pure water the saturation vapour pressure is greater than that over a plane surface, the excess vapour pressure increasing in inverse proportion to the radius of the drop. If therefore a small drop of water is introduced into an atmosphere saturated relative to a plane surface of water, it will at once evaporate, since a higher vapour pressure is necessary to keep it in equilibrium. Thus it is not correct to regard air, saturated in accordance with the customary definition, as containing the maximum possible amount of water vapour. It is possible to increase the water vapour content of clean air, free from all impurities, by spraying into it water in a sufficiently finely divided form.

§ 28. *Condensation on nuclei*

It is not possible for condensation to take place in the atmosphere by the casual grouping together of a number of molecules of water-vapour. An aggregation of 100 molecules of water-vapour would form a droplet with a radius of the order of 10^{-7} cm, and the equilibrium vapour pressure at the surface of such a droplet would be more than three times the saturation vapour pressure as normally defined. In normal conditions in the atmosphere the growth of the radius of a drop up to about 10^{-6} cm, in the absence of some central core on which the water could be deposited, would demand a higher degree of super-saturation than has yet been observed in natural conditions. There are present in the atmosphere small particles of hygroscopic substances. numbering from about 2,000 to 50,000 per cubic centimetre of air, and having radii of 10^{-5} to 10^{-6} cm. As can be seen from equation (2) above, the effect of the curvature on the equilibrium of water deposited on a nucleus of such a size is relatively small. It is moreover offset by the fact that the saturation vapour pressure over the solution of the hygroscopic nucleus is less than over pure water. Thus condensation begins when the vapour pressure in the atmosphere exceeds the equilibrium vapour pressure over a saturated solution of the nucleus by a small amount, this amount depending on the size of the nucleus. Condensation will then proceed rapidly until the stage is reached when all the salt has dissolved. From this stage onward the concentration of salt in the drop decreases, and the saturation vapour pressure in the air near the surface of the drop increases slowly, tending to approach the normal value for pure water. The precise physico-chemical processes involved in the initial condensation are not yet clearly understood, but there appears to be no doubt that in the atmosphere condensation always takes place on a hygroscopic nucleus of some kind. Condensation is always accompanied by the liberation of latent heat, and this retards the fall of temperature which produces super-saturation.

The nucleus, if sufficiently hygroscopic, may begin to absorb moisture from the atmosphere when the relative humidity is well below 100 per cent. It has frequently been observed that in fogs the air is well below saturation, and

J. S. Owens* has described occasions when nuclei began to gather moisture at relative humidities of 74 per cent.

The radii of droplets of water in country fogs usually lie between 4×10^{-4} and 3×10^{-3} cm, the average radius being probably about five or six times 10^{-4} cm. The droplets may number up to 1500 per cubic centimetre. Cloud droplets are normally greater than fog droplets, ranging up to 10^{-2} cm in radius, while the radii of rain drops range from about 10^{-2} cm up to 0·2 cm. Taking 10^{-5} cm as the radius of a hygroscopic nucleus which eventually forms a fog droplet of radius 10^{-3} cm, we see that the volume of the fog droplet is about one million times that of the original nucleus, so that in the normal fog droplet the solution of the nuclear salt is so weak that the presence of the salt can have no appreciable effect on the further growth of the drop. Further growth will require the cooling of the drop and its environment. Even the cooling of the drop alone will suffice to diminish the saturation vapour pressure at its surface and to produce further condensation on the drop.

When fogs form at the ground in calm conditions, the essential physical factor in their formation is the cooling of the ground and of the air in contact with it. The air is initially supposed to be unsaturated, but to contain nuclei on which condensation can take place. As the air cools its relative humidity increases, and, so far as nuclei of the same chemical composition are concerned, condensation starts first, and proceeds most rapidly, on the larger nuclei, both because the curvature effect is less for the greater drops, and because the concentration of the nuclear salt is greater initially in the drops formed around the larger nuclei. The smaller nuclei may remain free of condensation while the larger nuclei acquire water and grow into drops. It is in fact certain that the number of drops of water in a cubic centimetre of fog is much lower than the number of nuclei present, even as counted by the Aitken counter. In the Aitken apparatus as generally used, the air is near tenfold super-saturation, and this is sufficient to produce condensation on particles of radius 10^{-7} cm, which in the atmosphere could not be effective nuclei for condensation. Wilson† found that when samples of air are expanded and cooled slowly, fewer drops are formed than if the expansion and cooling are rapid. When the operation is performed slowly condensation starts on the larger nuclei at an early stage, removing some of the water vapour, and so decreasing the degree of super-saturation at each subsequent stage, so that the requisite degree of super-saturation for condensation on the smaller nuclei is not then attained. It is therefore concluded that particles counted in the Aitken counter are not all equally effective in the atmosphere.

It follows from this that it cannot be inferred from the fact that it is always possible to obtain condensation in the Aitken counter, with air taken from the atmosphere, that air is never free from effective nuclei on which condensation can take place. It seems possible that air in certain conditions contains only very small nuclei on which condensation would be possible only with a high degree of super-saturation. If there were only very small nuclei present in the air at any time, saturation could be surpassed without any formation of fog.

* *Proc. Roy. Soc.* A, **110**, 1926, p. 738.
† *Phil. Trans. Roy. Soc.* A, **189**, 1897, p. 263.

In the atmosphere nuclei of many kinds are present at the same time, e.g. sodium chloride from the evaporation of sea spray, and sulphur products from chimneys. The nuclei on which condensation will first take place will not of necessity be the largest, but rather those whose hygroscopic properties are most marked.

It must not be assumed that all non-gaseous constituents of the atmosphere can act as nuclei. Wigand* and Boylan† showed that ordinary dust particles, such as can be beaten from a carpet, or blown up from the floor of a coal-shed, do not act as nuclei, even in an atmosphere which is considerably supersaturated. It is mainly the water-soluble salts which act as nuclei, either dry, or in the form of hydrates containing water of crystallisation, or of solutions.

The formation of clouds is associated with convection or with the ascent of large masses of air over hills or over wedges of colder air. The cooling produced by expansion lowers the saturation vapour pressure, and the excess water-vapour is condensed on the nuclei, which grow progressively in size during the ascent, by the continued deposition of the surplus water-vapour from the atmosphere. It is usually supposed that this process goes on until the drop is large enough to fall through the ascending current. The distinction between cloud and rain is merely a distinction of size, and consequently of rate of fall of a drop through the air. (*Vide* Table X, p. 422.) But it is questioned by many writers whether this simple view of the formation of rain drops is even approximately true. Thus Bergeron‡ suggests that the direct condensation of further water on cloud droplets is never sufficient to produce a rain drop. He rejects the possibility of coagulation of small drops to form large drops, and puts forward the theory that rain is formed by the action of ice crystals falling through the cloud of small water drops. The vapour pressure being lower over the ice crystal than over a water drop, leads to the evaporation of the liquid water drops and condensation on the ice crystal, which in falling to the ground melts into a rain drop.

The question of coagulation of water drops in fog and cloud is not yet satisfactorily solved. While hydrodynamical arguments appear to negative the possibility, the observations of Köhler§ that the masses of droplets in fog and cloud are integral multiples of a certain unit, and that the concentrations of salt are integral multiples of a certain unit concentration, and also the observations of Defant‖ and Niederdorfer¶ that the masses of rain drops are integral multiples of a standard size, appear to demand the formation of large drops by the union of small drops of a uniform standard size.

Findeisen** states that the observations made over a period of some years on aircraft have confirmed Bergeron's views that rain of even moderate amount, and particularly rain in which the drops are large, predominantly originates as ice condensation.

Nothing has been said above concerning the other non-gaseous constituents of atmospheric air. The many investigations carried out in the field of atmo-

---

* *Met. Zeit.* **30**, 1913, p. 10.
‡ *Rept. Met. Assoc. U.G.G.I., Lisbon,* 1933, p. 156.
‖ *Wiener Ber.* **114**, 1905, p. 585.
** *Met. Zeit.* **55**, 1938, Heft 4, p. 121.

† *Proc. Roy. Irish Acad.* **37**, 1926, p. 58.
§ *Geofysiske Publ.* **2**, No. 6, 1922.
¶ *Met. Zeit.* **49**, 1932, p. 1.

spheric electricity show that the air is partially ionised. The small ions in the air consist of agglomerations of perhaps 10 molecules, carrying a single electronic charge, and having a radius of three to ten times 10^{-8} cm, and numbering 300 to 1000 per cm^3. The union of a small ion with an uncharged nucleus produces what is known as a large ion, of radius between 10^{-5} and 10^{-6} cm. Only about half the nuclei are charged by this process. We do not propose to enter into further details of the electrical aspect of atmospheric phenomena, which have no close relation with the thermodynamics of weather.

§ 29. *Supercooling of water in the atmosphere*

When damp air rises, so that water-vapour condenses into liquid drops, which are subsequently carried up to heights where the temperature is below o° C, the drops do not immediately freeze into ice. Clouds of liquid drops have been observed at temperatures as low as −40° C. The liquid water is then said to be *supercooled*.

Findeisen states (*loc. cit.*) that at cloud temperatures just below freezing-point, water drops predominate, but that at temperatures below −10° C, ice particles predominate.

It is readily possible to supercool water in the laboratory. If a small crystal of ice is introduced into the water, or if the side of the vessel is scratched vigorously, the water will partially freeze, and in the final state the vessel will contain ice and water all at o° C. Sufficient of the water will freeze to liberate enough latent heat to raise the whole contents of the vessel to o° C.

If, for example, the supercooled water is at −t° C, and a fraction *f* is frozen, the latent heat of fusion of ice being taken to be 80 g. cal. per gramme, we have

$$80f = t$$

or $$f = t/80.$$

When an aircraft flies through a cloud of supercooled water drops, any drop which strikes the leading edge of a wing will be partly frozen into ice, while the remaining portion of the drop will be splashed as liquid water over the wing, being rapidly frozen by contact with the cold wing and the cold air around it. The ice formed by the freezing of this water will be clear ice, but the ice formed on the leading edge of the wing will usually be white ice, since bubbles of air are trapped by the drop when frozen on impact.

§ 30. *Condensation at temperatures below the freezing-point*

Condensation at temperatures below the freezing-point may be in the form of either supercooled water or ice. It must be noted that the saturation vapour pressure over ice is less than that over water. (*Vide* Table V, p. 420.) In the presence of solid nuclei the condensation from air saturated relative to ice will be in the form of ice crystals. An ice crystal is initially a hexagonal prism, and its later growth may either be along the central axis, leading to the formation of a long thin prism, or along the six axes of the hexagon. In the latter case, the growth along the axes is frequently greater than in the space between them, resulting in the varied gem-like structures shown by photographs of snow crystals.* When the temperature of the air in which the condensation takes place is very low the amount of water-vapour is only sufficient to produce small crystals. Large crystals are formed when the condensation takes place at temperatures only slightly below freezing-point. The interlocking of numbers of such crystals produces a snowflake capable of falling to the ground.

Condensation in the form of water drops at temperatures below freezing-point has frequently been observed, in circumstances in which it is not possible to explain it as water condensed at temperatures above freezing-point and subsequently supercooled. Simpson† described an occasion on which a fog of water drops, showing rainbow or fog-bow, was observed in the Antarctic, at a temperature of about $-30°$ C. At the time hoar frost was being deposited on the fur of sleeping-bags and the wool of sweaters, which formed nuclei on which condensation in the solid state could occur. The atmosphere was therefore saturated relative to ice, but contained no solid nuclei on which water-vapour could condense in the solid form. Wegener reported that in similar conditions in Greenland the atmosphere was clear everywhere except downwind from the hut, from whose chimney smoke was at the time coming into the air.

§ 31. *Rain, snow and hail*

Rain will be formed when rapid condensation takes place in ascending air currents, at temperatures not much below the freezing-point. When the condensation takes place on solid nuclei, it will be initially in the form of snow. Large flakes of snow will only form when condensation takes place at a temperature of only a few degrees below zero. If the snow is formed at fairly high levels, with the temperature of the surface layers of the atmosphere some degrees above freezing-point, it will be melted before reaching the ground, and will reach the surface as rain. In middle latitudes a considerable proportion of the rain which reaches the ground originates as snow. When the ascending current in which condensation takes place rises to high levels in the atmosphere the precipitation may take the form of hail. Suppose, for example, that the level at which saturation is attained is at a temperature of $10°$ C, and that the air rises to the level at which the temperature is $-30°$ C. When the air rises above the level of $0°$ C it will contain water drops, which will remain as supercooled liquid drops until their temperature is probably about $-20°$ C. Experiments in the laboratory indicate that when water is supercooled, potential centres of crystallisation develop within the liquid, the number of such centres increasing as the degree of supercooling increases. Thus, as a current containing liquid drops ascends, the probability of some of the drops freezing increases. At the stage when the cloud contains both ice and water particles it is saturated relative to water, and therefore supersaturated relative to ice. There is therefore rapid condensation on an ice particle, which

* *Vide* Bentley and Humphreys, *Snow crystals*. (McGraw Hill Publishing Co.)
† Lecture to Royal Institution, March 2, 1923.

grows rapidly in size and falls through the ascending current, forming a potential centre of a hailstone. If it falls to the levels in which supercooled water drops are present it will from time to time collide with supercooled liquid drops. What precisely will occur at the collision is not clear. Some of the supercooled water will freeze on the ice particle while the rest will flow as liquid water over its surface, both water and ice initially at 0° C. The ice which is formed immediately on collision will trap some air within itself, and so will appear as white ice. The water, which flows in a thin film over the surface, will be rapidly cooled by radiation, conduction and evaporation to the surrounding colder air, and will freeze into a coating of clear ice. But it appears that the surface temperature of the hailstone during its passage through the region of supercooled water will not be much below 0° C at any time, and at the time when the hailstone falls through the level of the isotherm of 0° C, its surface temperature must be near 0° C, within very narrow limits. When the hailstone falls into the region of water drops above freezing-point, it will collect some water on its surface, the water having a higher temperature than the ice on which it rests.

Now the large hailstone formed in thunderstorms is usually a roughly spherical object, made of successive layers of white ice and clear ice with a core of soft white ice. Such a structure can be explained by the assumption that the ascending currents in which condensation takes place are intermittent, so that a hailstone which falls from the level at which it was first formed to a much lower level, is again caught up in an ascending current. Most of the white ice in any layer is deposited in the region where condensation is in the form of snow, while a layer of clear ice appears to be in the main deposited during the passage of the hailstone through the region of supercooled water drops. Possibly some of the water caught on the surface of the hailstone at temperatures above freezing-point is frozen (into clear ice) when the hailstone is again carried up to levels where the temperature is below 0° C. But so far as the present writer is aware, no completely satisfactory account of the mode of formation of a large hailstone has yet been given.

The formation of large hailstones demands very strong ascending currents, capable of maintaining the hailstone while it grows to a large size. It also demands that the clouds shall extend through a large range of height, so that the regions of snow and of supercooled water drops are extensive. There does not appear to be any special reason for supposing that an extensive region of water drops at temperatures above 0° C is essential.

For a discussion of the aerodynamics of hailstones the reader is referred to a paper by Bilham and Relf, *Q.J. Roy. Met. Soc.* **63**, 1937, p. 149, where it is shown that the maximum mass of hailstones must be about 1·5 lb. The thermodynamics of hailstone formation are discussed in a paper by Schumann in the same journal, **64**, 1938, p. 3.

§ 32. *Statement of the general meteorological problem*

The main problem to be discussed in connection with the thermodynamics of moist air is the variation of temperature produced by changes of pressure, which in the atmosphere are associated with vertical motion. When damp air ascends it must eventually attain saturation, and further ascent produces condensation, at first in the form of water drops, and as snow in the later stages.

Hertz* and Neuhoff† discussed the problem on the assumption that the products of condensation are carried up with the ascending current, and partake of the changes of temperature of the air. At any stage the amount of liquid water or snow depends both on the temperature of the mixture at that stage and on the initial condition of the air. The process is strictly reversible. For if the damp air and water drops or snow are again brought downward, the evaporation of water drops or snow uses up at each stage the same amount of latent heat as is liberated by condensation on the upward path.

It was pointed out by von Bezold‡ that the actual process in the atmosphere is not reversible, as some of the condensed water falls out as rain drops. If all the condensed water or ice falls out as rain or snow the changes of temperature are slightly altered, since there is no longer any interchange of heat between the air and the water or ice. The process which then takes place was named by von Bezold pseudo-adiabatic.

It has been shown by Fjeldstad§ and others that the differences which arise, according as we regard the products of condensation as falling out or being retained, are so small as to be negligible in practice. We shall therefore follow the method of Hertz and Neuhoff in what follows, retaining the products of condensation with the ascending air, since this leads to easier mathematical treatment than the pseudo-adiabatic process visualised by von Bezold. The results so derived will be applied to the discussion of the changes of temperature of ascending damp air, without further consideration whether the water drops and ice fall out, except at the hail stage. The errors which arise on account of the uncertainties as to what happens to the products of condensation are likely to be much smaller than the uncertainties of some of the other assumptions. The theoretical treatment must assume that there is no exchange of heat between the ascending air and the environment. This is equivalent to assuming that it is legitimate to neglect the effects of radiation, which are probably small during the relatively small time of ascent, and also the effects of turbulence, which are certainly not negligible, though it is impossible to correct for them in any way.

In the work of Hertz the changes of state of 1 gramme of mixture of air and water-vapour were considered. Neuhoff considered the changes of state of 1 gramme of dry air together with the appropriate admixture of water in its various forms—vapour, liquid and solid. This method is preferable in that it leads to simpler mathematical treatment. We shall therefore follow Neuhoff in the following pages.

Hertz and Neuhoff discussed four stages of the ascent of damp air: (a) the air unsaturated; (b) the air saturated and containing water drops, at a temperature above the freezing-point; (c) the stage in which all the water drops

* Met. Zeit. 1, 1884, pp. 421–31; also in English in Hertz, Misc. Papers (Macmillan).
† K. Preuss. Met. Inst. 1, 1900, p. 271; also in Abbe, Mechanics of Earth's Atmosphere, 3, 1910.
‡ Sitz. Berl. Akad. 1888.   § Geofys. Publikasjoner, 3, No. 13.

freeze into ice at $0°$ C; and (d) saturated air and ice at temperatures below $0°$ C. This scheme of events visualises that clouds at temperatures below the freezing-point are always composed of ice. But, as stated on p. 54, observation shows that clouds may be composed of supercooled water drops at temperatures as low as $-40°$ C. The idea that clouds are still composed of liquid drops at temperatures well below freezing-point has already been accepted and applied in § 29 above. If we could be certain of the temperature at which the water drops in a cloud would freeze, we could readily modify the treatment given by Neuhoff, by continuing stage (b) up to the level at which that temperature is attained. But as this temperature is uncertain, and probably different on different occasions, the treatment given by Neuhoff has been followed below. Further reference is made to this aspect of the subject in § 38, p. 63 below.

The initial conditions from which we start are defined by the pressure, temperature and humidity mixing ratio of unsaturated air. The analysis aims at the development of equations to represent the changes which take place as the air ascends and expands with diminishing pressure. Initially the water is present in the form of vapour, but in the later states the water may be present in the form of vapour, liquid or solid. The notation and constants used in the present chapter are given below:

$x=$ humidity mixing ratio, or weight in grammes of water-vapour associated with 1 gramme of dry air.

$y=$ weight in grammes of liquid water associated with 1 gramme of dry air.

$z=$ weight in grammes of ice associated with 1 gramme of dry air.

$\xi = x+y+z$. This is constant throughout the process.

Q or $dQ=$ an amount of heat in calories.

$T=$ absolute temperature.

$p=$ total pressure.

$e=$ partial pressure of water-vapour.

$c_p=$ specific heat of dry air at constant pressure $=0.2396$.

$c_v=$ specific heat of dry air at constant volume $=0.1707$.

$\gamma=$ ratio of specific heats of dry air $=1.403$.

$c_p'=$ specific heat of water-vapour $=0.4652$.

$c=$ specific heat of liquid water $=1.013$ (value at $0°$ C).

$c_i=$ specific heat of ice $=0.5$.

$\epsilon=$ specific gravity of water-vapour $=0.622$.

$R=$ constant of the gas-equation for dry air $=2.8703 \times 10^6$ in C.G.S. units.

$v=$ specific volume of dry air.

$\rho=$ true density of the gaseous mixture $=(1+x)/v$.

$L=$ latent heat of evaporation of water $=594.9-0.51\,(T-273)$.

$L_e=$ latent heat of fusion of ice $=79.7$.

$L_i=$ latent heat of sublimation of ice $=677$ (Fjeldstad's value).

$A=$ reciprocal of the mechanical equivalent of heat $=2.392 \times 10^{-8}$.

$M=\log_{10} e=0.4343$.

At each stage of the ascent, it is assumed that the entropy of the mixture of dry air, water vapour, liquid water and ice (of which all three phases will not be present together except in the freezing stage) remains constant. The equation for the entropy of a mixture of 1 gramme of dry air, x grammes water vapour, y grammes liquid water, and z grammes ice, is derived in § 50, p. 85 below. It is copied here for reference in the subsequent analysis.

$$\text{Total entropy} = \frac{(c_p + \xi c)}{M} \log T + \frac{Lx}{T} - \frac{L_e z}{T} - \frac{AR}{M} \log (p - e) \quad \ldots\ldots(4).$$

§ 33. Stage (a). Unsaturated air

During the stage when no condensation takes place the constitution of the mixture of air and water-vapour is unchanged, and the humidity mixing ratio has the constant value ξ. The specific heat of the damp air at constant pressure is $\frac{c_p + \xi c_p'}{1 + \xi}$.

Its pressure and temperature will be related by an equation

$$pT^{-m} = \text{const.} \quad \ldots\ldots(5)$$

as in equation (30), p. 38, with the difference that m will now be the appropriate value of $\gamma/\gamma - 1$ for the damp air. But

$$\frac{\gamma}{\gamma - 1} = \frac{c_p}{c_p - c_v} = \frac{c_p}{AR}$$

and so

$$m = \frac{c_p + \xi c_p'}{AR (1 + \xi/\epsilon)} = 3 \cdot 49 \times \frac{1 + 1 \cdot 94 \xi}{1 + 1 \cdot 608 \xi} \quad \ldots\ldots(6).$$

§ 34. Stage (b). Saturated air above 0° C, or the rain stage

Once saturation is attained, further cooling leads to condensation, and the quantity of water vapour and of liquid water will vary. We now have in equation (4) above

$$\xi = x + y, \quad z = 0,$$

Then

$$(c_p + \xi c) \log T + MLx/T - AR \log (p - e) = \text{const.} \quad \ldots\ldots(7).$$

Since $x = \epsilon e/(p - e)$, equation (7) is readily reduced to the form

$$\log (p - e) - 3 \cdot 93 \frac{L}{T} \frac{e}{p - e} - m_1 \log T = \text{const.} \quad \ldots\ldots(8)$$

where

$$m_1 = \frac{c_p + \xi c}{AR}.$$

This is the form which is normally used in the detailed computations, for further particulars of which reference should be made to Fjeldstad's paper.

§ 35. Stage (c). The freezing or hail stage

Hertz and Neuhoff assumed that as soon as the temperature falls below the freezing point the liquid drops begin to freeze. This does not normally happen in the atmosphere, as already stated in § 29, p. 54. The stage in which super-cooled water is present can be treated by means of equation (7) above, the rain stage being carried beyond the freezing point. For completeness, however, the freezing stage is included here. The treatment of the problem given below can be readily modified to deal with freezing at any other temperature than 0° C. In the initial stage $x = x_0$, $y = y_0$, $z = 0$, and $\xi = x_0 + y_0$; in the final stage $x = x_1$, $y = 0$, $z = z_1$, and $\xi = x_1 + z_1$, the temperature remaining unchanged at 273° A during the whole stage. From equation (4) above

$$\frac{Lx_0}{T} - \frac{AR}{M} \log (p-e)_0 = \frac{Lx_1}{T} - \frac{L_e z_1}{T} - \frac{AR}{M} \log (p-e)_1$$

$$= \frac{Lx_1 - L_e (\xi - x_1)}{T} - \frac{AR}{M} \log (p-e)_1$$

$$= \frac{(L+L_e)}{T} \frac{\epsilon e_1}{(p-e)_1} - \frac{AR}{M} \log (p-e)_1 - \frac{L_e \xi}{T}$$

$$\log (p-e)_0 - \frac{ML}{ART} \frac{\epsilon e_0}{(p-e)_0} - \frac{ML_e \xi}{ART} = \log (p-e)_1 - \frac{M(L+L_e)}{ART} \frac{\epsilon e_1}{(p-e)_1}$$

$$\dots \dots (9).$$

Inserting the values of the constants, we find

$$\log (p-e)_1 - \frac{59 \cdot 71}{(p-e)_1} = \log (p-e)_0 - \frac{52 \cdot 39}{(p-e)_0} - 1 \cdot 846 \xi \quad \dots \dots (10).$$

It will be noted that, as the temperature remains constant during the freezing stage, while the element of air under consideration ascends and expands, some of the liquid water must therefore evaporate in order to maintain saturation.

§ 36. Stage (d). Condensation below 0° C. The snow stage

In this stage, $y = 0$, $\xi = x + z$. The appropriate equation is most readily derived by re-writing equation (7), replacing c, the specific heat of water, by c_i, the specific heat of ice, and L by the latent heat of sublimation of ice. Fjeldstad showed (loc. cit.) that there was reason to regard the latent heat of sublimation of ice, L_i, as constant, and equal to 677 g.cal., and this value will be adopted here. Equation (7), with these modifications, then becomes

$$(c_p + \xi c_i) \log T + \frac{ML_i x}{T} - AR \log (p-e) = \text{const.} \quad \dots \dots (11)$$

or, if $\quad B = \frac{677M}{AR} \times 0 \cdot 622 = 2664, \quad m_2 = \frac{c_p + c_i \xi}{AR} = 3 \cdot 49 \, (1 + 2 \cdot 09\xi)$

$$\log (p-e) - m_2 \log T - \frac{Be}{(p-e) T} = \text{const.} \quad \dots \dots (12).$$

This is the form of the equation which is normally used in computation.

§ 37. *The Neuhoff diagram*

Equations (5), (8), (10) and (12) give the relations between p, T and x or e, and enable us to draw the adiabatic lines for the change of state of any mass of air for which the initial values of these variables are known. Very complete diagrams have been constructed by Hertz, Neuhoff and Fjeldstad, on the

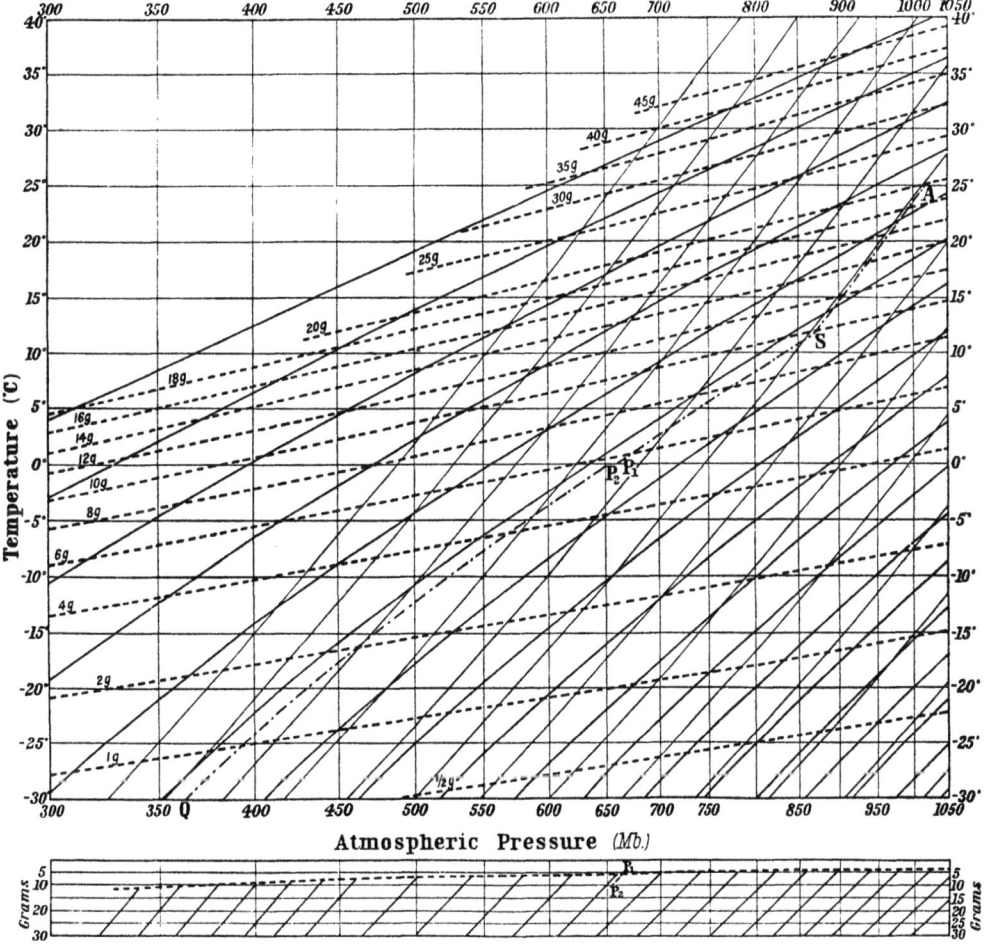

Fig. 17. The Neuhoff diagram.

assumption that all the condensed water and ice remain in suspension. In fig. 17 is reproduced a modified Neuhoff diagram, based on the computations of Fjeldstad. In this diagram the pressure in millibars is represented on a logarithmic scale along the horizontal axis. Temperature is represented on a linear scale along the vertical axis. There are three series of lines running across the diagram. The steepest series of lines are the adiabatic lines for dry air. The less steep series of continuous lines are the adiabatic lines for saturated air. The broken lines are the lines of equal values of the humidity mixing ratio x, and the numbers marked against these lines indicate the number of grammes of water-vapour which will saturate 1 kilogramme of the air. It will be noted that the dry adiabatic lines are very nearly straight lines; they would be perfectly straight if temperature had been represented on a logarithmic scale. The saturated adiabatic lines are more curved than the dry adiabatics, and the slopes of the two series of lines approach equality as the temperature diminishes. Note for example how small is the difference of slope of the two families of curves near the bottom right-hand side of the diagram.

The use of the diagram can be most readily understood from the example indicated by a broken line which starts at the top right-hand side of the diagram. This line traces out the history of a mass of air which is initially at a pressure of 1020 mb, with temperature 25° C and relative humidity 50 per cent, and which is set in upward motion through the atmosphere. The diagram shows that at pressure 1020 mb and temperature 25° C it requires 20 grammes of water-vapour to saturate the air, and as the relative humidity is 50 per cent, the kilogramme of dry air which we are considering has 10 grammes of water-vapour initially. During the early stages of its ascent the air will follow the dry adiabatic through the point A which represents its initial condition. It will follow this line until it reaches the vapour-content line of 10 grammes, at which point of its ascent it will have just reached saturation. Its further ascent will be along a saturated adiabatic. At the different stages of its ascent the water-vapour content of the air will be given by the figures shown against the broken lines, the remainder of the initial 10 grammes of water-vapour being carried up as liquid water. When it attains the level where the temperature is 0° C the water begins to freeze, and the temperature remains at 0° C until all the water is frozen. The latent heat liberated is used up in expanding the mixture. The mass of air we are considering reaches saturation at a pressure of 870 mb and temperature 11·5° C, represented on the diagram by the point S. The further ascent is along the saturated adiabatic through S, until the point P_1 is reached at which the temperature has just fallen to 0° C. At this stage of its ascent the air contains a little less than 6 grammes of water-vapour, and a little more than 4 grammes of liquid water. The small diagram at the bottom of the figure is used to estimate the range of pressure corresponding to the freezing stage. Take on the broken line of this small diagram the pressure corresponding to the pressure at which the air first reaches 0° C; in our example this is at the point P_1. Through this point take a line parallel to the sloping lines of the lower diagram and follow it down to the point at which it meets the line of water content equal to the original water content of the air, which in our example is 10 grammes. The latter point is at P_2 in our example. The pressure at P_2 gives the limit of the freezing stage, and indicates the pressure at which the air will begin to cool below 0° C, from the representative point P_2 in the diagram. After P_2 the representative point follows the saturated adiabatic P_2Q, and reaches $-30°$ C at a pressure of 362 mb, at which stage the original 10 grammes of water-vapour are represented by about 0·7 gramme of water-vapour, and about 9·3 grammes of ice.

If saturation is not attained until the level of $0°$ C has been surpassed, the freezing stage will not occur, and the adiabatic lines will cross the line of zero temperature without a break such as we have shown at $P_1 P_2$. But if saturation is attained at a higher temperature than $0°$ C, then the path followed by the adiabatics will consist of four parts—the dry adiabatic AS, the saturated adiabatic SP_1, the freezing state $P_1 P_2$ and the saturated adiabatic $P_2 Q$.

§ 38. *The effect of the loss of precipitated water and ice*

The equations which relate the changes of pressure in damp air to the variations of the other meteorological factors have been derived above in full, on the assumption that the products of condensation are retained with the air, none being precipitated as rain or snow. The mathematical treatment has been reproduced in detail rather for the sake of completeness than for its direct utility, since the practical problems of meteorology can be much more readily handled by the use of the diagram of fig. 17, than by means of the equations.

One questionable assumption in the treatment given above is that relating to the non-precipitation of the products of condensation. In the atmosphere the phenomena are not always in keeping with this assumption. Rain drops fall, while cloud particles float, or rather fall so slowly that their rate of fall is unimportant, and it appears that in nature the phenomena are intermediate between those assumed as the basis of the above treatment and the pseudo-adiabatic conditions discussed by von Bezold, in which all products of condensation are precipitated. The difference in the mathematical treatment of von Bezold and that of Hertz, Neuhoff and Fjeldstad arises through the variability of ξ in von Bezold's equations, leading to a term

$$\int (c_p + \xi c) \frac{dT}{T},$$

instead of $\qquad (c_p + \xi c) \int \frac{dT}{T} \quad \text{or} \quad (c_p + \xi c) \log T.$

Neuhoff (*loc. cit.*) investigated the difference between the results derived by his method and that of von Bezold, and found that while the pseudo-adiabatic changes always lead to higher pressures for the same temperatures when compared with the true adiabatic changes, the differences are so small as to be negligible, except near $0°$ C. Thus air initially saturated at a temperature of $20°$ C and pressure 760 mm attains a temperature of $0°$ C at a pressure of 465 mm in the pseudo-adiabatic system and 463 mm in the adiabatic system; it reaches a temperature of $-18°$ C at a pressure of 316 mm in the pseudo-adiabatic system, and 304 mm in the adiabatic system. The difference of 12 mm is almost entirely due to the hail stage, since on the adiabatic system the mixture first attains $0°$ C at a pressure of 463 mm, and begins to fall below $0°$ C at a pressure of 452 mm. It is in fact immaterial to the form of the saturation lines whether we regard the products of condensation as precipitated or retained, except at $0°$ C; the diagram of fig. 17 may be treated as adiabatic or pseudo-adiabatic, and if we wish to use it to discuss pseudo-adiabatic changes we simply omit the hail stage. In this event the representative point on the diagram crosses the line of zero temperature without side-stepping, and its further course is represented by an approximate continuation of the line SP_1 toward lower pressure. The adiabatics above and below 0° C are not strictly continuous in direction, there being a slight diminution of the lapse-rate just above zero, but the change of direction is so slight as to be scarcely noticeable in the diagram. The neglect of the hail stage means that the air reaches the bottom of the diagram at higher pressure in the pseudo-adiabatic system, at a point to the right of the point Q.

If we wish to consider the case in which condensation is in the liquid form until the temperature falls say to −20° C, we have to continue the saturated adiabatic followed by air down to 0° C, down to the temperature of −20° C. The difference in the slope of the water adiabatic and the ice adiabatic at temperatures below 0° C is so slight as to lead to no great difference in the temperatures deduced for different levels within the range of temperatures 0° C to −20° C, or even to considerably lower values than this.

The course of events described above is strictly reversible if none of the condensed liquid or solid products of condensation are precipitated, for then the air in descending is compressed and heated dynamically, and evaporation goes on continually until the air has returned to the state in which it was just saturated, with no liquid or solid content. Further descent beyond this stage brings the air to a temperature at which it is unsaturated, and consequently its further course is along the dry adiabatic which it followed in its initial ascent. If any of the water, ice or snow has fallen out of the air in which it was formed the result will be that on the reverse journey the point at which the air ceases to be saturated is attained at an earlier stage of the downward journey, and a larger part of the downward journey is along a dry adiabatic, resulting in the air reaching its initial pressure at a higher temperature than it had at the beginning. If, for example, the air which we took from 1020 mb and 25° C up to a pressure of 362 mb and temperature of −30° C should shed all its water or ice, and then be taken back to a pressure of 1020 mb, it would go the whole way along a dry adiabatic, and would attain the level of 1020 mb with a temperature of about 54° C, or 29° above the temperature which it had initially at that pressure. If we imagine air starting at sea level and being forced to rise over a ridge of mountains, the height to which the air forced to ascend being sufficient to produce considerable condensation and precipitation, the air on descending to its original level on the other side of the mountains would appear as a warm current (Föhn). This is but one of many important applications of the results derivable from the Neuhoff diagram.

There is a second questionable assumption underlying the mathematical treatment given above. It is that the conditions are strictly adiabatic, and that there are no exchanges of heat between the ascending air and its environment. This assumption is also inherent in von Bezold's pseudo-adiabatic treatment. It is equivalent to assuming that the effects of radiation and of turbulent mixing with the environment are both negligible. It is not possible to estimate arithmetically the effect of either, otherwise it would be possible to correct for them. But while it appears *a priori* that the effects of radiation are not likely to be very great, it is certain that on occasion the effects of turbulent mixing are very considerable, and that these effects produce a much greater uncertainty in the computations than the niceties of estimation of the amount of water or ice lost by precipitation.

§ 39. *The lapse-rate of ascending damp air*

The lapse-rate with height of damp air when displaced upward from its normal position can be readily derived without the use of the concept of entropy which is the basis of the derivation of equation (11) above. The condition of the air will be specified as in § 32 above, by the pressure p, the density ρ, the temperature T, the vapour-pressure e and the humidity mixing ratio x, and the amount of liquid water present at any stage will be denoted by $\xi - x$, where ξ is constant for any given element of mass of the damp air, being equal to the original vapour content of the mass when just saturated.

When unit mass of dry air, with the appropriate admixture of water-vapour and liquid water, is displaced upward through a distance dz, the increase in potential energy is balanced by the loss in internal energy of the air and water-vapour, and of any liquid water present. The internal energy of x grammes of water-vapour *plus* $(\xi - x)$ grammes of liquid water is equal to the internal energy of ξ grammes of liquid water *plus* the latent heat of x grammes of water-vapour, and is therefore equal to

$$\xi c T + Lx.$$

The change in this due to a vertical displacement dz is

$$\xi c\, dT + d\,(Lx).$$

The change in internal energy of 1 gramme of dry air is $c_p dT$. These two quantities are balanced by the increase in gravitational potential energy

$$g\,(1 + \xi)\,dz.$$

The equation which expresses the balance is

$$(c_p + \xi c)\,\frac{dT}{dz} + \frac{d}{dz}\,(Lx) + Ag\,(1 + \xi) = 0 \qquad \dots\dots(13).$$

For the consideration of the displacement of an element of air from its initial position, ξ becomes equal to x.

Substituting $x = \epsilon e/(p - e)$, and differentiating Lx, we find, after a slight approximation,

$$-\frac{\partial T}{\partial z} = \Gamma \times \frac{X + p}{Z + p} \qquad \dots\dots(14),$$

where

$$X = \frac{\epsilon}{AR}\,\frac{Le}{T}$$

$$Z = \frac{\epsilon}{c_p}\left\{ e\left(c + \frac{dL}{dT} \right) + L\,\frac{de}{dT} \right\} \qquad \dots\dots(15).$$

The details of the derivation of (14) have been given by Brunt.* This equation has been used to derive the lapse-rates for saturated air shown by the isopleths in fig. 18.

* *Q.J. Roy. Met. Soc.* **59**, 1933, p. 351.

The isopleths give the ratio of the saturated adiabatic lapse-rate to the dry adiabatic lapse-rate for different pressures and temperatures, and can therefore be used for any planetary atmosphere. The numbers shown against the lines in fig. 18 give this ratio, which for the earth's atmosphere can be interpreted as giving the saturated lapse-rate in degrees C per 100 metres.

Equation (13) above can also be derived by differentiation of equation (7), p. 50, with respect to T, and eliminating de/dT by the use of the Clausius-Clapeyron equation. (See § 62 below.)

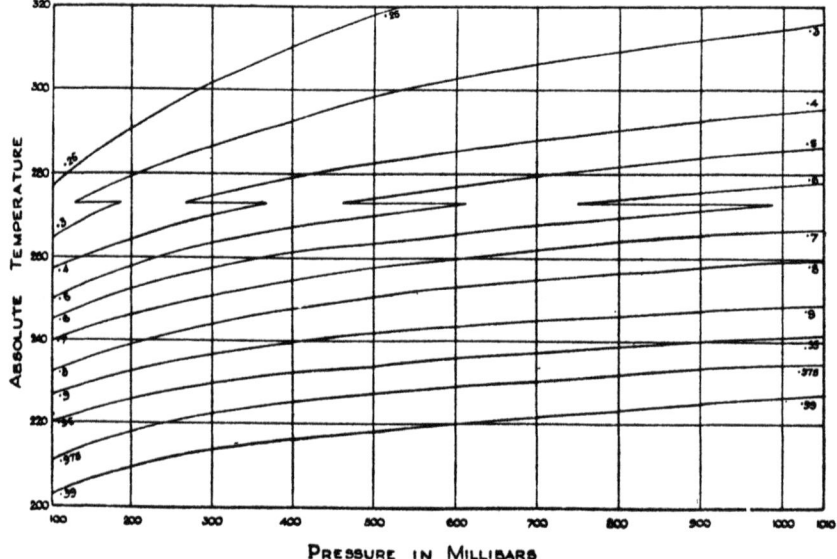

Fig. 18. Isopleths of the saturated adiabatic lapse-rate.

In the preparation of fig. 18. it was assumed that at temperatures below 0° C condensation will be in the form of ice. If it is required to allow for the precipitation being in the form of supercooled water, the lapse rate down to −20° C, or even lower, can be obtained with all the accuracy required in practice by the continuation of the curves in the upper part of fig. 18 for temperatures above 0° C into the lower portion.

§ 40. *Stability of saturated air*

If a small element of saturated air is displaced upward from its original position at a level z to a level $z+dz$, its temperature will fall by an amount $\left(\dfrac{\partial T}{\partial z}\right)_s dz$, where $\left(\dfrac{\partial T}{\partial z}\right)_s$ is the saturated adiabatic lapse-rate appropriate to the conditions of the air, as derived from equation (13) above. The stability or instability of the air is to be determined by the relative density of the displaced air and its environment. The displaced air has a temperature of

$$T+\left(\frac{\partial T}{\partial z}\right)_s dz,$$

and is surrounded by air whose temperature is

$$T + \frac{\partial T}{\partial z} dz.$$

But the density of the air will depend on the humidity as well as on the temperature of the environment.

If the whole environment is also saturated, the condition for stability is obviously

$$\frac{\partial T}{\partial z} > \left(\frac{\partial T}{\partial z}\right)_s \quad \text{or} \quad -\frac{\partial T}{\partial z} < -\left(\frac{\partial T}{\partial z}\right)_s \qquad \ldots\ldots(16),$$

i.e. the lapse-rate must be less than the saturated adiabatic for that level.

If the environment is unsaturated at the level $z + dz$, its density will be greater than that of saturated air at the same temperature and pressure, and instability will enter when the lapse-rate is rather less than the saturated adiabatic. In this case, stability requires that the lapse-rate shall be less than the saturated adiabatic by a finite amount. In practice it is not possible to take account of the details of the variation of water-vapour content of the atmosphere, and the result derived above is most usefully stated in the following form: An atmosphere whose lapse-rate is greater than the saturated adiabatic will be unstable for saturated air, and the degree of instability will be the greater the drier the environment. This appears at first sight contradictory, but a little consideration shows that it is in keeping with the fact that dry air is denser than damp air at the same temperature and pressure. It is to be used in the following way. Suppose the atmosphere is at all heights unsaturated, and that a small mass of air can be set in upward motion, until it attains saturation. The result we have derived above shows that once saturation is attained the air will be in an unstable position with a lapse-rate rather less than the saturated adiabatic above it, and that the further ascent of the saturated air will be facilitated by the fact that the air through which it moves is relatively dry.

The question of the stability of saturated air must always be discussed with regard to *upward* motion. If a mass of saturated air be set in downward motion it is heated at the dry adiabatic rate, unless it carries water drops in suspension. In the latter case the evaporation of the water drops keeps the rate of heating due to compression down to the saturated adiabatic rate.

If all condensed water drops are immediately precipitated, so that the saturated air carries no water drops, then saturated air is stable for downward motion when the lapse-rate is less than the dry adiabatic, but unstable for upward displacements unless the lapse-rate is less than the saturated adiabatic. The saturated adiabatic is the critical lapse-rate for both upward and downward motion only when the air carries with it a supply of water drops, which can evaporate when the air is dynamically heated by descent.

The fact that in a saturated atmosphere in which the lapse-rate is intermediate between the dry and saturated adiabatic any element of air is unstable for upward displacements, but stable for downward displacements, requires some consideration. The air removed upward may be replaced either by convergence of air laterally, or by the descent of an equal mass. If the replacement is by descent of an equal mass, this will impose some limitations on the amount of convection which can take place. Over a large area, we may suppose ascent to take place over portions of the area, and descent over other portions. Let S and S' be the areas over which air ascends or descends respectively. The rate at which potential energy is liberated by the ascent of unit mass is proportional to $\left(\Gamma_s + \dfrac{\partial T}{\partial z}\right)$, where Γ_s is the saturated adiabatic lapse-rate. Also the rate at which work must be done in order to force unit mass to descend is proportional to $\left(\Gamma + \dfrac{\partial T}{\partial z}\right)$. If the motion is to continue, the first of these expressions must exceed the second, so that, approximately

$$S\left(\Gamma_s + \frac{\partial T}{\partial z}\right) > S'\left(\Gamma + \frac{\partial T}{\partial z}\right).$$

When the lapse-rate only slightly exceeds the saturated adiabatic, convection will be in narrow columns, though the horizontal extent of the columns will probably be increased in time by the effect of turbulence. For a fuller discussion of this, reference should be made to a note by J. Bjerknes.*

* *Q.J. Roy. Met. Soc.* **64**, 1938, p. 325.

CHAPTER IV

THERMODYNAMICS OF THE ATMOSPHERE

§ 41. *The concept of entropy*

BEFORE we consider the thermodynamical concept of entropy, we shall find it useful to recall some of the main principles of thermodynamics. Heat, being a form of energy, can be made to do work, the work equivalent of the unit of heat, or, to give it its more familiar title, the mechanical equivalent of heat, amounting to $4\cdot18 \times 10^7$ ergs per gramme-calorie. This figure is usually denoted by $\mathcal{J}$, and its reciprocal by A. ($A = 2\cdot392 \times 10^{-8}$.)

The state of a gas at any instant, *assuming its constitution is unaltered*, is completely specified by any two of the variables p, v, T, which denote the pressure, specific volume and absolute temperature, respectively. If, in the discussion of meteorological problems, portions of the atmosphere could be treated as isolated entities, there would be no need to introduce anything further. In nature, however, there is no such isolation, and there is a continuous mutual interaction between each portion of the atmosphere and the surrounding portions. When heat is added to a portion of the atmosphere a part of the heat is used in warming the air, and the remainder in expanding the air against the pressure of its surroundings. The name of "working substance" which it is convenient to give to the portion of any gas or mixture of gases on which our attention is at the moment concentrated is itself a recognition of the importance of the environment in the changes of state of that mass.

The state of the working substance at any instant may be represented by a point in a diagram in which the ordinate and abscissa are the pressure and specific volume. Such a diagram is known as a p-v diagram or "indicator diagram". Any cycle of changes of the working substance may be represented by a continuous line in this diagram. In fig. 19 the work done by the body in going from a state denoted by $P(p, v)$ to the state denoted by a neighbouring point $P'(p+dp, v+dv)$ is $p\,dv$, and the total amount of work done in going

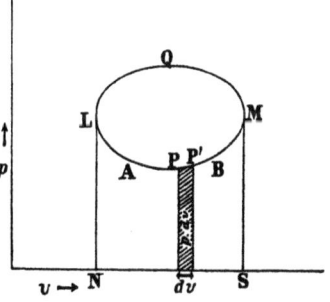

Fig. 19. Indicator diagram.

from A to B is $\int_A^B p\,dv$. If the working substance goes through a complete (Carnot) cycle of changes, starting at A, and eventually returning to A, the total amount of work done by the working substance is $\int p\,dv$, taken round the circuit, which is equal to the area enclosed by the curve. This amount of work done by the working substance is negative when the circuit is per-

69

formed in a counter-clockwise sense, as in the diagram (fig. 19) when the circuit is performed in the direction $APBQA$. For the work done by the working substance along the path $LABM$ is equal to the area $LAMSN$, and the work done in the path MQL is equal to the area $-MQLNS$. The total work done by the working substance is therefore negative and equal to the area enclosed by the loop, when the circuit is performed in the counter-clockwise sense. In this case work is done on the working substance by the surrounding medium. When work is done *by* the working substance, the circuit is performed in the reverse or clockwise sense, and energy is given out to the environment.

If we take a cycle bounded by two isothermals and two adiabatics, fig. 20, no heat is given to, or taken away from, the working substance in the parts of the cycle bounded by the adiabatics AD, BC, but in the parts of the cycle bounded by the isothermals AB, CD heat must be given to or taken away from the working substance in order to maintain the temperature constant. During the isothermal stages when heat is given to or taken away from the working substance, the temperature, and consequently the internal thermal energy, of the working substance remains constant. The thermal energy supplied is

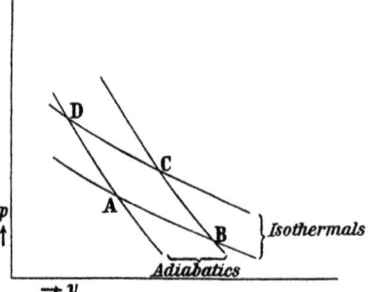

Fig. 20. Cycle bounded by iso-thermals and adiabatics.

passed on as work done upon the medium. The changes of temperature of the working substance take place during the adiabatic portions of the cycle, when no heat energy is supplied or extracted, and the changes of internal energy of the working substance are accounted for by the work done by the medium on the working substance during compression, or by the working substance on the medium during expansion.

When the working substance is expanding adiabatically it is doing work against the surroundings, and drawing upon its store of internal energy to do this. If in an adiabatic change the temperature falls from T_1 to T_2 the decrease of internal energy* is $c_v (T_1 - T_2)$, and this represents the work done in expanding against the pressure of the surroundings. When the working substance is expanding or contracting isothermally, an amount of heat must be given to or taken away from it, in just sufficient quantity to maintain the temperature constant.

An adiabatic is most conveniently specified by the corresponding potential temperature θ (see p. 38). The equation to the adiabatic may be written (see equations (31) and (32), p. 38)

$$pv^\gamma = D\theta^\gamma \qquad\qquad \dots\dots(1),$$

where D is a constant depending on the standard pressure with respect to which θ is defined. $(D = R^\gamma / p_0^{\gamma-1}.)$

If the working substance is kept at temperature T while it moves from the

* See p. 37, after equation (23).

adiabatic θ_1 to the adiabatic θ_2, the amount of work done by the working substance is*

$$\int p\,dv = RT \int dv/v = \frac{RT}{M}\log\frac{v_2}{v_1} = \frac{RT}{M}\log\frac{p_1}{p_2}$$

$$= \frac{c_p T}{AM}\log\frac{\theta_2}{\theta_1} \qquad \text{......(2)},$$

by equation (32), p. 38.

The amount of heat which must be given to the working substance in order to enable it to perform this work is $\dfrac{c_p T}{M}\log\dfrac{\theta_2}{\theta_1}$. The quantity of heat necessary in order to enable the working substance to go isothermally from one adiabatic to another is thus proportional to the absolute temperature of the isothermal, i.e. to T.

Now consider the heat necessary to take the working substance from a point A, defined by p, v, T, θ, to a point B defined by p', v', T', θ', along any specified path in the indicator diagram. A point in the indicator diagram is completely specified by any two of the variables p, v, T, or θ. Divide the path into any number of very small arcs of which PQ is one. Let PR and QR be the adiabatic through P and the isothermal through Q respectively. Then the amount of heat which must be communicated to the working substance in order to take the representative point along PRQ is

$$\frac{c_p T}{M}\log\frac{\theta_Q}{\theta_P} \qquad \text{......(3)}.$$

This differs from the amount of heat required to take the point along the direct arc PQ only by the area of the small triangle PRQ, which is a small quantity of the second order.

Thus if θ, θ_1, θ_2, θ_3, θ_4, ... be the potential temperatures, and T, T_1, T_2, T_3, T_4, ... the absolute temperatures, at successive points along the curve AB, the total quantity of heat which must be communicated to the working substance in order to take it along the line AB is equal to

$$\frac{c_p}{M}\left\{ T_1\log\frac{\theta_1}{\theta} + T_2\log\frac{\theta_2}{\theta_1} + T_3\log\frac{\theta_3}{\theta_2} + ... + T'\log\frac{\theta'}{\theta_n} \right\} \qquad \text{......(4)},$$

or in the limit in which the small arcs are infinitesimally small the amount of heat which must be given to the working substance is

$$c_p \int T\frac{d\theta}{\theta} \qquad \text{......(5)}.$$

Either expression shows that the total amount of heat is dependent on the path followed. Thus the total quantity of heat does not of itself determine the final state of the system.

It will be seen, however, that if the quantity of heat Q communicated to the working substance in the small arc QR is divided by T, then

$$\frac{Q}{T} = \frac{c_p}{M}\log\frac{\theta_Q}{\theta_P} \quad \text{and} \quad \Sigma\frac{Q}{T} = \frac{c_p}{M}\log\frac{\theta_B}{\theta_A} \qquad \text{......(6)}.$$

* All logarithms are to base 10.

Thus $\Sigma \frac{Q}{T}$ depends only on the initial and final states of the working substance.

For this reason the quantity $\Sigma \frac{Q}{T}$, or, to take the more general expression, $\int \frac{dQ}{T}$, has come to be regarded as a fundamental concept in thermodynamics. It is called the *Entropy*, and is usually denoted by the Greek letter ϕ. It is defined by the equation

$$d\phi = \frac{dQ}{T} \quad \text{or} \quad \phi = \int \frac{dQ}{T} \qquad \ldots\ldots(7).$$

Conversely, the amount of heat communicated in a process which produces a change of entropy $d\phi$ is dQ, where

$$dQ = T\,d\phi \qquad \ldots\ldots(8).$$

We might therefore represent the changes of a system, in so far as the communication of heat is concerned, by an indicator diagram in which the abscissa and ordinate are T and ϕ. In an elementary complete cycle, represented by an elementary rectangle of sides $d\phi$ and dT, the total amount of heat communicated is $dT\,d\phi$, and in any closed cycle the total heat communicated is equal to the area enclosed in the T-ϕ diagram.

Since in any reversible process the change of entropy in going from A to B is

$$\phi_B - \phi_A = \frac{c_p}{M} \log \frac{\theta_B}{\theta_A},$$

we may therefore define the entropy ϕ by the equation

$$\phi = \frac{c_p}{M} \log \theta \qquad \ldots\ldots(9),$$

where θ is the potential temperature. We may, if we choose, add a constant on the right-hand side of this equation, but as we are never concerned with the absolute magnitude of the entropy, but only with its changes, the addition of a constant is of no particular significance.

It is of fundamental importance to bear in mind the difference between reversible and irreversible physical processes, with regard to the changes of entropy. If the working substance is taken from a state A to a state B, and then brought back again to its original state, then the entropy will return to its original value at A, when the cycle is completed, if, and only if, the process is reversible. If any irreversible action occurs, as for example the conduction of some of the heat to a colder body, or the condensation and falling out of water drops, then the process ceases to be reversible, and the entropy of the whole system is increased thereby.

Consider the changes in entropy when the body goes through a complete (Carnot) cycle. The working substance takes in an amount of heat Q_1 at a temperature T_1, the heat being used in expansion against the pressure of the environment. The entropy increases by an amount Q_1/T_1. During the isothermal compression at temperature T_2 the working substance gives out an amount of heat Q_2 and loses an amount of entropy Q_2/T_2. It follows from equation (6) above that

$$Q_1/T_1 = Q_2/T_2.$$

For the amount of heat involved in the transfer of the working substance iso-thermally from the adiabatic θ_1 to another adiabatic θ_2 is $\dfrac{c_p}{M} T \log \dfrac{\theta_2}{\theta_1}$. Thus Q/T is the same for all isothermal transfers from one adiabatic to another. In any reversible cycle the gain of entropy along one isothermal is equal to the loss of entropy along the other isothermal, so that the initial and final values of the entropy are equal. This statement is readily generalised to the statement that in any reversible cycle the entropy returns to its original value.

The value of the concept of entropy lies in the fact that it is a function depending only on the state of the working substance, and independent of the sequence of changes by which that state was attained. In this respect it should be contrasted with the quantity of heat $\int dQ$ which must be given to the working substance to take it from the state A to the state B. The substance can be taken from the one state to the other by an infinite number of possible sequences, and to each of these possible sequences will correspond a different value of $\int dQ$, but the change of entropy in going from A to B will be the same for all the possible paths, provided that the changes which take place are all reversible.

Though entropy does not in fact give any additional information when we know already the pressure and temperature of each constituent of the system, it is frequently of great value in physical discussion. The fundamental vari-ables of the observer are undoubtedly pressure and temperature, but one is tempted to say that entropy is one of the fundamental variables of nature. Entropy is frequently described as a mathematician's device, enabling him to integrate his differential equations; but it is probably much truer to describe it as a physicist's device for sorting out the most important physical factors involved in any particular process. An example of this was given in § 34, in the derivation of equation (7) for the changes of state of saturated air. The argu-ment which was used was based on the fact that in a reversible process the entropy is a one-valued function of the state of the working substance. In general, if $d\phi$ can be expressed as a function of any variables x_1, x_2, x_3, etc. in the form

$$d\phi = f_1 dx_1 + f_2 dx_2 + f_3 dx_3 + \text{etc.} \qquad \ldots\ldots(10),$$

the right-hand side is a complete differential. Thus having specified the quan-tity of heat dQ involved in an elementary change of state we may divide it by T and so find $d\phi$, which must be a complete differential. It might conceivably have been possible to derive equation (7), p. 59 without making use of the con-cept of entropy, if it were possible to state with certainty the order in which the different processes of condensation, heating and expansion take place. The use of entropy takes charge of this troublesome stage of the problem by sorting out the essential from the non-essential facts. The quantity of heat dQ involved in the elementary change discussed in § 34, when expressed as a function of T and x, is not a perfect differential, the physical reason being that in a complete cycle $\int dQ$ is not zero, being equal to the amount of work done, so that $\int dQ$ is not a single-valued function of the state of the working substance. A variety of different paths in the indicator diagram join any two points selected, and though the final state is clearly defined by the position of the appropriate point in the diagram, different paths represent the conversion of different amounts of heat into work done on the environment, so that $\int dQ$ is not specified by the initial and final points of the path.

§ 42. *The effect of conduction or mixing on the total entropy*

The effect of transfer of heat by conduction or mixing is to increase the total entropy of the system. For suppose a quantity of heat Q is conveyed from a body at temperature T_1 to a body at temperature T_2. (T_2 must obviously be lower than T_1 if the transfer is to be physically possible.) The first body loses an amount of entropy Q/T_1 and the second gains an amount of entropy equal to Q/T_2. The net gain of entropy of the system is

$$Q/T_2 - Q/T_1 = \frac{Q}{T_1 T_2}\,(T_1 - T_2) \qquad \qquad \ldots\ldots(11),$$

which is positive since $T_1 > T_2$.

It should be noted, however, that while the total entropy of a system is increased by mixing, that is, the mean entropy of the unit mass within the system is increased by mixing, the effect of mixing masses of air of different potential temperatures is that the potential temperature of the mixture is the mean potential temperature of the original masses unmixed. For let a quantity m_1 of air at temperature T_1, pressure p_1, and potential temperature θ_1 be mixed with a quantity m_2 of air of temperature T_2, pressure p_2, and potential temperature θ_2, without any gain or loss to the environment. Let the resultant pressure of the final mixture be p. We may suppose the mixing to be carried out in two steps. First the two masses are brought adiabatically to the pressure p, when they will have temperatures $\theta_1 \left(\dfrac{p}{p_0}\right)^{\frac{\gamma-1}{\gamma}}$ and $\theta_2 \left(\dfrac{p}{p_0}\right)^{\frac{\gamma-1}{\gamma}}$ respectively $\left(\dfrac{\gamma-1}{\gamma} = 0{\cdot}288\right)$. Next let the two masses mix. The internal energy of the mixture will now be the sum of the internal energies of the constituents of the mixture, and the final temperature of the mixture will therefore be

$$\frac{m_1 \theta_1 + m_2 \theta_2}{m_1 + m_2} \left(\frac{p}{p_0}\right)^{\frac{\gamma-1}{\gamma}}$$

The potential temperature of the mixture will therefore be $\dfrac{m_1 \theta_1 + m_2 \theta_2}{m_1 + m_2}$, which is equal to the mean potential temperature of the constituents.

§ 43. *Formulae for entropy*

Provided we are dealing with a substance whose constitution is not variable, and whose specific heat at constant pressure is c_p, we may define entropy by either of the equations

$$\phi = \frac{c_p}{M} \log \theta \qquad \qquad \ldots\ldots(12),$$

$$\phi = \frac{c_p}{M} \log T - \frac{AR}{M} \log p \qquad \qquad \ldots\ldots(13).$$

An arbitrary constant may be added to the right-hand side of either of these equations.

We may start from any convenient point for the zero of entropy. If, for example, we wish to take T_0 and p_0 as the starting points for temperature and pressure,

$$\phi = \frac{c_p}{M} \log \frac{T}{T_0} - \frac{AR}{M} \log \frac{p}{p_0}.$$

Convenient values are $T_0 = 100°$, and $p_0 = 1000$ mb. Then the equations are

$$\phi = 2 \cdot 303 \times 10^7 \log \frac{T}{100} - 0 \cdot 661 \times 10^7 \log \frac{p}{1000} \qquad \ldots\ldots(14).$$

The entropy is then given in C.G.S. units, the appropriate values having been inserted for c_p and R to ensure this. Since 1 joule $= 10^7$ ergs, the entropy in joules per gramme is given when the factors 10^7 are omitted in the above expression,

$$\phi = 2 \cdot 303 \log \frac{T}{100} - 0 \cdot 661 \log \frac{p}{1000} \qquad \qquad \ldots\ldots(15)$$

measured in joules per gramme.

§ 44. *Efficiency of a heat engine*

If a reversible engine working between temperatures T_1 and T_2 takes in a quantity of heat Q_1 at temperature T_1, and rejects a quantity of heat Q_2 at temperature T_2, the changes from T_1 to T_2 and from T_2 to T_1 being adiabatic at potential temperatures θ and θ', then the efficiency of the engine is the ratio $(Q_1 - Q_2)/Q_1$, which measures the fraction of the heat put into the engine at the higher temperature which is converted into work.

But since the quantities Q_1 and Q_2 are proportional to the absolute temperatures at which the isothermal stages of the cycle are performed (equation (3) above), the efficiency is also equal to

$$(T_1 - T_2)/T_1.$$

For the perfect reversible engine the efficiency increases as the temperature T_2 of the cold source diminishes. If the cold source were at absolute zero, the efficiency of the engine would be unity.

The condition for reversibility might be written

$$Q_1/T_1 = Q_2/T_2.$$

The engine is not reversible when heat is lost by some such agency as conduction to some external source, or precipitation of condensed water. The amount of heat finally rejected is then less than in the absence of such agencies, and

$$Q_2/T_2 < Q_1/T_1.$$

It follows from the definition of entropy given above that in such a case the entropy of the working substance is finally greater than at the beginning of the cycle.

§ 45. Entropy-temperature diagrams or T-φ diagrams

If a working substance is taken through a series of changes in a reversible manner, the state of the working substance, being thus at each instant fixed by the temperature and entropy, can be represented in a diagram with temperature and entropy as co-ordinates. It is convenient to measure entropy along the vertical axis, and temperature along the horizontal axis. The amount of heat communicated to the working substance in order to take it from the state represented by the point P to the state represented by the point P' is $T d\phi$, or the area $PNN'P'$ (fig. 21) and the total amount of heat given to the working

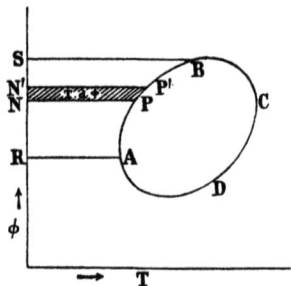

Fig. 21. T-φ diagram.

substance in going from state A to state B is the area $ARSB$, or $\int_A^B T d\phi$. The total amount of heat is positive in the transit from A to B as drawn in the diagram, since the entropy increases steadily. If the path is followed in the reverse order from B to A, $\int_B^A T d\phi$ is negative, and heat is taken from the working substance.

If the representative point executes a closed curve, indicating that the working substance, after going through a complete cycle, has returned to its original state, the total amount of heat communicated to the working substance is equal to the area enclosed by the closed curve $ABCD$, and it is readily seen that if the cycle is performed in the clockwise direction, i.e. in the direction $ABCD$, the net amount of heat communicated to the working substance is negative, or, in other words, heat is extracted from the working substance. When the cycle is performed in the counter-clockwise direction, as along $ADCB$, the net amount of heat communicated to the working substance is positive. In the first case work is done on the environment by the working substance, and in the second case work is done by the environment on the working substance. The amount of work done in either case is measured by the area enclosed by the curve in the T-ϕ diagram.

Examples of T-ϕ diagrams corresponding to certain particular cases are given in any textbook of thermodynamics, and the reader is referred to such books for further information as to the general use of T-ϕ diagrams. The T-ϕ diagram has one outstanding advantage in that the isothermals and adiabatics are straight lines in these diagrams, being parallel to the axes of co-ordinates, instead of being curved as in the usual p-v diagram. But in all thermodynamic diagrams the saturated adiabatics are curved lines.

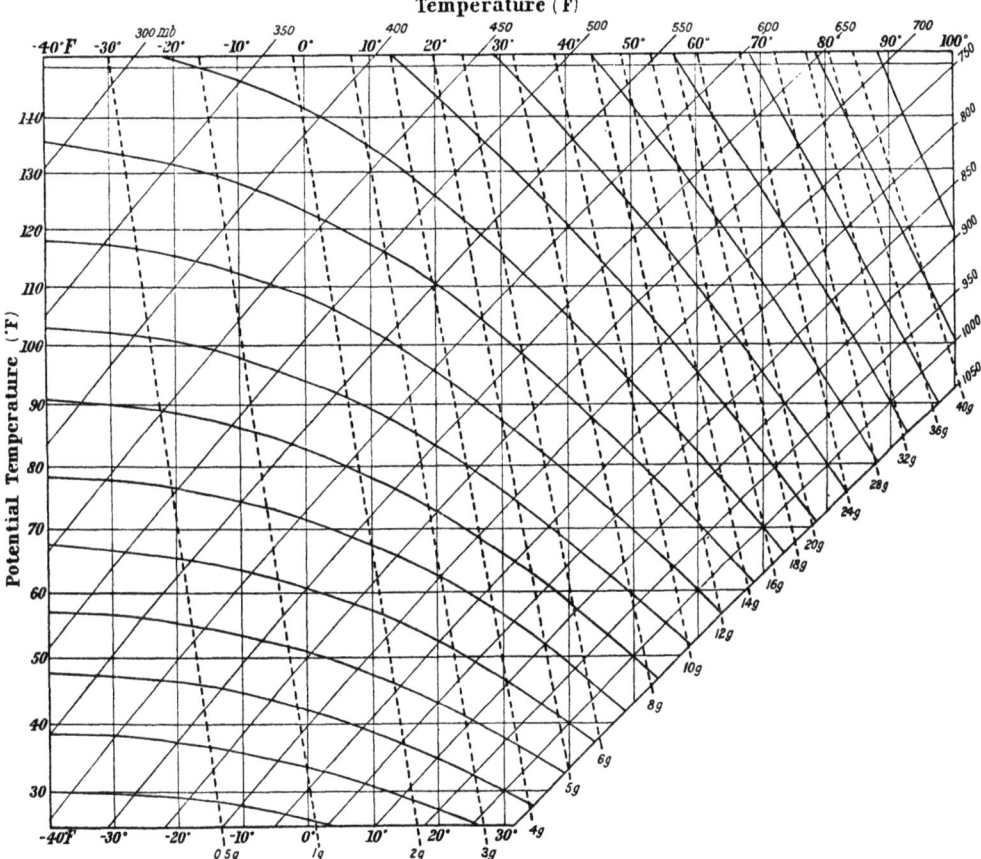

Fig. 22. Tephigram.

§ 46. The tephigram

The T-ϕ diagram has been adapted for meteorological use by Shaw in the *Manual of Meteorology*, **3**. On the tephigram (T-ϕ-gram) the temperature in ° F is measured along the horizontal axis, and the entropy along the vertical axis (see fig. 22). Shaw identifies entropy with $(c_p \log \theta)/M$, where θ is the potential temperature, and the linear entropy scale is actually a logarithmic scale of potential temperature. The adiabatics or pseudo-adiabatics for saturated air are shown as curved lines, and the pressure lines are shown sloping downward from right to left while the broken lines indicate the number of grammes of water-vapour necessary to saturate 1 kilogramme of

dry air at the temperature shown. The values of entropy read off for points on
the saturated adiabatics are values of the entropy of dry air at the corresponding temperature, evaluated by the formula

$$\phi = \frac{c_p}{M} \log \theta + \text{const.}$$

$$= \frac{c_p}{M} \log \frac{T}{T_0} - \frac{R}{M} \log \frac{p}{p_0} \qquad \ldots\ldots(16),$$

where c_p is taken in dynamic units, and $c_p/M = 2\cdot303 \times 10^7$. It should be borne
in mind that in all discussion of tephigrams hitherto published, the effect of
water-vapour on the entropy and density of the air has been disregarded. Nor
should it be overlooked that the discussions of adiabatic descent or ascent of
air also neglect the possible effects of radiation, absorption and turbulence,
whose magnitude has not been estimated in a single instance.

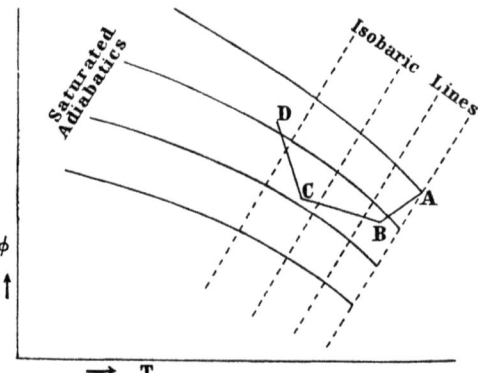

Fig. 23. Tephigram, schematic.

Subject to the above limitations the tephigram makes it possible to see at
a glance the state of the air as regards stability. The dry adiabatics are horizontal lines, while the isotherms are vertical lines (see fig. 22). When the lapse-
rate exceeds the dry adiabatic the curve in the tephigram slopes downward to
the left. When the lapse-rate is less than the dry adiabatic the curve slopes
upward to the left, while an inversion gives a line sloping upward to the right.
Isothermal conditions are represented by a portion of the curve running
vertically, while a lapse-rate between the saturated and the dry adiabatic will
slope upward to the left at an angle less than the slope of the saturated
adiabatic.

Thus if *ABCD* in fig. 23 represents the observed conditions in the atmosphere, *AB* represents a super-adiabatic lapse-rate, *BC* a lapse-rate less than
the dry adiabatic, and *CD* a lapse-rate less than the saturated adiabatic.

Figs. 22, 23 have been drawn so that temperature increases from left to
right, as in the Neuhoff diagram, and in accordance with recent British practice.

§ 47. *The use of the Neuhoff diagram or the tephigram to evaluate the energy liberated during the ascent of air*

In the Neuhoff diagram the co-ordinates are temperature on a linear scale and pressure on a logarithmic scale. Let the conditions at height z in the atmosphere be represented by p, ρ and T. Let an element of air of unit mass, which is set in motion in a manner that we need not at the moment specify, have density ρ and temperature T' when it is at height z. It will have the same pressure as its environment at the same level.

Whether the moving air or its environment is damp or not will not be considered, and any portion of air will be regarded as satisfying the same gas-equation

$$p = R\rho T,$$

R being the appropriate constant for the air.

The difference involved in neglecting the effect of water-vapour on density is slight.

The acceleration of the element of moving air is

$$g\frac{\rho - \rho'}{\rho'} \quad \text{or} \quad g\frac{T' - T}{T} \qquad \ldots\ldots(17).$$

In moving through a distance dz it does an amount of work dW, where

$$dW = \frac{g\,(T' - T)\,dz}{T} = \frac{g\rho\,(T' - T)\,dz}{\rho T} = -\frac{R\,(T' - T)\,dp}{p} \qquad \ldots\ldots(18),$$

or
$$dW = -R\,(T' - T)\,d\log p \qquad \ldots\ldots(19).$$

In fig. 24 (*a*), if AQC be the line indicating the changes of temperature of the moving air (this may be either a dry or saturated adiabatic, or partly one, partly the other), and if $ABCD$ represents the conditions in the environment, then the moving air at Q is surrounded by air whose condition is represented by P at the same pressure. Let $P'Q'$ be a subsequent position of PQ, separated by a very small vertical distance $d\log p$. Then

$$dW = R \times \text{area } PQQ'P'.$$

Thus the amount of work done, or energy released, in an elementary displacement is equal to R times the area $PQQ'P'$ drawn as shown in the figure.

If the air at A is in an unstable condition, and is set in vertical motion along the adiabatic AQC, it will attain a level where it can again be in equilibrium at C, and the amount of energy liberated per gramme of ascending air is R times the area enclosed between the curve of observed temperatures and the curve of temperature of the ascending air. The latter will be an adiabatic line of some kind. It may be a dry or a saturated adiabatic, or it may be a dry adiabatic along the lower part of its course and a saturated adiabatic along the remainder of its course. The proof given above is perfectly general, and applies to any of these possibilities.

A result which is precisely similar can be derived for the tephigram. By definition

$$\phi = \frac{c_p}{M} \log T - \frac{AR}{M} \log p + \text{const.} \qquad \ldots\ldots(20),$$

or

$$d\phi = c_p \frac{dT}{T} - AR \frac{dp}{p} \qquad \ldots\ldots(21).$$

As in equation (18)

$$dW = -(T' - T) R \frac{dp}{p}.$$

In fig. 24 (b) the line $ABCD$ represents the condition of the environment, or, in other words, it is the curve which shows the actual observations of temperature, and AQC represents the change of state of the moving mass. PP'' is

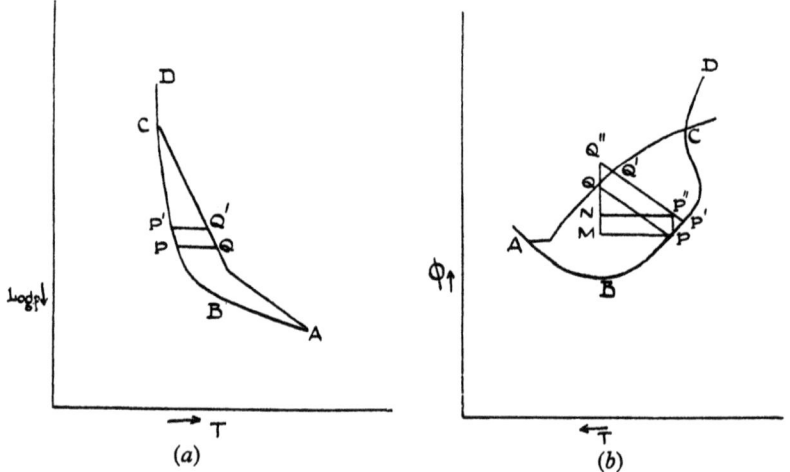

Fig 24. Energy in the Neuhoff diagram and the tephigram.

the change of entropy when the pressure changes by dp while the temperature remains constant. Thus

$$PP'' = R \frac{dp}{p},$$

$$dW = -(T' - T) R \frac{dp}{p} = \text{area } PP''NM = \text{area } PP''Q''Q$$

$$= \text{area } PP'Q'Q,$$

since the ends of the strip, the two small triangles $PP'P''$ and $QQ'Q''$, may be neglected as small quantities of the second order. Thus the energy released by the ascent of the air from Q to Q' is equal to the area $PQQ'P'$, and by integration it follows that the energy released by the motion of the ascending air from A to C, where it is again in equilibrium with its environment, is measured by the area $APCQA$.

The use of the tephigram in the way suggested is to be regarded as an approximation, as it takes no account of the water-vapour in the atmosphere, in the estimate of the buoyancy of a moving mass. No method has been devised which can take account of the water-vapour in the atmosphere for all purposes. The tephigram in effect takes account only of the density of the dry air, though it takes account of the effect of the presence of water-vapour on the temperature changes which are produced by the ascent of the moist air. Strictly speaking, equilibrium of a mass of air in the atmosphere is determined by its density and not by its entropy, and the tephigram therefore fails to give a criterion which will apply to all cases. In practice, the error will be very small for motions on a large scale, and no serious error is to be feared if the tephigram is used in the manner suggested.

A clear distinction should be drawn between the entropy of damp air, as derived from the formulae which have been developed above, and the entropy which is represented in the tephigram. We may write (see § 49 below)

$$\text{Entropy of moist air} = \text{Entropy of dry air}$$
$$+ \text{Entropy of water-vapour}$$
$$= \frac{c_p}{M} \log T - \frac{AR}{M} \log (p - e)$$
$$+ \frac{xc}{M} \log T + Lx/T.$$

It is an approximation to the first part of the entropy, given by

$$\frac{c_p}{M} \log T - \frac{AR}{M} \log p$$

which is represented in the tephigram, and the water-vapour is disregarded as having no effect on the stability of the atmosphere.

The tephigram gives precisely the same results as the Neuhoff diagram, both for estimating the stability of the atmosphere, and for estimating the amount of energy which can be released by the ascent of air. The use of the word entropy in this connection may mislead the uninitiated, and a word of warning is given here, that all methods which base their criteria on the neglect of the density effect of water-vapour must of necessity lead to the same results.

It has been suggested that the entropy of any mass of air will indicate the height at which it will be in equilibrium with its environment, but the entropy of dry air alone, which Shaw calls realised entropy, increases with height, and so the entropy measured by the tephigram as realised entropy is not constant for a given mass of air. Ascending air will come to rest when two conditions are realised, namely, the pressure and the density of the ascending air must both attain equality with those factors for the environment. It is assumed that the ascending air will at all stages take up the pressure of the environment. This is readily taken into account by plotting temperature against pressure, or against a function of temperature and pressure such as entropy. The graph of observed conditions is plotted on a diagram on which the adiabatics of dry and saturated air are shown, so that it is possible to determine the height at which the ascending air disturbed from its position as part of the normal environment will again attain the same temperature as the environment. This is not strictly the level at which the ascending air will attain equilibrium. For equilibrium is determined by equality of density, and density is partly determined by the water-vapour content. Neither the tephigram nor the Hertz or

Neuhoff diagram can take account of the effect of water-vapour on density, though it would be possible to construct a diagram which would do so. This could be done by showing in the graph, not the observed temperatures, but temperatures at which the air would, when saturated, have the same density as the actual air of the environment.* But such a diagram would of necessity give the same general results as the tephigram or Neuhoff diagram, since the main large-scale variations of density are produced by variations of temperature, the main effect of the variations of water-vapour content being to change the height at which the air when disturbed from its equilibrium position will attain saturation.

To put the matter into a few words, there is no special virtue in any one of the standard methods of plotting temperature observations, and whether we adopt the tephigram, the T-log p diagram of Neuhoff, or any other method, is more a matter of personal inclination than a matter of real scientific significance. The tephigram and the Neuhoff diagram have the great advantage that, as shown in fig. 24 above, it is possible to evaluate on these diagrams the amount of energy which is liberated by the ascent of unit mass of air, by the direct computation of an area. This advantage is definitely of importance, but any diagram which will enable us to carry out this computation will have all the virtues of any other.

The tephigram has one great advantage, from the point of view of mere convenience, that the isothermals and dry adiabatics are straight lines at right angles to each other, and as a result the diagram is usually more compact than is the Neuhoff diagram, on which the area between the pseudo-adiabatics and the curve of observed temperatures is usually a long and narrow area.

To most of its users the tephigram is a diagram in which the co-ordinates are temperature and pressure, along axes which are not rectangular, but are so arranged that isothermals and dry adiabatics are vertical and horizontal respectively. That the vertical co-ordinate is entropy is not in practice taken into account by the average user of the tephigram.

§ 48. *Latent heat of evaporation*

The latent heat of evaporation of water is defined as the quantity of heat which is required to evaporate 1 gramme of liquid water in contact with its saturated vapour. The heat is used partly in raising the internal energy of the water, and partly in doing the work of expansion against the external pressure.

Let
$v_1 = $ specific volume of liquid water.
$v_2 = $ specific volume of water-vapour.
$E_1 = $ internal energy of 1 gramme of liquid water.
$E_2 = $ internal energy of 1 gramme of water-vapour.
$L = $ latent heat.
$e = $ pressure of saturated vapour.
$p = $ total pressure.
$A = $ reciprocal of mechanical equivalent of heat.
$\epsilon = $ ratio of densities of water-vapour and dry air.

---

* Such a temperature might be called the "saturated virtual temperature".

The first six of these quantities are to be defined as the values appropriate to the temperature T of both liquid and vapour.

If the water is evaporated under the pressure of its own vapour, then it is seen that the latent heat of L is given by

$$L = E_2 - E_1 + Ae\,(v_2 - v_1) \qquad \ldots\ldots(22).$$

The specific volume of liquid water, v_1, is negligible by comparison with that of water-vapour, v_2, and equation (26) then reduces to

$$L = E_2 - E_1 + ART/\epsilon \qquad \ldots\ldots(23).$$

The question arises whether equation (23) still holds when the evaporation takes place into air.

Let a volume V be occupied by air and vapour at a total pressure p, and let an additional quantity m of water be evaporated, thereby increasing the volume to $V + v$. Then by definition

$$mL = mE_2 - mE_1 + Apv \qquad \ldots\ldots(24).$$

In the final stage let p_1 be the partial pressure of the air and water-vapour originally present, and p_2 the partial pressure of the added vapour. Then

$$p = p_1 + p_2,$$
$$p_1\,(V + v) = pV \text{ by Boyle's law,}$$
$$(p_1 + p_2)\,(V + v) = p\,(V + v).$$

Hence
$$p_2\,(V + v) = pv.$$

But p_2 and $V + v$ are the partial pressure and volume of a mass m of water-vapour at the temperature T. Hence

$$pv = p_2\,(V + v) = mRT/\epsilon \qquad \ldots\ldots(25).$$

Substituting this value of pv in equation (24), we again obtain equation (23), which is therefore true in the case of evaporation into air.

§ 49. *The entropy of moist air*

In discussing the thermodynamics of moist air we shall require to write down the entropy of 1 gramme of water-vapour at a given temperature T. By the second law of thermodynamics, as we have seen in the earlier parts of the present chapter, the entropy will be independent of the manner in which the water-vapour is formed, as it depends only on the final state of the vapour.

Consider first the formation of 1 gramme of water-vapour, saturated at temperature T. Its entropy must equal that of 1 gramme of liquid water at the same temperature *plus* the added entropy required to convert it into vapour. Thus we have the result

Entropy of 1 gramme of saturated water-vapour at temperature T

$$= \text{entropy of 1 gramme of liquid water}$$
$$+ \frac{\text{Latent heat at temperature } T}{T}$$
$$= \int \frac{c\,dT}{T} + \frac{L}{T} \qquad \ldots\ldots(26),$$

where c is the specific heat at temperature T of liquid water in contact with its own saturated vapour. The error involved in assuming a constant value for c is very slight, and the entropy may be written, with sufficient accuracy

$$\phi = \frac{c}{M} \log T + L/T \qquad \ldots\ldots(27).$$

We have already made use of this expression in § 34 in the discussion of the thermodynamics of ascending moist air.

If we have to consider saturated air whose humidity mixing ratio is x, then the entropy of $(1+x)$ grammes of the moist air is equal to the entropy of 1 gramme of dry air, *plus* the entropy of x grammes of water-vapour. It therefore amounts to

$$\frac{c_p}{M} \log T - \frac{AR}{M} \log (p-e) + \frac{xc}{M} \log T + Lx/T$$

or

$$\frac{(c_p+xc)}{M} \log T - \frac{AR}{M} \log (p-e) + Lx/T \qquad \ldots\ldots(28).$$

It has been assumed that the specific heat of water has a mean value c; this assumption is not strictly necessary, and the term which we have written in the last equation as $xc \log T$ might have been written as $x\int\frac{c\,dT}{T}$. Actually the error involved in writing $c \log T$ instead of the integral is usually very small, as the changes of entropy of moist air are predominantly due to the changes in the value of the term Lx/T. Moreover, the expression which has been derived above for the entropy of moist air will usually only be used over a range of temperature which does not involve a wide variation in the value of c. Over such a restricted range we may take the mean value of c appropriate to that range.

Next consider the entropy of 1 gramme of unsaturated water vapour at temperature T, having the dew-point T_d. It is now necessary to bear in mind that the change of entropy involved in the conversion of 1 gramme of water into 1 gramme of water vapour at the same temperature is L/T, if, and only if, the liquid water is at all stages of the process in contact with its saturated vapour, since the latent heat is only defined with this limitation (see § 48 above). It follows that in order to evaluate the entropy of unsaturated water-vapour it is necessary to postulate the initial formation of the water-vapour as saturated vapour, i.e. at the dew-point temperature T_d. The change of entropy in raising the temperature from T_d to T can be computed by the use of equation (13), p. 75, with the substitution of c_p' for c_p. During this change the external pressure p is unchanged. Hence

Entropy of 1 gramme of water-vapour at temperature T_d

$$= \frac{c}{M} \log T_d + L_d/T_d$$

Entropy of 1 gramme of water-vapour at temperature T

$$= \frac{c}{M} \log T_d + L_d/T_d + \frac{c_p'}{M} \log T/T_d$$

$$= \frac{c}{M} \log T + L_d/T_d - \frac{(c-c_p')}{M} \log T/T_d \qquad \ldots\ldots(29).$$

It will be seen that in the treatment of the problem of the ascent of damp air in Chapter III no uncertainty arises as to the use of equation (27), since we were there concerned only with the ascent of saturated air. In the unsaturated stage, the difficulty was avoided by the use of an adiabatic equation similar to that for dry air.

§ 50. *The entropy of a mixture of damp air, water and ice*

Consider a mixture containing 1 gramme of dry air, x grammes of water-vapour, y grammes of liquid water, and z grammes of ice. Let its temperature be T, and its vapour-pressure e. Also let

$$\xi = x + y + z.$$

The entropy of 1 gramme of dry air is ϕ_1, where (see § 43, p. 75)

$$\phi_1 = \frac{c_p}{M} \log T - \frac{AR}{M} \log (p - e).$$

The entropy of x grammes of water-vapour (saturated) $+ y$ grammes of liquid water is ϕ_2, where (see § 49 above)

$$\phi_2 = (x + y) \frac{c}{M} \log T + Lx/T \qquad \text{......(30)}.$$

The entropy of z grammes of ice, ϕ_3, is equal to the entropy of z grammes of water at the same temperature, *minus* the entropy involved in its conversion from water into ice.

$$\phi_3 = \frac{zc}{M} \log T - L_e z/T.$$

Hence the total entropy is

$$\phi_1 + \phi_2 + \phi_3 = \frac{(c_p + \xi c)}{M} \log T + Lx/T - L_e z/T - \frac{AR}{M} \log (p - e) \quad \text{......(31)},$$

where M is the modulus of logarithms, 0·4343.

For an ascending mass of damp air which carries with it all the products of condensation, the changes which take place are reversible, and the total entropy, as represented in (31) above, remains constant.

§ 51. *The thermodynamics of the wet- and dry-bulb hygrometer*

When it has attained a steady state the wet-bulb thermometer is not gaining or losing heat, so that the heat required to evaporate water from the bulb must be supplied by the cooling of the air which comes into contact with the wet bulb. It is usually assumed that the air in contact with the wet bulb becomes saturated at the temperature of the wet bulb. This assumption is not justifiable *a priori*, and it must be judged by the results which follow from it rather than on strictly theoretical grounds.

We shall use the following notation:

T = absolute temperature of the dry-bulb thermometer.
T' = absolute temperature of the wet-bulb thermometer.
x = humidity mixing ratio of the normal air.
x' = humidity mixing ratio of air saturated at the wet-bulb temperature T'.
e = vapour-pressure of the normal air.
e' = vapour-pressure of air saturated at temperature T'.
c_p = specific heat at constant pressure of dry air.
c_p' = specific heat at constant pressure of water-vapour.
L = latent heat of water-vapour at temperature T.
L' = latent heat of water-vapour at temperature T'.
ϵ = ratio of densities of water-vapour and dry air at same temperature and pressure.
p = total pressure.

In the original air of the environment 1 gramme of dry air is associated with x grammes of water-vapour at temperature T. The assumption which we have mentioned above means that $(1+x)$ grammes of the original air, in cooling from T to T', yield up sufficient heat to evaporate $(x'-x)$ grammes of water, and that the result is to produce $(1+x')$ grammes of air saturated at temperature T'. The evaporation takes place at the temperature T' of the wet bulb, and the equation which gives the heat exchange is

$$(c_p + xc_p')\,(T-T') = L'\,(x'-x) \qquad \ldots\ldots(32).$$

The first factor on the left-hand side of this equation is very frequently reduced to its first term c_p, since c_p'/c_p is about 2, and x seldom exceeds 0.025. Equation (32) then becomes

$$c_p\,(T-T') = L'\,(x'-x)$$

or

$$T + \frac{L'x}{c_p} = T' + \frac{L'x'}{c_p} \qquad \ldots\ldots(33).$$

This is the form used by August and Apjohn, and also by Normand[*] in his discussion of the thermodynamics of the wet-bulb temperature. But equation (32), when expanded without any approximation, reads

$$c_p T + L'x + c_p'x\,(T-T') = c_p T' + L'x' \qquad \ldots\ldots(34).$$

If T'' is the temperature of absolutely dry air whose wet-bulb temperature is T', then equation (34) must hold for $T=T''$ and $x=0$, so that

$$c_p T'' = c_p T' + L'x' \qquad \ldots\ldots(35).$$

Combining equations (34) and (35) we find

$$c_p T + L'x + c_p'x\,(T-T') = c_p T' + L'x' = c_p T'' \qquad \ldots\ldots(36).$$

The temperature T'' thus defined is known as the *equivalent temperature*, and it is clear from equation (35) that T'' is a function of the wet-bulb temperature only.

* *Mem. Indian Met. Dept.* **23**, 1921, p. 1.

When in equation (32) we substitute $x = \epsilon e/(p-e)$, and $x' = \epsilon e'/(p-e')$, we find

$$\frac{\epsilon \, (e'-e) \, pL'}{(p-e) \, (p-e')} = (T-T') \left(c_p + \epsilon c_p' \, \frac{e}{p-e} \right) \qquad \ldots \ldots (37).$$

But $\epsilon c_p'/c_p = 1 \cdot 21$, and when e is not a large fraction of p we may neglect its variation from unity. The right-hand side of equation (37) then becomes $c_p \, (T-T') \, p/(p-e)$, and the equation may be written

$$e' - e = Bp \, (T-T') \qquad \ldots \ldots (38),$$

where

$$B = (1 - e'/p) \, \frac{c_p}{L'\epsilon} \qquad \ldots \ldots (39).$$

For the detailed application of this formula to psychrometry, reference should be made to a paper by Whipple,* in which this form was first given. The usual historic form of the equation has unity in place of the factor $1 - e'/p$.

It is not clear that the use of the latent heat L' on the right-hand side of equation (32) is strictly justified, except by the fact that the equation in this form appears to give consistent results in the measurement of humidity.

§ 52. *Some thermodynamical propositions relating to the wet-bulb temperature*

In § 51 above we have stated the assumptions which are made in Normand's work on the thermodynamics of the wet-bulb temperature. It is necessary to bear these very clearly in mind in what follows, if confusion between assumption and deduction is to be avoided. Normand deduced the following propositions from the assumptions:

Proposition I. The heat content of the air is equal to the heat content of the same air saturated at the wet-bulb temperature, *minus* the heat content of the additional liquid water required so to saturate it.

Proposition II. The entropy of air is equal to the entropy of the same air saturated at the wet-bulb temperature, *minus* the entropy of the additional liquid water required so to saturate it.

The first of these propositions appears a self-evident truth when stated as above. It is in fact no more than a re-statement of the assumptions made above as to what happens at the wet bulb, and equation (32) above is the mathematical statement of these assumptions. Any attempt at a mathematical proof of proposition I, which would of necessity use equation (32) in some form, would be to argue in a circle.

Proposition II is not quite in the same category. The process which we visualise is not reversible, since heat is taken from air at a temperature between T and T', and taken up by water at a temperature T', so that there is a gain of entropy. The simplest method of approaching this proposition is to evaluate the gain of entropy. The system which we have to consider consists initially of $(1 + x)$ grammes of moist air, plus $(x' - x)$ grammes of liquid water, and at the final stage consists of $(1 + x')$ grammes of moist air at temperature T'. An

* *Proc. Phys. Soc.* **45**, 1933, p. 307.

amount of heat $(c_p + xc_p')(T - T')$ is taken from the air and absorbed by water at a temperature T', being used up in evaporation. The gain of entropy by the water is equal to

$$(c_p + xc_p')\frac{T - T'}{T'}.$$

The loss of entropy of the initial damp air is

$$(c_p + xc_p')\log\frac{T}{T'}.$$

The total gain of entropy of the system is

$$(c_p + xc_p')\left(\frac{T - T'}{T'} - \log\frac{T}{T'}\right) \qquad\qquad ......(40).$$

But $\quad \log T/T' = \log\left(1 + \frac{T - T'}{T'}\right) = \frac{T - T'}{T'} - \frac{1}{2}\left(\frac{T - T'}{T'}\right)^2 + ...$ etc.

Substituting in equation (40) above, we find that the net gain of entropy is

$$(c_p + xc_p')\left[\frac{1}{2}\left(\frac{T - T'}{T'}\right)^2 + \text{higher powers of } \frac{T - T'}{T'}\right] \qquad ......(41).$$

Thus the net gain of entropy of the system is very closely equal to $\frac{1}{2}\left(\frac{T - T'}{T'}\right)$ times the loss of entropy of the original air. This is a small fraction, and may be neglected, and the process which we have been considering may therefore be treated as isentropic. Proposition II may thus be taken as verified to a high degree of approximation.

Proposition I has a number of interesting applications to the thermodynamics of moist air. It follows from the original assumption on which the derivation of equation (32) was based that $(1 + x)$ grammes of moist air at temperature T, having a wet-bulb temperature T', can be cooled to the temperature T' by the evaporation into it of the necessary $(x' - x)$ grammes of liquid water. Being then saturated, it cannot be cooled beyond T', and the wet-bulb temperature may accordingly be defined as the lowest temperature to which air can be cooled by the evaporation of water into it.

If we now start off with a given mass of air whose temperature is T, and whose wet-bulb temperature is T', then we know that T' is the lowest temperature to which the air can be cooled by evaporation. But the cooling down to the temperature T' obviously need not be carried out in one step. Suppose that an amount of water is evaporated into the air, but that it is not sufficient to cool it down to T'. In the intermediate stage then reached, the maximum cooling down to T' will still be attainable by the further evaporation of the right amount of additional water. The limit attainable by cooling will still be the same, and hence the wet-bulb temperature is still T'. Hence it follows that if air is cooled by the evaporation of water into it, the wet-bulb temperature remains unaltered during the process, and the air cannot be cooled below the wet-bulb temperature by this means, since it will be saturated when it reaches that temperature, and any further evaporation will be impossible.

There is implied in the above discussion the assumption that the evaporation all takes place at the temperature T', or that the liquid water is introduced into the system at a temperature T'. Any error introduced by the variations in the temperature of the water from T' will be slight, on account of the smallness of the specific heat of water by comparison with its latent heat.

This result can be derived mathematically from equation (32), though the general argument used above, if it has been properly stated, should be more convincing than any equation. The heat content of $(1+x)$ grammes of moist air is

$$c_p T + L'x + xc_p' (T - T') + xcT' \qquad \ldots\ldots(42).$$

By equation (32) this is equal to

$$(c_p T' + L'x') + xcT' \qquad \ldots\ldots(43).$$

Thus the heat content of the air, *minus* the heat content of the water vapour when reduced to liquid water at temperature T', is a function of the wet-bulb temperature T' only. Now evaporate a further quantity x_2 of water. The total heat content of the air will then amount to

$$(c_p T' + L'x') + (x + x_2) cT' \qquad \ldots\ldots(44).$$

It is seen at this stage that the part of this expression which remains when the heat content of the equivalent amount of water is subtracted is unaltered. It follows therefore that the wet-bulb temperature is unchanged.

It can also be deduced from equation (43) that when portions of air of different temperatures, but having the same wet-bulb temperatures, are mixed, the mixture has that same wet-bulb temperature. If for example two masses m_1 and m_2, whose humidity mixing ratios are x_1 and x_2, each having the wet-bulb temperature T', are mixed together, it follows from (43) that the total heat content of the mixture is

$$(m_1 + m_2) (c_p T' + L'x') + (m_1 x_1 + m_2 x_2) cT' \qquad \ldots\ldots(45)$$

and the heat content of 1 gramme of the mixture is

$$(c_p T' + L'x') + \frac{m_1 x_1 + m_2 x_2}{m_1 + m_2} cT' \qquad \ldots\ldots(46).$$

Hence the wet-bulb and the equivalent temperature of the mixture are both the same as for the separate constituents of the mixture.

§ 53. *Variation of dry- and wet-bulb temperatures, and of dew-point in air ascending adiabatically*

Normand has further shown that the dry adiabatic through the dry-bulb temperature T, the saturated adiabatic through the wet-bulb temperature T', and the dew-point line through the dew-point T'', all meet in a point. For if air saturated at T' be taken up adiabatically from B (fig. 25) until the humidity mixing ratio reaches the value x at the point D, the air represented by D will have $(x' - x)$ grammes of liquid water mixed with it. Let this water be removed as rain, and let the air then return to ground level, following the dry adiabatic DA. The process has been adiabatic, and therefore the initial entropy of the air saturated at T' is equal to the entropy of the air at A, *plus* the entropy of the water removed at D. Hence it follows from the second proposition of p. 87 that the wet-bulb temperature is the same in the initial and final conditions. Further, since T' is the wet-bulb temperature of air with temperature A and humidity mixing ratio x, it follows that A represents the initial dry-bulb temperature T. The line DC represents the temperatures and pressures at which x grammes of water-vapour will saturate the air, and so C must be the dew-point of the initial air. Thus the proposition is proved. Hence if unsaturated air be raised adiabatically, it will attain saturation at the level where the dry adiabatic through T meets the saturated adiabatic through T'. This gives a simple method of finding the saturated adiabatic which ascending air will eventually follow, and provides a very simple method of completing the tephigram or Neuhoff diagram.

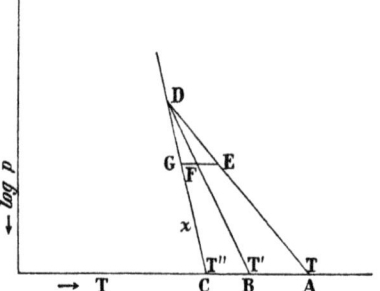

Fig. 25. Normand's temperature-height diagram.

If in fig. 25 EFG is drawn horizontally to intercept the three lines through D, the air, when it has reached the height appropriate to this level, will have dry-bulb temperature, wet-bulb temperature and dew-point temperature represented by the points E, F and G respectively.

It follows directly from the results which we have stated above that all masses of air which have at ground level the same wet-bulb temperature, though varying dry-bulb temperature, will ultimately follow the same saturated adiabatic when they ascend.

The proposition whose proof we reproduce above from Normand (*loc. cit.*) is not mathematically exact. The temperature at which the water is removed from the system at D is not equal to the wet-bulb temperature but the error involved is very slight, less in fact than the errors of observation of temperature would introduce.

§ 54. *The wet-bulb potential temperature*

It has been shown that the wet-bulb temperature has the property of remaining invariant during the evaporation of water into the air. Also when a mass of damp air rises or falls in the atmosphere, the wet-bulb temperature follows a saturated adiabatic during the whole of the ascent or descent, whether there is actual condensation taking place or not. Consider a mass of air rising through the atmosphere, having not yet attained saturation. Let the ascent be checked at some stage, and let some additional water be evaporated into it. Up to the time of the addition of the water, the wet-bulb temperature followed a saturated adiabatic. During the evaporation the wet-bulb temperature remained constant, and during any subsequent ascent, the wet-bulb temperature must follow the same saturated adiabatic, since the evaporation did not move the wet-bulb temperature from that adiabatic. If then we introduce the concept of a wet-bulb potential temperature, θ', which we define as the wet-bulb temperature that the air will take up when brought adiabatically to a standard pressure, then the wet-bulb potential temperature so defined will have the property that it remains invariant during all adiabatic and pseudo-adiabatic changes, whether accompanied by condensation or evaporation or not. Loss or gain of water has no effect on the wet-bulb potential temperature, which only varies when heat is given to the air from outside itself, either by mixing or by radiation or conduction, or is taken from it by the same agencies. We may contrast with this property of the wet-bulb potential temperature the property of ordinary potential temperature, which remains invariant for adiabatic changes of pressure only so long as there is neither condensation nor evaporation taking place. The wet-bulb potential temperature is therefore a much more potent instrument in determining the differences of origin of different air masses.

A direct application of the wet-bulb potential temperature to the phenomenon of the Föhn was mentioned by Normand in the paper to which reference has frequently been made in the present chapter. It follows from the preceding discussion that if a mass of air rises over a mountain range, and deposits most of its original water-vapour content on the windward side as rain, when after passing over the crest it is again brought down on the other side to its original level, it will be warmed at the dry adiabatic rate. The dry-bulb temperature will accordingly be much higher on the lee side than on the windward side of the mountain, but from the preceding discussion it follows that the wet-bulb temperature will be the same on the two sides. Thus the wet-bulb temperature might be used as a criterion for establishing the original identity of the warm dry air on the lee side with the cooler damp air on the windward side of the mountain.

Normand quotes the following records of conditions at Berkeley, California, during a short-lived Föhn wind, in which an easterly wind from the Berkeley Hills replaced the usual westerly sea winds for a period of about half an hour. The temperature rose $12° F$ in less than 10 minutes, while the relative humidity dropped below 50 per cent with remarkable suddenness. The variations are shown in the small table below:

Time (hours)	$7\frac{1}{2}$	$7\frac{3}{4}$	8	$8\frac{1}{2}$	9
Air temperature (°F)	52	63	56	56	61
Relative humidity (%)	99	49	90	95	85
Wet-bulb temperature (°F)	52	53	54	55	58·5

The wet-bulb temperature rose during the morning at the normal diurnal rate, showing no sign of an abrupt change with the onset of the easterly wind. It may be concluded that the easterly wind might have had the same origin as the westerly wind, but had been subjected to a series of adiabatic changes by ascent and subsequent descent.

When air subsides its temperature increases, and it is marked by extreme dryness when the subsidence has proceeded through a considerable range of height. When rain falls through the subsiding air and is wholly or partly evaporated, the humidity of the subsiding air may be maintained at a high level. The air is at the same time cooled as a result of the evaporation, and so subsiding air through which rain has fallen is no longer distinguishable by high temperature. From what has been said above, however, it follows that the wet-bulb potential temperature of the subsiding air is unaltered, and the best criterion for identifying subsiding air is therefore its wet-bulb potential temperature.

For the computation of the wet-bulb potential temperature we may use the Neuhoff diagram or the tephigram, marking each saturated adiabatic with the temperature at which it crosses the isobar of 1000 mb. The wet-bulb temperature plotted on the diagram will give a point for which we can interpolate a value of θ' from those corresponding to the two neighbouring saturated adiabatics.

§ 55. *Practical use of wet-bulb potential temperature*

The curve obtained by plotting the wet-bulb potential temperature θ' against the pressure, as given by an upper air ascent at one station, frequently shows some marked peculiarity which can be used to identify the level of this air at a later stage of its history in which upper air conditions are again observed. This method was used by Hewson[*] to investigate subsidence over the North American Continent, and the ascent of air at warm fronts.

For example fig. 26 shows the variation of θ' with pressure as observed at San Diego, U.S.A. (Station NNI), on Feb. 20, 1935. It was found from a consideration of the upper winds over the western half of the United States that the air above San Diego on the 20th was approximately over Cheyenne on the 21st. The curve showing θ' against pressure at this station (CX) is also shown in fig. 26, and the similarity of the CX curve to the NNI curve when shifted through about 180 mb is very striking, and may be held to indicate the identity of the two air masses shown, except for the loss of a shallow layer of surface air from San Diego, which was probably diverted by the marked irregularities of the ground over which the air passed.

Fig. 27, also due to Hewson (*loc. cit.*, I), shows a pair of ascents made, so far as it is possible to judge from the upper winds, in the same air mass. Comparing the pressures at which equal values of θ' occur over the two stations we find that while air initially at 500 mb had subsided to approximately 700 mb, the air initially at 700 mb had only reached a pressure of about 820 mb. Thus the subsidence was accompanied by marked divergence, the amount of this divergence being found to agree very closely with the amount computed from the wind distribution, which on this occasion was sufficiently clearly defined to make an estimate of divergence possible. Fig. 28 shows the θ', p, distribution in air over Sealand on Feb. 15, 1935, and in approximately the same air, so far as could be estimated from pilot balloon ascents, over Breslau on the next day. On the 15th, Sealand was in the warm sector of a depression,

---

[*] The application of wet-bulb potential temperature to air mass analysis. I. *Q.J. Roy. Met. Soc.* **62**, 1936, p. 387; II. *ibid.* **63**, 1937, p. 7; III. *ibid.* **63**, 1937, p. 324; IV. *ibid.* **64**, 1938, p. 407.

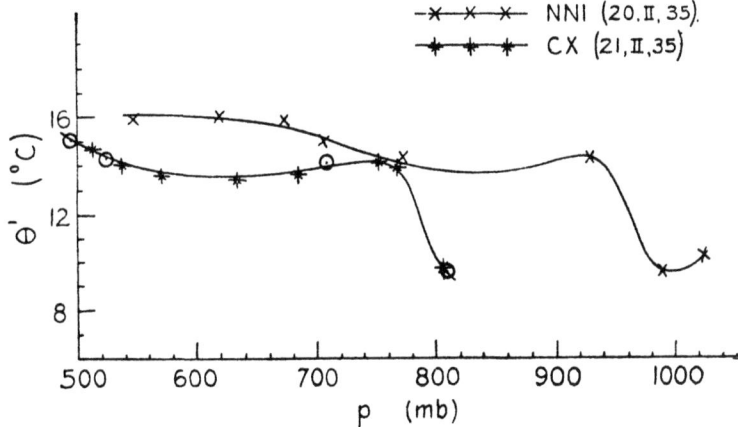

Fig. 26. Distribution of wet-bulb potential temperature in an
air-mass on two successive days.

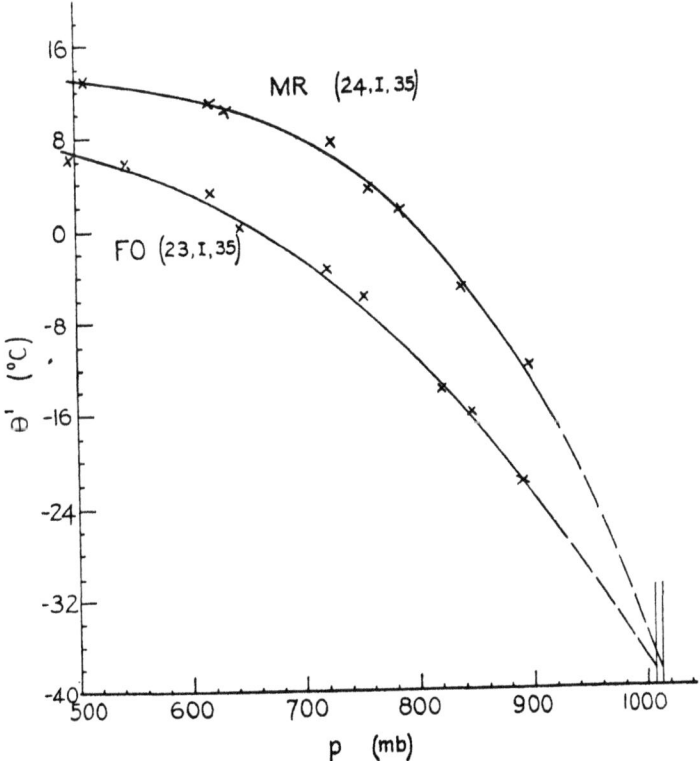

Fig. 27. Estimating subsidence by use of wet-bulb potential temperature.

but by the time the air had reached Breslau, it had climbed up over the wedge of cold air in advance of the warm front. The lower cold air at Breslau extended up to a pressure of about 740 mb and above this level was the warm air, in which the distribution of θ' shows a close resemblance to that at Sealand, except that there is here evidence that the warm air has undergone considerable divergence in climbing up over the warm front.

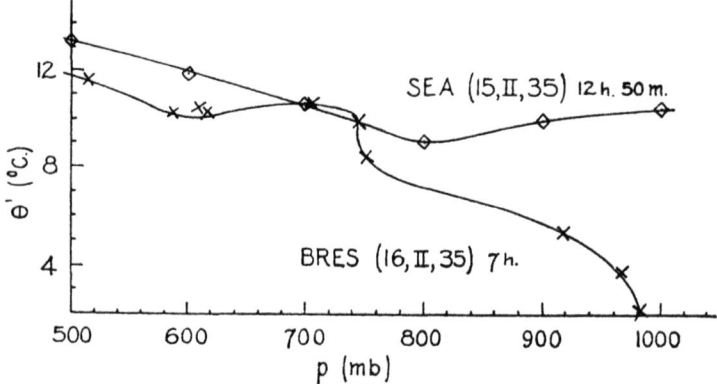

Fig. 28. Estimating ascent of air at a warm front.

§ 56. *Equivalent temperature and equivalent potential temperature*

In § 51 above the following equation was derived:

$$c_p T + L'x + c_p'x\,(T - T') = c_p T' + L'x' = c_p T'' \qquad \ldots\ldots(47).$$

This equation may be taken as defining the equivalent temperature T'' as the temperature of absolutely dry air which has the wet-bulb temperature T' of the air under consideration. The total heat content of $(1+x)$ grammes of moist air at temperature T and wet-bulb temperature T' may be written in any of the following forms:

$$c_p T + L'x + c_p'x\,(T - T') + xcT' = c_p T' + L'x' + xcT'$$
$$= c_p T'' + xcT' \qquad \ldots\ldots(48).$$

If then we regard the heat content of $(1+x)$ grammes of moist air as made up of the heat content of x grammes of liquid water at temperature T' *plus* the remainder of any of the expressions in (48), we may regard that remainder as measuring the "effective" heat content of the moist air. The part xcT' can never become effective in the same sense that the remainder can, since this amount will be carried away by the liquid water when this is condensed. What we have here called the effective heat is the only part that can become available for any other purpose, and since this part of the heat content is measured by the equivalent temperature, we should expect that the equivalent temperature should have important applications in the thermodynamics of the atmosphere.

If the mass of air under consideration could be carried upward adiabatically until all its water-vapour had been condensed and precipitated, and then brought down again to its original level and to its original pressure, its temperature when it returned to its original pressure would be T'', if we could assume that the condensed water had all been carried away at a temperature T'. The difference is not negligible. A simple example will serve to show the order of magnitude of the difference. From the tephigram of fig. 22 it is seen that a mass of air saturated at $53°$ F and 1000 mb will, when raised sufficiently high to lose all its water-vapour, assume a potential temperature of $97°$ F, so that when brought back down to a pressure of 1000 mb it will have a temperature of $97°$ F. But the computation of T'' from equation (47) above yields $T'' = 89°$ F.

The use of the equivalent potential temperature has not yet been thoroughly justified in practice. Most writers on the subject use numbers of approximations in their equations, and no analysis of the subject can be regarded as satisfactory the until accuracy of the approximations has been justified. Computation of the equivalent temperature corresponding to different points on a saturated adiabatic in fig. 22 shows that the equivalent potential temperature computed on the basis of equation (47) is not constant along the saturated adiabatic.

The name "effective heat content" suggested above is not a generally accepted term, and is introduced here in order to avoid repetition of a long definition. It would have been preferable to call it "realised heat", but that would have been liable to confusion in view of the use by Shaw of the term "realised entropy" to indicate the entropy of the dry air alone.

The effective heat content can be measured by either the wet-bulb temperature or by the equivalent temperature T'', which is a function of T' only, as is seen from equation (47). A simple table could be drawn up to give the equivalent temperature in terms of the wet-bulb temperature only for a given pressure, or, simpler still, the wet-bulb thermometer could be graduated so as to give, for a selected pressure, the equivalent temperature by direct reading. But it must be noted that the relation between T'' and T' involves the pressure, and a table which gave the relation between T'' and T' at one standard pressure would not hold at another pressure.

If a mass of moist air is carried upward adiabatically, its dry-bulb and wet-bulb temperatures will follow a dry adiabatic and a saturated adiabatic respectively. Imagine that at the same time a second mass of completely dry air, whose wet-bulb temperature is initially T', the same as that of the first mass, at the ground, is also taken upward adiabatically. Its wet-bulb temperature will follow the same saturated adiabatic as that of the other mass, and its dry-bulb temperature will follow a dry adiabatic. But the temperature of the second mass is the equivalent temperature of the first mass. Thus when a mass of moist air is taken upward adiabatically from the ground, its equivalent temperature follows a dry adiabatic, which from our previous consideration we can state to be that dry adiabatic which at great heights is asymptotic to the saturated adiabatic that passes through the surface wet-bulb temperature.

We are thus led to define an equivalent potential temperature θ'', which is the equivalent temperature attained by a mass of air brought adiabatically to a standard pressure. It should be noted that θ'' is the temperature of absolutely dry air which has the wet-bulb temperature θ'. The variable θ'' is related to the entropy of moist air in a manner similar to the relation of the entropy of dry air to its potential temperature, by a relationship which we shall now derive.

It has been shown in § 52 that if S is the entropy of 1 gramme of water at temperature T', ϕ the entropy of $(1 + x)$ grammes of moist air at temperature T, and ϕ_s the entropy of $(1 + x')$ grammes of air saturated at temperature T', then

$$\phi = \phi_s - (x' - x)\, S \qquad \text{......(49)},$$

or $\qquad \phi = c_p \log T' - AR \log (p - e) + L'x'/T' + xS + \text{const.} \qquad \text{......(50)}.$

But $L'x'/c_p T'$ is normally a fairly small fraction of unity on account of the small value of x', and we may write

$$\log (1 + L'x'/c_p T') = L'x'/c_p T'$$

neglecting higher powers of $L'x'/c_p T'$. Substituting this in equation (50), we find

$$\phi = c_p \log \left(T' + \frac{L'x'}{c_p} \right) - AR \log (p - e) + xS + \text{const.} \qquad \text{......(51)}$$

$$= c_p \log T'' - AR \log (p - e) + xS + \text{const.} \qquad \text{......(52)}.$$

But since, as explained above, θ'' is the potential temperature of absolutely dry air at temperature T'', equation (52) becomes, approximately,

$$\phi = c_p \log \theta'' + xS + \text{const.} \qquad \text{......(53)}.$$

This is the generalised form of the entropy-potential temperature relationship.

§ 57. *Alternative definition of equivalent potential temperature*

The definition of equivalent potential temperature usually given by continental and American writers differs somewhat from that given above, though the actual value obtained for it in any given circumstances differs little in the two methods of approach.

If a mass of air is taken up pseudo-adiabatically until it has condensed all its water-vapour content and shed it as precipitation, and is then brought adiabatically down to a standard pressure, the temperature which it finally attains is called the equivalent potential temperature. Further the temperature which the air then attains at any pressure p on the *downward* journey is defined as the equivalent temperature of the air on its *upward* journey at pressure p. Thus to ovaluate the equivalent potential temperature of a mass of air in any given condition at pressure p, we require to take the mass of air up to a height at which all its water-vapour is condensed and lost as precipitation, and then to bring it adiabatically back to the original pressure.

It is easy to see by referring to the tephigram of fig. 22 that the equivalent potential temperature is the ordinary potential temperature along that horizontal dry adiabatic which is asymptotic to the saturated adiabatic followed by the air mass when it has become saturated.

Rossby[*] related the equivalent potential temperature θ'' to the potential temperature, by using a modification of equation (7), p. 59, neglecting the term $c\xi$. Then

$$c_p \log T - AR \log (p-e) + \frac{Lx}{T} = \text{constant along a pseudo-adiabatic}$$

or

$$c_p \log \theta_d + \frac{Lx}{T} = \text{constant} \qquad \ldots\ldots(54)$$

where θ_d is the potential temperature computed from the partial pressure of the dry air. Now θ'' is the value of θ when $x = 0$

$$\therefore \quad \log \theta'' = \log \theta_d + \frac{Lx}{c_p T} \text{ for saturated air} \qquad \ldots\ldots(55)$$

or

$$\theta'' = \theta_d \, e^{\frac{Lx}{c_p T}}.$$

For unsaturated air it is clear that

$$\theta'' = \theta_d \, e^{\frac{L_0 x}{c_p T_0}} \qquad \ldots\ldots(56)$$

where T_0 is the temperature at the level where x is sufficient to saturate the air, and L_0 is the corresponding latent heat. Rossby writes the last equation in the form

$$\log \theta'' = \log \theta_d + (\log \theta_s'' - \log \theta_d) \frac{RH}{100} \qquad \ldots\ldots(57).$$

It is not clear how great will be the error involved in the simplifications made by Rossby in the derivation of this result, but it is not likely to lead to any appreciable error in the use of the equation over a moderate range of pressure and temperature.

If Normand's theory of the invariance of the wet-bulb potential temperature during ascent or descent of air be accepted, the equivalent potential temperature can be obtained very simply from the Neuhoff diagram or the tephigram. Each saturated adiabatic can be labelled with a figure indicating the equivalent potential temperature of air which eventually rises along that adiabatic. But since the saturated adiabatic along which the air will pass when saturated is the one which passes through the point representing the wet-bulb temperature, the equivalent potential temperature of the air is the value corresponding to the point representing the wet-bulb temperature. The value required is read off in the same way as the wet-bulb potential temperature. If Normand's theorem is accepted, it is obviously preferable to work with wet-bulb potential temperature than with equivalent potential temperature, since this is more closely related to the quantities observed. The results derived must be the same, whichever of the two concepts be applied, since equivalent potential temperature is a one-valued function of the wet-bulb potential temperature.

[*] Rossby, "Thermodynamics applied to air-mass analysis", *Mass. Inst. Tech., Met. Papers*, **1**, No. 3, 1932.

§ 58. *Potential instability and the variation of wet-bulb potential temperature or of equivalent potential temperature with height*

If a layer of air has lower wet-bulb potential temperature at the top than at the base, this will still be true when the layer has been elevated until it is all saturated, and it will follow that the lapse-rate within the layer is then greater than the saturated adiabatic. Such a layer is said to be *potentially unstable*, the instability being realised when the layer is raised sufficiently to bring it to saturation. The condition for this, that the wet-bulb potential temperature shall decrease with height, is that the lapse-rate of the wet-bulb temperature within the layer in its original state, shall exceed the corresponding saturated adiabatic.

Conversely, a layer will remain stable when elevated until it is saturated, if the wet-bulb potential temperature increases with height, i.e. if the lapse-rate of the wet-bulb temperature is less than the saturated adiabatic.

The atmosphere will have potential instability when the equivalent potential temperature decreases with height.

§ 59. *The Rossby diagram*

The tephigram has the disadvantage that the dew-point lines (i.e. the x-lines) are very unequally spaced. Rossby (*loc. cit.*) produced a diagram in which the vertical co-ordinate is the logarithm of the dry air potential temperature (log θ_d), and the horizontal component is x, now represented on a linear scale. Thus the dew-point lines are vertical, the isentropics horizontal. There are three other sets of lines on the diagram:

(*a*) the isobars, which slope downwards from right to left of the diagram, much as they do in the tephigram of fig. 22, but with a very marked increase in the downward slope as they approach low values of x;

(*b*) the isotherms, which are slightly curved, sloping upward towards the right in the diagram;

(*c*) the lines of equal equivalent potential temperature, sloping upward to the left of the diagram, each labelled with the value of the potential temperature at which it crosses the vertical axis $x=0$ (this is in agreement with equation (56) above).

In the practical use of the Rossby diagram the partial (dry air) potential temperature, and the humidity mixing ratio x, are used as working co-ordinates, and when these have been evaluated, a point is marked in the diagram in the corresponding position. Clearly, an unsaturated mass of air rising or falling through the atmosphere will be represented by the same point during the whole of the motion, provided it remains unsaturated. A point in the diagram normally represents air having the appropriate values of θ_d, x and equivalent potential temperature. The scales for pressure and temperature cannot be taken to apply to the air unless it is saturated. Alternately we may say that the pressure and temperature scales give the level, and the conditions at that level, where the air first attains saturation during adiabatic ascent.

The user of the tephigram can visualise the Rossby diagram as derived from the tephigram by first distorting it in order to bring the *x*-lines vertical, and then further distorting it to equalise the spacing of these lines.

A layer of air which is thoroughly mixed, so that it has everywhere the same potential temperature and the same humidity mixing ratio, will be represented by a point in the diagram. In general, the more nearly homogeneous a layer of air is as regards the potential temperature and humidity mixing ratio, the shorter the line in the Rossby diagram which represents it.

In general potential temperature increases with height in the atmosphere, and the direction of increasing height along the line in the Rossby diagram representing the results of an upper air ascent is usually the direction of increasing potential temperature.

A layer in which equivalent potential temperature decreases as potential temperature increases will become unstable when raised sufficiently to become saturated, and the pressure reading corresponding to the lowest point of the line representing this layer in the diagram will give the pressure at which saturation will be produced. The difference between the latter value and the observed pressure at the base of the layer will give the range of pressure through which the layer must be elevated in order that its base shall become saturated.

It has been found that air masses over the North American continent show sharp distinctions when plotted in this way. Even in summer, when the contrasts between air masses are reduced to a minimum, their representation in the Rossby diagram brings out sharply the distinctions between air masses.[*] Over the British Isles, where so much of the air comes from the ocean, the distinction is much less marked.

§ 60. *Latent instability*

Normand[†] has applied the name *latent instability* to denote a state which, white stable for small displacements, is unstable for sufficiently large displacements. In fig. 29 let *ABCD*, *A'B'C'D'*, represent the distribution of temperature and of wet-bulb temperature respectively. The air is everywhere unsaturated, and the lapse rate being everywhere less than the dry adiabatic, there is stability at all the levels shown.

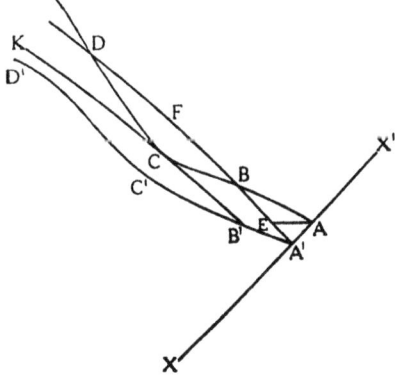

Fig. 29. Tephigram showing latent instability.

If an element of the surface air is displaced upward, it becomes saturated at *E*, and if its displacement continues to a higher level, its representative point will follow the saturated adiabatic *EBFD*. As soon as the air has passed the level *B*, it will at all subsequent levels up to *D* be warmer than its surroundings, and will rise freely. To raise the air to *B* will

* *Vide* Willett, "American air mass properties", *Mass. Inst. Tech.*, *Papers in Physical Oceanography and Meteorology*, **2**, No. 2, 1933.

† Normand, "On instability from water vapour", *Q.J. Roy. Met. Soc.* **64**, 1938, p. 47.

require the application of an amount of work measured by the area AEB, and during the ascent from B to D, an amount of energy measured by the area $BCDFB$ will be liberated. If, as in this example, the amount of work which has to be applied to elevate the air is less than the amount of energy released during the further free ascent of the air, the latent instability can be realisable, since the net release of energy is positive. When the net release of energy is negative, i.e. when the work which has to be done in the initial elevation of the air is greater than the amount of energy released during the later free ascent, the nominal latent instability cannot be realised.

An examination of fig. 29 shows that the surface air has latent instability if the saturated adiabatic through A' intersects the curve of temperature $ABCD$. The saturated adiabatic KCB' has been drawn to touch $ABCD$ at C. It is the saturated adiabatic with the lowest possible value of θ' that can be drawn to touch $ABCD$. It intersects $A'B'C'D'$ at B', and it is readily seen that air above the level of B' cannot have latent instability.

It is found that latent instability is as a rule developed before the occurrence of thunderstorms and dust storms of the convective type in India,* and it is also a frequent occurrence before thunderstorms and heavy rain in the British Isles.

§ 61. *The general conditions of stability in the atmosphere*

In the light of the preceding paragraphs we can now summarise the conditions of stability as follows.

(*A*) A dry atmosphere is stable, for small or large displacements of individual particles, if the potential temperature increases with height, i.e. if the lapse-rate is less than the dry adiabatic. It is unstable, even for small displacements of individual particles, if the potential temperature decreases with height, i.e. if the lapse-rate exceeds the dry adiabatic.

(*B*) A saturated atmosphere is stable for small or large displacements of individual particles, if the lapse-rate is everywhere less than the saturated adiabatic, and unstable if the lapse-rate exceeds this limit. Strictly speaking, a saturated atmosphere in which the lapse-rate exceeds the saturated adiabatic, is stable for large or small *downward* displacements, but since in practice disturbances in the atmosphere cannot be restricted to downward motion, it is necessary to regard a saturated atmosphere as unstable if the lapse-rate exceeds the saturated adiabatic.

(*C*) In an unsaturated atmosphere the conditions for stability or instability for small displacements are the same as those for dry air. But when portions of an unsaturated atmosphere are given large vertical displacements which may lead to their becoming saturated, it is necessary to consider carefully what will then occur. If the saturated adiabatic through the point representing the initial wet-bulb temperature of the particle does not cut across the line representing the dry-bulb temperature then no upward displacement of the particle can bring it to a level in which it is as warm as its environment, so that the atmosphere is then stable for all displacements, no matter how great. If

* Sohoni and Paranjpe, *Mem. Indian Met. Dept.* **26**, No. 7, 1937.

the saturated adiabatic through the point representing the wet-bulb temperature of any small element cuts across the line representing the dry-bulb temperature distribution, the element of air is said to have *latent instability*, but this latent instability will not be realised unless the potential energy ultimately liberated exceeds the work done in raising the element until it is unstable. (See § 60, and fig. 29.)

(*D*) A layer of dry air is stable if the lapse-rate within it is less than the dry adiabatic, and unstable if the lapse-rate exceeds the dry adiabatic. The stability or instability is not altered by any adiabatic vertical displacement of any amount, since the difference between the lapse-rate and the dry adiabatic does not alter its sign. (See § 23, p. 44.)

A layer of saturated air is stable if the lapse-rate is less than the saturated adiabatic, and unstable if the lapse-rate exceeds the saturated adiabatic. It will retain its original condition of stability or instability if displaced vertically through any distance, since the temperature of each element will follow a saturated adiabatic, retaining its wet-bulb potential temperature unchanged.

A layer of air which is initially unsaturated will be stable for *small* disturbances of any kind, if the lapse-rate is less than the dry adiabatic. When displaced vertically through a sufficient range of height to bring it all to saturation, it will then be unstable if the wet-bulb potential temperature decreased with height in the original state of the layer.

§ 62. *The Clausius-Clapeyron equation*

It is possible to relate the latent heat of evaporation to the rate of change of the saturation vapour-pressure with temperature. The relationship is most readily derived by considering a Carnot cycle, as represented in fig. 30. The pressure

Fig. 30. Cycle for the Clausius-Clapeyron equation.

e is the vapour-pressure, and the lines of equal *e* are therefore isothermal lines. Let AB be the isothermal of T, the abscissae A and B representing the specific volumes of 1 gramme of water and of water-vapour respectively. Let CD be the isothermal $T+dT$, the points C and D representing water-vapour and water respectively, as before.

Let the latent heat of vaporisation be L at temperature T, and $L+dL$ at temperature $T+dT$. Now take 1 gramme of water round the cycle $ABCDA$, starting at A. To take the representative point from A to B, an amount of heat L must be communicated to the water, which at B is 1 gramme of water-vapour, under a pressure e. Next the water-vapour must be taken from B to C, and during this transit it must be kept saturated at each stage. Let s_2 be the specific heat of water-vapour when constrained to change its state in this way. Then an amount of heat $s_2 dT$ must be communicated to the water-vapour during the path BC. Next, by extraction of a quantity of heat $L+dL$ the water-vapour is condensed, yielding 1 gramme of liquid water, the representative point moving from C to D, along the isothermal $T+dT$. Then by extraction of a quantity of heat $c\,dT$, the water is brought back to the original point A. Here c is the specific heat of liquid water at the temperature T. Let v_1 and v_2 be the specific volumes of liquid water and water-vapour respectively.

The total amount of heat communicated to the working substance is equal to the work done by the substance on the environment, and since the cycle has been performed in the clockwise sense, the work done on the environment is negative, and equal in magnitude to the area of the cycle in the diagram. Hence

$$L + s_2 dT - L - dL - c\,dT + A\,(v_2 - v_1)\,de = 0$$

or

$$s_2 - c - \frac{dL}{dT} + A\,(v_2 - v_1)\,\frac{de}{dT} = 0 \qquad \ldots\ldots(58).$$

But since the cycle is reversible, the change of entropy in going round it is zero. Hence

$$\frac{L}{T} + s_2 \frac{dT}{T} - \frac{L + dL}{T + dT} - c\,\frac{dT}{T} = 0,$$

$$\frac{s_2 - c}{T}\,dT = -\frac{L}{T} + \frac{L}{T}\left(1 + \frac{dL}{L}\right)\left(1 + \frac{dT}{T}\right)^{-1}$$

$$= \frac{dL}{T} - \frac{L\,dT}{T^2},$$

or

$$s_2 - c = \frac{dL}{dT} - \frac{L}{T} \qquad \ldots\ldots(59),$$

or

$$s_2 - c = T\frac{d}{dT}\left(\frac{L}{T}\right).$$

Comparing (58) and (59) we find

$$\frac{L}{T} = A\,(v_2 - v_1)\,\frac{de}{dT} \qquad \ldots\ldots(60),$$

where e is the saturation vapour-pressure of water-vapour at the temperature T.

This equation is known as the Clausius-Clapeyron equation. Neglecting v_1, which is small by comparison with v_2, and substituting for v_2 from the equation

$$ev_2 = R'T \qquad \ldots\ldots(61)$$

we can write equation (60) in the form

$$\frac{1}{e}\frac{de}{dT} = \frac{L}{AR'T^2} \qquad \ldots\ldots(62).$$

The boiling point of water is the temperature at which the saturation vapour-pressure is equal to the external pressure. If, then, the boiling point can be determined accurately, a table of vapour-pressures will give the corresponding external pressure at the time. It is, however, more convenient to use equation (62) to evaluate de/dT, which gives the change of pressure per degree of depression of the boiling point.

The use of this method of determination of pressure demands very accurate thermometers. If the method is to yield the height correct to within 10 feet, the boiling point temperature must be correct to the nearest hundredth of a degree Centigrade.

§ 63. *The functional relation of saturation vapour-pressure to temperature*

In the preceding sections we have deduced a relationship between the latent heat and the saturation vapour-pressure of water-vapour. If L is a constant or a known function of the temperature T, the equation (62) above can be integrated. From equation (62),

$$\frac{L}{T^2}\,dT = \frac{AR'}{e}\,de \qquad\qquad \ldots\ldots(63).$$

For temperatures below freezing point, and for vapour over ice, L may be treated as a constant, whose value is 677 gramme-calories, or $677 \times 4\cdot18 \times 10^7$ ergs. Integrating the last equation, we find

$$AR' \log e = \text{const.} - L/T \qquad\qquad \ldots\ldots(64).$$

An equation of this form was suggested by Young in 1820, from a consideration of experimental results. When the appropriate values of the constants are inserted, equation (64) gives for the saturation pressure over ice

$$\frac{1}{273} - \frac{1}{T} = 1\cdot629 \times 10^{-4} \log \frac{e}{e_0},$$

where e_0 is the value of e at 273° A. This equation gives a remarkably close fit to the observations. A graphical comparison is shown in the *Manual of Meteorology*, **3**, fig. 92.

Whipple[*] integrated equation (63) on the assumption that

$$L = L_v - (c - c_p')\,t,$$

where t is the Centigrade temperature, and L_v the latent heat at 0° C. He found the following results for temperatures below 0° C:

(a) For vapour-pressure over water,

$$\log \frac{e}{e_0} = 10\cdot78 \frac{t}{273 + t} - 5\cdot01 \log \frac{t + 273}{273};$$

(b) For vapour-pressure over ice,

$$\log \frac{e}{e_0} = 9\cdot95 \frac{t}{273 + t} - 0\cdot445 \log \frac{t + 273}{273};$$

(c) For relative humidity r over ice,

$$\log \frac{r}{100} = 4\cdot56 \log \frac{t + 273}{273} - 0\cdot83 \frac{t}{t + 273}.$$

In these formulae e_0 is the saturation vapour-pressure at 0° C.

Whipple found that the results given by his equations gave very close agreement with the vapour-pressures observed by Washburn. The close agreement shows that over the range of temperatures considered the variation of $c - c_p'$ with temperature is negligible.

[*] *Monthly Weather Review*, 1927, p. 131.

For temperatures above 0° C covered by meteorological observations the variations of L are slight. If these variations are neglected, equation (59) can again be integrated to give the saturation vapour-pressure over water. The result is quoted from Shaw, *Manual of Meteorology*, **3**, p. 239,

$$\frac{1}{273} - \frac{1}{T} = 1 \cdot 844 \times 10^{-4} \log \frac{e}{e_0}.$$

Shaw shows that the fit of the observations to the formula is again very close.

We shall not in practice require to use these formulae, as it will be found more convenient to use the tabulated values of the vapour-pressures, but the results are of very great interest, in that they demonstrate the power of the second law of thermodynamics to measure relationships which otherwise would appear to be incapable of theoretical treatment.

CHAPTER V

RADIATION

§ 64. *Radiation of light and heat*

THE phenomena of radiation consist in the transmission of energy from one body to another through the intervening medium. The transmission takes place along straight lines, and with a velocity which, though large compared with the velocities of motion of matter in bulk with which we are familiar, is yet determinable by careful experiment. The process of transfer by radiation is analogous to wave motion, and a wide range of wave-lengths is possible. We can build into one spectrum the visible rays known as light rays, the infra-red or heat rays, X-rays, and the Hertzian waves used in wireless telegraphy; but in meteorology we are only concerned with a relatively narrow range of wave-length, as will be seen when we come later to discuss solar and terrestrial radiation. The units used in measuring wave-lengths are $\mu = 10^{-3}$ mm $= 10^{-4}$ cm, $\mu\mu = 10^{-6}$ mm $= 10^{-7}$ cm, and the Ångström unit 10^{-8} cm.

A clear distinction can be drawn between the phenomena of radiation and molecular conduction. All bodies radiate energy, the amount radiated depending on their temperature. But as any radiating body may also absorb radiation from surrounding bodies, its net gain or loss of energy can only be determined when the temperature of each portion of the environment is known. By contrast with this, the conduction of heat requires a slope of temperature along which the heat travels from high to low temperature, and the net flow of heat by conduction can be evaluated when the slope of temperature is known. Radiation requires no such slope of temperature, and does not of necessity warm the medium through which it passes. It is true that it is possible to have conduction of heat through a medium without warming the medium, but this demands the very special condition of a constant gradient of temperature across the medium.

From the point of view of meteorology the distinction which is usually drawn between light and heat is a convenience rather than a fact of physical significance. The effect of the absorption of light, like that of heat, is to warm the medium which absorbs it, the absorbed energy being converted into molecular motion. The distinction which we draw between light and heat is mainly due to the limitation of the susceptibility of the human eye to one octave of wave-length; but it is a convenient distinction to have in mind, since air (and to a less extent water) is almost completely transparent to light rays.

§ 65. *Kirchhoff's law*

The rate at which a body sends out radiation from unit surface may be called the "emissive power", and the rate at which unit surface absorbs radiation falling upon it may be called "absorptive power". These two quantities are closely related, the relation as expressed by Kirchhoff being that "at a given temperature the ratio between the absorptive and emissive power for a given wave-length is the same for all bodies". It should be noted that two distinct physical facts are involved in this statement. In the first place it gives a qualitative rule which connects the radiation and absorption for a given substance; if a body emits radiation of a given wave-length at a given temperature, it will also absorb radiation of the same wave-length at that temperature. In the second place it gives a quantitative rule which establishes a relationship between different bodies.

If E_λ, A_λ are respectively the emission and absorption of a particular wave-length λ for a particular body at temperature T, and $E_\lambda/A_\lambda = c_\lambda$, then Kirchhoff's law states that c_λ is a function of λ and T only, and is independent of the nature of the body. This does not preclude the possibility of both E_λ and A_λ being zero, and the body being transparent to radiation of wave-length λ. Many bodies, notably gases and vapours, only absorb in limited ranges of wave-length, and their emission is therefore limited to the same ranges of wave-length.

It also follows from Kirchhoff's law that a good absorber is a good radiator, and a bad absorber a bad radiator. Highly polished surfaces, which reflect nearly all short-wave radiation falling upon them, are poor radiators of short waves. But a body which is transparent to one range of wave-lengths may readily absorb other wave-lengths. Thus water-vapour is nearly transparent to the short-wave radiation coming from the sun, but absorbs readily the radiation emitted at terrestrial and atmospheric temperatures within a wide range of wave-lengths.

§ 66. *Black-body radiation*

Experiment shows that there is an upper limit to the amount of radiation which can be emitted by unit surface of a body at a given temperature. A body which radiates for every wave-length the maximum amount of radiation possible at a given temperature is known as a "black body" or "perfectly black", or a "perfect radiator". The nearest approach to black-body radiation is the radiation which passes out through a small cavity in a solid body at a uniform temperature.

§ 67. *Grey radiation*

A body which emits for each wave-length a fixed proportion of the black-body radiation at the same temperature is called a "grey" body, and its radiation is known as "grey" radiation.

§ 68. *Planck's law*

The distribution of the energy in the spectrum of a black body has been represented by Planck by means of a formula

$$E_\lambda = \frac{c_1 \lambda^{-5}}{e^{\frac{c_2}{\lambda T}} - 1} \qquad\qquad \text{......(1)},$$

where E_λ is the energy emitted per unit area per unit time within unit range of wave-length centred on λ, and c_1 and c_2 are constants. The equation may be written

$$\frac{E_\lambda}{T^5} = \frac{c_1 (\lambda T)^{-5}}{e^{\frac{c_2}{\lambda T}} - 1} = f(\lambda T) \qquad \ldots\ldots(2).$$

The function $f(\lambda T)$ is zero for $\lambda = 0$ and for $\lambda = \infty$, and has a maximum at some finite value of λT. If this value of λT be a, then the wave-length at which E_λ is a maximum for a given temperature is given by Wien's law

$$\lambda_m = a/T \qquad \ldots\ldots(3).$$

The value of a as given by Lummer and Pringsheim[*] is 2940, when λ is measured in μ, and T in degrees absolute. A slightly lower value, 2890, was given by Coblentz. The maximum intensity in the spectrum of the radiation from a black body at a temperature of 300° A is at about 10μ, while for a body at a temperature of 200° A the maximum is at about 15μ. These are roughly the limits of temperature within the earth's atmosphere. Conversely, by observing the wave-length of the maximum intensity of the radiation we may determine the temperature of the body which emits the radiation, by the use of the equation

$$\lambda_m T = 2940 \qquad \ldots\ldots(4).$$

Further, by measuring the maximum value of E_λ we can determine T by using the relationship

$$\left(\frac{E_\lambda}{T^5}\right)_{\max} = \text{const.} = f(a).$$

These two methods have been employed to determine the temperature of the sun, and give approximate agreement, which may be taken as evidence that the sun radiates approximately as a black body. It is found that the maximum intensity in the solar spectrum is at about 0·5μ, corresponding to an effective temperature of 5600° A.

Since E_λ/T^5 is a function of λT, if a diagram is drawn to represent E_λ/T^5 as a function of λ for any one temperature, it can be used to represent E_λ for any other temperature by making the scale of λ inversely proportional to T, and taking the scale of E_λ proportional to T^5. Thus in fig. 31 the emission E_λ is plotted against λ for a variety of temperatures, the appropriate horizontal scale being used according to the temperatures shown against the various scales. The unit μ used in the scale of wave-length is $1/1000$ mm, or 10^{-6} metre.

In fig. 31 the common horizontal scale is in reality λT, the vertical scale measuring E_λ/T^5. It will be seen that if the curves of radiation intensity be drawn for two different temperatures, on the same absolute scale of wave-length and intensity, the curve for the higher temperature lies completely above the curve for the lower temperature. This is most readily seen from equation (1), which shows that the value of E_λ for any selected value of λ increases as T increases. For as T increases the denominator decreases so that

[*] *Verh. Deut. Phys. Ges.* **1**, 1889, pp. 23 and 215.

the expression as a whole increases. Hence a curve representing the variation of E_λ with λ rises along its whole length as the temperature increases.

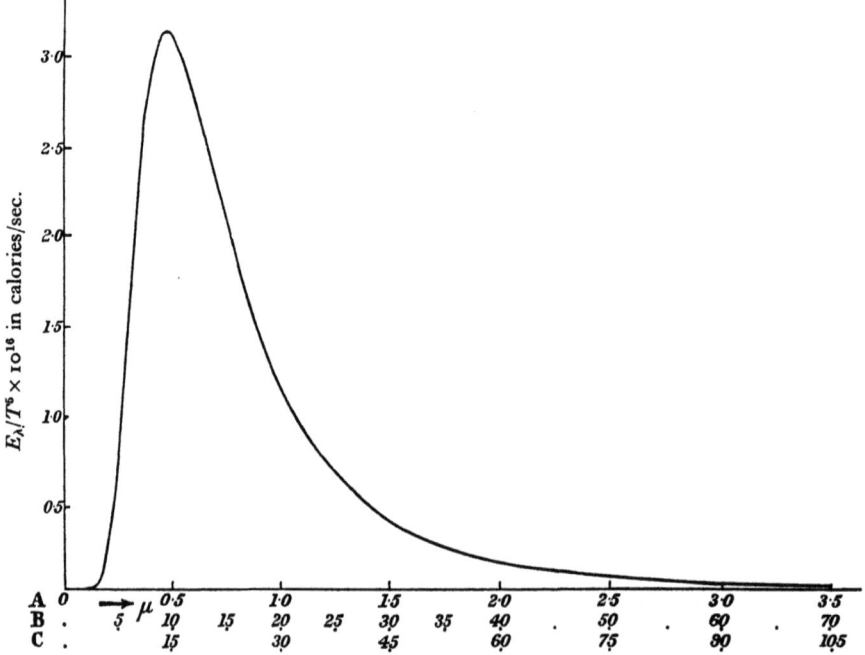

Fig. 31. Theoretical curve of distribution of black-body radiation. Scale A corresponds to $T = 6000°$ A, B to $T = 300°$ A, and C to $T = 200°$ A.

§ 69. *Stefan's law*

Stefan's law states that the amount of energy radiated per unit time from unit surface of a black body is σT^4, where σ is known as Stefan's constant, and T is the absolute temperature. In the usual C.G.S. units σ amounts to $5·709 \times 10^{-5}$ ergs per cm² per second, or $5·709 \times 10^{-12}$ watts per cm², or 82×10^{-12} gramme-calories per cm² per minute.

This result may be derived from equation (1), from which it follows that

Total radiation from a perfect radiator at temperature T

$$= \int_0^\infty E_\lambda d\lambda = c_1 \int_0^\infty \frac{\lambda^{-5}}{e^{\frac{c_2}{\lambda T}} - 1} \, d\lambda$$

$$= c_1 T^4 \int_0^\infty \frac{(\lambda T)^{-5}}{e^{\frac{c_2}{\lambda T}} - 1} \, d(\lambda T)$$

$$= c_1 T^4 \int_0^\infty \frac{x^{-5}}{e^{c_2/x} - 1} \, dx$$

$$= \sigma T^4,$$

where σ is a constant. Values of σT^4 are given in Table VIII, p. 421.

§ 70.   Range of wave-lengths in radiation from bodies at different temperatures

The distribution of complete or black-body radiation with wave-length is shown in fig. 31, with a uniform scale of λT. It can readily be seen that on account of the steep slope of the curve on the side of low wave-lengths the amount of radiation below a limit $\lambda T = 1000$ is negligible, being considerably less than 0·1 per cent of the total radiation. The slope of the curve is far less steep on the side of high wave-lengths, and the limit is not so readily set. Only 0·1 per cent of the total radiation will be neglected if we stop at an upper limit set by $\lambda T = 54,000$, and 1 per cent will be neglected if the upper limit is set by $\lambda T = 24,000$. For our present purposes we shall assume that the spectrum does not extend beyond the limit (set by $\lambda T = 24,000$) which contains 99 per cent of the total radiation.

Thus black-body radiation at 6000° A, approximately the temperature of the sun, is contained within the limits of 0·17μ and 4μ, and has its maximum intensity in the blue-green at about 0·5μ. Black-body radiation at terrestrial temperatures of roughly 300° A will be contained within the limits 3μ and 80μ, having its maximum intensity at 10μ; and black-body radiation at stratospheric temperatures of about 200° A will be contained within the limits 4μ and 120μ, having its maximum intensity at 15μ. Thus black-body radiation at atmospheric and terrestrial temperatures is in the far infra-red part of the spectrum. Since this range of wave-lengths is widely separated from that of incoming solar radiation, it is customary to call it "long-wave" radiation, to distinguish it from direct solar radiation, which is called "short-wave" radiation.

§ 71.   The spectral distribution of solar radiation

The outer surface of the sun may be regarded as approximately a black body at a temperature of about 5600° A. Its radiation should therefore be contained within the limits of wave-length 0·15μ to 4μ. An appreciable amount of this radiation should be contained in the ultra-violet range, below 0·4μ, while a still larger amount is contained in the range beyond 0·7μ, in the infra-red region. Roughly half the total energy in the sun's radiation is contained within the range 0·4 to 0·7μ, so that about half of the total solar energy is in the form of light.

Close investigation of the intensity of the radiation of different wave-lengths indicates that there are many dark lines in which little or no radiation is present. Many of these dark lines are absorption lines due to the absorption, in the outer atmosphere of the sun, of radiation coming from deeper layers. There are, however, some dark lines due to absorption in the earth's atmosphere. The blue end of the spectrum is sharply limited at 0·3μ, even when photographed from considerable heights above the earth's surface. Wigand in 1913 succeeded in photographing the solar spectrum from a height of 9 km, and found the same sharp limitation at 0·3μ which had been noted in spectra photographed from the earth's surface. It was first suggested by Hartley that the absence of all wave-lengths below 0.3μ was due to absorption by ozone. This question was studied in greater detail by Fabry and Buisson in 1912, and these writers brought forward evidence that the ozone is localised at heights inaccessible to direct observation. More recent studies of the day to day variation in the amount of ozone absorption by Dr G. M. B. Dobson and his collaborators* have brought out some striking relationships between the amount of ozone at heights of about 10–40 km and the distribution of pressure at the earth's surface.

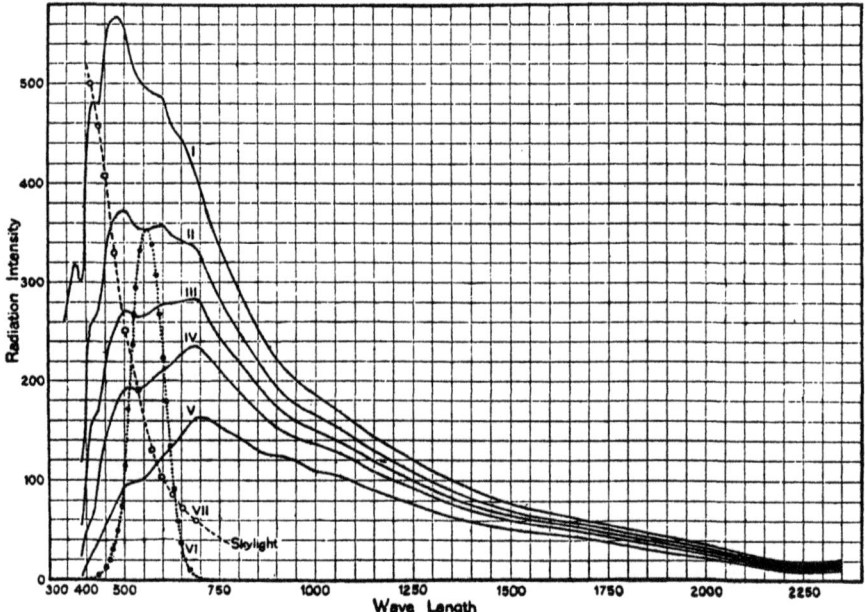

Fig. 32. Observed intensity distribution of solar radiation.

Within the limits of the appreciable solar spectrum the main lines of atmospheric absorption are those due to oxygen at 0.69μ and 0.76μ, those due to water-vapour at 0.72μ, 0.81μ, 0.93μ, 1.13μ, 1.42μ, 1.89μ, and the two wide bands due to water-vapour centred at about 2.01μ and 2.05μ. Except for the two lines due to oxygen referred to above, the simple gases of the atmosphere do not absorb any radiation, and carbon dioxide does not absorb within the limits of wave-length of the solar beam. The oxygen lines are so narrow that they represent only a very minute loss of energy from the solar beam.

Fig. 32 shows in curve II the spectral distribution of solar radiation as received at Washington with zenith sun and cloudless sky, and in curve I the distribution as estimated outside the atmosphere in the latitude of Washington, after allowing for the absorption and scattering of the atmosphere. Curves III, IV, V give the distributions for solar altitudes 30°, 19·3°, and 11·3° (air-masses

* *Vide* references, footnote, p. 20 above.

2, 3, 5), while curve VI gives the relative brightness of the parts of the spectrum.

According to Fowle,* on a clear day, with the sun in the zenith, the total loss from the solar beam due to absorption in the atmosphere down to sea level is only about 6 to 8 per cent of the incident radiation. Thus on a clear day by far the greater part of the solar radiation passes unchanged through the earth's atmosphere, and is either absorbed by the earth's surface, or is reflected upward by various parts of the earth's surface. Clouds will also reflect a large part of the incident radiation. It is estimated that 0·43 of the solar radiation which reaches the outer limit of the earth's atmosphere is lost by reflexion or absorption, so that only 0·57 of the incident radiation becomes available for heating the earth's atmosphere and surface. We shall return to this aspect of the question in § 74 below.

§ 72. *The solar constant*

The beam of incoming solar radiation is liable to suffer loss by reflexion at the surfaces of clouds, by absorption by the gaseous constituents of the atmosphere and by scattering by small particles suspended in the atmosphere. All three sources of loss are diminished when observations are carried out at a high level station, and a correction can be made for the residual loss by absorption and scattering in the uppermost layers of the atmosphere. The most exhaustive observations of the intensity of the incoming solar radiation are those carried out by the Smithsonian Institution at the high level observatories on Table Mountain, California, Mt. Montezuma, Chile, and Mt. St Catherine, Sinai.

The intensity of the solar radiation at the outer limit of the earth's atmosphere is known as the "solar constant". The mean value deduced from a long series of observations is 1·94 gramme-calories per cm^2 per minute, or 136 kilowatts per square dekametre. The measured values show a variation from day to day, with a total range of a few per cent of the whole. The reality of these variations is by no means generally accepted, though there is some evidence that high or low values occur at the same time in the observations made at Mount Wilson and at Calama. Notable depressions in the measured values occurred in 1902 and 1912, associated with the volcanic eruptions of Pelée, Santa Maria, and Colima in 1902, and of Katmai in 1912. Large variations also followed earlier volcanic eruptions, but those mentioned gave the most marked depressions in the measured values of the solar constant. It should be added that as methods of observation have been improved, the magnitude of the variations in the day to day observations of the solar constant has diminished, and this is in itself an argument against the reality of these variations. Pettit gives (in the *Third Report on Solar and Terrestrial Relationships*) a curve showing the variation in the ratio of the intensity of ultra-violet radiation at about 0·3μ to the intensity in the green at about 0·5μ, and this curve shows very clearly an annual variation, suggesting that the effects of the earth's atmosphere have not been entirely eliminated. It is possible that the residual

* *Ann. Astroph. Obs.* 4, p. 274.

error is mainly attributable to the estimate of the ultra-violet intensity, but the result leaves us in some doubt as to whether the variations of atmospheric turbidity have been entirely eliminated in the evaluation of the solar constant.

§ 73. *The variation of insolation with season and latitude*

The amount of solar radiation reaching unit area of any part of the earth's surface in one day depends upon

    (*a*) the solar constant,
    (*b*) the transparency of the atmosphere,
    (*c*) the latitude of the place, and
    (*d*) the time of the year.

No simple formula will summarise these effects. Angot,* starting from the assumption that the atmosphere is transparent, has calculated a table giving for intervals of latitude, for each month of the year, the insolation or total amount of radiation reaching unit surface of the earth per month. In this table the unit is the amount of energy that would be received on unit area on the equator in one day, at the equinox, with the sun at its mean distance, the atmosphere being assumed completely transparent. This unit amounts to 458·4 times the solar constant, or 889 gramme-calories per cm², taking the solar constant to be 1·94 gramme-calories per cm² per minute. Angot's table is reproduced in the table below. It will be noted that the figures for the Southern hemisphere are not exactly equal to those for the Northern hemisphere shifted through 6 months, on account of the varying distance of the earth from the sun. The totals for the whole earth show a maximum in Dec.-Jan., and a minimum in June-July, the times of perihelion and aphelion respectively.

Table 2

Calculated insolation reaching the earth

Lat.	Jan.	Feb.	Mar.	Apr.	May	June	July	Aug.	Sept.	Oct.	Nov.	Dec.	Year
N 90	0·0	0·0	1·9	17·5	31·5	36·4	32·9	21·1	4·6	0·0	0·0	0·0	145·4
80	0·0	0·1	5·0	17·5	30·5	35·8	32·4	20·9	7·4	0·6	0·0	0·0	150·2
60	3·0	7·4	14·8	23·2	30·2	33·2	31·1	24·9	16·7	9·0	3·8	1·9	199·2
40	12·5	17·0	23·1	28·6	32·4	33·8	32·8	29·4	24·3	18·4	13·4	11·1	276·8
20	22·0	25·1	28·6	30·9	31·8	32·0	31·8	30·9	28·9	25·8	22·5	20·9	331·2
Equat.	29·4	30·4	30·6	29·6	28·0	27·1	27·6	28·6	30·1	30·2	29·5	28·9	350·3
S 20	33·8	32·2	29·0	24·9	21·2	19·6	20·5	23·7	27·7	31·1	33·3	34·1	331·2
40	34·8	30·4	23·9	17·4	12·5	10·4	11·6	15·8	21·9	28·5	33·6	36·0	276·8
60	33·0	25·3	16·0	8·1	3·3	1·7	2·7	6·5	13·6	22·6	31·1	35·3	199·2
80	34·2	20·5	6·3	0·3	0·0	0·0	0·0	0·0	3·8	16·0	.31·0	38·1	150·2
90	34·7	20·7	3·2	0·0	0·0	0·0	0·0	0·0	1·0	15·6	31·5	38·7	145·4

§ 74. *The earth's albedo*

On p. 111 we quoted as the total reflecting power of the earth, or the earth's albedo, the figure 0·43, a value due to Aldrich.† This figure is based on the assumption that with a cloudless sky a fraction 0·08 is reflected from the

---

    * *Ann. Bur. Cent. Met.*, Paris, 1883 1ère partie, pp. B 136–61.
    † *Smithsonian Misc. Coll.* **69**, No. 10, 1919; also *Ann. Astroph. Obs.* **4**, 1922, p. 379.

earth's surface, and 0·09 from the atmosphere, giving a total loss of 0·17. For totally overcast skies the reflexion factor is taken as 0·78, and for a mean cloud amount 0·52, which is estimated to be the mean cloud amount of the earth, Aldrich computes that a fraction 0·43 of the incoming radiation is reflected by the atmosphere, the clouds and the earth's surface.

Ångström[*] gave a linear relationship between the albedo a and the cloud amount c, in the form

$$a = 0.70c + 0.17 (1 - c).$$

The difference between the albedo of the whole earth deduced from Ångström's equation and the value estimated by Aldrich is very slight, and the value 0·43 for the whole earth may be taken as a reasonably accurate value.

Different natural surfaces have widely different factors of reflexion for solar radiation.[†] Thus snow reflects from 70 to 80 per cent; water an amount varying from 2 per cent with the sun at an altitude of $47°$ to 71 per cent with the sun at an altitude of $5\frac{1}{2}°$; grass reflects from 10 to 33 per cent; rock 12 to 15 per cent; dry mould 14 per cent, and wet mould 8 to 9 per cent. Thus when the ground is not snow-covered the differences in reflective power are not of great importance. The greatest variation which has to be taken into account is the difference between clear and clouded skies, since with overcast skies the reflexion factor is 0·78, whereas the mean reflexion factor of the earth's surface is of the order of 0·1, except when the ground is covered with snow, when it becomes 0·7 to 0·8.

§75. *The coefficient of absorption*

The law of absorption of radiation as usually stated is known as Beer's law. If I_0 is the intensity of the incident radiation, the intensity of the radiation after it has passed over a path containing m units of mass of the absorbing substance per cm² of cross-section is reduced to I, where

$$I = I_0 e^{-km}$$

and k is defined as the coefficient of absorption.

The reduction of intensity in a path containing dm of absorbing substance is given by

$$dI/I = -k \, dm.$$

This equation might equally well be taken as defining the coefficient of absorption.

In practice, the exponential form is not as convenient as the form

$$I = I_0 \times 10^{-0.4343km}$$

and it is also convenient to substitute another symbol α' for $0.4343k$, so that the equation reads

$$I = I_0 \times 10^{-\alpha'm}.$$

The two symbols k and α' might be distinguished by calling k the "Napierian" coefficient of absorption, and α' the "decimal" coefficient of absorption.

---

[*] *Beitr. Geophys. Leipzig*, **61**, 1926, h. 1.
[†] Ångström, *Geog. Ann.* 1925, h. 4.

In the following pages we shall use either k or α', taking the one which happens to be most convenient in each problem to be considered.

§ 76. *Absorption of long-wave radiation in the atmosphere*

It was stated in § 71 that within the limits of the wave-lengths included in the beam of radiation from the sun, the amount of atmospheric absorption is relatively slight, being limited to certain bands mainly due to water-vapour, but in part also to ozone and oxygen. Radiation from, and absorption by, bodies at atmospheric and terrestrial temperatures, as we have seen from fig. 31, will lie within a totally different range of wave-lengths, say 3μ to 100μ. Within these limits there is no appreciable absorption by any of the simple gases of the atmosphere. Ozone, however, has a strong absorption band at $9-10\mu$. Carbon dioxide* has only one absorption band above 4μ, a narrow intense band centred at $14\cdot7\mu$, and extending from 12μ to about $16\cdot3\mu$. By far the most important absorbing constituent of the atmosphere is water-vapour, which shows very marked absorption within a very wide range of wave-lengths. It was for the latter reason that many writers treated water-vapour as a "grey" radiator, which absorbed all wave-lengths in the same proportion. It was first clearly demonstrated by G. C. Simpson† that this assumption led to erroneous conclusions, and that it was necessary to take account of the actual observed nature of the absorption spectrum of water-vapour in order to explain atmospheric phenomena.

§ 77. *The absorption spectrum of water-vapour*

The absorption spectrum of water-vapour has been the subject of many researches, of which the most notable are those of Hettner‡ and of Weber and Randall.§ Earlier workers had found marked absorption bands centred at $2\cdot7\mu$ and $6\cdot7\mu$, and a wide band extending from about 14μ upwards. Rubens and Hettner‖ mention the occurrence of very strong absorption at 50μ, $58\cdot5\mu$, 66μ and 79μ.

Hettner's observations have been used to compute the decimal coefficient of absorption α', up to a wave-length of 34μ, and the results are shown graphically in fig. 33. If the amount of precipitable water per cm² of cross-section of the path is m, the transmission is $10^{-\alpha'm}$, where α' is the ordinate read off from fig. 33. In fig. 33 the inset diagram is a reproduction, with an enlarged vertical scale, of the portion of the spectrum between 9μ and 19μ. The diagram can be used to compute the fraction of the incident radiation which will be transmitted through any given amount of water-vapour. Thus at $19\cdot5\mu$, $\alpha' = 50$, and the fraction of the incident radiation transmitted through 1 mm of precipitable water is 10^{-5}, while a fraction 10^{-10} is transmitted through 2 mm of precipitable water.

* Rubens and Aschkinass, *Ann. Phys. u. Chem.* **64**, 1898, p. 584.
† *Memoirs R. Met. Soc.* **3**, No. 21.
‡ *Ann. Phys.* **55**, 1918, p. 476.
§ *Physical Review*, **40**, 1932, p. 835.
‖ *Berlin Sitzber. Ak. Wiss.* 1916, p. 167.

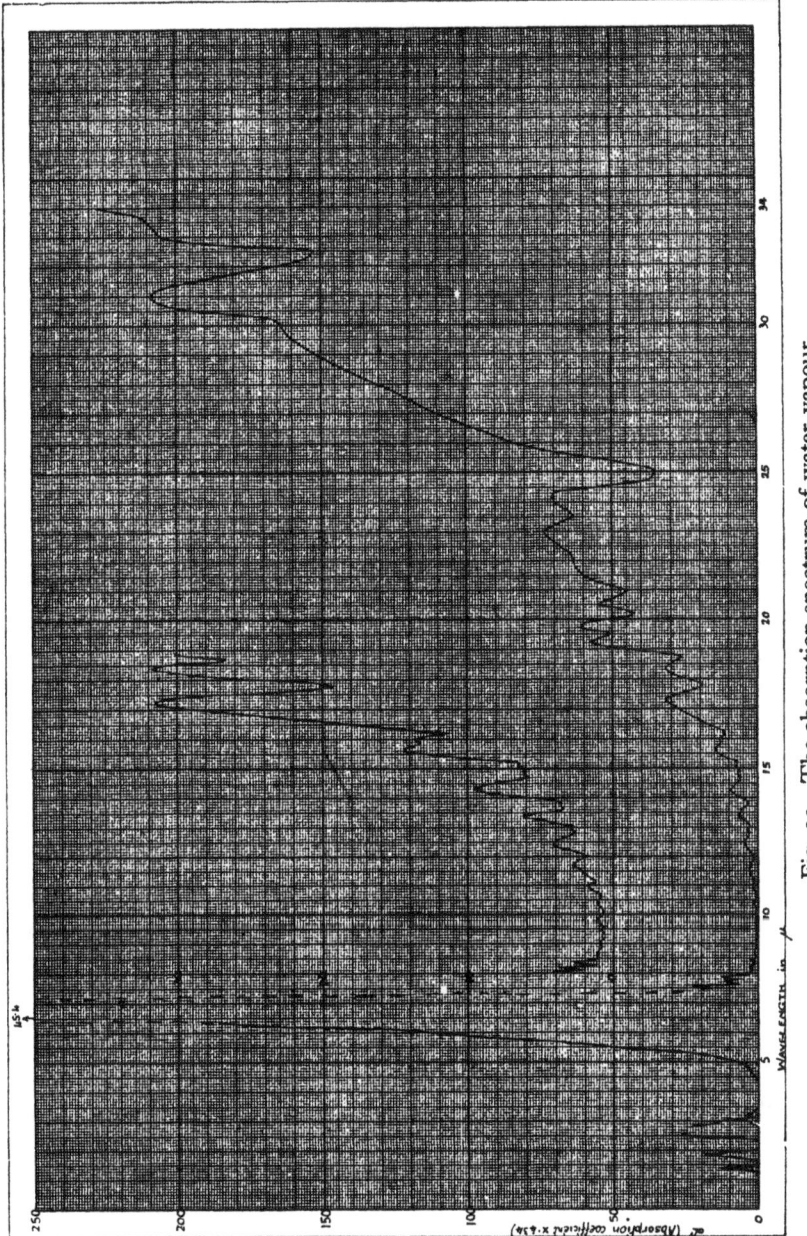

Fig. 33. The absorption spectrum of water-vapour.

Apart from some very narrow lines around 1μ, the main regions of absorption shown by fig. 33 are:

(a) Bands centred at $1\cdot37\mu$, $1\cdot84\mu$, and $2\cdot66\mu$;

(b) A very intense band centred at $6\cdot26\mu$;

(c) A wide band beginning at about 9μ, rising rapidly, and except for a series of oscillations, maintaining the same tendency for increasing absorption with increasing wave-length up to the limit of 34μ at which Hettner's observations terminate. There is reason to suppose that this intense absorption continues up to about 80μ, and possibly still further.

Hettner found no appreciable absorption between $3\frac{1}{2}\mu$ and $4\frac{1}{2}\mu$. Between $8\frac{1}{2}\mu$ and $9\frac{1}{2}\mu$ the absorption measured by Hettner is so small that it is questionable whether it is to be regarded as real. Also Fowle, in his observations of the transmission of radiation through water-vapour, found no evidence of absorption in this region, and it is probable that when the density of the water-vapour is as small as in the earth's atmosphere, the region $8\frac{1}{2}\mu$ to $9\frac{1}{2}\mu$ is effectively transparent.

Simpson (*loc. cit.* fig. 1) gave a curve of absorption by $0\cdot3$ mm of precipitable water. His results are readily derived by the use of fig. 33 above, with $m = 0\cdot03$ gm. Only one-tenth of the incident radiation will pass through the column for all wave-lengths for which α' is greater than about 33. Thus all radiation of wave-length greater than about 19μ, or between $5\cdot5\mu$ and 7μ, will be absorbed practically completely by the water-vapour in the column. Simpson assumed that the absorption was negligible in the range $8\frac{1}{2}\mu$ to 11μ, and taking the absorption by CO_2 in the region 13μ to 17μ as added to the effect of the absorption by water-vapour, he was able to reduce the main features of atmospheric absorption by a column containing $0\cdot3$ mm of precipitable water in the form of vapour to

(a) effectively complete absorption from $5\cdot5\mu$ to 7μ, and from 14μ upwards;

(b) complete transparency from $8\cdot5\mu$ to 11μ and below about 4μ; and

(c) intermediate regions of incomplete absorption in the ranges of wave-length 7–$8\cdot5\mu$ and 11–14μ.

Simpson's simplifications of the water-vapour spectrum enabled him to give for the first time a reasonable explanation of a number of observed phenomena, and the final justification of his hypotheses is their success in application.

Weber and Randall's observations in steam at atmospheric pressure yield the values of α' shown graphically in fig. 34a, and their observations in saturated air at $22\cdot5°$ C, yield the values of α' shown in fig. 34b. The differences in figs. 33 and 34 a and b are considerable. Weber and Randall's measurements not only make the coefficients of absorption of water-vapour much smaller than those of Hettner, but they also make the water-vapour spectrum appear even more complex than it was pictured by Hettner. Fig. 34 shows the band which Simpson assumed to be transparent from $8\frac{1}{2}\mu$ to 11μ as transparent to at least $12\frac{1}{2}\mu$. It indicates that water-vapour radiation from shallow layers of damp air will only begin to be really effective from about 17μ, and for the wave-lengths above 17μ, 1 mm of precipitable water will suffice to absorb over $9/10$ of the incident radiation.

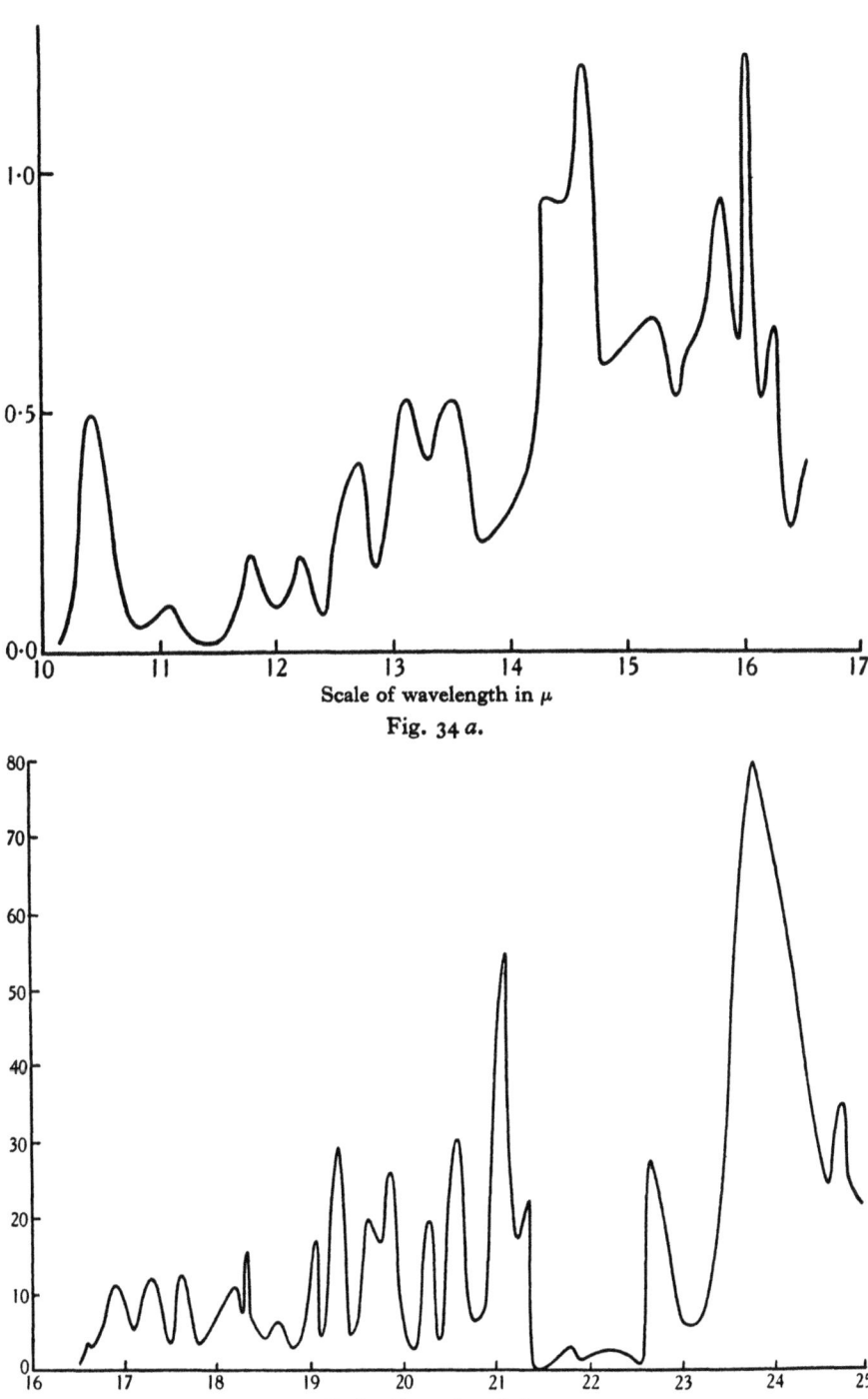

Scale of wavelength in μ

Fig. 34 *a*.

Scale of wavelength in μ

Fig. 34 *b*.

Fig. 34. Absorption spectrum of water-vapour according
to Weber and Randall's measurements.

If we assume 1·2 mm of precipitable water, instead of the 0·3 mm of precipitable water assumed by Simpson, to act effectively as a complete absorber, we see from fig. 34 that this amount will absorb practically the whole of the radiation in wave-lengths above 17μ. We could then increase fourfold the depth l of the layer which acts effectively as a black body as computed on the basis of Hettner's observations. (See § 85, p. 129.) But this leads to difficulties which are discussed in the next chapter. The most serious difficulty which has to be faced in any meteorological application of laboratory measurements of the coefficients of absorption of water-vapour lies in the fact that the precise mode of variation of the form of an individual line in the spectrum with variations of temperature, air pressure, and partial pressure of the water-vapour, is not known.

§ 78. *The effect of air pressure, partial vapour pressure, and temperature, on the water-vapour spectrum*

The general question of the effect on the form of the lines in the water-vapour spectrum, of variations in temperature and total and partial pressure, has been briefly discussed by Elsasser,* but there still remains a considerable degree of uncertainty as to the precise nature of these effects. Dennison† gave for the form of a spectral line the following relationship between the coefficient of absorption and the frequency ($\lambda = c/\nu$), c being the velocity of light, and ν_0 the frequency at the centre of the line,

$$k_\nu = \frac{1}{\pi} \frac{a}{(\nu - \nu_0)^2 + a^2} \qquad \ldots\ldots(5).$$

The effect of variations of temperature or pressure will be to vary the value of a. But

$$\int_{-\infty}^{+\infty} k_\nu \, d\nu = \frac{1}{\pi} \int_{-\infty}^{+\infty} \frac{a}{(\nu - \nu_0)^2 + a^2} \, d\nu$$

$$= \frac{1}{\pi} \left[\tan^{-1} \frac{\nu - \nu_0}{a} \right]_{-\infty}^{+\infty} = 1 \qquad \ldots\ldots(6).$$

Hence if k_ν is plotted against ν it will always have the same area between it and the axis $k = 0$. When the line is widened, the intensity of the absorption in the central core, $1/\pi a$, is diminished, with a corresponding increase in intensity in the wings. At $\nu - \nu_0 = \pm a$, $k = 1/2\pi a$, so that the absorption coefficient falls to one-half the central value at a frequency distance a from the centre. The quantity $2a$ is conveniently called the half-width of the line.

The theory of radiation shows that a spectral line is never infinitesimally narrow, though the natural width is only a small fraction of an Ångström unit. There is also a widening produced by the Doppler effect, or the motion of the molecules. This widening, which is proportional to the square root of the absolute temperature, is small in the earth's atmosphere. More important than either of the preceding effects is that produced by the perturbation of the

* *Monthly Weather Review*, **65**, 1937, p. 323.
† *Physical Review*, **31**, 1928, p. 503.

radiating molecule by the impacts of other molecules, whether similar or not to the radiating molecule. If, for example, we consider a radiating molecule of water-vapour, subject to impacts with air molecules, it can be shown that

$$a = N_0 \sigma^2 p \sqrt{\frac{1 + m_w/m_a}{2\pi m_w RT}} \qquad \ldots\ldots(7)$$

where N_0 is Avogadro's number, σ the effective diameter for an impact between an air and a water molecule, m_w and m_a the molecular weights of water-vapour and air respectively, T the absolute temperature, and R the gas constant.

When numerical values are inserted in the above,

$$a = 1\cdot4 \times 10^7 \frac{p}{\sqrt{T}}$$

where p is in millibars and T in degrees absolute.

Let $p = 1000$ millibars, $T = 300°$ A. Then the half-width of a spectral line, expressed as a frequency difference, $2a$, is

$$2\cdot8 \times 10^7 \times 100/\sqrt{3} = 1\cdot61 \times 10^9.$$

Let this half-width correspond to a wave-length difference of $\Delta\lambda$. Then since

$$\lambda = c/\nu, \quad \Delta\lambda = -\frac{c}{\nu^2}\Delta\nu = \frac{2a\lambda^2}{c}.$$

If the unit in terms of which λ and $\Delta\lambda$ are expressed is μ,

$$\Delta\lambda = 5\cdot4 \times 10^{-6}\lambda^2.$$

We are interested in values of λ of the order of magnitude of 10μ. For

$$\lambda = 10\mu, \quad \Delta\lambda = 5\cdot4 \times 10^{-4}\mu.$$

Hence the widths of absorption lines in the spectrum, on this theory, will be very small, far too small to be resolved by any spectrograph at present available. Weber and Randall used slit widths equivalent to $0\cdot04\mu$ in the region near 10μ, and $0\cdot08\mu$ in the longer wave-lengths, while Hettner used much greater slit widths. But even with Weber and Randall's instrumental equipment it is impossible to resolve a line of the width of many times $5\cdot4 \times 10^{-4}\mu$. The Doppler widening is negligible in these conditions, but is of the same order as the collision effect at low pressures.

In an atmosphere of steam alone, it is necessary to modify equation (7) above by substituting m_w for m_a, and increasing σ to allow of the larger perturbation of a radiating molecule by a similar molecule. For this reason, observations of absorption in steam should not be applied to the discussion of absorption in water-vapour of low density mixed with air.

The final conclusion to which we are forced is that the phenomena of water-vapour absorption are so complex and the experimental observations subject to so many difficulties, that at present it is not possible to apply any of the observed values to discuss radiation phenomena in the atmosphere, with any degree of confidence in the results obtained. It is certain that over a considerable range of wave-lengths, Beer's law of absorption (§ 75, p. 113), is no longer applicable. In thin layers of an atmosphere, when the density of water-vapour is small, the absorption is largely limited to the cores of the lines, in which the energy is entirely absorbed by relatively small amounts of water-vapour, and the total loss of energy by absorption is not sensitive to changes in the length of the path. But in an extensive atmosphere the wings of the lines become effective, and with increasing mass of the atmosphere the total emission in any wave-length, even between the emission lines, tends to increase as the mass of the atmosphere increases, asymptotically approaching black-body radiation. This was verified by Dennison (*loc. cit.*) for the bands of HCl.

We may thus anticipate that the downward long-wave radiation from the water-vapour in the atmosphere will contain not only radiation within the central parts of the spectral lines originating at low levels, but also radiation in the regions between the lines, of amount determined by the mean temperature of the atmosphere, and possibly by the square root of the total amount of water-vapour above the place of observation. Whether this will reach the maximum black-body radiation cannot be stated with certainty, though in view of the reasonably close agreement which Simpson found between the observations of W. H. Dines and the estimated radiation, which amounted to complete black-body radiation within a wide range of wave-lengths, it appears that the downward radiation from the atmosphere must approach black-body radiation within wide ranges of wave-length. It is certainly impossible to account for the magnitude of the long-wave radiation from the atmosphere as radiation within narrow lines separated by wide intervals of complete transparency.

§ 79. *The absorption spectrum of liquid water*

The absorption spectrum of liquid water has been investigated in some detail by Rubens and Ladenberg* up to wave-lengths of 18μ, using films of water and glycerine, and by Aschkinass† using pure water, up to a wave-length of 7μ. The curve of absorption coefficients plotted against wave-length has the same general form as the curves in figs. 33 or 34, except that the values are much greater. Using the same notation as is used above, referring to a unit of 1 cm of liquid water we obtain the values of α' plotted in fig. 35, which are therefore directly comparable with those in figs. 33 or 34. The only striking differences of form between figs. 33 or 34 and 35 are the intensity of the band at 3μ in the liquid water spectrum, and the relatively smaller intensity of the band at 6μ by comparison with the intensity in the region beyond 12μ. It is seen that even for those wave-lengths for which it is most transparent to long waves, water of a thickness of 1 mm only transmits 10^{-20} of the incident unreflected radiation, and that a layer of water of 0·1 mm thickness only transmits 1/100th. In the regions of greater absorption such as the band at 6μ a layer of thickness 0·02 mm (20 microns) only transmits 1/100th of the unreflected radiation, and even a layer of 10 microns only transmits 1/10th of the incident radiation.

---

* *Verhandl. Phys. Ges.* **11**, 1909, p. 16.
† *Wied. Ann.* **55**, 1895, p. 401.

The broken curve of fig. 35 gives the reflecting power of a water surface for normal incidence. In the wave-lengths at which atmospheric radiation is most intense, say 5μ to 15μ, the value is everywhere small, so that only a few per cent of the incident radiation is reflected. Hence we may assume that films or drops of water of a thickness of $0\cdot1$ mm or more are within a few per cent of being black-body radiators.

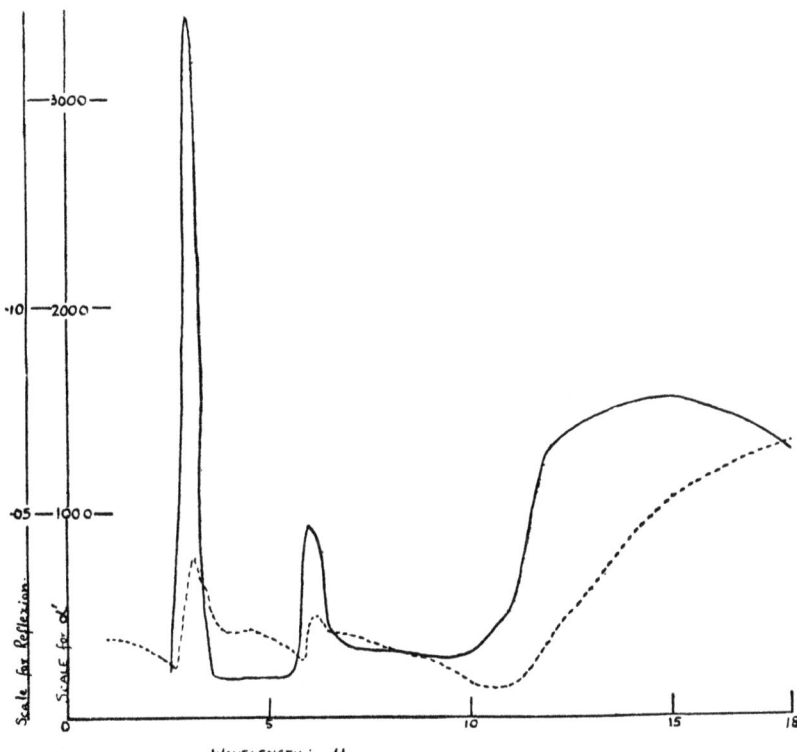

WAVE-LENGTH in μ

Fig. 35. The absorption spectrum of liquid water (continuous line), and the reflecting power of liquid water (broken line).

The absorption curve of fig. 35 is based on observations on thin plane films of water, but presumably the results may be applied to water drops also. Thus individual water drops of $\frac{1}{2}$ mm in diameter may be treated as black-body radiators for all practical purposes. Even individual small drops of 10μ in diameter, such as are found in fogs and clouds, may be regarded as almost equivalent to black-body radiators, and a layer of fog or mist containing the equivalent of $\frac{1}{2}$ mm of water per cm² of cross-section may be treated as a perfect black body within the limits of a few per cent. It is, however, clear that in a fog near the ground there will be a certain amount of diffuse scattering or reflexion of long-waved radiation produced in the same manner as the scattering of light in the sky, by particles whose dimensions are less than the wave-lengths of the radiation. W. H. Dines* found that the radiation from a

* M.O., *Geoph. Mem.* No. 18, 1921.

fog was equivalent to the complete black-body radiation at the same temperature, but his measurements could not separate the true radiation from the fog particles from the radiation originating at the ground and undergoing diffuse scattering in the fog.

According to the *Meteorological Glossary*, the amount of suspended water in a cloud or fog is of the order of a few grammes of water per (metre)^3. Thus 1 (metre)^3 contains say 2 cc of liquid water, and the length of path which contains 0·1 mm of liquid water per cm^2 of cross-section is 50 metres. Thus a layer of cloud or fog a few metres thick can be treated as a black-body radiator.

The computation of the thickness of drop-laden air which is capable of functioning as a black body neglects the effect of reflexion and scattering by the drops. A figure of a few per cent was quoted above for the reflecting power of water (in sheets), but it is not strictly justifiable to use the same figure for a layer of a few metres in thickness, containing small water drops, since any portion of a beam of radiation passing through these few metres would meet a large number of reflecting surfaces, and the number would be greater the smaller the drops. Thus a cloud or fog is nearer to being a black body when the drops are large than when the drops are small. There appears to be no method of estimating the effect mentioned.

The coefficients of absorption of liquid water for wave-lengths less than 2μ, such as occur in the direct solar beam, are everywhere small, and the values derived for wave-lengths less than $0·5\mu$ can be attributed to Rayleigh scattering by the molecules. The following table gives the values of α' for a range of wave-lengths below $1·2\mu$ referred to 1 cm of water:

Wave-length (μ)	0·779	0·865	0·945	1·19
α'	0·118	0·128	0·234	0·842

Thus the absorptive power of water is very small for light waves. In a cloud or fog, the incident solar rays are scattered by irregular reflexion by the drops, and the total path through the water drops in the upper layers of the cloud or fog will be many times greater than the equivalent depth of liquid water. It is therefore possible that there is an appreciable amount of absorption of solar rays in the upper region of a cloud sheet or fog.

§ 80. *Diffuse reflexion and scattering*

The scattering of light by small obstacles was studied in great detail by the late Lord Rayleigh.* When a beam of radiation passes through a transparent medium containing in suspension small particles whose refractive index differs from that of the medium, some of the radiation is taken from the direct beam, and sent out in all directions from the particles. Rayleigh showed that the incident beam I_0 is reduced to $I_0 e^{-sx}$ in a distance x, s being the "coefficient of scattering". The coefficient s is proportional to λ^{-4}, and is therefore much greater for blue than for red light. Hence blue light is much more readily scattered than red light. Rayleigh explained the blue colour of the sky as due to the effect of scattering by the molecules of the air. He also showed that true scattering can only take place when the diameter of the obscuring particles is

---

* *Scientific Papers*, **1**, p. 91. *Vide* also L. V. King, *Phil. Trans.* A, ccxii, 1913, p. 375.

less than the wave-length of the incident light. Thus tobacco smoke, which consists of liquid drops whose diameter is about 0.2μ, scatters light of all wave-lengths, but scatters the blue light much more than the red light, and so appears blue by scattered light; while the sun appears red when viewed through a smoke fog or haze, the light from the blue end of the spectrum being removed from the solar beam by scattering.

When the diameter of the particles is larger than the wave-length of the incident light true scattering no longer occurs, and the effect of the particles is of the nature of diffuse reflexion, which is equally effective for all wave-lengths. The last statement is borne out by the fact that the light reflected from a cloud when the sun is behind the observer is pure white, and that the sun appears white when viewed through a fog of water drops, whose diameter is of the order of 10μ.

Solar radiation is scattered by molecules or other sufficiently small particles in the air, and suffers loss by diffuse reflexion by larger particles. The larger particles such as cloud droplets occur in greater quantities at low levels than at high levels in the atmosphere, and hence observations of the intensity of solar radiation are made at stations at high level, in places removed as far as possible from the effects of low cloud and of local sources of pollution. At such stations the observations of the intensity of solar radiation are only affected by scattering, and possibly a small amount of absorption, in the upper atmosphere. Schuster showed that on clear days the loss of radiation from the solar beam of incoming radiation at Mount Wilson could be accounted for by the effect of molecular scattering alone.

The importance of the effects included in the terms scattering and diffuse reflexion will be seen when we come later to discuss the effects of cloud sheets on temperature (p. 144). The magnitude of the effects of scattering upon radiation of short wave-lengths is shown by the low amount of ultra-violet light measured in large towns as compared with the amounts measured in the country.

It is important to realise the difference in the effects of small particles or molecules and cloud droplets. The former, by scattering a larger amount of light of the shorter wave-lengths, diminishes the proportion of these wave-lengths in the direct beam, and so causes the direct beam to be richer in red rays, and the scattered light to be richer in blue rays, than the original incident light. On the other hand, cloud droplets do not produce any change in the relative intensities of the different wave-lengths, and when a cloud appears red the colour is not to be explained as a result of the cloud droplets showing a selective scattering effect, but as due to the loss of the blue light by scattering from the beam which illuminates the cloud.

The scattering of light by volcanic dust is capable of producing a decided attenuation of the incoming solar radiation, and may give rise to marked temperature effects.

Footnote to §§ 77, 78. A further discussion of the water-vapour spectrum has been given by Elsasser in *Q.J. Roy. Met. Soc.* Canadian Supplement, 1940.

CHAPTER VI

RADIATION IN THE TROPOSPHERE

§ 81. *General survey of radiation phenomena in the atmosphere*

THE incoming solar radiation suffers depletion in at least five ways:

(1) Absorption by oxygen* at heights well above 100 km, and by the ozone layers at heights of about 10–40 km.

(2) Absorption by the other constituents of dry air.

(3) Scattering by molecules of dry air and water-vapour.

(4) Absorption by water-vapour.

(5) Scattering and diffuse reflexion and absorption by solid and liquid particles suspended in the atmosphere.

Item (1) will be discussed more fully in the next chapter. Item (2) is not of any considerable importance, as the atmospheric gases at ordinary atmospheric temperatures absorb practically no radiation. ·Item (3) has been estimated to be sufficient in itself to account for the blue colour of the sky. Schuster† and Fowle‡ have shown that scattering by the molecules of dry air is in itself sufficient to account for the loss of solar radiation in the uppermost layers of the atmosphere, within which water-vapour absorption is negligible. The corresponding coefficients of transmission can be evaluated by the use of the formulae of Rayleigh and King. Item (4), the absorption by water-vapour, has already been mentioned in § 76, and is discussed in some detail below. The amount of the loss due to the phenomena grouped under item (5) is highly variable, according to the conditions prevailing at the time of observation. The amount of dust in the lower atmosphere varies with time and place, and may comprise not only particles small enough to exert an effect which is greater for blue light than for red light in accordance with Rayleigh's theory, but also larger particles which do not come within the scope of Rayleigh's theory, and produce diffuse reflexion and absorption rather than scattering. Again, droplets of water in fog and cloud, being large by comparison with the wave-length of the incident radiation, also produce reflexion rather than true scattering. A cloud layer is capable of reflecting back to the sky about four-fifths of the incident radiation, and therefore the main factor in determining the net inflow of solar radiation is cloud amount.

The radiation which reaches the earth's surface is in part reflected by the sea, or diffusely reflected and scattered by the different parts of the earth's surface, such as trees, grass and other details of the earth's structure. Over land, the portion which is not reflected or scattered is absorbed by the earth's

---

* *Vide* Chapman, *Q.J. Roy. Met. Soc.* **60**, 1934, p. 135.
† *Nature*, **38**, 1909, p. 97.
‡ *Astroph. Journ.* **38**, Nov. 1913.

surface, except for the fraction which is used in evaporating water from the surface. Over the sea, the amount lost by reflexion back to the sky is great, and varies from about 2 to over 70 per cent; the amount of energy used in evaporation is probably greater than over land; and as the residual penetrates to considerable depths before it is completely absorbed, the changes of temperature of the surface of the sea are much less than the changes of temperature of the surface of the land.

Considerable variations are, however, possible in the temperature of the sea surface where there are large variations in depth, from quite shallow water to deep water. Where the water is shallow the radiation which penetrates through warms the sea floor, and this in turn warms the shallow water above it to a higher temperature than that of deep water. The high temperature of shallow pools on the sea shore bear out this view in a striking manner.

While the sun is high, the surface of the earth is continually heated by the incoming radiation, and the earth's surface radiates back to the sky an amount approximately in accordance with Stefan's law. The radiation from the earth's surface is long-waved radiation, much of which is readily absorbed by the water-vapour in the atmosphere. The water-vapour radiates in turn, both upward and downward, an amount of energy dependent on its temperature, and the result is to yield a complexity of streams of radiation which require careful consideration. The phenomena are further complicated by convection currents which carry upward streams of warm air, and so act as carriers of heat.

The amount of incoming radiation from the sun reaches its maximum when the sun is on the meridian at noon, but the rise in temperature of the earth's surface continues for some time after this, since the loss of energy by radiation is not at noon sufficient to balance the incoming radiation. The balance between incoming and outgoing radiation is reached some time after noon, but after that time the loss by radiation from the earth's surface exceeds the incoming radiation, and the temperature steadily falls during the rest of the day, and during most of the night. The fall is only checked by the beginning of incoming radiation shortly before sunrise. The diurnal variation of temperature should thus show a maximum shortly after noon, and a minimum shortly before sunrise. The curve of variation will not be a pure sine-curve, but its form will vary with the time of the year, though its main features will remain unchanged.*

The air temperature will follow the same general course as the soil temperature, but with a lag in the times of occurrence of the maximum and minimum. The relationship cannot be simple, on account of the complexity of the phenomena of radiation and turbulence, but we should expect to find the diurnal variation of temperature decreasing with height, as is actually the case. The transfer of heat upward through the atmosphere is discussed in §§ 85, 137 and 140 below.

* H. L. Wright, *Mem. R. Met. Soc.* **4**, No. 31; also fig. 13, p. 23 above.

§ 82. *Water-vapour as a controlling agent in atmospheric absorption and radiation*

Before proceeding to discuss the various streams of radiation which must be considered, we must form a mental picture of the radiative mechanism of the atmosphere. The main result of § 81 above is that so far as true absorption in the atmosphere is concerned the incoming solar beam passes through the atmosphere almost undiminished. The light reflected and scattered by molecules of dry air and water-vapour, water drops, etc., will remain short-wave radiation. The long-wave radiation from the atmosphere itself, and the absorption in the atmosphere of long-wave radiation from the earth's surface, are so nearly completely due to water-vapour that we may, at least in a preliminary survey, neglect the radiation and absorption of all other gaseous constituents of the atmosphere. This is equivalent to regarding the atmosphere as consisting of water-vapour only, so far as radiative effects are concerned. The effect of the addition of dry air to the water-vapour is to weight the water-vapour with an added specific heat, without modifying any of its other pro-perties. Since the establishment of a balance between incoming and outgoing radiation depends on the temperatures of the radiating bodies, it may be concluded that the temperature distribution at which equilibrium will be finally established will not be affected by the addition of the dry air to the water-vapour atmosphere, but that the time taken to establish any given change of temperature will be thereby increased. The mathematical treatment of the problem is not much simplified by the suggestion made here, since the water-vapour atmosphere will be endowed with a specific heat whose magni-tude will be subject to wide variations with height, and with time at any given level, but it will nevertheless be worth bearing in mind the relative functions of dry air and water-vapour in the radiative processes of the atmosphere.

§ 83. *The heat balance of the atmosphere*

On the average the intensity of the incoming solar beam at the outer limit of the atmosphere is 1·94 gramme-calories per cm² per minute, upon a surface placed at right angles to the beam. Of this amount 6 to 8 per cent is absorbed in the atmosphere, 9 per cent is reflected back from the atmosphere, 8 per cent is reflected back from the earth's surface, and a considerable fraction is scat-tered by the molecules of the atmosphere, or suffers diffuse reflexion by par-ticles in suspension in the atmosphere. The short-wave radiation which reaches the ground thus comprises two categories: (*a*) the direct solar beam, and (*b*) the radiation diffused from the sky. Ångström* has estimated that if Q is the total short-wave radiation $(a+b)$, and D the diffuse radiation from the sky (b), the fraction D/Q has a minimum of 0·25 in May, and a maximum of 0·8 to 0·9 in winter (at Stockholm). He further estimates that on a completely overcast day the total Q amounts to one-fourth of its value on a clear day, and

* *Vide* "Radiation and Climate", *Geog. Ann.* 1925, p. 122.

that on a day when the maximum possible number of hours of sunshine is N and the actual number of hours of sunshine is n, the total value of Q is given by

$$Q = Q_0 \left(0 \cdot 25 + 0 \cdot 75 n / N\right),$$

where Q_0 is the total incoming radiation on a clear day.

Over land the short waves which reaoh the earth's surface are in part reflected by the surface. The loss by reflexion does not usually exceed about 10 per cent, except over a snow surface, which is capable of reflecting about 80 per cent of the incident short-wave radiation. A fraction, whose magnitude is not readily estimated, is used in evaporating water from the surface of grass, leaves, etc., and in high latitudes a portion is used in melting ice and snow, and the remainder is absorbed.

Over the sea the phenomena are somewhat different. The fraction reflected is very slight when the sun is high, but increases rapidly as the sun's zenith distance increases. Schmidt* has estimated that on the average for the whole year the percentage r of the total incoming radiation which is reflected at a water surface in different latitudes is as shown in the following table:

Lat.	0	10	20	30	40	50	60	70	80	90
r	3·3	3·5	3·6	4·2	4·6	5·3	6·2	8·0	11·5	13·5

Ångström† has given estimates of evaporation based on observations extending over ten days of August 1905, over Lake Vassijäure, the period being selected so that there was practically no difference in the temperature of the air and water, and no progressive change in the temperature of the water, during the period. His estimate of the evaporation amounted to a little less than 2 mm of water per day, corresponding to nearly 120 gramme-calories per cm², or approximately one-third of the incoming net radiation from sun and sky. Lake Vassijäure being in latitude 68° N, we thus find that of the incoming radiation from sun and sky, 8 per cent is reflected from the water surface, and 33 per cent is used in evaporation, the remaining 59 per cent being absorbed by the water and re-radiated to the atmosphere as long waves. Ångström's figure for evaporation is rather less than that of 2·3 mm per day estimated by Wallén for Lake Hjälmaren during August, but it is in close agreement with Witting's estimate for Botten Bay, which is in almost the same latitude as Lake Vassijäure, while Lake Hjälmaren is farther south. It is of interest to compare these estimates with the estimate given by Wüst† of the total evaporation during the year from sea and land, his respective figures being 267 and 112, in units of 1000 km³ of water. The total surface of the oceans being $3 \cdot 68 \times 10^{18}$ cm², over which there has to be distributed 267×10^{18} cm³ of evaporation, the depth of water evaporated per annum is 267 ÷ 3·68, or 72·6 cm, yielding an average of 2 mm per day over the whole earth. The area of the earth's land surface is $1 \cdot 45 \times 10^{18}$ cm², so that the total evaporation per annum amounts to 112 ÷ 1·45 cm, or 77·2 cm, or an average of about 2·1 mm per day. At this stage we do not wish to emphasise these actual magnitudes, but the general agreement of the figures given by different workers appears to

* *Ann. Hydrog. u. Marit. Met.* 1915, h. 3–4.   † *Geog. Ann.* 1920, h. 3.
‡ *Zeit. Ges. Erdkunde*, Nos. 1–2, 1922.

justify the adoption of Ångström's figures as giving the order of magnitude of the consumption of heat in evaporation.

The part of the incoming solar radiation which reaches the earth's surface and is not reflected or used up in evaporation is absorbed by the surface. As a result the temperature of the earth's surface increases steadily while the amount of incoming radiation is increasing during the morning. The maximum temperature of the earth's surface occurs some time after noon, the lag after noon depending on the nature of the surface. Thus at a depth of 1 cm in sand the maximum occurs at 12 h 30 m, while just inside the soil under a grass covering the maximum occurs some three hours after noon. The surface of the earth radiates to the atmosphere, its radiation being of long wave-length, and therefore readily absorbed by the water-vapour in the atmosphere. The water-vapour in turn sends out long-wave radiation, a portion of which is directed downward to the earth's surface, where it is absorbed. A considerable fraction of the radiation from the earth's surface falls within the band of transparency of water-vapour ($8\frac{1}{2}\mu$ to 11μ), and the radiation within this band of wave-lengths will go upward through the atmosphere without being absorbed.

The net outward flow of long-wave radiation from the earth, which is the difference between the radiation from the earth's surface and the long-wave radiation of the atmosphere, is of the same order of magnitude by night and by day. But as it is much easier to observe this net radiation by night than by day, on account of the absence at night of the diffuse radiation of short wave-length from the sky, night observations have been more frequent than day observations. Moreover, the nocturnal cooling is bound up with the practical problem of forecasting the occurrence of frost, a problem of considerable economic importance.

Observations of the radiation of the atmosphere at night have been carried out by Dines, Ångström, Asklöf and others, and the amount of this radiation is found to be of the order of three-fourths of the full black-body radiation at the temperature of the surface. Since the earth sends out practically the full black-body radiation appropriate to its own temperature, it follows that the net flow of radiation outward from the earth (outgoing − incoming) is of the order of one-fourth of the black-body radiation. This net outflow of radiation is mainly the radiation of the earth's surface in the transparent band from $8\frac{1}{2}\mu$ to 11μ.

In discussing radiation of long waves at ordinary terrestrial temperatures, we may treat as black-body radiators the solid ground, the sea surface, the surface of snow, a cloud sheet thick enough to cast a shadow, and a fog. These assumptions are found to fit the facts with great accuracy.

§ 84. *The equations of radiative transfer of heat*

In the first place we shall consider the variation of the upward and downward beams of radiation of a particular wave-length λ. Let the upward and downward beams of radiation of this wave-length be A_λ and B_λ at a level where the temperature is T, and let E_λ be the black-body radiation of this wave-length at temperature T. Let the level be specified by the amount τ of the water-vapour *above* this level. The beams entering and leaving a thin layer containing an amount of water-vapour $d\tau$ are shown in fig. 36. The upward beam

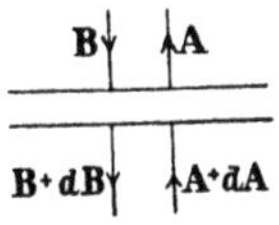

Fig. 36. Diagram of radiative transfer.

loses an amount $k_\lambda d\tau A_\lambda$ by absorption, but gains an amont $k_\lambda d\tau E_\lambda$ from the upward radiation of the layer itself. Hence

$$\frac{dA_\lambda}{d\tau} = k_\lambda \, (A_\lambda - E_\lambda) \qquad\qquad \dots\dots(1),$$

$$\frac{dB_\lambda}{d\tau} = k_\lambda \, (E_\lambda - B_\lambda) \qquad\qquad \dots\dots(2).$$

In the atmosphere, under normal conditions, when the temperature decreases steadily with height, the upward beam A_λ is greater than E_λ at all heights, except at the ground, when the two are equal for all wave-lengths radiated and absorbed by water-vapour. For long-wave radiation B_λ is zero at the outer limit of the atmosphere, and remains less than E_λ at all lower levels. The difference between A_λ or B_λ and E_λ will be small at all levels for those values of λ for which k_λ is great, and will be greatest at all levels for those values of λ for which k_λ is small, i.e. for those wave-lengths to which water-vapour is transparent.

It should be noted that, for a parallel beam, equations (1) and (2) are true whether the radiation is black-body, grey, or selective.

The nature of the beams ΣA and ΣB will thus be very complicated, and neither beam can be represented as the radiation of water-vapour of a particular temperature with any great accuracy.

Some simplification of the problem is, however, possible along the lines of Simpson's discussion as amplified by Brunt.* But at this stage, the serious difficulty that arises through the diffuse nature of the streams of radiation will be left out of account. The upward and downward streams of radiation are not, in fact, parallel streams of radiation, but consist of an infinite number of beams in widely varying directions. For reasons stated in § 93 below, a reasonable approximation is obtained by treating the beams as parallel, and doubling the coefficient k_λ. For the present we shall limit our discussion to that of parallel beams without considering too closely the appropriate magnitude of the constant k in equations (1) and (2) above.

§ 85. *An equation for radiative transfer*

The maximum intensity of the radiation from the earth's surface is at about 10μ, and that of black-body radiation from a body at the temperature of the stratosphere is at about $12\frac{1}{2}\mu$. Thus the existence of the band from $8\frac{1}{2}\mu$ to 11μ,

---

* *Proc. Roy. Soc.* A, **124**, 1929, p. 201.

in which water-vapour is effectively transparent, is of predominating importance in meteorology. As soon as radiation within this band of wave-lengths leaves the earth's surface, it is effectively lost as far as the atmosphere is concerned, if the sky is clear. The treatment which follows here attempts to take these facts into account.

We shall assume with Simpson that a column of air which contains $0\cdot3$ mm of precipitable water as vapour will completely absorb all radiation of wave-lengths between $5\cdot5\mu$ and 7μ, and of wave-lengths greater than 14μ. We shall neglect the partial absorption between 4μ and $5\cdot5\mu$, between 7μ and $8\cdot5\mu$, and between 11μ and 14μ. It is considered that a reasonable first approximation to the facts is obtained by making this approximation; and the success with which Simpson applied these simplifications justifies their use as a first approximation.

The name "W-radiation" has been suggested (Brunt, *loc. cit.*) for radiation restricted to the wave-lengths within which water-vapour absorbs and radiates, and the term will be adopted here in order to save continual re-statement. It is also assumed that dry air does not absorb or radiate long waves to an appreciable extent.

The length l of the column of air which contains $0\cdot3$ mm of precipitable water, or $0\cdot03$ gramme of water-vapour per cm², is readily deduced. Let e be the vapour-pressure in mb, ρ_w the density of the water-vapour, T the temperature, and R' the gas-constant for water-vapour. Then

$$e = R'\rho_w T, \quad \text{where} \quad R' = 4\cdot62 \times 10^3.$$

By the definition of l
$$l\rho_w = 0\cdot03,$$

$$l = 0\cdot03/\rho_w = 0\cdot03 \times 4\cdot62 \times 10^3 \times T/e = 139\,T/e \text{ cm} = 1\cdot39\,T/e \text{ metres} \quad \ldots\ldots(3).$$

If we take $T = 280°$ A, $\quad l = 380/e$ metres.

In the lower atmosphere e is normally of the order of 10 mb, and is often considerably greater than this. Hence l is normally of the order of 40 metres, and it is legitimate to treat the layer of thickness l as having uniform temperature equal to the mean temperature of the layer, so far as the computation of the radiation from the layer is concerned. We shall adopt this method as the basis of computation of the upward and downward streams of radiation in the atmosphere.

Beginning at the ground we divide the whole atmosphere into layers of varying thickness such that each contains $0\cdot3$ mm of precipitable water. The thickness of each layer is $139\,T/e$ cm, where T and e represent the mean absolute temperature and the mean vapour-pressure in millibars, within each layer. The thickness of each layer is thus proportional to the appropriate value of T/e. Let AB in fig. 37 be the upper boundary of the rth layer. The amount

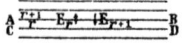

Fig. 37. Diagram of transfer of W-radiation.

of W-radiation moving upward across AB is equal to E_r, the radiation from the rth layer, since no radiation reaching AB can by hypothesis have originated below CD, the lower boundary of the rth layer. Similarly the amount of W-radiation moving downward across AB is equal to E_{r+1}, the radiation from the $(r+1)$th layer. Let the net upward flux of radiation across AB be F_r. Then

$$F_r = E_r - E_{r+1} \qquad \dots\dots(4).$$

If T_r be the mean temperature of the rth layer, then with the usual notation

$$F_r = E_r - E_{r+1} = -\Delta E_r = -\frac{\Delta E}{\Delta T}(T_{r+1} - T_r)$$

$$= -\frac{\Delta E}{\Delta T}\left\{\frac{1}{2}\left(\frac{\Delta T}{\Delta z}l\right)_{r+1} + \frac{1}{2}\left(\frac{\Delta T}{\Delta z}l\right)_r\right\} \qquad \dots\dots(5),$$

$$= -\frac{\partial E}{\partial T}\frac{\partial T}{\partial z}l = -\frac{139T}{e}\frac{\partial E}{\partial T}\frac{\partial T}{\partial z} = -j\frac{\partial T}{\partial z} \qquad \dots\dots(6).$$

The use of the differential coefficient $\partial E/\partial T$ in place of the corresponding finite-difference ratio is justified by the slow rate of change of this quantity with T, as shown in the next paragraph. The use of the differential coefficient $\partial T/\partial z$ in place of the corresponding finite-difference ratio is justified if we neglect small quantities of the first order, and also by the fact that in practice $\partial T/\partial z$ is determined by the measurement of the difference of temperature at points separated by a finite difference of level. In this way we obtain the expression $-j\partial T/\partial z$ for the net upward flux of heat, j being a quantity which normally changes only very slowly with height, as is seen from a consideration of the factors involved in it.

The value of j depends upon the assumption that 0·3 mm of precipitable water absorbs completely the W-radiation passing through it. Should the estimate of 0·3 mm be too small, the value of j is proportionately too small. Thus any uncertainty as to the true value of j is due only to uncertainty as to the truth of the estimate of 0·3 mm for the quantity of precipitable water which will completely absorb W-radiation. The value of $\partial E/\partial T$ can be readily deduced from a table of values of E given by Simpson (loc. cit. p. 8):

Temperature	200° A	220° A	270° A	295° A
$\frac{\partial E}{\partial T} \times 10^3$	1·6	2·0	3·0	3·5

E is measured in gramme-calories per cm² per min.

Within these limits of temperature $\partial E/\partial T$ appears to be almost accurately a linear function of temperature

$$\frac{\partial E}{\partial T} \times 10^3 = 3\cdot0 + 0\cdot02\,(T - 270) \qquad \dots\dots(7).$$

The vapour-pressure e varies from day to day at a given place, but shows no marked diurnal variation. It varies to some extent with height. According to Hann* the relative values of e at the ground, at 0·5 km and at 1 km are 1·0, 0·83 and 0·68. But observations frequently show a much slower decrease with height than this. Thus Steiner† found in his discussion of kite ascents

---

* Lehrbuch der Meteorologie, 4th edn., p. 243.
† Wiss. Abhandl. Luftwarte Rostock, 1926, h. 1.

made at Rostock, that vapour-pressure showed extremely little variation with height up to 600 metres.

If we assume a temperature of 275° A and a vapour-pressure of 5 mb, corresponding to relative humidity of about 70 per cent, the value of j in equation (6) is found to be

$3 \times 10^{-3} \times 139 \times 275 \div 5 = 23$, the units being the gramme-calorie and minute.

We can use this figure to compute the upward flux of radiation corresponding to any given value of $\partial T/\partial z$. Let us assume the lapse-rate of temperature to be adiabatic, so that $\partial T/\partial z = -10^{-4}$. Then the vertical flow of radiation is $2 \cdot 3 \times 10^{-3}$ gramme-calories per minute. To compare this with the average amount of radiation coming in from the sun, we take the solar constant to be 2 gramme-calories per square centimetre per minute. Allowing for averaging over day and night, and over all latitudes, and assuming Aldrich's value of the albedo 0·43, we find that the amount of incoming energy which has to be transported out again by radiation, turbulence, or otherwise, is about 0·275 gramme-calories per square centimetre per minute. Thus with the assumptions made above in evaluating j, and assuming an adiabatic lapse-rate, except for the radiation contained within the transparent band from $8\frac{1}{2}\mu$ to 11μ, which passes out through the atmosphere without absorption, less than 1/100th of the incoming energy can be disposed of by passing outward as W-radiation through the lowest layers. While it is true that the lapse-rate may from time to time exceed many times the adiabatic, this only occurs during a part of the day, and moreover the value of j deduced above is based on a low figure for vapour-pressure. With larger vapour-pressures j is diminished. It can therefore be accepted that under normal conditions only a very small fraction of the incoming solar radiation passes back through the lowest layers of the atmosphere as W-radiation.

It will be noted that the flow of energy by radiation is always from high temperature to low temperature, so that radiation tends to produce isothermal conditions. There is resemblance between the results here derived and the equations for conduction of heat in solids.

The net flux of heat across unit area of a horizontal surface at height $z + dz$ is

$$-j \frac{\partial T}{\partial z} - \frac{\partial}{\partial z}\left(j \frac{\partial T}{\partial z}\right) dz.$$

A disc of air of unit horizontal area and thickness dz will therefore gain a quantity of heat $-\frac{\partial}{\partial z}\left(j \frac{\partial T}{\partial z}\right) dz$ per minute. If c_p be the specific heat of air at constant pressure, and the unit of time t be 1 second, this gain of heat must be equal to $60 \rho c_p \frac{\partial T}{\partial t} dz$. Thus the equation of radiative transfer of heat may be written

$$\rho c_p \frac{\partial T}{\partial t} = \frac{1}{60} \frac{\partial}{\partial z}\left(j \frac{\partial T}{\partial z}\right) = \frac{139}{60} \frac{\partial}{\partial z}\left(\frac{T}{e} \frac{\partial E}{\partial T} \frac{\partial T}{\partial z}\right) \qquad \ldots\ldots(8).$$

We have seen that in the lowest layers of the atmosphere j will vary slowly with height, while $\partial T/\partial z$ is shown by observation to be subject to variation within wide limits, a lapse-rate 10 times the dry adiabatic lapse-rate being by no means uncommon near the ground. Thus in an investigation of the transfer of heat by radiation in the lowest layers we may treat j as constant in equation (8), giving T and e the values corresponding to the level under discussion. Equation (8) then reduces to

$$\frac{\partial T}{\partial t} = \frac{139}{60\rho c_p} \frac{T}{e} \frac{\partial E}{\partial T} \frac{\partial^2 T}{\partial z^2} = K_R \frac{\partial^2 T}{\partial z^2} \qquad \ldots\ldots(9).$$

This equation is similar to that for conduction of heat in a solid with a constant K_R replacing the thermometric conductivity κ, where

$$K_R = \frac{j}{60\rho c_p} = \frac{139}{60\rho c_p} \frac{T}{e} \frac{\partial E}{\partial T} \qquad \ldots\ldots(10).$$

This value of K_R is adjusted to give a unit of time of 1 second in equation (9).

Again, taking $T = 280°$ A, $e = 5$ mb, we find

$$K_R = \frac{23}{60\rho c_p} = \frac{23}{60 \times 0.00125 \times 0.24} = 1.3 \times 10^3 \qquad \ldots\ldots(11).$$

For the reasons given above, in discussing j, this estimate of K_R is likely to be rather higher than the normal values, on account of the low value assumed for e. For larger values of e, K_R is reduced in inverse proportion. The constant K_R might, on the analogy with the conduction of heat in solids, be called the *radiative diffusivity*. Even allowing for the fact that the value of 1.3×10^3 is likely to be rather high, it is very much in excess of the corresponding coefficient for the molecular conduction of heat, whose value is about 0.16 in the same units.

The derivation of equation (6) is based on the assumption that there is a complete layer of air whose water-vapour content is equivalent to 0.3 mm of precipitable water, above and below the level across which the flux of radiation is to be stated. It cannot therefore be applied without further consideration to heights above the ground of less than l as defined above.

If, instead of Hettner's coefficients of absorption for water vapour, we adopt the corresponding coefficients yielded by Weber and Randall's observations, it will no longer be possible to assume that the W-radiation is completely absorbed in a layer containing 0.3 mm of precipitable water. Some higher value of the water-vapour content, of the order of four times Simpson's value, will have to be adopted, if we allow, as was not done above, that the coefficients of absorption may be doubled in order to compensate for the diffuseness of atmospheric radiation. (See § 93.) The layer containing 1.2 mm of precipitable water will have a depth of the order of 300 metres, even under surface conditions, while K_R in equations (9), (10) and (11) above is also increased four-fold. The method outlined above for the computation of the radiative transfer of heat will then be a rougher approximation to the truth, since the total W-radiation from a layer 300 metres thick will usually differ from the corresponding radiation at the mean temperature of the layer.

§ 86. The diurnal variation of temperature as affected by radiation

We have seen in equation (9) above that the transfer of heat by radiation is analogous to the conduction of heat in a solid, but with a much larger coefficient of radiative diffusivity substituted for the usual coefficient of conduction.

The theory as given above is not strictly applicable down to the surface. Let us as a first approximation neglect this limitation, and consider how the diurnal variation of temperature will vary with height, if the transfer of heat is entirely by radiation in accordance with equation (9).

Let the temperature at the earth's surface be given by a pure diurnal sine-curve

$$T = T_0 + A \sin qt \qquad \text{......(12).}$$

The solution of equation (8) which agrees with this relation at the ground is

$$T = T_0 - \beta z + A e^{-bz} \sin (qt - bz) \qquad \text{......(13),}$$

where b is a constant given by

$$b^2 = q/2K_R \quad \text{and} \quad q = 2\pi/(24 \times 60 \times 60).$$

The term βz is included to allow for the mean lapse-rate β from the ground upwards.

Equation (13) gives the diurnal variation of temperature at any height z. The amplitude $A e^{-bz}$ falls off exponentially, and the time of maximum at height z occurs at bz/q seconds later than at the surface.

The observations on the Eiffel Tower may be compared with the results shown in equation (13). Taking $K_R = 650$, corresponding to a vapour-pressure of 10 mb, we find $b = 0.000236$. At the top of the Eiffel Tower

$$z = 300 \text{ metres} = 3 \times 10^4 \text{ cm}, \quad \text{and} \quad bz = 7.08.$$

Hence the ratio of the diurnal range at the top and bottom of the Tower should be e^{-7} or about 10^{-3}. If we take $K_R = 2600$, $b^2 = q/2K$, $bz = 7.08 \div 2 = 3.5$. The ratio of the diurnal range of temperature at the top and bottom of the Eiffel Tower will then be

$$e^{-3.5} = 10^{-1.5} = 1/30 \text{ approximately.}$$

The observed ratio is between 1/3 and 2/3, according to the time of year. We therefore conclude that radiation is only of slight importance in the spread of heat upward to any considerable distance above the ground, as has indeed been seen in § 85. As will be seen later in Chapter XIII, turbulence is a far more effective agent in effecting the transfer of heat upward through the atmosphere.

At very small heights above the surface of the ground, turbulence is unable to develop effectively, and the transfer of heat is there mainly by radiation. Hence on sunny days, with a large amount of incoming radiation, the surface heat is transferred only slowly up to small heights above the ground, and the result is the formation of very large lapse-rates in the immediate neighbourhood of the earth's surface. Many efforts have been made to study these large lapse-rates in detail, but the instrumental difficulties are serious. The theoretical difficulties are equally serious. Efforts to give a detailed theory have been made by Malurkar and Ramdas,* and by Brunt,† but neither of these efforts is wholly successful, owing to the difficulty of dealing with the discontinuous nature of the phenomena near the surface of the earth.

---

* *Indian Journ. Phys.* **6**, pt 6, 1932.   † *Proc. Roy. Soc.* A, **130**, 1930 p. 98.

§ 87. *Limitations of the discussion of §§ 85 and 86*

The difficulties in the way of a theoretical discussion of radiative phenomena are threefold: (*a*) In the first place, since the absorption is mainly due to water-vapour, it is important to be able to specify the distribution of water-vapour with height. This cannot be done with any degree of accuracy, since the distribution is irregular at any instant, and is subject to wide variations from one hour to the next. (*b*) In the second place, the radiation which has a general upward (or downward) component of motion is not a parallel beam, but is made up of an infinite number of beams distributed through all possible directions in space, whose intensities follow no simple law. In a first approximation it is necessary to simplify the treatment to a parallel beam, though it is not necessary to assume that the beam is normal to the earth's surface. If the direction of the parallel beam makes an angle ψ with the vertical, equations (1) and (2), p. 129, only require modification by substitution of $d\tau \sec \psi$ for $d\tau$, or, what comes to the same thing, the substitution of $k \sec \psi$ for k. (*c*) In the third place, the coefficient of absorption k is a complex function of the wave-length, and is not capable of representation by any simple mathematical law.

The third of these difficulties led to the development given above in § 85, following the lines adopted by Simpson. It is realised that the representation of the transport of heat by radiation as strictly analogous to the transport of heat by conduction in a solid is only a first approximation. It fails to take account of the parts of the inward and outward streams of radiation which are of such wave-lengths as to pass through the atmosphere with only very slight absorption. But referring back to equation (9), p. 133, we see that the coefficient K_R aims at taking account only of those parts of the streams of radiation which can produce changes of temperature, and the neglect of the wave-lengths which are slightly absorbed is therefore of no practical importance. Roberts[*] has computed the outward stream of radiation through the atmosphere with certain simplifying assumptions, and representing the stream by a formula

$$A_1 \frac{\partial T}{\partial z} + A_2 \frac{\partial^2 T}{\partial z^2} + A_3 \frac{\partial^3 T}{\partial z^3} + \dots,$$

has assumed that the coefficient A_1 is the radiative diffusivity. He finds, as is indeed obvious *a priori*, that the contributions to A_1, due to wave-lengths which are only very slightly absorbed, are very great, and concludes from this that the transport of heat by radiation is not analogous to transport by conduction in a solid. The treatment of § 85 above assumes that the transport of heat by radiation is in part analogous to conduction, and is in part a steady stream passing through the atmosphere without change. The real physical difference between the two points of view is not great, but the method of § 85 has the advantage that it leads to equations which can be applied to concrete problems.

A much more serious limitation of our method is the increase of l, the depth of the layer containing 0·3 mm of precipitable water, at heights where the

[*] *Proc. Roy. Soc. Edin.* **50**, 1930, p. 225.

vapour-pressure is small. It is not then permissible to treat the layer as having uniform temperature equal to the mean temperature of the layer.

If Weber and Randall's absorption coefficients are used, a much larger value must be adopted for the depth of the layer which absorbs practically completely all incident radiation, and, particularly at heights where the vapour pressure is small, the treatment of the layer as having a uniform temperature equal to the mean temperature of the layer will not be permissible.

§ 88. *The interchange of long-wave radiation between the atmosphere and the earth's surface. Nocturnal radiation with clear skies*

It has been stated above in § 83 that the total amount of long-wave radiation from the atmosphere to the earth's surface is of the order of $\frac{1}{2}$ to $\frac{3}{4}$ of the black-body radiation at the temperature of the earth's surface. Methods have been developed for measuring the net loss of heat from the earth's surface by long waves. Some of these methods are only applicable at night, in the absence of direct and diffuse solar radiation, but other methods, notably that of W. H. Dines,[*] can be used both by day and by night. The observations are, however, easier to carry out at night, and hence the net exchange of long-wave radiation between the atmosphere and the earth's surface has been more extensively studied by night than by day. The net outward flow of long waves by day cannot differ as to order of magnitude from that which occurs at night, since the water-vapour content of the atmosphere is not subject to any considerable diurnal variation.

Simpson showed (*loc. cit.*) that the net outward flow of long waves from the earth's surface can be placed between certain limits on the basis of his assumptions as to the nature of the water-vapour spectrum. He compared the actual observations of W. H. Dines and L. H. G. Dines[†] with the conclusions he was able to draw on theoretical grounds, and found close agreement, the mean observed values falling well within the prescribed limits.

Ångström[‡] found that the formula

$$R/\sigma T^4 = A - B \; 10^{-\gamma e} \qquad\qquad \ldots\ldots(14)$$

represented with fair accuracy the relation between R, the total long-wave radiation downward from the atmosphere, and the total black-body radiation at the surface temperature T; A, B, and γ are constants, for which the latest value given by Ångström are 0·25, 0·32 and 0·052, when e is in millibars.

Brunt[§] showed that many series of mean values of R could be represented with close fidelity by a formula

$$R/\sigma T^4 = a + b \sqrt{e} \qquad\qquad \ldots\ldots(15).$$

High coefficients of correlation were found between $R/\sigma T^4$ and $\sqrt{e}$ in monthly mean observations at Benson, both when the sky was divided into 6 zones, and when the whole hemisphere was considered. Other series of grouped means yielded high coefficients of correlation between $R/\sigma T^4$ and $\sqrt{e}$, and

* *Met. Mag.* Oct. 1920.   † *Mem. R. Met. Soc.* **2**, No. 11.
‡ *Geog. Ann.* **2**, 1920, p. 253.   § *Q.J. Roy. Met. Soc.* **58**, 1932, p. 389.

though series of individual observations showed considerable deviations from the regression line of equation (15), the series of observations examined were rather more closely represented by equation (15) than by equation (15), although the latter contains one more adjustable constant.

But while it is possible to find values of the coefficients a and b in equation (14) above which will yield a good fit to any series of observations of radiation of long waves from the earth to the atmosphere, the values of a and b differ for different series of observations. The following table shows the values of a and b, and the correlation coefficient between the ratio $R/\sigma T^4$ and $\sqrt{e}$, for different series of observations, reduced to refer to vapour-pressures in millibars in each case.

	a	b	r	Range of e in mb
Dines (Benson)	0·52	0·065	0·97	7–14
Asklöf (Upsala)	0·43	0·082	0·83	2–8
Ångström (Bassour)	0·48	0·058	0·73	5–15
Boutaric (France)	0·60	0·042	—	3–11
Robitsch (Lindenberg)	0·34	0·110	1·0	3–22
Ramanathan and Desai (Poona)	0·47	0·061	0·92	8–18
Mean	0·44	0·080	0·92	

The precise reason for the wide differences in the values of a and b is by no means obvious, but it is probable that it is mainly due to instrumental causes. The methods of observation adopted by the different observers whose results are represented in the above table were widely different, and the constants of some of these instruments were possibly not known with certainty. It is also probable that all the instruments do not treat radiation from all zones of the sky in the same manner.

Ramanathan and Ramdas[*] have given a theoretical derivation of equation (14), using mean values of the absorption coefficients of water-vapour within certain ranges of wave-length, taken from Weber and Randall's experimental values, a method whose validity is doubtful. Pekeris[†] has given a proof that the radiation from a large mass of gas should be proportional to the square root of its mass, and has justified equation (15) on that basis. Baur and Phillips,[‡] using mean values of Hettner's coefficients with a small correction for pressure for specified ranges of wave-length, evaluated the downward radiation from the atmosphere in different latitudes, and showed that the results were represented by

$$R/\sigma T^4 = 0·60 + 0·042 \sqrt{e} \qquad \ldots\ldots(16)$$

with a coefficient of correlation 0·99, the values of a and b being almost exactly equal to those derived for Boutaric's observations, as shown in the table above.

Neither formula can be strictly justified theoretically, and the choice between the two rests on convenience in use. When $e = 0$, $R/\sigma T^4 = A - B$, or a, according to the equation used. This would appear to indicate that a dry atmosphere would radiate an amount of energy equal to about one-half the

---

[*] *Proc. Ind. Acad. Sci.* **1**, 1935, p. 822.   [†] *Astroph. Journ.* **79**, 1934, p. 441.
[‡] *Gerlands Beitr.* **45**, 1935, p. 82.

black-body radiation at the surface temperature, or considerably more than is radiated by the water-vapour in the atmosphere. All the available evidence points to the falsity of such a supposition. Equations (14) and (15) are best regarded as empirical formulae, which, within the limits of vapour-pressure which occur in the atmosphere, give a reasonable representation of the average amount of radiation in given conditions. No such formula can give accurate representation of individual observations, which frequently show wide variations of R for the same values of T and e.

The results derived by Baur and Phillips, referred to above, and the close agreement between the values of the sky radiation computed by Simpson on the basis of Hettner's coefficients, and the observed sky radiation at Benson, appear to indicate that the use of Hettner's coefficients of absorption lead to values which do not deviate widely from the truth. Computations of R similar to those of Baur and Phillips, using Weber and Randall's coefficients of absorption, would be useful for comparison with observed values of sky radiation. The smaller the assumed coefficients of absorption, the deeper the layer from which the water-vapour radiation reaching the ground emanates, and the less its amount. This method would afford a useful check on any observations of the water-vapour spectrum, but would require that the comparison with observed atmospheric radiation be limited strictly to those occasions when the sky is clear of cloud and haze of any kind.

§ 89. *Nocturnal radiation: conditions within the surface layers of the ground*

Since the earth's surface radiates effectively as a black body, the amount of heat which it sends into the atmosphere is independent of the nature of the ground. The amount of energy which the earth's surface gains from the atmosphere depends upon the distribution of temperature and humidity in the lower layers of the atmosphere. Thus the net outward radiation at night from the earth's surface depends only on atmospheric conditions, and on the temperature of the earth's surface. But the temperature changes which are produced in the earth by a given amount of radiation will depend upon the readiness with which the loss of heat from the surface is compensated by conduction of heat upwards from lower layers of the earth. The precise nature of this dependence can be readily ascertained.

We shall neglect the effect of the transfer of heat by conduction from the air to the earth's surface, as there is no obvious method of allowing for this. Further, we are here concerned mainly with conditions during night inversions and in such conditions the transfer of heat is from air to ground, tending to prevent the fall of temperature of the ground. Thus any formula which we derive for estimating the drop of temperature at the ground will give an overestimate, or a limit to the maximum fall of temperature at the surface.

The surface layers of the earth will be assumed to have a specific conductivity of heat κ_1. Values of κ_1 have been derived by Johnson and Davies[*] and by

---

[*] *Q.J. Roy. Met. Soc.* **53**, 1927, p. 45. See also Johnson, *M.O. Geoph. Mem.* No. 46.

Wright,* the mean of their determinations being $\kappa_1 = 4 \cdot 7 \times 10^{-3}$ in C.G.S. units. The equation of conduction of heat through the ground is then

$$\frac{\partial T}{\partial t} = \kappa_1 \frac{\partial^2 T}{\partial z^2} \qquad \ldots\ldots(17),$$

where z is the depth measured positive downwards.

If the net loss of heat by radiation from the ground to the atmosphere is R_N, then

$$R_N = \kappa_1 \rho_1 c_1 \left(\frac{\partial T}{\partial z}\right)_{z=0} \qquad \ldots\ldots(18),$$

where ρ_1 and c_1 are respectively the density and specific heat of the ground. For the loss of heat from the surface outwards is equal to the flow of heat from below to the surface. Further, we have seen that to a first approximation the net radiation R_N remains constant throughout the night, since

$$R_N = \sigma T^4 \left(1 - a - b \sqrt{e}\right) \qquad \ldots\ldots(19).$$

The vapour-pressure has only a slight diurnal variation, and since the fall of temperature during the night is only a relatively small fraction of T, we may assume as a first approximation that R_N is constant. By equation (18) this involves a constant value of $\partial T/\partial z$ at $z = 0$.

Differentiating equation (17) once with respect to z and replacing $\partial T/\partial z$ by S, we find

$$\frac{\partial S}{\partial t} = \kappa_1 \frac{\partial^2 S}{\partial z^2} \qquad \ldots\ldots(20).$$

This now has to be solved with the boundary condition

$$S = \frac{R_N}{\rho_1 c_1 \kappa_1} \text{ at } z = 0.$$

The appropriate solution is given in any textbook on the subject, in the form

$$S = \frac{2}{\sqrt{\pi}} \frac{R_N}{\rho_1 c_1 \kappa_1} \int_{\frac{z}{2\sqrt{\kappa_1 t}}}^{\infty} e^{-u^2} du \qquad \ldots\ldots(21).$$

Integrating equation (21) we readily find

$$T = T_1 - \frac{2}{\sqrt{\pi}} \frac{R_N}{\rho_1 c_1 \kappa_1} \left\{ \sqrt{\kappa_1 t} \cdot e^{-\frac{z^2}{4\kappa_1 t}} - z \int_{\frac{z}{2\sqrt{\kappa_1 t}}}^{\infty} e^{-u^2} du \right\} \qquad \ldots\ldots(22).$$

Thus the temperature at $z = 0$ is given by

$$T = T_1 - \frac{2}{\sqrt{\pi}} \frac{R_N}{\rho_1 c_1 \sqrt{\kappa_1}} \sqrt{t} \qquad \ldots\ldots(23).$$

In this equation R_N is to be measured in gramme-calories per second and t in seconds. It is readily seen that we can take R_N to indicate the net radiation in gramme-calories per minute, and t the time in hours.

Before we consider the application of this result to any practical problem, we must consider the conditions under which it has been derived. Referring back to equation (21) we find that at $z = 0$, S or $\partial T/\partial z$ is constant at all times

* Mem. R. Met. Soc. 4, No. 31.

except at $t=0$, when it is zero, through the lower limit of the integral becoming infinite. Thus the solutions represented by equations (22) and (23) correspond to the case when the outward radiation is initially zero at $t=0$, and then instantaneously jumps to the value R_N. Such a change in physical conditions cannot happen in nature, but something approaching it probably happens at sunset, on a clear evening. It is known that the incoming short-wave radiation from sun and sky becomes negligibly small just before sunset, and that just before this happens there is a very rapid fall in the amount of the incoming radiation. Thus while the change from a net flow of radiation inwards through a complete balance, to a net flow outwards at the rate R_N, does not take place instantaneously, it takes only a very short time to be accomplished, and so equation (23) should give a reasonable approximation to the fall of temperature of the ground during a clear night, t being measured from sunset, except for small values of t, when the infinite rate of fall of temperature given by equation (23) at $t=0$ would in practice be replaced by a steep but finite rate of fall.

A nearer approximation to the conditions pre-supposed in the deduction of the above equations occurs on occasions when the sky, after being overcast with low cloud, suddenly clears. It is shown later in § 90 that there is only a very slight loss of heat from the ground when the sky is overcast with low cloud, and so the clearing of the cloud brings on suddenly the net loss R_N appropriate to a clear night. On such occasions, therefore, we may use equation (23) with fair confidence to forecast the fall of temperature in a time t after the disappearance of the cloud, assuming that there is no wind and therefore no convection of heat to or from the ground.

There do not appear to be any published observations available which make it possible to test the hypotheses adopted above with regard to the lapse-rate of temperature in the ground, and some doubt inevitably remains as to the possible effects of the conditions prevailing during the daytime upon the temperature changes during the night.

Returning to equation (23) we find that the fall of temperature after sunset is (approximately)

$$\frac{2}{\sqrt{\pi}}\frac{R_N}{\rho_1 c_1 \sqrt{\kappa_1}}\sqrt{t}=\frac{2}{\sqrt{\pi}}\frac{\sigma T^4\left(1-a-b\sqrt{e}\right)}{\rho_1 c_1 \sqrt{\kappa_1}}\sqrt{t}\qquad\ldots\ldots(24).$$

The minimum temperature of the night is derived by making t here equal to the number of hours from sunset to sunrise. For dry soil we may adopt as approximate values

$$\rho_1=2\cdot5,\qquad c_1=0\cdot2,\qquad \kappa_1=4\cdot7\times10^{-3}.$$

Johnson gives (loc. cit. Fig. 15 a) the mean curve for clear days and nights in June, and taking the mean value of T at $287°$ A, and assuming the factor $1-a-b\sqrt{e}$ to be equal to its mean value at Benson in June, $0\cdot225$, we find that the fall of temperature in the seven hours from sunset to sunrise should be

$$\frac{2\times0\cdot56\times0\cdot225\times2\cdot65}{\sqrt{\pi}\times2\cdot5\times0\cdot2\times\sqrt{4\cdot7\times10^{-3}}}\text{ or } 11°\text{ C.}$$

The mean fall of temperature shown by Johnson's curve is about $9°$ C, which is in very close agreement with the theoretically derived value, but is lower than the latter, as we should expect, since the effect of wind will be to diminish the fall of surface temperature by the downward transfer of heat through the inversion to the ground.

The figures for the specific heat and diffusivity of dry soil used above are Patten's estimates.* If we assume that the corresponding values given by Patten for light soil containing 20 per cent of water will be appropriate for December, giving a value of $\rho_1 c_1 \sqrt{\kappa_1}$ five times greater than for dry soil, the mean fall of temperature between sunset and sunrise during clear nights in December as calculated from equation (23) is $3 \cdot 3^\circ$ C, as compared with the value $2 \cdot 9^\circ$ C which Johnson found by observation. Both the June and December values are mean temperatures as observed in a Stevenson screen, and only a rough mean value of the factor $(1 - a - b \sqrt{e})$ has been used in each case. The agreement in each of the two months is sufficiently close to show that the main physical factors have been taken into account.

The chief difficulty in the way of using formula (23) for forecasting the night minimum temperature lies in the uncertainty of the value of the coefficient $\rho_1 c_1 \sqrt{\kappa_1}$. It has already been mentioned that the addition of 20 per cent of water to dry soil increases this factor five-fold.

Fig. 38 shows two examples of thermograms for clear nights: (*a*) on Salisbury Plain; and (*b*) at a station in Malaya, at a height of 1090 feet above sea level. The agreement of the form of the night portion of these curves with the parabolic form predicted on theoretical grounds is in each case very close.

Equation (23) indicates that the amplitude of the variation of temperature is inversely proportional to $\rho_1 c_1 \sqrt{\kappa_1}$. This factor therefore indicates the nature of the dependence of the variations of temperature upon the type of ground surface. For snow rough values of the constants are

$$\rho_1 = 0 \cdot 1, \quad c_1 = 0 \cdot 5, \quad \rho_1 c_1 \kappa_1 = 0 \cdot 00025.$$

For snow $\rho_1 c_1 \sqrt{\kappa_1} = 0 \cdot 004$; for soil $\rho_1 c_1 \sqrt{\kappa_1} = 0 \cdot 035$.

The temperature variation over the surface of snow may thus be many times that over a surface of soil, if radiation and conduction are the only factors which are active. This accords with experience, the lowest temperatures which are measured being associated with a snow covering over the ground. While snow causes extremely low temperatures at its surface it is at the same time a protection to anything which it covers, in that it conducts heat so slowly that any object covered by it retains its own heat. During any prolonged spell of clear weather with snow on the ground the temperature does not recover during the day, since 80 per cent of the solar radiation is reflected upwards at the snow surface, leaving only a small fraction to be absorbed and to become available for raising the surface temperature.

This is in accordance with the extraordinary results found by Simpson† in his discussions of the diurnal range of temperature on the Barrier in the Antarctic, where the small diurnal change in daily insolation corresponding to the sun oscillating between 10° and 35° above the horizon produced a diurnal variation of temperature with an average amplitude of 20° F.

* Washington, D. C., *Bull. U.S. Dept. Agri. Bur. Soils*, No. 59, 1909.
† *British Antarctic Expedition*, 1910–13, *Meteorology*, **1**, p. 56.

It has been assumed that the transfer of heat in the soil can be represented by a single uniform set of constants ρ_1, c_1 and κ_1, which have the same values at all depths below the surface. This is known to be a very rough approximation to the facts (*vide* Wright, *loc. cit.*), but it appears to be a sufficiently good approximation to enable us to derive the form of the curve of change of the surface temperature.

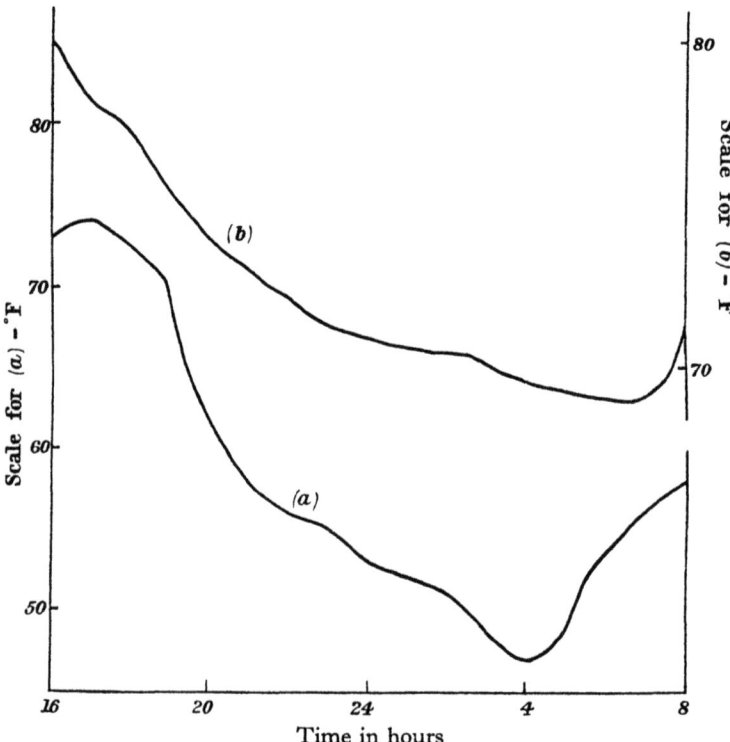

Fig. 38. Diurnal variation of temperature on clear nights

§ 90. *Nocturnal radiation with cloudy skies*

When the sky is overcast with cloud the radiation conditions are enormously modified, on account of the fact that the cloud sheet absorbs and radiates effectively as a black body. The effect can be readily understood from a consideration of the rate of change of the intensity of the beam of radiation of one wave-length λ, in the vertical direction.

In § 84 it was found that the upward beam of radiation of wave-length λ is equal to the black-body radiation E_λ at the ground, and is greater than E_λ at heights above the ground; while the downward beam is zero at the outer limit of the atmosphere, and in general is less than E_λ at all levels in a clear atmosphere. This statement is obviously true in a cloudless atmosphere, if the temperature decreases steadily with height. A cloud sheet whose thickness is at least of the order of say 50 metres can be assumed to radiate like a black body (*vide* § 79). The presence of a sheet of cloud therefore modifies very seriously the flow of radiation. At the base of the cloud the upward moving radiation of all wave-lengths is completely absorbed, and the downward moving radiation leaving the cloud base is equal to the complete black-body radiation at the temperature of the cloud base. Similarly at the top of the cloud, all the downward moving radiation is completely absorbed, and the upward moving radiation leaving the top of the cloud is equal to the complete black-body radiation at the temperature of the top of the cloud.

There are three important features to be observed in relation to the presence of a sheet of cloud:

(*a*) *At the top of the cloud.* The downward moving radiation of long wave-length will be only *W*-radiation, and at each level less (to a varying extent in different wave-lengths) than the corresponding black-body radiation. It will be completely lacking in radiation of wave-lengths corresponding to the transparent bands. The long-wave radiation leaving the top of the cloud will therefore exceed the downward beam from above, in all wave-lengths, so that there is a net loss of heat at the upper surface. The net loss will be greater, the lower the cloud. The upper surface of the cloud acts in effect as an elevated ground surface.

(*b*) *At the base of the cloud.* The above argument is simply reversed, the upward beam coming into the cloud from below exceeding the black-body radiation in all wave-lengths. There is therefore a net gain of heat by the lower surface of the cloud, and the net gain will obviously increase as the height of the cloud increases.

(*c*) *At the ground.* The chief effect of the interposition of a cloud sheet between the ground and outer space is to put into the downward beam radiation of the wave-lengths corresponding to the transparent band, which are not present when the sky is clear. The net loss of heat by the ground is therefore considerably diminished, approximately by an amount equal to the radiation from the cloud in the wave-lengths of the transparent band, and the net nocturnal radiation from the ground will bear a rough proportionality to the height of the cloud.

Ångström[*] has given the results of observations of (*a*) net radiation received from the ground and the intervening atmosphere, by a horizontal black body, and (*b*) the net radiation lost by a horizontal black body to the atmosphere above, the observations being carried out in a free balloon. The figures for the first of these were as follows:

Height (metres)	975	1350	2000	3000	4000
Net gain of radiation	0·015	0·000	0·008	0·019	0·036

The net radiation is in gramme-calories per cm² per minute. The observations were commenced at about 10 p.m. on July 3, 1922, so that there was in all probability an inversion at the ground, and the radiation leaving the ground would be initially less than the equivalent black-body radiation at the tem-

[*] *Beitr. Phys. fr. Atmos.* **14**, h. 1–2, 1928.

perature of air a short distance above the ground. Thus the conditions were not similar to those we have supposed, and the net radiation measured is at all levels less than would have been measured if there had been no inversion. From 1350 m upwards there is, however, a very rough proportionality of net radiation to excess of height above 1350 m. Similar results were obtained during the night of June 5–6, 1923, when, probably with an inversion at the ground, negative values of net gain of radiation were observed up to nearly 1200 m, above which the values were positive, becoming 0·036 at 2000 m and 0·055 at 2750 m, the greatest height at which measurements were obtained. On this occasion there was a varying amount of cloud about 2600 m to 2850 m. Above this level observations were made of the net loss of radiation of a black body to the atmosphere above it. This value was 0·226 at 3300 m, diminishing to 0·216 at 4400 m. We shall make further use of these results later in discussing the effects of radiation on cloud sheets.

Ångström's observations are sufficient to show that our conclusion of a rough proportionality between $(\Sigma A - \Sigma E)$ and the height is justified as to order of magnitude; and we therefore conclude that in a similar way $(\Sigma E - \Sigma B)$ is proportional to depth below the cloud level. The special form of the last relation which we require is that the net loss of radiation from the ground is roughly proportional to the height of the cloud. Asklöf's observations* made in the period March to June, 1918, gave the following average values, for different types of cloud covering:

Cloud	Net radiation	Average height, Upsala
Nb., St., or St.-Cu.	0·023	1·5 km
A.-Cu.	0·039	2·8 ,,
Ci.-St.	0·135	6·4 ,,
Clear sky	0·169	—

Thus the net loss of heat from the ground with a sky covered with high cloud is almost as great as with clear skies, while when the sky is covered with low cloud, the net loss of heat from the ground is only about one-seventh the value observed for clear skies. Consequently the fall of temperature during the night with a sky overcast with high cloud should be nearly as great as with a clear sky, but should be only of the order of one-seventh of this amount, if the sky is overcast with low cloud. N. K. Johnson† reproduces a chart showing temperature variations during the night March 15–16, 1923, in which the drop of temperature during the night is shown to be about 1° F, the sky being overcast with St.-Cu. and Nb. This is of about the right order of magnitude to agree with Asklöf's observations.

Ångström has also suggested that the net loss of radiation from the ground may be represented by

$$R_m = (1 - 0·09m) R_0 \qquad \ldots\ldots(25),$$

where R_m is the observed net loss of radiation from the ground when m tenths of the sky are covered with cloud, R_0 being the net loss of radiation to a clear sky in the same circumstances of temperature and humidity; but it is clear that such a rule cannot deal with clouds of different height, and a different

* *Geog. Annaler, Stockholm,* **2**, 1920, p. 253.
† M.O., *Geoph. Mem.* No. 46.

formula, with a constant appropriate to the height, should be employed for each type of cloud.

During the daytime the net loss of heat from the ground by long-wave radiation is practically the same as it would be at night with the same atmospheric conditions of temperature and humidity. But the surface of the earth gains heat by absorption of the short-wave radiation in the direct beam of sunlight, or in the diffuse solar radiation from the sky. When the sky is overcast, the amount of diffuse radiation which reaches the earth's surface is very much diminished, but it is still sufficient to outweigh the net loss by long-wave radiation from the ground. W. H. Dines and L. H. G. Dines* give tables of mean monthly values of diffuse radiation from overcast skies which can be directly compared with their mean values of net loss by long waves, and for each month the gain from short waves outweighs the net loss by long waves. We therefore conclude that the surface temperature should rise during the day, even with overcast skies, but the rise will be small when the sky is overcast with a thick sheet of low cloud. The record for an overcast day and night, March 15–16, 1923, reproduced by Johnson (*loc. cit.*) indicates a rise of only about 2° F during the day on the 15th.

§ 91. *Conditions which favour the nocturnal cooling of the ground*

In the derivation of equation (23), which represents the nocturnal cooling of the ground by radiation alone, the effect of convection of heat to the ground was assumed negligible. If the wind remains strong and turbulent during the night the cooling of the surface of the earth by radiation is impeded, as fresh air is then being continually brought into contact with the ground, and a flow of heat from the air to the ground prevents the temperature of the ground from falling rapidly. The outward stream of radiation from the ground is not appreciably affected by this, and the effect of the convection is to distribute the loss of heat through a much deeper layer of air, so that an inversion of temperature cannot form.

The conditions which favour the formation of inversions at the ground are:

(*a*) clear skies, or only high cloud;

(*b*) absence of wind;

(*c*) low vapour-pressure in the atmosphere;

(*d*) low thermal conductivity and specific heat of the ground.

These four conditions are in fact well known from experience. Their effects are to some extent stated mathematically in some of the preceding sections. As to the last of these conditions, it has been shown that the higher value of the coefficient $\rho_1 c_1 \sqrt{\kappa_1}$ in winter when the ground is waterlogged accounts for the smaller diurnal variation of temperature in winter than in summer, and far outweighs the effect of lower vapour-pressure and longer nights in winter.

Our lack of knowledge of the physical constants of soils, and in particular, of the precise nature of the effect of a grass covering, makes it at present im-

* *Mem. R. Met. Soc.* **2**, No. 11.

possible to use the equation given above to forecast night minimum temperatures. There is a considerable literature dealing with the what is known as the "frost problem", which is concerned with the forecasting of night minimum temperatures, and a vast number of empirical formulae have been proposed by various writers. Details of such formulae will be found in papers by Angström,* Ellison,† Warren Smith,‡ and Pick.§

* *Geog. Ann.* 1920, h. 1; 1921, h. 3; 1923, h. 4.
† *Monthly Weather Review*, **51**, 1928, p. 485.
‡ *Ibid.* Supplement No. 16.
§ *Met. Mag.*, various papers during 1927 to 1930.

CHAPTER VII

RADIATIVE EQUILIBRIUM AND THE STRATOSPHERE

§ 92. *Statement of the problem*

THE most striking problem to which we have to apply considerations of radiative processes is the existence of an upper region of the atmosphere, known as the stratosphere, within which the temperature is practically constant, or even increases slowly with height, while below it, in the troposphere, there is a definite lapse-rate whose mean value is about 6° C per km at all heights and in all latitudes. The change from a definite fall of temperature with height to constant temperature is abrupt, and any satisfactory theory must be capable of explaining both the sudden change of lapse-rate at the tropopause, and the occurrence of constant or increasing temperature above that level.

It was first suggested by Gold* and Humphreys† that the conditions within the stratosphere can be explained as due to a complete balance between radiation and absorption, leading to a constancy of temperature, dynamical disturbances of the temperature at these levels being regarded as negligible on account of the stability of the air at these levels preventing convection or turbulence on any considerable scale. Both of these writers regarded the stratosphere as controlled by the long-wave radiation coming up from the lower levels of the atmosphere. Strictly speaking, account should be taken of the incoming beam of solar radiation of short wave-length, as well as of the outgoing beam of radiation of long wave-length. But as the incoming beam passes down through the atmosphere with relatively little attenuation, we shall leave out of consideration, at least in a first survey of the question, the possible absorption of the direct solar beam in the atmosphere, on its way downward. In effect this means that we shall regard the earth's surface as the source of heat of the atmosphere.

Humphreys points out that the temperature changes in the stratosphere between winter and summer are such as to justify the assumption that the conditions in the upper air are determined by radiation from the lower atmosphere, since the ratios of the absolute temperatures in winter and summer are approximately the same at all heights. He therefore proceeds to consider the effect of radiation from the earth and the lower atmosphere, leaving out of consideration the absorption of incoming solar radiation in the stratosphere. The radiation which comes up from below is effectively (says Humphreys) the radiation from a black body at a temperature of 259° A, and he therefore replaces the earth and the lower atmosphere by a black body at this temperature, sending out radiation of the appropriate amount. The curvature of the earth may be neglected in the discussion of the narrow range of heights

* *Proc. Roy. Soc.* A, **82**, 1909, p. 43.   † *Astroph. Journ.* **29**, 1909, p. 14.

10-2

with which we are concerned, so that we may replace the spherical black body by an infinite plane black body.

Now consider two such surfaces, parallel and directly facing each other, at a finite distance apart, each having a temperature T_2. If any object is placed midway between the planes, in what is effectively an enclosed space (since the lateral boundaries of the surfaces are very distant), it will take up the temperature T_2. If now one of the parallel planes is removed, the state of the object, which is kept in the same position, is akin to the state of the stratosphere and the problem is now to find the temperature at which the object will be in equilibrium. Since it now absorbs and radiates only half the amount it absorbed and radiated when it was placed between the two planes, it follows that if the radiation is proportional to T^n, its final temperature T_1 will be given by the equation

$$2T_1{}^n = T_2{}^n, \quad \text{or} \quad T_1 = 2^{-\frac{1}{n}} T_2.$$

There is therefore a minimum temperature T_1 below which the temperature of the upper atmosphere may not fall on account of the radiation from the lower atmosphere. Assuming the radiation of the upper atmosphere to be roughly continuous, and therefore to follow approximately Stefan's law, we find for the temperature of the stratosphere $T_1 = 2^{-\frac{1}{4}}T_2$; and since the effective temperature T_2 is about 259° A, it follows that $T_1 = 218°$ A. This value is in close agreement with the mean observed temperature of the stratosphere, which is estimated as 219° A for Europe.

At first sight Humphreys' proposed solution of the problem appears complete and accurate, but on closer examination it appears to offer several points for criticism. What Humphreys *proves* is that an isothermal state is possible in the first system he postulates, i.e. when the atmosphere is irradiated in both directions by radiations of its own temperature, and there is no net flow of radiation in any direction. The second system which he considers, with one radiating plane only, is irradiated from one side only by radiation of a higher temperature than its own, having a net outflow of heat upwards, and it is not possible to deduce what will happen in the second system from what happens in the first system. Thus the extension of the argument of constancy of temperature in an enclosed space to the unenclosed space outside one plane is not justifiable. Humphreys does not prove that the amount of radiation which will reach the element of mass exposed to one plane only will be the same at all points, and it is in fact clear that this cannot be so in practice in the system he considers. The thermodynamical system which he considers is also open to objection. He regards the stratosphere, whose temperature is everywhere T_1, as irradiated from below by a beam of intensity $2T_1{}^n$, which is equivalent to the net outward flow of radiation necessary to compensate the inward flow of solar radiation. The stratosphere absorbs an amount C of this beam, and re-radiates it, $\frac{1}{2}C$ outward to space, and $\frac{1}{2}C$ downward to the earth. The net outward flow is thus diminished by $\frac{1}{2}C$, which the earth gains at the expense of the outward beam. A further objection will be seen later when we come to discuss the problem mathematically.

§ 93. *The difficulties in the way of mathematical treatment*

It has been shown in § 78 that in the present state of our knowledge of the spectrum of water-vapour it is not possible to apply the laboratory observations of the spectrum to the discussion of the absorption and radiation of energy in the stratosphere, with any degree of confidence. There is a further difficulty in the fact that radiation in the atmosphere cannot be regarded as consisting of parallel beams directed vertically upward and downward, but is diffuse. Emden* has shown that in an isothermal atmosphere of infinite mass the total absorptive power of a layer for diffuse black-body radiation is twice the absorptive power for normal incidence of a parallel beam. Emden found that no simple law could be derived for other atmospheres, and he finally concluded that a reasonably close approximation to actual phenomena is obtained by treating the radiation as a parallel beam, and allowing for the diffuse nature of the radiation by doubling the coefficient of absorption appropriate to normal incidence. This is the method followed by Schwarzschild† and by Milne,‡ in their work on stellar atmospheres. It should be noted, however, that Roberts§ found that the diffuse radiation from the earth's surface followed approximately the simple law of absorption, having $1\frac{1}{2}$ times the coefficient of absorption of a parallel beam.

Gold‖ has discussed the radiative processes in the atmosphere, allowing for the diffuseness of the radiation and assuming, in the later stages, $T^4 \propto p$, as an approximation to the theoretical law $T^{3.5} \propto p$, which holds for an adiabatic atmosphere. Gold's assumption is only strictly correct for a tri-atomic atmosphere.

So long as it was considered possible to treat the radiation from water-vapour as grey radiation, it was possible to write down the total radiation from a layer of optical thickness $d\tau$ as $\sigma T^4 k\, d\tau$. If we abandon this assumption, and assume the radiation to be strictly selective, it is no longer possible to express the total radiation in such a simple form.

All these difficulties make any rigorous treatment impossible, and the brief discussion which follows is given rather as an indication of what is likely to be true, than as a valid treatment of the subject.

§ 94. *Emden's solution for grey radiation*

Emden assumed that the stratosphere is in radiative equilibrium. Each element of air radiates in unit time as much energy as it absorbs from the incoming and outgoing beams of long-wave radiation. Emden assumed that the radiation from an element of water-vapour in the atmosphere was in all wave-lengths a fixed fraction of black-body radiation, so that the coefficient of absorption was the same in all wave-lengths, but for the moment we shall

---

* *Sitzber. Math.-Phys. Kl. Akad. Wiss. München*, 1913, h. 1.
† *Göttingen Nachrichten*, 1906, p. 41.
‡ *Phil. Mag.* **144**, 1922, p. 871.
§ *Proc. Roy. Soc. Edin.* **50**, 1930, p. 225.
‖ *Proc. Roy. Soc.* A, **82**, 1909, p. 43.

make no assumption of this kind. Equations (1) and (2) of § 84 are

$$\frac{dA_\lambda}{d\tau} = k_\lambda (A_\lambda - E_\lambda) \qquad \ldots\ldots(1),$$

$$\frac{dB_\lambda}{d\tau} = k_\lambda (E_\lambda - B_\lambda) \qquad \ldots\ldots(2).$$

It is obviously unnecessary to assume complete grey radiation. We need only assume that in all wave-lengths in which it radiates, water-vapour has the same coefficient of absorption k, and that in other wave-lengths it is transparent, with $k=0$.

Summing (1) and (2) for all wave-lengths, we have

$$\frac{dA}{d\tau} = k (A - E) \qquad \ldots\ldots(3),$$

$$\frac{dB}{d\tau} = k (E - B) \qquad \ldots\ldots(4),$$

where $A = \Sigma A_\lambda$, $B = \Sigma B_\lambda$, $E = \Sigma E_\lambda$. Thus A, B, and E denote the total upward beam, the total downward beam, and the total selective radiation at temperature T, respectively.

The condition of radiative equilibrium is

$$k (A + B) = 2kE \qquad \ldots\ldots(5).$$

From (3), (4) and (5), we find

$$\frac{d}{d\tau} (A - B) = k (A + B - 2E) = 0 \qquad \ldots\ldots(6).$$

Hence $\qquad\qquad A - B = \text{const.} = A_0 \qquad \ldots\ldots(7),$

where A_0 is the amount of long-wave radiation leaving the atmosphere at its outer boundary. Adding (3) and (4), we find,

$$\frac{d}{d\tau} (A + B) = k (A - B) = kA_0,$$

$$2E = A + B = kA_0\tau + \text{const.} = A_0(1 + k\tau) \qquad \ldots\ldots(8).$$

From (7) and (8)

$$\left.\begin{aligned} A &= A_0 (1 + \tfrac{1}{2}k\tau) \\ B &= \tfrac{1}{2}A_0 k\tau \\ E &= \tfrac{1}{2}A_0 (1 + k\tau) \end{aligned}\right\} \qquad \ldots\ldots(9).$$

Now E will be a function of T, which with black-body or grey radiation is equal to σT^4, but which with selective radiation is not of necessity expressible as a simple function of T. But it will be a function of T which will increase as T increases, since by Planck's law the radiation in each wave-length increases as T increases (see equation (1), p. 105). We can therefore conclude from the last of equations (9) that as τ decreases with height, T will decrease with height, but that since the magnitude of τ is everywhere small in the stratosphere, T will decrease slowly with height.

To a first approximation, the temperature in the stratosphere is determined by the amount of the radiation coming into the stratosphere from below. The less the amount of this radiation, the lower will be the value of A_0, of the temperature, and of the lapse-rate of temperature. Thus it should be in the cloudiest regions that the effects of other factors would show most clearly the tendency to produce an increase of temperature with height in the stratosphere.

We are therefore led to the conclusion that the temperature of the stratosphere cannot remain constant with increasing height, still less increase with increasing height, if its temperature is controlled by long-wave radiation from the lower atmosphere. There is, however, one further result of interest which can be derived from the analysis of the above paragraph. Equation (9) states that the value of E, and therefore the temperature, will increase as the net flow of radiation outwards increases. It is seen from equation (6) of § 85 above that the flow of radiation outwards at a particular temperature is proportional to the lapse-rate, and inversely proportional to the vapour-pressure. Though it was not stressed at the time, this result is true for each separate wave-length of W-radiation, and is thus not dependent on any obviously doubtful assumption. Gold* has suggested that the condensation of water-vapour governs very closely the actual lapse-rate in the troposphere. It has been shown that the saturated lapse-rate at a given temperature diminishes with the pressure, and is therefore greater when that temperature occurs at low levels in the atmosphere than when it occurs at high levels.† If then Gold's view (which is partly confirmed by the examination of the tephigrams reproduced by Sir Napier Shaw, in the *Manual of Meteorology*, **3**) is accepted, we should expect the lapse-rate at a given temperature to be greater when that temperature occurs at low levels than when it occurs at high levels. Thus the net outward flow across a given isothermal surface is greater where that surface is low in the atmosphere than where it is high, and so the net outward flow of heat across the upper troposphere is greater in high latitudes than at the equator. But equation (9) above indicates that the temperature in the upper stratosphere increases with the net outward flow, and so the temperature in the stratosphere should increase from the equator towards higher latitudes. This argument might be affected by an increase of vapour-pressure at a given temperature in the same sense as the increase of lapse-rate, in such a way as to diminish the ratio of the two, and lead to a decrease with increasing latitude of the net outflow of radiation across the upper troposphere. This is a very improbable occurrence, since the relative humidity is more or less uniform through the whole atmosphere, and certainly shows no definite tendency, so far as observations are available to test it, to increase from equator to pole.

It is thus established that while it is not possible to explain any vertical distribution of temperature, except a slow decrease of temperature with height, by the effect of long-wave radiation from the lower atmosphere, it is possible to explain rationally the latitude variation of temperature in the stratosphere by this effect, taking into account the controlling effect of the condensation of water-vapour on the actual lapse-rate in the troposphere.

* M.O., *Geoph. Mem.* No. 5.   † See fig. 18, p. 66.

The next logical step is to consider what other factors can exert a control over the temperature distribution in the stratosphere, the most obvious factor being the absorption of ultra-violet radiation by ozone.

§ 95. *The effect of absorption of ultra-violet radiation by ozone on the temperature distribution in the stratosphere*

Fabry and Buisson* have given an account of the methods by which was established the now generally accepted view that the sharp termination of the solar spectrum at 0.2890μ is due to absorption by ozone at levels of 10 to 40 km above the earth's surface. Even when photographed from heights of 9 km above the earth's surface, the spectrum showed no extension beyond this point. Laboratory observations have shown that ozone has an intense absorption band extending from 0.23μ to 0.32μ, with a pronounced maximum at 0.255μ. This is known as the Hartley band, from its discoverer. Ozone also has other absorption bands, one in the red and orange, with a maximum at 0.61μ, another in the infra-red, centred at 4.8μ, and a third at 9 to 11μ. Absorption by the Hartley band accounts for the sharp limitation of the solar spectrum on the ultra-violet side, and it is estimated that from 4 to 6 per cent of the total energy in the incoming solar beam is taken from it by this means, though the total amount of ozone is so small that if concentrated in a uniform layer at normal temperature and pressure, it would only make a layer 2–3 mm in thickness.

The details of the methods of observation of the quantity and amount of the ozone are given in papers by Dobson and others† (see footnote, p. 20). The mean height is about 22 km, and the amount of ozone measured may vary 25 per cent above or below the mean value. The ozone amount shows a marked annual variation, with a maximum at the spring equinox, and a minimum at the autumnal equinox; it also shows a marked variation with latitude, the amount increasing from equator to pole. Perhaps the most remarkable feature of all is the occurrence of very high values in the western quadrants of depressions, and of very low values in the western quadrants of anticyclones. It appears that, so far as is at present known, the ozone amount in the rear of depressions is greater than the normal amount anywhere over the earth's surface. Why the ozone should show this association with depressions is as yet a mystery, whose solution may reverse many of the ideas at present accepted concerning the dynamics of the atmosphere.

The energy absorbed by the ozone is in the ultra-violet, but at normal atmospheric temperatures the ozone could only re-radiate this energy in the relatively weak band between 9 and 11μ. It could only maintain radiative equilibrium, and re-radiate at the same rate as it absorbs, if it had a very much higher temperature than would allow it to persist without dissociating. We are therefore forced to conclude that the re-radiation is performed mainly by another constituent of the atmosphere, most probably water-vapour.

---

* *Mem. Sci. Phys., Acad. Sci. Paris*, Fasc. XI, 1930.

† See also *Nature*, **127**, 1931, p. 668; *Proc. Roy. Soc.* A, **145**, 1934, p. 416; *Q.J. Roy. Met. Soc.* **62**, 1936, Supplement; Penndorf, *Veröff. Geoph. Inst. Leipzig*, Bd. VIII, h. 4.

Since Dobson and his collaborators showed that the centre of gravity of the ozone in the upper atmosphere is to be found at a height of about 22 km, several computations have been made of the temperature distribution in the layer in which the ozone occurs. Gowan* made a series of computations assuming a variety of conditions as to temperature and humidity up to 55 km above the ground. They all agreed in giving a region of maximum temperature at a height of about 50 km, and a fairly rapid increase of temperature with height above 30 km. Penndorf† derived similar results, the region of maximum heating shown by his computations occurring just below 50 km. Penndorf gave a definite warning that the actual numerical results were not to be taken too seriously, but added that it was certain that an increase of temperature with height in the stratosphere could be explained by the absorption of ultra-violet rays by ozone.

There are several uncertainties involved in such work as that of Gowan and Penndorf. In the first place, the amount of water-vapour in the stratosphere is uncertain. If it comes up from the troposphere, mixed with dry air churned up with it, then the proportion of water-vapour to dry air cannot exceed the value at the tropopause. As a result, relative humidity must decrease with height in the stratosphere, being directly proportional to the pressure if the stratosphere is isothermal. In the second place, the form of the water-vapour spectrum, in such conditions as hold at heights of 10 to 40 km above the ground, is not yet known, nor is the precise form of the ozone spectrum in these conditions known. The distribution of ozone appears to be the most definitely known of all the variables concerned. In the third place, the spectral distribution of the radiation coming up into the stratosphere is also uncertain, though it must obviously contain a large amount of energy in the bands around 10μ, to which water-vapour is transparent. The main band in which ozone radiates and absorbs long waves is almost centred over this transparent band, which must therefore be an important factor in the heat economy of the ozone layers.

We may however regard the distribution of temperature in the higher stratosphere as explained qualitatively, but not quantitatively, by the presence of ozone. Fig. 12, p. 18, indicates that the increase of temperature with height in the stratosphere is most marked in the tropics. Now ozone has a well-marked latitude variation, the total ozone being least in the tropics, and greatest in high latitudes. It also has a clearly defined annual variation, with a maximum in spring, and a minimum in autumn, the range of variation being small in the tropics, and greatest in high latitudes, in either hemisphere. A summary of these facts has been given by Dobson and Meathem,‡ who also showed that the amount of ozone in the region above the north-westerly winds in the rear of a depression is greater than the normal amount of ozone in any latitude, while slightly westward of the centre of an anticyclone the amount of ozone is well below the average value appropriate to the latitude. Meathem§ showed that the centre of maximum ozone is usually some 200 miles from the centre

---

* *Q.J. Roy. Met. Soc.* **62**, 1936, Supplement, p. 34.
† *Ibid.* p. 37; also *Veröff. Geoph. Inst. Leipzig*, Bd. VIII, h. 4.
‡ *Q.J. Roy. Met. Soc.* **60**, 1934, p. 265.   § *Ibid.* **63**, 1937, p. 289.

of a depression, in a direction between south and south-west. He also found a high correlation between total ozone and the potential temperature at 18 km. No satisfactory explanation of these facts has yet been evolved.

A theory of the formation of ozone has been given by Chapman,* who attributes the formation of ozone to the dissociation of oxygen by ultra-violet radiation, and the subsequent dissociation of ozone to the same cause. Chapman's theory leads to the result that above about 80 km the ratio of atomic oxygen to molecular oxygen should increase without limit, and that above 60 km the ratio of ozone to molecular oxygen must decrease indefinitely with increasing height. In the extreme outer layers of the atmosphere oxygen tends to exist in the atomic rather than in the molecular form.

We may refer in passing to two zones of ionisation in the high atmosphere, one at 100 km, and known as the Kennelly-Heaviside layer, and another at 200 km, known as the Appleton layer. The ionisation in both may probably be ascribed to ultra-violet radiation, the lower being due to the ionisation of nitrogen molecules, and the upper to the ionisation of atomic oxygen.†

Chapman‡ has discussed the absorption by oxygen in the range from 0.13μ to 0.175μ. It appears that within this range oxygen absorbs even more strongly than ozone does in the Hartley band from 0.23μ to 0.32μ. Chapman states that this absorption takes place at heights well above 100 km.

§ 96. *The heat balance of the atmosphere*

When allowance has been made for the earth's albedo the mean inflow of solar radiation to the earth's surface, averaged over all latitudes, and over day and night, is 0.278 gramme-calorie per cm² per minute. Since the earth is not getting any hotter or colder on the average, this amount must be re-radiated out to space. Before leaving the subject of radiation we must make some effort to form a picture of the processes by which the earth and its atmosphere balance the account between incoming and outgoing radiation.

Fig. 39, reproduced from Simpson's paper,§ shows (curve I) the effective incoming solar radiation in hundredths of gramme-calories per cm² per minute in different latitudes, with a mean value of 0.278. Here allowance has been made for the reflexion of solar radiation from the atmosphere, the earth's surface, and the appropriate cloud amount for each latitude. Simpson has computed for each latitude the total amount of outgoing radiation with clear and cloudy skies. To follow the details of his computation we have to bear in mind the three categories into which he divided radiation at terrestrial and atmospheric temperatures:

(*a*) Wave-lengths in which water-vapour is transparent to radiation— $8\frac{1}{2}\mu$ to 11μ.

* *Mem. R. Met. Soc.* **3**, No. 26, 1930.
† *Vide* Chapman, Bakerian Lecture, *Proc. Roy. Soc.* A, **132**, 1931, p. 353.
‡ *Q.J. Roy. Met. Soc.* **60**, 1934, p. 127.
§ *Mem. R. Met. Soc.* **3**, No. 21, 1928.

(b) Wave-lengths in which water-vapour amounting to 0·3 mm of precipitable water will completely absorb all radiation—$5\frac{1}{2}\mu$ to 7μ, and $> 14\mu$.

(c) Wave-lengths in which water-vapour absorption is intermediate between (a) and (b)—7μ to $8\frac{1}{2}\mu$, and 11μ to 14μ.

With a clear sky the radiation which gets out into space will belong in part to each of these categories. The radiation of category (a) will have originated at the earth's surface, and its amount can therefore be computed from the known mean temperatures of the earth's surface in different latitudes. Simpson assumes that the stratosphere contains at least 0·3 mm of precipitable water, and therefore the radiation of category (b) which gets out into space will have originated in the stratosphere, and its amount can therefore be computed from radiation tables such as are given by Simpson (loc. cit.). Of the radiation of category (c) which gets out into space we cannot make any equally direct statement. It will be intermediate in amount between the radiation of this kind at the temperature of the stratosphere and that at the temperature of the earth's surface, and we shall not be far out if we follow Simpson and take a value roughly intermediate between the two. Fig. 40, reproduced from Simpson's memoir, shows the method which he followed, assuming ground temperature 280° A, and stratosphere temperature 218° A. In the horizontally hatched region from $8\frac{1}{2}\mu$ to 11μ the radiation to space is that at the surface temperature. From $5\frac{1}{2}\mu$ to 7μ, and above 14μ, the radiation to space is equal to the vertically hatched area which measures the radiation at 218° A. In the intermediate regions, from 7μ to $8\frac{1}{2}\mu$, and

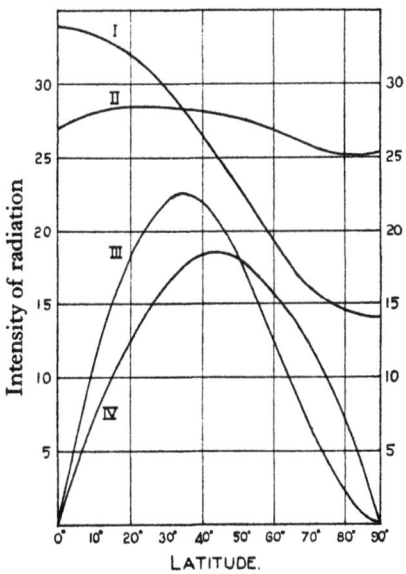

Fig. 39. Solar and terrestrial radiation.

from 11μ to 14μ, the amount of radiation is the diagonally hatched area. The upper boundaries of these areas cannot be specified with absolute accuracy, but as Simpson says, "...there is very little latitude; for any reasonable curve will divide the area $JDEK$ into two nearly equal parts". Thus a reasonable approximation is obtained by taking the area $DPQK$ equal to the mean of $JPQK$ and $DPQE$. On this basis, Simpson was able to compute the total radiation passing out into space through cloudless skies.

When the sky is overcast, since for all practical purposes a cloud sheet radiates like a black body, the only difference in the computation is that the surface of the earth is replaced by the upper surface of the cloud, and the temperature of the earth's surface has to be replaced by the temperature of the cloud, which Simpson assumes to be 261° A in all latitudes.

Simpson takes no account of the variation of cloud amount with latitude in his second memoir, but uses a mean cloud amount of 5/10 for all latitudes, so that his final figure for the total radiation from the whole earth is the mean of the figures for clear and cloudy skies. The values so obtained give a mean outgoing radiation of 0·271 cal. per cm² per min., which is within a few per cent of the mean incoming radiation estimated by using Aldrich's value of the albedo. A correction can be made so as to bring the mean incoming and the mean outgoing radiation into agreement, and the values plotted in curve II of fig. 39 are the values so corrected.

A comparison of curves I and II in fig. 39 shows that from the equator to about latitude 35° the outgoing radiation is less than the incoming radiation, while in higher latitudes, the outgoing radiation exceeds the incoming radia-

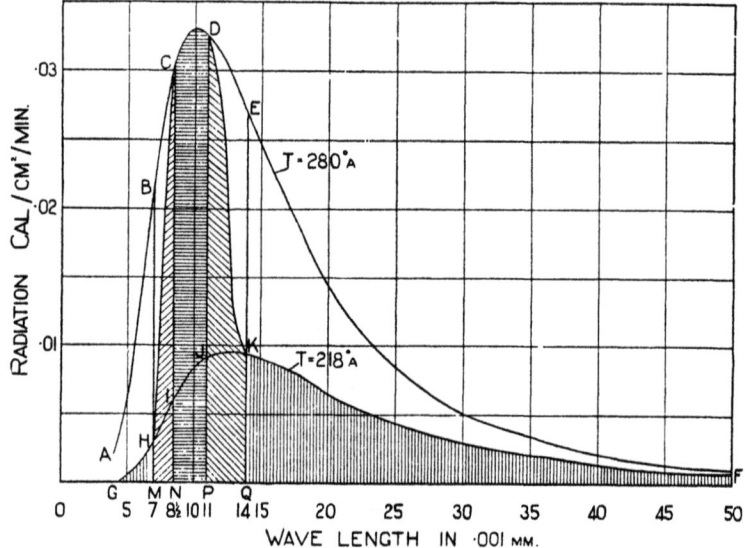

Fig. 40. The heat balance of the atmosphere.

tion. There is therefore a horizontal transfer of heat from low to high latitudes, through the agency of the general circulation. In fig. 39 curve III gives the total horizontal heat transfer across circles of latitude, and curve IV the horizontal heat transfer per cm of circle of latitude.

There is one feature of Simpson's work which should be emphasised, that while he makes use of Hettner's observations to show that 0·3 mm of precipitable water as vapour will suffice to absorb practically completely all radiation of certain wave-lengths, and is transparent to certain other wave-lengths, he makes no use of the actual values of the coefficients of absorption evaluated from Hettner's observations, which we have represented graphically in fig. 33 above. This is a point which should not be overlooked, as it lends great weight to the close agreement between the theoretical deductions and the observations which we find in Simpson's "Further Studies".

The outstanding result of this research by Simpson is that the amount of energy radiated from the earth and its atmosphere to space varies surprisingly little with latitude. It emphasises the heat transporting function of the general circulation, and puts this on a quantitative basis. In his third memoir, Simpson* has worked out similar data, using this time monthly, and not annual, means of the temperatures.

An earlier study of the same question by Mügge,† based on the assumption that the radiation from water-vapour could be treated as grey radiation, yielded the result that at the poles twice as much energy is radiated into space, per unit area of the earth's surface, as is radiated at the equator. Simpson‡ repeated Mügge's work, with a different series of assumptions as to the distribution of temperature and water-vapour, and found that while the intensity of outgoing radiation was strikingly constant between 50° N and 50° S, towards the poles there was a slight falling off in the intensity of the outgoing radiation. Simpson's investigation of the balance between incoming and outgoing radiation, which has been briefly described above, was the first in which allowance was made for the existence of transparent bands in the water-vapour spectrum.

Baur and Phillips§ made similar investigations of the radiation to space, of the downward radiation from the atmosphere, and of the transport of water-vapour by the general circulation. They assumed different mean values of the coefficients of absorption of water-vapour in the ranges 4μ to $8\frac{1}{2}\mu$, $8\frac{1}{2}\mu$ to 13μ, and 13μ to 50μ, and in effect used integrals of equations (1) and (2), p. 129 above, in the form

$$A = A_0 e^{k\tau} - e^{k\tau} \int E e^{-k\tau} d\tau,$$

$$B = B_0 e^{-k\tau} + e^{-k\tau} \int E e^{k\tau} d\tau.$$

The evaluation of the integrals on the right-hand side of these two equations is only possible when T and τ can be expressed as functions of the same variable. Baur and Phillips adopted an exponential law for the variation of vapour-pressure with height. The rather tedious integrations, which they handled with great skill, yielded a curve which differed from Curve II, fig. 39, obtained by Simpson, in yielding rather higher values for the outgoing radiation in low latitudes, compensated by rather lower values than Simpson's in high latitudes. To some extent the differences are to be accounted for by the different coefficients of absorption used in the two investigations, but the larger value which Baur and Phillips obtained for the outgoing radiation in low latitudes can be largely explained by the fact that they assumed much lower values for the water-vapour content of the stratosphere in low latitudes, so that they ascribed the outgoing atmospheric radiation to a lower, and therefore warmer, radiating layer than was assumed by Simpson.

At the present time it is not possible to pronounce a final judgment on the value of such work as that of Simpson and of Baur and Phillips, to which reference has been made above, in view of the considerations raised in § 77 above with regard to the present state of our knowledge of the water-vapour spectrum.

The computations referred to above were made with a view to estimating the intensity of the outgoing radiation in different latitudes. There is another

* *Mem. R. Met. Soc.* **3**, No. 23, 1929.
‡ *Mem. R. Met. Soc.* **2**, No. 16, 1927.
† *Zeit. für Geophysik*, **2**, 1926, p. 63.
§ *Gerlands Beiträge*, **45**, 1935, p. 82.

aspect of the problem which merits consideration. If the mean distribution of temperature and water-vapour, and the mean flux of radiant energy, be known or computed for different heights, it is possible to compute the rate of change of temperature, if any, at each level.* On any view of the structure of the spectrum of water-vapour, it is difficult to escape the conclusion that with a clear sky, the radiation loss from the middle troposphere must lead to a steady fall of temperature. The bearing of such changes of temperature on dynamical and thermodynamical processes in the atmosphere must be of fundamental importance.

§ 97. *The vertical transport of heat in the atmosphere*

In Chapter VI, § 85, it was shown that the net outflow of heat by radiation is proportional to $T\dfrac{dE}{dT}\dfrac{1}{e}\dfrac{dT}{dz}$, where E is the total W-radiation at temperature T. It was shown that in the lower atmosphere at ordinary temperatures, the fraction of the net outflow of energy represented by the net outflow of W-radiation is very small. Where the vapour-pressure is small, as in very high latitudes, and still more, at great elevations above the earth's surface, the thickness of the layer containing 0·3 mm of precipitable water is so great that it is no longer strictly permissible to assume that the radiation from such a layer is equal to the radiation from a layer of equal thickness whose temperature is the mean temperature of the actual layer. Nevertheless, it must still be a first approximation to the radiation from such a layer, and we shall assume that it is at least sufficiently near to the truth to give an idea of the order of magnitude of the net outflow from the atmosphere.

An idea of the change with height of the net outward flux of heat by W-radiation can be readily got by taking say the July observations of temperature for the British Isles given in Table 1, p. 19. Unfortunately it is not possible to assign a mean vapour-pressure to each of these heights, since the observations of humidity are not as reliable as those of temperature. We shall assume a mean relative humidity of 60 per cent at all heights, so that the vapour-pressure is 60 per cent of the saturation vapour-pressure at the corresponding temperature. The following table shows the computation of the net outflow of heat at each level:

	T	e_s	e	$dE/dT \times 10^3$	$dT/dz \times 10^4$	Net flux
Ground	289	18·2	10·9	3·4	0·6	$7·5 \times 10^{-4}$
2 km	278	8·7	5·2	3·1	0·5	$11·5 \times 10^{-4}$
4 km	267	3·7	2·2	2·9	0·6	$29 \ \times 10^{-4}$
6 km	255	1·26	0·76	2·6	0·7	$85 \ \times 10^{-4}$
9 km	234	0·14	0·09	2·2	0·75	$384 \ \times 10^{-4}$
10 km	226	0·056	0·034	2·1	0·6	$1170 \ \times 10^{-4}$

The net outflow of heat by W-radiation is $\dfrac{139}{e} T \dfrac{dE}{dT}\dfrac{dT}{dz}$. The total outflow from the ground of W-radiation is about 0·31 gramme-calorie at the temperature assumed above, and so the net flow which we have computed in the

* Mügge and Möller, *Met. Zeit.* **49**, 1932, p. 95.

table amounts at low levels to only a small fraction of the total, but becomes at higher levels an increasingly great portion of the total flow. But as the flow of heat outward increases steadily with height, it is clear that either the upper troposphere must be continually cooling, or the main transport of heat in the troposphere is carried out by some other mechanism than radiation. The mechanism is obviously convection, which takes large quantities of heat upwards to middle levels of the atmosphere. In low latitudes, in particular, the increase of the vapour-pressure e would more than counterbalance the increase of $T (dE/dT)$, and the water-vapour in the atmosphere acts as a blanket on the outward flow of radiation, and by keeping the energy at low levels, gives the general circulation of the atmosphere time to carry it away to high latitudes. The increase in the outflow of heat with height is slow at first, but increases rapidly above 4 km, showing that the effects of condensation are greater in the middle levels of the atmosphere, as we might have anticipated from the greater frequency of clouds at those levels.

The absence of any condensation in the atmosphere would involve a direct cooling of the middle and upper troposphere, leading to a diminution of the lapse-rate in the upper troposphere, and an increase in the middle troposphere, until some kind of equilibrium was attained. There should therefore be a very definite difference in the vertical temperature distribution in regions of condensation and those of continued clear weather, if other factors could be allowed for. Note that this certainly is not true of the difference between the cyclone and anticyclone, according to W. H. Dines. The anticyclone is warmer in the upper troposphere relative to the cyclone that it is at the ground, and it is certain that dynamical factors are more important than radiational factors in the anticyclone.

The above discussion is admittedly based on rather rough approximations, and all the values of net flux given in the last column may require correction by a factor due to the smallness of the thickness of the layer which acts effectively as a black body. It does however give an idea of the way in which the effects of radiation may have a bearing on the genesis of instability in the atmosphere.*

* See also Brunt, *Scientia*, Jan. 1937.

CHAPTER VIII

THE GENERAL EQUATIONS OF MOTION

§ 98. *The general equations of motion in spherical polar co-ordinates*

THE equations of motion in three dimensions can be most readily derived by an extension of the equations for two dimensions. In two dimensions the position of a point P (fig. 41 a) may be fixed by means of the length r and the

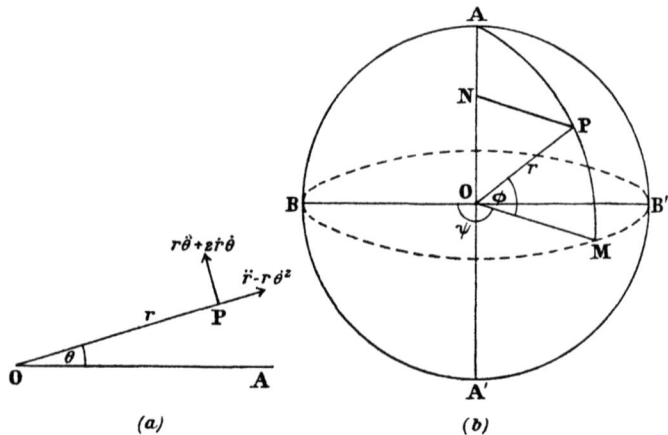

Fig. 41. Polar co-ordinates.

angle AOP, or θ, O being a fixed point, and OA a fixed direction. The accelerations of the point P are

$$\ddot{r} - r\dot{\theta}^2 \text{ along } OP$$

and

$$r\ddot{\theta} + 2\dot{r}\dot{\theta} \quad \text{or} \quad \frac{1}{r}\frac{d}{dt}(r^2\dot{\theta}) \qquad \qquad \text{......(1)}$$

at right angles to OP, where the dots represent differentiation with respect to time t.

For the case of motion relative to the earth we adopt a system of spherical co-ordinates. In fig. 41 b let O be the centre of the earth, and AOA' the axis of rotation of the earth. The equatorial plane is shown in the figure by a dotted circle BMB'. We require to fix the position of the point P. Let the sphere whose centre is at O and radius OP cut the equatorial plane in the circle BMB', and the axis of the earth at the points A, A'. Through P draw the great circle $APMA'$ cutting the equatorial section in M. This great circle is the meridian through P. Let the angle between its plane and the plane $ABA'B'$ fixed in space be ψ. Also let the angle POM be denoted by ϕ. Then

ϕ is the geocentric latitude of P, and r, ϕ and ψ are the spherical co-ordinates of P. If λ be the longitude of P, taken as positive when measured to East, then

$$\psi = \lambda + \omega \qquad \qquad \dots\dots(2),$$

where ω is the angular velocity of rotation of the earth. For ψ is the rate of rotation in space of the meridian plane through P, and λ is its rate of rotation relative to the standard meridian plane fixed in the earth, and ω is the rate of rotation in space of the standard meridian plane.

In the diagram (fig. $41\,b$) PN is drawn perpendicular to AA', and all the points P, A, A', B, B' and M are on the sphere whose centre is O, and whose radius is r. The motion of P is made up of

(i) a motion in the meridian plane APA', and

(ii) the effect of the rotation of this plane about the axis AA', or in other words, that part of the motion of P due to the rotation of PN about N.

Each of these effects is derivable directly from tne polar equations for two dimensions. From (i), P has accelerations $\ddot{r} - r\dot{\phi}^2$ along OP and $r\ddot{\phi} + 2\dot{r}\dot{\phi}$ directed towards North along the tangent at P to the meridian PA. From (ii),

P has accelerations $-r\cos\phi\,.\,\dot{\psi}^2$ along NP and $r\cos\phi\,.\,\ddot{\psi} + 2\dot{\psi}\dfrac{d}{dt}(r\cos\phi)$

along the tangent to the parallel of latitude drawn at P towards East.

Note that under (ii) we are only concerned with the accelerations which involve the rotational terms in ψ, and the radial term is therefore not included. Collecting the terms under (i) and (ii), and resolving the acceleration along NP into two components, directed along OP and to North respectively, we find the accelerations of P along three rectangular axes to be as follows:

Along OP, $\qquad\qquad\qquad \ddot{r} - r\dot{\phi}^2 - r\cos^2\phi\,.\,\dot{\psi}^2 \qquad\qquad \dots\dots(3).$

Perpendicular to OP and to East,

$$r\cos\phi\,.\,\ddot{\psi} + 2\dot{\psi}\,(\dot{r}\cos\phi - r\sin\phi\,.\,\dot{\phi}) \qquad\qquad \dots\dots(4).$$

Perpendicular to OP and to North,

$$r\ddot{\phi} + 2\dot{r}\dot{\phi} + r\sin\phi\cos\phi\,.\,\dot{\psi}^2 \qquad\qquad \dots\dots(5).$$

These expressions give the accelerations in three directions of any body whose co-ordinates are r, ϕ, ψ. The accelerations are relative to a system of axes fixed in space. To obtain the motion relative to the earth, we need only substitute

$$\left.\begin{array}{c} \psi = \lambda + \omega \\ \ddot{\psi} = \ddot{\lambda} \end{array}\right\} \qquad\qquad \dots\dots(6).$$

It will be seen that ψ only enters into the accelerations in the form $\dot{\psi}$ or $\ddot{\psi}$.

Now take axes x, y and z at P, the axis of x being horizontal, and directed towards East, the axis of y being horizontal and directed towards North, and the axis of z being vertical, or along OP continued. These axes are "moving axes" whose directions change with the motion of P. Let the total forces acting on unit mass at P have components X', Y' and Z' along the axes thus specified. Then we may equate these components of force to the expressions which we have derived above for the accelerations of P. In rewriting these expressions we substitute for $\dot{\psi}$ and $\ddot{\psi}$ from (6)

$$r \cos \phi \, . \, \ddot{\lambda} + 2 \, (\dot{\lambda} + \omega) \, (\dot{r} \cos \phi - r \sin \phi \, . \, \dot{\phi}) = X' \qquad \ldots \ldots (7),$$

$$r \ddot{\phi} + 2 \dot{r} \dot{\phi} + r \sin \phi \cos \phi \, (\dot{\lambda} + \omega)^2 = Y' \qquad \ldots \ldots (8),$$

$$\ddot{r} - r \dot{\phi}^2 - r \cos^2 \phi \, (\dot{\lambda} + \omega)^2 = Z' \qquad \ldots \ldots (9).$$

These equations can also be readily derived by the use of Lagrange's equations. The reader unfamiliar with the use of Lagrange's equations may, however, find it useful to follow the above derivation of these equations from simple beginnings.

Equations (7), (8) and (9) must meet all cases of motion, including the special case of a particle at rest on the earth's surface. For a particle at rest,

$$\dot{r} = \dot{\phi} = \dot{\lambda} = \ddot{r} = \ddot{\phi} = \ddot{\lambda} = 0.$$

Substituting these values in (7), (8) and (9), we find

$$X' = 0,$$

$$Y' = r \sin \phi \cos \phi \, . \, \omega^2,$$

$$Z' = -r \cos^2 \phi \, . \, \omega^2.$$

Thus it is not possible for a particle to remain permanently at rest on a spherical earth, without the action of a force Y' whose magnitude is given by the second of these equations; it requires a force $r \sin \phi \cos \phi \, . \, \omega^2$ directed towards North to keep it in equilibrium. In the absence of such a force any body free to move on the earth's surface would move towards the equator.

It has hitherto been assumed that the earth may be regarded as a sphere, and that the radius vector OP will always represent the vertical at P. Actually the earth is not a sphere, but a spheroid, whose polar axis is slightly less than the equatorial axis. The angle between the radius vector OP and the vertical at P fixed by the normal to the geoidal surface is $700'' \sin 2\phi$. This very small angle is the difference between the geocentric and the astronomical latitudes. If now we alter our system of axes, so that the z-axis becomes the true vertical at P, and the y-axis becomes the true horizontal drawn towards North, while the x-axis remains as before, we have to rotate our original system of axes about the x-axis through an angle which even at its maximum is only slightly greater than ten minutes of arc. But there can be no component of force Y' along the new y-axis. The term $r \sin \phi \cos \phi \, . \, \omega^2$ is then in fact compensated by a component of gravitational attraction. On account of the spheroidal form of the earth, the gravitational attraction of the earth on a material point outside its own mass is not directed exactly towards the centre of the earth, and when the force of attraction is resolved along the vertical direction and the horizontal line directed to North, there is a component along the latter direction which is exactly equal to the constant $r \sin \phi \cos \phi \, . \, \omega^2$ of equation (8). The component along the true vertical at the point under consideration is combined with the constant term $-r \cos^2 \phi \, . \, \omega^2$ of equation (9), and the sum is what is usually known as g. We cannot by direct observation separate the two parts of g.

Let the external forces acting at P, other than gravitation, have components X, Y, Z along the axes as modified; the direction of the axis of z is then the true vertical. Then we may write equation (7), (8) and (9) in the form

$$r \cos \phi . \ddot{\lambda} + 2 (\dot{\lambda} + \omega) (\dot{r} \cos \phi - r \sin \phi . \dot{\phi}) = X \qquad ......(10),$$

$$r\ddot{\phi} + 2\dot{r}\dot{\phi} + r \sin \phi \cos \phi . \dot{\lambda} (\dot{\lambda} + 2\omega) = Y \qquad ......(11),$$

$$\ddot{r} - r\dot{\phi}^2 - r \cos^2 \phi . \dot{\lambda} (\dot{\lambda} + 2\omega) = Z - g \qquad ......(12).$$

The quantities X, Y, Z are related to X', Y', Z' by the relations

$$X = X' \qquad ......(13),$$

$$Y = Y' - r \cos \phi \sin \phi . \omega^2 \qquad ......(14),$$

$$Z = Z' + g + r \cos^2 \phi . \omega^2 \qquad ......(15).$$

We have made use of the fact that the earth is a spheroid in order to account for the second term on the right-hand side in equation (14), but the angle between the true vertical and the radius vector from the earth's centre is so small that we need not take the difference into account in what follows, and we may now without risk of obtaining erroneous results assume equations (10), (11) and (12) to apply to motion relative to a spherical earth.

Equation (10) can be readily integrated when $X = 0$, or the motion is under "no forces". Multiplying the equation throughout by $r \cos \phi$, it becomes

$$\frac{d}{dt} \{r^2 \cos^2 \phi (\dot{\lambda} + \omega)\} = 0 \qquad ......(16),$$

from which it follows that

$$r^2 \cos^2 \phi (\dot{\lambda} + \omega) = \text{const.} \qquad ......(17).$$

This equation states that the moment of momentum about the axis of the earth remains constant in any motion during which there is no West-East component of force. Hence if the West-East component of velocity of an element of mass is known in any one latitude, equation (17) enables us to compute it for any other latitude attained by that element, provided there is no West-East component of force acting on the element. It cannot be too strongly emphasised that the question whether a specified element of mass originally in some specified latitude will subsequently attain some other specified latitude has to be answered from other considerations. Neglect to take account of this has frequently led writers on meteorology into giving estimates of velocity in the earth's atmosphere which are far in excess of observed velocities.

§ 99. *Expression of the general equations of motion in Cartesian co-ordinates*

We have already defined a convenient set of axes of co-ordinates as follows:

> x horizontal and drawn to East,
>
> y horizontal and drawn to North,
>
> z vertical.

Let u, v, w be the component velocities along these three axes. We then have the following relations:

$$u = r \cos \phi . \dot{\lambda},$$

$$v = r\dot{\phi},$$

$$w = \dot{r}.$$

Also

$$\frac{du}{dt} = r \cos \phi . \dot{\lambda} + \dot{r} \cos \phi . \dot{\lambda} - r \sin \phi . \dot{\phi}\dot{\lambda},$$

$$\frac{dv}{dt} = r\ddot{\phi} + \dot{r}\dot{\phi},$$

$$\frac{dw}{dt} = \ddot{r}.$$

By means of these relations we can re-write equations (10), (11) and (12) in the form

$$\frac{du}{dt} - v\dot{\lambda} \sin \phi + w\dot{\lambda} \cos \phi + 2\omega \, (w \cos \phi - v \sin \phi) = X \qquad \ldots\ldots(18),$$

$$\frac{dv}{dt} + w\dot{\phi} + u\dot{\lambda} \sin \phi + 2\omega u \sin \phi \qquad\qquad = Y \qquad \ldots\ldots(19),$$

$$\frac{dw}{dt} - u\dot{\lambda} \cos \phi - v\dot{\phi} - 2\omega u \cos \phi \qquad\qquad = -g + Z \ldots\ldots(20).$$

But as the point moves over the earth, the axes of reference of equations (18), (19) and (20) rotate, and the rates of rotation are readily seen to be $\dot{\phi}$, $\dot{\lambda} \cos \phi$ and $\dot{\lambda} \sin \phi$. The first three terms on the left-hand side of these equations therefore give the accelerations along axes fixed in the earth, but instantaneously coinciding with the moving axes.* Hence when the motion is referred to axes *fixed in the earth*, the equations of motion become

$$\frac{du}{dt} + 2\omega \, (w \cos \phi - v \sin \phi) = X \qquad\qquad \ldots\ldots(21),$$

$$\frac{dv}{dt} + 2\omega u \sin \phi \qquad\qquad = Y \qquad\qquad \ldots\ldots(22),$$

$$\frac{dw}{dt} - 2\omega u \cos \phi \qquad\qquad = Z - g \qquad \ldots\ldots(23).$$

The terms in ω represent the effect of the rotation of the earth. If u is positive, or the motion has a component directed towards East, the term $2\omega u \cos \phi$ effectively diminishes g, while when u is negative, or the motion has a component towards West, the rotational term effectively increases g. Thus westward moving air should behave as though it were denser than eastward moving air of the same temperature and pressure.

The velocity u in the second term on the left-hand side of equation (23) is the horizontal velocity directed towards East. If axes other than those directed towards East and North are used, account must be taken of this in

* See, for example, Routh's *Rigid Dynamics*, **1**, p. 205.

equation (23), u being replaced by the appropriate expression for velocity to East.

Equations (21), (22) and (23) indicate that the effect of the rotation of the earth can be taken into account by adding in the equations of motion accelerations $-2\omega (w \cos \phi - v \sin \phi)$, $-2\omega u \sin \phi$, $2\omega u \cos \phi$, parallel to the axes of x, y and z respectively. Alternatively we may say that the effect of the earth's rotation is a "deviating force" whose magnitude per unit mass has the components stated above. Since the direction cosines of the earth's axis are o, $\cos \phi$, $\sin \phi$, it follows that the deviating force is perpendicular to the earth's axis. Further, when the components of the deviating force are multiplied by u, v, w, respectively, the sum of the products is zero, and hence the deviating force is perpendicular to the resultant velocity. The direction of the total deviating force on any particle is therefore parallel to the plane of the equator, and is perpendicular to the direction of motion of the particle.

If axes x, y', z' be taken, with the x-axis drawn to East as before, the y'-axis parallel to the earth's axis of rotation, and the z'-axis perpendicular to the other two, and in the plane of the circle of latitude, the velocities relative to the new set of axes are

$$u, \quad v' = v \cos \phi + w \sin \phi, \quad w' = w \cos \phi - v \sin \phi.$$

The components of the deviating acceleration along these axes are

$$-2\omega (w \cos \phi - v \sin \phi) \qquad\qquad = -2\omega w'$$

$$-2\omega u \sin \phi \cos \phi + 2\omega u \cos \phi \sin \phi = 0$$

$$2\omega u \cos^2 \phi + 2\omega u \sin^2 \phi \qquad = 2\omega u.$$

The resultant acceleration is therefore in a plane parallel to the plane of the equator, and its magnitude is 2ω times the resultant velocity in this plane.

§ 100. *The equations in hydrodynamical form*

In equations (21), (22) and (23) the terms du/dt, dv/dt, dw/dt represent the accelerations of an element of mass, or, in other words, they represent "differentiation following the motion". In the discussion of hydrodynamical and aerodynamical questions it is often more convenient to express by u, v, w the component velocities at a given point (x, y, z) at a time t. The complete form of equations (21), (22) and (23) may then be written

$$\frac{\partial u}{\partial t} + u\frac{\partial u}{\partial x} + v\frac{\partial u}{\partial y} + w\frac{\partial u}{\partial z} + 2\omega (w \cos \phi - v \sin \phi) = X \qquad \ldots\ldots(24),$$

$$\frac{\partial v}{\partial t} + u\frac{\partial v}{\partial x} + v\frac{\partial v}{\partial y} + w\frac{\partial v}{\partial z} + 2\omega u \sin \phi \qquad\qquad = Y \qquad \ldots\ldots(25),$$

$$\frac{\partial w}{\partial t} + u\frac{\partial w}{\partial x} + v\frac{\partial w}{\partial y} + w\frac{\partial w}{\partial z} - 2\omega u \cos \phi \qquad\qquad = -g + Z \ldots\ldots(26).$$

§ 101. *The special case of horizontal motion*

When the motion is purely horizontal, or $w = 0$, the equations (21) and (22) reduce to

$$\frac{du}{dt} - 2\omega v \sin \phi = X \qquad \dots\dots(27),$$

$$\frac{dv}{dt} + 2\omega u \sin \phi = Y \qquad \dots\dots(28).$$

The terms involving ω represent the effect of the rotation of the earth. These yield accelerations $2\omega v \sin \phi$ along the x-axis, and $-2\omega u \sin \phi$ along the y-axis. Thus the resultant acceleration is $2\omega \sin \phi \times$ (resultant velocity), and is directed at right angles to the motion, and to the right-hand side of the line of motion. For if, in fig. 42, OP represents the total velocity V whose components are u, v, and OQ represents the total acceleration due to the ω terms, $QN = 2\omega u \sin \phi$, $ON = 2\omega v \sin \phi$; also since $\dfrac{QN}{ON} = \dfrac{u}{v}$, the angle $QON =$ angle OPM, and therefore OQ is at right angles to OP, and its magnitude is $2\omega \sin \phi \times V$.

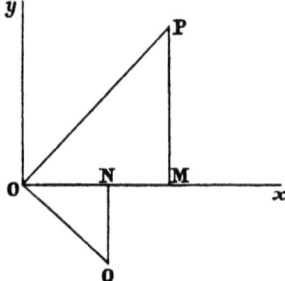

Fig. 42. The deviating force in two dimensions.

The result as stated is true for all latitudes. In the Southern hemisphere the latitude ϕ is to be regarded as negative, and the acceleration can be more usefully described as $2\omega \sin \phi \times V$ to the left of the direction of motion, and at right angles to the latter.

Thus, so far as horizontal motion is concerned, the effect of the rotation of the earth is to produce an acceleration directed at right angles to the motion, of magnitude $2\omega \sin \phi$ times the velocity. The equations (27) and (28) will always represent this fact whether the axes are drawn to East and North or not. In fact, the limitation of the direction of the axes is no longer necessary so far as these two equations are concerned. In contrast with this, it should be noted that equations (21), (22), (23) are only true for motion referred to axes drawn as specified above. For the term in $2\omega w \cos \phi$ in equation (21) represents an acceleration directed along the West-East line, of magnitude proportional to the vertical velocity; and if we use axes not directed to East and North, this acceleration must be resolved along the new axes, giving a component along both, instead of along one axis only, as in equation (21). Note also that equation (23) shows that there is a vertical acceleration proportional to the velocity along the West-East line.

§ 102. *The deviating force of the earth's rotation*

The acceleration of $2\omega \sin \phi . V$, where V is the total velocity in the horizontal plane, is frequently referred to as the "deviating force due to the earth's rotation". Since it is always directed at right angles to the direction of motion, it cannot produce any change of velocity, but only changes of direction of motion. If, therefore, a small mass is set in motion in any latitude, with a total velocity V, it will continue to move with the same velocity, though the direction of its motion will not remain constant.

Let c be the radius of curvature of the path at the point P, where the velocity is V. The acceleration towards the centre of curvature is V^2/c, and this is equal to the deviating force $2\omega V \sin \phi$. Hence $c = V/2\omega \sin \phi$ is the radius of curvature of the path, and since the speed remains unaltered, the radius of curvature c also remains unaltered, and the path is a circle. It is assumed in the above that the variation of ϕ may be neglected. The circle is known as the *circle of inertia*. The same result is readily derived by the direct use of equations (27) and (28) above. Multiplying these equations by u and v respectively, and adding the products, we find

$$\frac{d}{dt}(u^2 + v^2) = 0, \quad \text{or} \quad u^2 + v^2 = V^2 = \text{const.}$$

If initially the particle is at the origin (0, 0), and has components of velocity u_0, v_0, we may integrate equations (27) and (28) directly,

$$u - u_0 - 2\omega \sin \phi \cdot y = 0, \quad \text{or} \quad u = u_0 + 2\omega \sin \phi \cdot y,$$

$$v - v_0 - 2\omega \sin \phi \cdot x = 0, \quad \text{or} \quad v = v_0 - 2\omega \sin \phi \cdot x.$$

Hence $\qquad V^2 = u^2 + v^2 = (u_0 + 2\omega \sin \phi \cdot y)^2 + (v_0 - 2\omega \sin \phi \cdot x)^2$

or $\qquad \left(x - \frac{v_0}{2\omega \sin \phi}\right)^2 + \left(y + \frac{u_0}{2\omega \sin \phi}\right)^2 = \frac{V^2}{(2\omega \sin \phi)^2}.$

The particle therefore moves in a circle with uniform velocity V, the circle having a radius $V/2\omega \sin \phi$. In deriving the equation to this path we have neglected the variation of latitude along the path. If account is taken of the variation of ϕ, the path deviates from a circle, though only very slightly, except at the equator. For a full discussion of the form of paths of particles moving over the earth's surface under no forces, the reader is referred to a paper by F. J. W. Whipple.[*]

The radius of the circle of inertia is proportional to the velocity of projection, and inversely proportional to $\sin \phi$. For a velocity V of 10 metres per second the radius of the circle is 69 km at the poles, 90 km in latitude 50°, 138 km in latitude 30°. A mass started moving horizontally poleward with a velocity of 10 metres per second in latitude 8° would execute approximately a circle whose radius would be equal to 4° of latitude.

The fact that there is no deviating force at the equator, if the motion is purely two-dimensional, as here discussed, should be contrasted with the result stated on p. 165 at the end of § 99. Treating the motion as three-dimensional, we find that the circle of inertia should always lie in a plane parallel to the plane of the equator. Any element of air projected with a velocity of 1 metre per second in any direction parallel to the plane of the equator will then execute a circle of radius 7 km, independently of the latitude in which the motion takes place, always provided the atmosphere is adiabatic, so that hydrostatic reactions may be excluded. The time taken to execute the complete revolution in the circle of inertia is 12 sidereal hours.

* *Phil. Mag.* **33**, 1917, p. 457.

§ 103. *The order of magnitude of the terms in the equations of motion*

It is important to keep in mind an estimate of the order of magnitude of the different terms in equations (21), (22) and (23). We shall take the component velocities u and v to be both 20 metres per second, giving a total velocity of 28 metres per second. This is distinctly greater than the mean velocity of motion observed in the earth's atmosphere, and is therefore a reasonably safe guide to the terms which may be neglected.

$$g \text{ is of the order of 1000 cm/sec}^2,$$

$$2\omega u \sin \phi = 2\omega v \sin \phi = 0 \cdot 29 \sin \phi \text{ cm/sec}^2,$$

$$2\omega u \cos \phi = 0 \cdot 29 \cos \phi \text{ cm/sec}^2.$$

The vertical component of velocity, w, is usually small, and of the order of magnitude of 1 metre/sec, and seldom exceeds 5 metres/sec, except possibly in thunderstorms. Thus in general it is possible in equation (21) to neglect $2\omega w \cos \phi$ by comparison with the term $2\omega v \sin \phi$, but care may be necessary in special cases. The term $2\omega u \cos \phi$ is normally negligible by comparison with g, but it will be seen that it may be of importance at a surface of discontinuity of density or velocity (see § 121 below).

§ 104. *The equation of continuity*

Let ρ represent the density of a fluid, which we regard as a function of the co-ordinates x, y, z, and of the time t. Consider the flow into a small parallelepiped whose edges are parallel to the axes of co-ordinates, and of length dx, dy, dz. The rate of flow of mass across a face set parallel to the yz plane is $\rho u \, dy \, dz$, and the net flow per unit time out of the parallelepiped across the two faces parallel to the yz plane is

$$\frac{\partial}{\partial x}(\rho u) \, dx \, dy \, dz ;$$

similar considerations apply to the other faces, and the result is that the net flow of mass per unit time out of the parallelepiped is

$$\left\{\frac{\partial}{\partial x}(\rho u) + \frac{\partial}{\partial y}(\rho v) + \frac{\partial}{\partial z}(\rho w)\right\} dx \, dy \, dz.$$

This is equal to the rate of diminution of mass within the element of volume, and is therefore equal to

$$-\frac{\partial}{\partial t}(\rho \, dx \, dy \, dz) = -\frac{\partial \rho}{\partial t} \, dx \, dy \, dz.$$

Hence the condition that fluid is nowhere being annihilated or created within the region is

$$\frac{\partial \rho}{\partial t} + \frac{\partial}{\partial x}(\rho u) + \frac{\partial}{\partial y}(\rho v) + \frac{\partial}{\partial z}(\rho w) = 0 \qquad \text{......(29)}$$

or

$$\frac{1}{\rho}\frac{d\rho}{dt} + \frac{\partial u}{\partial x} + \frac{\partial v}{\partial y} + \frac{\partial w}{\partial z} = 0 \qquad \text{......(30),}$$

where d/dt represents total differentiation following the fluid, or

$$\frac{d}{dt} = \frac{\partial}{\partial t} + u\frac{\partial}{\partial x} + v\frac{\partial}{\partial y} + w\frac{\partial}{\partial z}.$$

In the special case of incompressible fluid

$$\frac{\partial u}{\partial x} + \frac{\partial v}{\partial y} + \frac{\partial w}{\partial z} = 0 \qquad \qquad \ldots\ldots(31).$$

Equation (29) or (30) is known as the "equation of continuity". In polar co-ordinates this equation becomes

$$\frac{\partial \rho}{\partial t} + \frac{1}{r\cos\phi}\frac{\partial}{\partial\lambda}(\rho u) + \frac{1}{r}\frac{\partial}{\partial\phi}(\rho v) - \frac{\rho v}{r}\tan\phi + \frac{\partial}{\partial r}(\rho w) + \frac{2}{r}\rho w = 0 \quad \ldots\ldots(32).$$

§ 105.   *The forces X, Y, Z*

So far no attempt has been made to specify the nature of the forces X, Y, Z. These forces were defined as the external forces acting on an element of mass of the body whose motion is being considered. They must clearly include the effects of pressure, friction and turbulence. The effect of pressure distribution can be readily introduced into our equations, and the detailed discussion of the effects of friction and turbulence will be postponed to a later chapter.

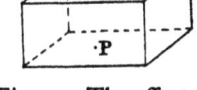

Fig. 43. The effect of pressure gradient.

In a fluid the pressure may vary from point to point, but the variations are continuous. In three-dimensional space, defined by Cartesian co-ordinates x, y, z, the pressure p will be a continuous function of x, y, z.

Take a parallelepiped with edges dx, dy, dz parallel to the axes of co-ordinates, of which the z-axis is vertical. Let P, fig. 43, the centre of the lower face of the parallelepiped, be (x, y, z), and let P', the centre of the upper face, be $(x, y, z+dz)$. If p is the pressure at P, this may be regarded as the mean pressure over the lower face. The pressure at P' will be $p + \frac{\partial p}{\partial z}dz$, and this may be regarded as the mean pressure over the upper face. The total pressure on the lower face is $p\,dx\,dy$ acting upward, and the total pressure on the upper face acting downward is $\left(p + \frac{\partial p}{\partial z}dz\right)dx\,dy$. Thus the net pressure on the element of volume along the direction of the axis of z is $-\frac{\partial p}{\partial z}dx\,dy\,dz$, acting upward. This is equivalent to a force $-\frac{\partial p}{\partial z}$ per unit volume, or $-\frac{1}{\rho}\frac{\partial p}{\partial z}$ per unit mass. Similarly the resultant pressures parallel to the axes of x and y are respectively $-\frac{\partial p}{\partial x}$, $-\frac{\partial p}{\partial y}$ per unit volume, or $-\frac{1}{\rho}\frac{\partial p}{\partial x}$, $-\frac{1}{\rho}\frac{\partial p}{\partial y}$ per unit mass. Hence the parts of the forces X, Y, Z, which are due to the variation of pressure from point to point of the fluid, may be written $-\frac{1}{\rho}\frac{\partial p}{\partial x}$, $-\frac{1}{\rho}\frac{\partial p}{\partial y}$, $-\frac{1}{\rho}\frac{\partial p}{\partial z}$. The negative signs denote that the effect of the pressure distribution is equivalent to a force tending to push the fluid towards the region of lowest pressure, as might have been inferred from general considerations. The resultant force is at right angles to the isobaric surfaces. When the isobars are drawn for particular levels, the resultant of the two terms $-\dfrac{1}{\rho}\dfrac{\partial p}{\partial x}$, $-\dfrac{1}{\rho}\dfrac{\partial p}{\partial y}$ is normal to the isobars. The name "pressure gradient" is usually given to the resultant of $-\dfrac{\partial p}{\partial x}$, $-\dfrac{\partial p}{\partial y}$.

Should pressure be given as a function of spherical polar co-ordinates, r, ϕ, λ, it is a simple exercise to show that the components of the effective pressure forces to East, North and vertically upward are

$$-\frac{1}{\rho r \cos \phi}\frac{\partial p}{\partial \lambda}, \quad -\frac{1}{\rho r}\frac{\partial p}{\partial \phi}, \quad -\frac{1}{\rho}\frac{\partial p}{\partial r}.$$

If all frictional and viscous forces may be neglected, equation (23) may be written

$$\frac{dw}{dt} - 2\omega u \cos \phi = -g - \frac{1}{\rho}\frac{\partial p}{\partial r} \qquad \ldots\ldots(33).$$

If there is no motion, all the terms on the left-hand side of this equation vanish, and we obtain the statical equation of p. 34,

$$\frac{\partial p}{\partial r} = -g\rho \qquad \ldots\ldots(34).$$

In equation (33) the term $2\omega u \cos \phi$ is usually negligible by comparison with g, but dw/dt may be considerable in very strong convection currents, and it is not then permissible to assume that the variation of pressure with height is given by equation (34).

We may sum up the results derived in the present chapter in the following equations, where X, Y, Z are the components of the forces not due to pressure and gravity:

$$\frac{du}{dt} + 2\omega (w \cos \phi - v \sin \phi) = -\frac{1}{\rho}\frac{\partial p}{\partial x} + X \qquad \ldots\ldots(35),$$

$$\frac{dv}{dt} + 2\omega u \sin \phi \qquad\qquad = -\frac{1}{\rho}\frac{\partial p}{\partial y} + Y \qquad \ldots\ldots(36),$$

$$\frac{dw}{dt} - 2\omega u \cos \phi \qquad\qquad = -g - \frac{1}{\rho}\frac{\partial p}{\partial z} + Z \qquad \ldots\ldots(37),$$

where d/dt may be replaced by

$$\frac{\partial}{\partial t} + u\frac{\partial}{\partial x} + v\frac{\partial}{\partial y} + w\frac{\partial}{\partial z}.$$

To these we add the equation of continuity in the form in which it is most commonly used:

$$\frac{1}{\rho}\frac{d\rho}{dt} + \frac{\partial u}{\partial x} + \frac{\partial v}{\partial y} + \frac{\partial w}{\partial z} = 0 \qquad \ldots\ldots(38).$$

Equations (35), (36) and (37) are referred to axes drawn to East, to North, and vertically upward respectively. In practice it is frequently convenient to use axes oriented in some other direction. Let us, for example, take a new axis of x drawn at an angle β with the East-West line, as shown in fig. 44. Then the relation between the new co-ordinates x', y' and the original co-ordinates are as shown below:

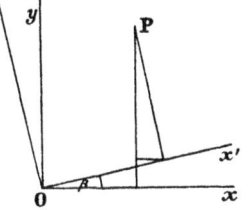

$$x = x' \cos \beta - y' \sin \beta,$$

$$y = x' \sin \beta + y' \cos \beta,$$

$$u = u' \cos \beta - v' \sin \beta,$$

$$v = u' \sin \beta + v' \cos \beta.$$

Fig. 44. Co-ordinate axes in any direction.

Substituting these values in equations (35), (36) and (37), we find the corresponding form referred to the axes x and y, with the axis of z vertical.

Only two points call for special care in the process of substitution. Equation (35) indicates that the effect of vertical motion is to give an acceleration $2\omega w \cos \phi$ along the West-East line. It will now be necessary to resolve this along the new axes of x' and y', yielding components $-2\omega v \cos \phi \sin \beta$ along the axis of y', and $2\omega w \cos \phi \cos \beta$ along the axis of x'. Similarly equation (37) indicates a vertical acceleration $2\omega \cos \phi \times$ velocity to East, which is now represented by $u' \cos \beta - v' \sin \beta$. Hence equations (35), (36) and (37) are now replaced by the following, the accents being now omitted:

$$\frac{du}{dt} + 2\omega \left(w \cos \phi \cos \beta - v \sin \phi \right) = -\frac{1}{\rho} \frac{\partial p}{\partial x} + X \qquad \ldots\ldots(39),$$

$$\frac{dv}{dt} + 2\omega \left(-w \cos \phi \sin \beta + u \sin \phi \right) = -\frac{1}{\rho} \frac{\partial p}{\partial y} + Y \qquad \ldots\ldots(40),$$

$$\frac{dw}{dt} - 2\omega \cos \phi \left(u \cos \beta - v \sin \beta \right) = -g - \frac{1}{\rho} \frac{\partial p}{\partial z} + Z \qquad \ldots\ldots(41).$$

These three equations agree with the preceding set of equations when $\beta = 0$.

§ 106. *Barotropic and baroclinic fluids*

The field of distribution of pressure in a fluid is conveniently given by a system of isobaric surfaces, which are such that on any one surface the pressure is the same at all points. If surfaces are drawn for each unit of pressure, the whole fluid considered is thereby divided into a set of isobaric sheets.

Similarly the distribution of density may be represented by a system of surfaces of equal density, or *isopycnic surfaces*, and if isopycnic surfaces are drawn for equal or unit intervals of density, the fluid is divided into a set of isopycnic sheets.

V. Bjerknes has discussed the properties of such surfaces,* and has distinguished the case in which the surfaces intersect from the case in which they do not intersect. In the former case the fluid is said to be *baroclinic*, in the latter case *barotropic*. It is thus seen that barotropy is a special case of degeneration of the two families of equiscalar surfaces into one family, the

* *Geofys. Publ.* **2**, No. 3.

isobaric surfaces then being also isopycnic. The physical distinction between barotropic and baroclinic fluids is that in the barotropic fluid pressure is a function of density alone, while in the baroclinic fluid pressure is not determined by density alone, and is in part determined by other factors such as temperature or water-vapour content.

Instead of drawing surfaces of equal density we might draw surfaces of equal specific volume, to which Bjerknes gives the name of *isosteric* surfaces. Any isosteric surface is also isopycnic, but the spacing of surfaces separated by unit difference of density and of specific volume will of course be different. If we wish to consider only the form of these surfaces without specifying the variable, we may, following Bjerknes, call them "equisubstantial" surfaces.

When the fluid is baroclinic, the isobaric and equisubstantial surfaces have definite lines of intersection. The two sets of surfaces will divide the fluid into a system of unit tubes. In a barotropic fluid neither curves of intersection nor unit tubes can be specified. It may be added that the fluid of "classical" hydrodynamics, which is normally assumed to be homogeneous and incompressible, is obviously barotropic.

§ 107. *Circulation and vorticity*

The circulation around a closed curve is defined as the line-integral of the velocity around the curve. For an element ds of the curve, the tangential component of the velocity v is evaluated, and the sum of the products $v_T ds$ for the whole curve is evaluated. The quantity so defined, $\int v_T ds$, is known as the circulation around the circuit. The circulation, which we shall denote by C, is readily seen to be given by

$$C = \int (u\,dx + v\,dy + w\,dz) \qquad \ldots\ldots(42),$$

where the integral is taken around the closed curve.

The simplest form of circulation which we can consider is uniform motion around a circle of radius r, with tangential velocity v. The circulation is then seen to be

$$C = 2\pi r v.$$

Stokes[*] has shown that the motion of a small element of fluid for a short time may be regarded as made up of four parts:

(a) a motion of bodily translation,

(b) a motion of solid rotation of the whole element,

(c) a volume expansion or contraction, and

(d) a shear.

This classification, which is only true to the first order of small quantities, has permeated the whole mechanics of the continuum. The reader who desires to consider its complete analytical discussion is referred to Love's *Treatise on Elasticity*, or to Lamb's *Hydrodynamics*, Chapter III.

We shall here consider briefly the item (b) of the above classification, since

* *Collected Papers*, **1**, p. 80; or *Camb. Phil. Trans.* **7**, 1845

a large part of meteorology is concerned with the nature and growth of circulation and rotation. The rotation as a whole of the small element, which is assumed to be sufficiently small to permit our assuming that its rotation may be regarded as uniform, is usually specified by the *vorticity*, which is equal to twice the angular velocity of rotation. Being a vector, it may be specified by the components about three rectangular co-ordinate axes. The adoption of twice the angular velocity as the vorticity is merely a useful convention which avoids the necessity for the continual recurrence of a factor 2 in a large number of hydrodynamical equations.

Consider first the simple case of a circular cylinder rotating as a solid with angular velocity $\frac{1}{2}\zeta$. Let the outer boundary have the radius R. The vorticity has everywhere the value ζ. The circulation around the boundary will be $2\pi R \times \frac{1}{2}\zeta R = \pi R^2 \zeta$, or: Circulation = area × vorticity. This result is also readily seen to be true for an area of any form when the fluid rotates as a solid.

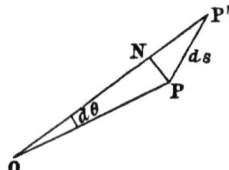

Fig. 45. Circulation and vorticity.

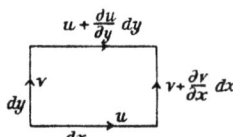

Fig. 46. Vorticity in Cartesian co-ordinates.

In fig. 45 let O be any point within the circuit, and let PP' be a small element of the circuit. Let $PP' = ds$, and let the angle POP' be $d\theta$. The velocity of rotation of the point P about O is $\frac{1}{2}\zeta$. Hence the contribution to the circulation by the small element PP' is

$$PP' \times \tfrac{1}{2}r\zeta \times \cos NPP' = \tfrac{1}{2}r^2\zeta \times P\hat{O}P' = \text{vorticity} \times \text{small area } POP'.$$

Integrating around a complete circuit will give the same result as before that the circulation is equal to area × vorticity.

The last relationship may be taken as the definition of vorticity, that it is such that the circulation around a small plane circuit is equal to the product of the component of the vorticity about an axis perpendicular to the element and the area of the element. This definition enables us to derive a number of general relationships of interest. When we come to the stage at which it is necessary to make actual computations we shall find it necessary to use Cartesian co-ordinates. The natural development of the formulae is by the use of polar co-ordinates, but the transformation to Cartesians is now readily made. In fig. 46 the small rectangle in the xy plane, whose sides are parallel to the co-ordinate axes, is assumed to be so small that the angular velocity about the axis of z has the same value over the whole area. The value of the circulation is

$$C = \left(\frac{\partial v}{\partial x} - \frac{\partial u}{\partial y}\right) dx\,dy = \zeta\,dx\,dy,$$

where ζ is the component of vorticity parallel to the axis of z. Hence

$$\zeta = \frac{\partial v}{\partial x} - \frac{\partial u}{\partial y}$$

Similarly
$$\xi = \frac{\partial w}{\partial y} - \frac{\partial v}{\partial z} \quad\quad \dots\dots(43).$$

$$\eta = \frac{\partial u}{\partial z} - \frac{\partial w}{\partial x}$$

These equations give positive vorticity and positive circulation for counter-clockwise motion, which, in the Northern hemisphere, is the direction of rotation of the earth.

Any fluid motion which has vorticity is called "rotational" motion, and fluid motion which has no vorticity is known as "irrotational" motion. The simplest case of rotational motion is rotation as a solid. The earth's atmosphere rotates with the earth, and the mean vorticity of the atmosphere, apart from local variations, is 2ω, where ω is the angular velocity of the earth.

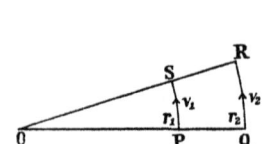

Fig. 47. Irrotational symmetrical
motion.

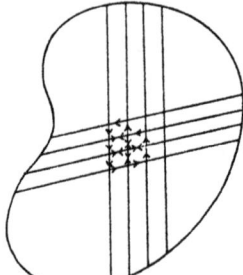

Fig. 48. Circulation and vorticity
in any circuit.

The simplest form of symmetrical motion which shall be irrotational is readily found as follows. Take a sector of the circular distribution. Let r_1 and r_2 be two radii, and v_1 and v_2 the corresponding velocities. Take the circulation around the circuit $PQRS$ in fig. 47. The sides PQ and RS being at right angles to the flow do not contribute to the circulation. The total circulation around $PQRS$ is therefore.

$$(v_1 r_1 - v_2 r_2) \times \text{angle } POS,$$

and since this must be zero by hypothesis, $v_1 r_1 = v_2 r_2$, and the product vr must be constant. The circulation is in fact constant around all concentric circles. This distribution, in which the transverse velocity is everywhere inversely proportional to the distance from the centre, is called the "simple vortex", or the "vr vortex". It is, however, irrotational, and has everywhere zero vorticity, except at the centre, where the velocity is infinite.

It will be seen from the above discussion that it is possible to have cyclic motion without vorticity in the main part of the field. The so-called "vr vortex" is of the nature of an infinite cylinder in the vertical direction. It extends out to an infinite distance laterally, and has infinite velocity at the centre. In nature the vr vortex can be readily set up through a finite region if these infinities can be avoided. The infinities at the ends of the cylinder are avoided if the fluid has upper and lower boundaries, as for example in a bowl of water, in which the bottom of the bowl and the free surface of the water form the necessary boundaries. The infinite horizontal extent is also removed by the boundaries formed by the side of the bowl, and the infinity of velocity at the centre may be evaded in two ways, either by the core being empty, as is the case when water whirls rapidly in a bowl, or by the provision of an inner region in which there is a continuous, but not necessarily constant, distribution of vorticity.

Around any closed plane curve the circulation is equal to the integral $\int \zeta_n dS$, where dS is an element of surface, and ζ_n the component vorticity perpendicular to the element dS. This is readily seen from fig. 48. The circulation around the outer boundary is equal to the sum of the circulations around all the elementary circuits, or in other words to the sum of the products $\zeta_n dS$ corresponding to each element, and when the elements of area are taken sufficiently small, the sum may be replaced by the integral $\int \zeta_n dS$, which is therefore equal to the circulation around the outer circuit.

Fig. 48 can be applied to the general case in which the circuit is not a plane circuit, and the surface of which it is a boundary is no longer part of a plane. The elementary circuits into which the surface is divided must now be sufficiently small to be treated as plane elements. The circulation around a small circuit enclosing an area dS is $dS \times$ vorticity about the normal to dS. If the direction cosines of the normal to dS be l, m, n, and if the components of vorticity be ξ, η, ζ, the circulation about the elementary circuit is

$$(l\xi + m\eta + n\zeta)\, dS.$$

Hence the circulation C about the circuit which forms the outer boundary of the surface is given by

$$C = \int (u\, dx + v\, dy + w\, dz) = \iint (l\xi + m\eta + n\zeta)\, dS,$$

where dS is a small element of any surface bounded by the circuit. This equation defines the close relation which subsists between circulation and vorticity.

§ 108. *Vortex lines, filaments and sheets*

A line drawn from point to point so that its direction is everywhere that of the instantaneous direction of the axis of rotation of the fluid is called a "vortex line". If through every point of a small closed curve we draw the corresponding vortex line, these lines mark out a tube, which is called a "vortex tube", and the fluid within the tube is called a "vortex filament". The circulation around the boundary of any cross-section of the tube, taken normal to its length, is $\Omega\sigma$, where Ω is the total vorticity $\sqrt{\xi^2 + \eta^2 + \zeta^2}$, and σ is the area of the small cross-section.

A vortex filament may be surrounded by fluid in irrotational motion, but the field of the vortex filament must be regarded as permeating the whole fluid.

A surface of discontinuity of velocity may be regarded, from the point of view of pure kinematics, as a surface or sheet of infinite vorticity, the axis of the vorticity being along the lines of relative flow. If, for example, the motion be represented by components of velocity, u, v, w, on one side of the sheet, and u', v', w', on the other side, the components of vorticity are proportional to $u-u'$, $v-v'$, $w-w'$, and the vortex lines at any point are given by

$$\frac{dx}{u-u'} = \frac{dy}{v-v'} = \frac{dz}{w-w'}.$$

The concept of a surface of discontinuity as a vortex sheet is a purely mathematical idea, and does not of necessity assist in the discussion of meteorological problems.

§ 109. *The generation of circulation in the atmosphere*

Kelvin established a theorem that if

$$C = \int (u\,dx + v\,dy + w\,dz),$$

then

$$\frac{dC}{dt} = \int \left(\frac{du}{dt}\,dx + \frac{dv}{dt}\,dy + \frac{dw}{dt}\,dz\right) \qquad \dots\dots(44),$$

where the integrals are both taken around a contour c composed of the same moving particles of fluid. By direct differentiation it is seen that

$$\frac{dC}{dt} = \int \left(\frac{du}{dt}\,dx + \frac{dv}{dt}\,dy + \frac{dw}{dt}\,dz\right) + \int \left(u\,\frac{d\,(dx)}{dt} + v\,\frac{d\,(dy)}{dt} + w\,\frac{d\,(dz)}{dt}\right).$$

Since

$$\frac{dx}{dt} = u, \quad \frac{dy}{dt} = v, \quad \frac{dz}{dt} = w,$$

this reduces to

$$\frac{dC}{dt} = \int \left(\frac{du}{dt}\,dx + \frac{dv}{dt}\,dy + \frac{dw}{dt}\,dz\right) + \frac{1}{2}\int \frac{\partial}{\partial s}\,(u^2 + v^2 + w^2)\,ds.$$

Since the velocity is a single-valued function of position, the second term on the right-hand side of this equation vanishes, and the equation reduces to (44) above.

Substituting in equation (44) the values of du/dt, dv/dt and dw/dt from equations (35), (36) and (37) above,

$$\frac{dC}{dt} = \int -\frac{1}{\rho}\left(\frac{\partial p}{\partial x}\,dx + \frac{\partial p}{\partial y}\,dy + \frac{\partial p}{\partial z}\,dz\right) + \int (X\,dx + Y\,dy + Z\,dz)$$

$$- \int g\,dz - 2\omega \int \{(w\cos\phi - v\sin\phi)\,dx + u\sin\phi\,.\,dy - u\cos\phi\,.\,dz\}$$

$$= -\int \frac{dp}{\rho} + \int (X\,dx + Y\,dy + Z\,dz) - \text{terms in } 2\omega \qquad \dots\dots(45).$$

The terms in ω are readily evaluated as follows:

In fig. 49 consider the circuit f which is the projection of the original circuit c upon the plane of the equator. In the plane of the equator take axes x', y', where x' is parallel to the original axis of x, and y' is the intersection of the original yz plane with the equator. Let u', v' be the components of velocity in the equatorial plane of the original velocities u, v, w.

Then
$$dx' = dx, \quad dy' = \sin \phi . dy - \cos \phi . dz,$$
$$u' = u, \quad v' = \sin \phi . v - \cos \phi . w,$$
$$(w \cos \phi - v \sin \phi) \, dx + u \sin \phi . dy - u \cos \phi . dz = -v' dx' + u' dy'.$$

The terms in ω in (45) may therefore be written as
$$-2\omega \int (u' dy' - v' dx') \qquad \qquad \dots \dots (46),$$

where the integral is taken around f, the projection of the original contour on the plane of the equator. We take the circulation to be positive in the counter-

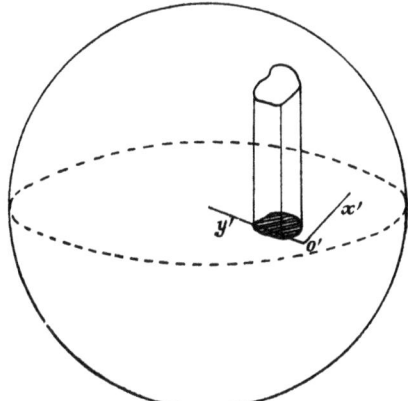

Fig. 49. Circulation in the earth's atmosphere.

clockwise sense in the plane of the equator. It is seen that if F denote the area of the contour f, (46) reduces to
$$-2\omega \frac{dF}{dt} \qquad \qquad \dots \dots (47),$$

and writing
$$W = \int (X \, dx + Y \, dy + Z \, dz),$$

equation (45) becomes
$$\frac{dC}{dt} = -\int \frac{dp}{\rho} + W - 2\omega \frac{dF}{dt} \qquad \qquad \dots \dots (48).$$

In (48) F is to be taken as positive when the circuit c is traversed in such a sense that the circuit f is traversed in the counter-clockwise direction.

Equation (48) shows that changes in circulation may be due to one of three causes. Of these, W represents the effects of friction, which is normally only effective in decreasing circulations which are already in existence, and does not actually generate circulations. It is shown below that the term $-\int \frac{dp}{\rho}$ represents the effects of inequalities of density in producing shearing motion, and that it does not represent a factor capable of producing cyclic circulations. The third term in equation (48) shows that cyclic circulation is generated around a circuit, in the cyclonic sense when there is convergence of air inward across the circuit, and in the anticyclonic sense when there is divergence of air outward across the circuit. This factor is considered more fully in Chapter xv.

The term $-\int \dfrac{dp}{\rho}$ taken around a closed circuit disappears if, and only if, the fluid is barotropic, and p is a function of ρ only. If p is a function of other variables, in particular of temperature, $-\int \dfrac{dp}{\rho}$ will not in general vanish around a closed circuit.

Let any surface be drawn through the circuit c as its boundary, and divide the area enclosed by c, by isobaric and isopycnic lines, into small parallelograms, such as $ABCD$, fig. 50. The value of $-\int \dfrac{dp}{\rho}$ taken around the circuit c is

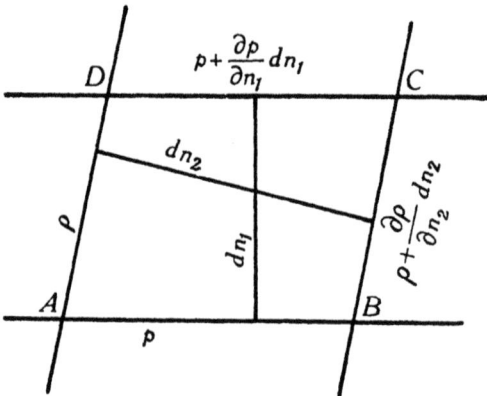

Fig. 50. Line and surface integrals for $-\int \dfrac{dp}{\rho}$.

the sum of the values of $-\int \dfrac{dp}{\rho}$ taken around all the elementary circuits. Let the distances between opposite sides of the parallelogram $ABCD$ be dn_1 and dn_2, and let the angle ABC be χ.

The contributions to $-\int \dfrac{dp}{\rho}$ along AB and CD are zero; along BC and DA they are $\dfrac{-\dfrac{\partial p}{\partial n_1} dn_1}{\rho + \dfrac{\partial \rho}{\partial n_2} dn_2}$ and $\dfrac{1}{\rho}\dfrac{\partial p}{\partial n_1} dn_1$ respectively.

The value of $-\int \dfrac{dp}{\rho}$ taken around the small circuit $ABCD$ is, to the second order of small quantities,
$$\frac{1}{\rho^2} dp\, d\rho.$$

$ABCD$ is approximately a parallelogram, whose area, to the second order of small quantities, is
$$dn_1 dn_2 / \sin \chi.$$

Hence the value of $-\int \dfrac{dp}{\rho}$ taken around $ABCD$ is, correctly to the second order of small quantities,

$$\frac{1}{\rho^2}\,dp\,d\rho = \frac{1}{\rho^2}\frac{\partial p}{\partial n_1}\frac{\partial \rho}{\partial n_2}\,dn_1\,dn_2 = \frac{1}{\rho^2}\frac{\partial p}{\partial n_1}\frac{\partial \rho}{\partial n_2}\sin\chi \times \text{area } ABCD.$$

χ is positive when the rotation of the gradient of density into coincidence with the gradient of pressure is in the counter-clockwise direction. The total value of the integral $-\int \dfrac{dp}{\rho}$ taken around any circuit c may be regarded as equivalent to the sum of the contributions

$$\frac{1}{\rho^2}\frac{\partial p}{\partial n_1}\frac{\partial \rho}{\partial n_2}\sin\chi \qquad \qquad \dots\dots(49)$$

per unit area of any surface bounded by the circuit. This expression was first given by Silberstein.*

The result thus derived shows that in any region where the surfaces of equal pressure and equal density intersect there is a tendency for a circulation to be set up about axes which lie along the intersections of these families of surfaces, in such a sense as to bring the gradient of density into line with the gradient of pressure.

The same result has been derived by V. Bjerknes† in a slightly different way. The whole of space is regarded as divided into a system of tubes by families of isobaric and isosteric surfaces which are separated respectively by unit difference of pressure and of density. The result expressed in (49) above is equivalent to saying that the rate of increase of circulation around the circuit c is measured by the number of unit tubes or solenoids which pass through the closed curve c. Here we prefer to follow the expression (49) for this effect, instead of the expression in terms of solenoids given by Bjerknes.

The result which is represented in equation (49) has been derived by a transformation of equations (35), (36) and (37) above. It may be represented by the diagram of fig. 52, which indicates that the result may be described as a tendency of the lighter air to seek lower pressure, and of the heavier air to seek higher pressure. This result might in fact have been inferred from equations (35), (36) and (37), in which the effect of the pressure distribution is represented by terms such as $-\dfrac{1}{\rho}\dfrac{\partial p}{\partial x}$. In a given gradient of pressure the effect is inversely proportional to the density, and so is greater for the lighter than for the denser air. The lighter air will therefore show a greater tendency to drift across the isobars into low pressure. Before considering further the interpretation of expression (49) we shall investigate some transformations of this expression into terms of other variables.

In meteorology density is not directly observed, and it is convenient to express the above result first in terms of the gradients of pressure and temperature, and finally of pressure and potential temperature, which is the most useful form.

---

* *Bull. Acad. Sci. Cracow*, 1896; *Vectorial Mechanics* (Macmillan), p. 167.
† *Geofys. Publ.* 2, No. 3.

At any point in the atmosphere let p, ρ, T and θ be the pressure, density, absolute temperature and potential temperature respectively. Then

$$p = R\rho T \qquad \qquad \text{......(50)},$$

where R is the gas-constant for dry air. Neglecting the effect of water-vapour, and taking logarithms in (50) and differentiating,

$$\frac{dp}{p} = \frac{d\rho}{\rho} + \frac{dT}{T} \qquad \qquad \text{......(51)}.$$

In the notation of vector analysis, equation (51) leads at once to

$$\frac{\text{grad } p}{p} = \frac{\text{grad } \rho}{\rho} + \frac{\text{grad } T}{T} \qquad \qquad \text{......(52)},$$

the vector relation between the gradients of pressure, density and temperature.

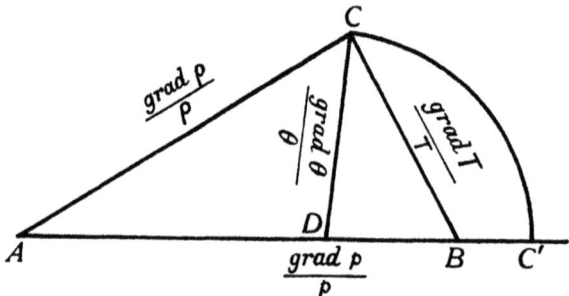

Fig. 51. The geometrical relations between the gradients of p, ρ, θ and T.

In fig. 51 AB, AC, CB represent $\dfrac{\text{grad } p}{p}$, $\dfrac{\text{grad } \rho}{\rho}$ and $\dfrac{\text{grad } T}{T}$, in magnitude and direction. The potential temperature θ, defined as the temperature which a mass of air at pressure p will take when brought adiabatically to a standard pressure p_0, is given by

$$\theta = T \left(\frac{p_0}{p} \right)^{\frac{\gamma - 1}{\gamma}} \qquad \qquad \text{......(53)},$$

where γ is the ratio of the specific heats of air. From (53)

$$\frac{d\theta}{\theta} = \frac{dT}{T} - \frac{\gamma - 1}{\gamma} \frac{dp}{p},$$

whence

$$\frac{\text{grad } \theta}{\theta} = \frac{\text{grad } T}{T} - \frac{\gamma - 1}{\gamma} \frac{\text{grad } p}{p} \qquad \qquad \text{......(54)}.$$

In fig. 51, $DB = \dfrac{\gamma - 1}{\gamma} AB$, and from (54) $\dfrac{\text{grad } \theta}{\theta}$ is represented by CD. Fig. 51 is a true vector diagram, connecting the total gradients of pressure, density, temperature and potential temperature. It can also be treated as referring to the components of these gradients in any plane.

In (49) above, the quantity $\frac{1}{\rho^2}\frac{\partial p}{\partial n_1}\frac{\partial \rho}{\partial n_2}\sin\chi$, which is the rate of increase of circulation per unit area, may be regarded as a vector perpendicular to $ABCD$, and is equal to the vector product

$$\frac{1}{\rho^2}\operatorname{grad} p \times \operatorname{grad} \rho \qquad \ldots\ldots(55),$$

where the symbol $\times$ denotes vectorial multiplication.

From (52) and (54)

$$\frac{\operatorname{grad} p}{p}\times\frac{\operatorname{grad}\rho}{\rho} = -\frac{\operatorname{grad} p}{p}\times\frac{\operatorname{grad} T}{T} = -\frac{\operatorname{grad} p}{p}\times\frac{\operatorname{grad}\theta}{\theta}\ \ldots\ldots(56).$$

From (55) and (56)

$$\frac{1}{\rho^2}\operatorname{grad} p \times \operatorname{grad}\rho = -\frac{1}{\rho T}\operatorname{grad} p \times \operatorname{grad} T = -\frac{1}{\rho\theta}\operatorname{grad} p \times \operatorname{grad}\theta\ \ldots\ldots(57).$$

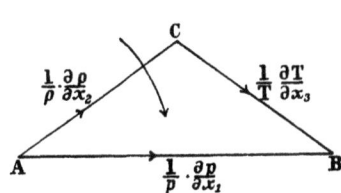

Fig. 52. The direction of apparent spin.

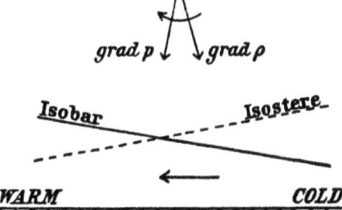

Fig. 53. Land and sea breezes.

It was shown above that in any region where the gradients of pressure and density are not parallel there is a growth of circulation or vorticity tending to rotate the density gradient towards coincidence with the pressure gradient. Since the density is a function of the pressure, it cannot be assumed that the gradient of density rotates (about A in fig. 51) without change of magnitude. Neglecting all effects due to radiation, conduction of heat, or mixing, we may assume that the changes of state of the fluid are adiabatic, so that each element retains its original potential temperature, and the gradient of potential temperature rotates with the fluid, with its scalar magnitude unchanged. Then in fig. 51 the point C revolves about D, moving towards C' on AB. If the inertia of the fluid carries C beyond C' to the other side of C', the forces called into play tend to check its motion, and to carry the point C back towards C'. If therefore the initial conditions be represented by fig. 47, the phenomena produced may be of the nature of an oscillation of the line CD about the mean position $C'D$.

The argument used above assumes that the pressure gradient remains unchanged during the motion. In the atmosphere this is a reasonable assumption to make, since we may regard the pressure distribution as superposed from above.

The phenomena of land and sea breezes afford an interesting and simple illustration of the results derived above. Suppose the atmosphere is initially at rest, with pressure uniform at mean sea level. When now the land becomes heated during the day, the rate of decrease of pressure with height is less over land than over the sea, and the isobaric surfaces slope upward from sea to land. The isosteric surfaces slope in the opposite direction, upward from land to sea. The gradients of pressure and density are as shown in fig. 53, and the tendency will be therefore to produce a circulation from sea to land. This is no more than the tendency of the cold air to push under the warm air, and in the idealised conditions we have supposed to exist the sea breeze will be normal to the shore line. So long as the land remains warmer than the sea the motion will continue. Fig. 53 only represents the initial motion of the air, and we are not concerned at this stage with the subsequent history of the air masses thus set in motion. At night when the land becomes colder than the sea the process is reversed.

The result stated in (49) above is a perfectly general result which appears to make no assumptions of any kind as to the nature of the fluid, and it therefore appears at first sight that the growth of cyclic motion in the atmosphere might be explained in this way. But in the atmosphere it is possible to have cyclic motion with no vorticity in the main body of fluid, as in the so-called simple vortex, and also to have vorticity without cyclic motion, as for example when there is a gradient of velocity in one direction. It is not at first sight easy to decide whether the apparent growth of circulation deduced above is purely of the nature of a shearing of the elements of fluid considered, with no tendency to produce cyclic motion.

It is important to bear in mind that the motions produced by the effect of the term $-\int \frac{dp}{\rho}$ may involve convergence or divergence which may lead to an effect in the opposite sense, as measured by $2\omega \frac{dF}{dt}$.

Take a specially simplified problem of motion in a horizontal plane, in a field of pressure represented by equidistant straight isobars, drawn parallel to the axis of x. The equations of motion may be written

$$\left.\begin{aligned}
\frac{du}{dt} - 2\omega \sin \phi . v &= -\frac{1}{\rho}\frac{\partial p}{\partial x} = 0 \\
\frac{dv}{dt} + 2\omega \sin \phi . u &= -\frac{1}{\rho}\frac{\partial p}{\partial y}
\end{aligned}\right\} \quad \ldots\ldots(58).$$

Consider the motion of an element of air of density ρ, whose component velocities at time $t=0$ are $u=u_0$, $v=v_0$. Multiplying the second of the above equations by $i\,(\sqrt{-1})$, and adding to the first, we find

$$\frac{d}{dt}(u+iv) + 2\omega \sin \phi . i (u+iv) = -\frac{i}{\rho}\frac{\partial p}{\partial y} \qquad \ldots\ldots(59).$$

This equation is readily integrated, giving

$$u+iv = Ae^{-2\omega \sin \phi i(t-\beta)} - \frac{1}{2\omega \sin \phi}\frac{1}{\rho}\frac{\partial p}{\partial y},$$

or $\quad u+iv = \left\{u_0+iv_0 + \frac{1}{2\omega \sin \phi}\frac{1}{\rho}\frac{\partial p}{\partial y}\right\} e^{-2\omega \sin \phi\, it} - \frac{1}{2\omega \sin \phi}\frac{1}{\rho}\frac{\partial p}{\partial y}$ $\ldots\ldots(60).$

This equation indicates that the motion of the air as defined by the equations and conditions stipulated above will be an oscillation of the velocity about the geostrophic value $-\dfrac{1}{2\omega \sin \phi}\dfrac{1}{\rho}\dfrac{\partial p}{\partial y}$, except in the case when the initial velocity is equal to the geostrophic value, in which case the velocity remains constant. Equation (60) shows that the constant value about which the velocity oscillates is inversely proportional to the density ρ.

Now suppose that initially a mass of air is moving with the appropriate geostrophic velocity along the straight isobars postulated above, and that a difference of temperature is produced in any manner. The warmer and therefore the less dense portion of the air will acquire a velocity which oscillates about a higher mean value than that of the denser portion. The initial effect is obviously to cause the lighter air to move faster inwards across the isobars towards lower pressure, but the eventual result is to cause the lighter air to get further and further ahead downwind. Thus the result which we have expressed in (49) above is only applicable in the initial state, and the final result of a difference of density may be exactly opposite to the initial effect. It is thus not possible to explain the growth of large-scale horizontal circulations in the atmosphere by the reactions due to the gradients of pressure and density not being parallel to each other.

§ 110. *Velocity potential*

When the motion is irrotational, or the vorticity is zero,

$$\frac{\partial u}{\partial z}=\frac{\partial w}{\partial x}, \quad \frac{\partial u}{\partial y}=\frac{\partial v}{\partial x}, \quad \frac{\partial w}{\partial y}=\frac{\partial v}{\partial z}.$$

These three equations are the conditions that

$$(u\,dx+v\,dy+w\,dz)$$

should be a complete differential. Call this $-d\phi$. Then

$$u=-\frac{\partial \phi}{\partial x}, \quad v=-\frac{\partial \phi}{\partial y}, \quad w=-\frac{\partial \phi}{\partial z}.$$

If the fluid is incompressible, the equation of continuity then reduces to the form

$$\frac{\partial^2 \phi}{\partial x^2}+\frac{\partial^2 \phi}{\partial y^2}+\frac{\partial^2 \phi}{\partial z^2}=0 \qquad \ldots\ldots(61).$$

In two-dimensional flow, we may also imagine a function $\psi\,(x, y)$ which gives the form of the stream lines. Along the stream lines

$$\frac{\partial \psi}{\partial x}\,dx+\frac{\partial \psi}{\partial y}\,dy=0 \qquad \ldots\ldots(62)$$

and we may then take $\qquad v=\dfrac{\partial \psi}{\partial x}, \quad u=-\dfrac{\partial \psi}{\partial y}.$

The equation of continuity for two-dimensional flow is

$$\frac{\partial u}{\partial x}+\frac{\partial v}{\partial y}=0,$$

if the fluid is incompressible; this is automatically satisfied by the velocities deduced from ψ as shown above. The vorticity is given by

$$2\zeta = \frac{\partial^2 \psi}{\partial x^2} + \frac{\partial^2 \psi}{\partial y^2} \qquad \qquad \dots \dots (63).$$

If the flow is irrotational, $\nabla^2 \psi = \dfrac{\partial^2 \psi}{\partial x^2} + \dfrac{\partial^2 \psi}{\partial y^2} = 0.$

In two-dimensional polars equation (63) becomes

$$\frac{1}{r} \frac{\partial}{\partial r} \left(r \frac{\partial \psi}{\partial r} \right) + \frac{1}{r^2} \frac{\partial^2 \psi}{\partial \theta^2} = 2\zeta \qquad \qquad \dots \dots (64).$$

The radial velocity is given by $\partial \psi / r \partial \theta$, and the transverse velocity by $\partial \psi / \partial r$. If ζ is a known function of r and θ, the equation is soluble.

We have only collected some of the more important results here. For further details see Lamb's *Hydrodynamics*, or Glauert's *Aerofoil and Airscrew Theory*.

§ 111. *The stresses in a viscous fluid*

If at a point P in the fluid we imagine three planes to be drawn perpendicular to the axes of x, y, z respectively, the stresses exerted across the first of these planes are

$$p_{xx}, \quad p_{xy}, \quad p_{xz}.$$

The stresses across the second plane are

$$p_{yx}, \quad p_{yy}, \quad p_{yz},$$

and the stresses across the third plane are

$$p_{zx}, \quad p_{zy}, \quad p_{zz}.$$

If the fluid is frictionless $p_{xx} = p_{yy} = p_{zz} = -p.$

Taking moments about the centre of an element $dx\,dy\,dz$, we find

$$p_{xy} = p_{yx}, \quad p_{yz} = p_{zy}, \quad p_{zx} = p_{xz}.$$

These three relationships reduce the number of components of stress to six. The detailed development of these formulae will be found in Lamb's *Hydrodynamics* (6th edition, art. 326):

$$\left. \begin{aligned} p_{xx} &= -p - \tfrac{2}{3}\mu \left(\frac{\partial u}{\partial x} + \frac{\partial v}{\partial y} + \frac{\partial w}{\partial z} \right) + 2\mu \frac{\partial u}{\partial x} \\[4pt] p_{yy} &= -p - \tfrac{2}{3}\mu \left(\frac{\partial u}{\partial x} + \frac{\partial v}{\partial y} + \frac{\partial w}{\partial z} \right) + 2\mu \frac{\partial v}{\partial y} \\[4pt] p_{zz} &= -p - \tfrac{2}{3}\mu \left(\frac{\partial u}{\partial x} + \frac{\partial v}{\partial y} + \frac{\partial w}{\partial z} \right) + 2\mu \frac{\partial w}{\partial z} \end{aligned} \right\} \qquad \dots \dots (65),$$

$$\left. \begin{aligned} p_{yz} &= p_{zy} = \mu \left(\frac{\partial w}{\partial y} + \frac{\partial v}{\partial z} \right) \\[4pt] p_{zx} &= p_{xz} = \mu \left(\frac{\partial u}{\partial z} + \frac{\partial w}{\partial x} \right) \\[4pt] p_{xy} &= p_{yx} = \mu \left(\frac{\partial v}{\partial x} + \frac{\partial u}{\partial y} \right) \end{aligned} \right\} \qquad \dots \dots (66),$$

where μ is the coefficient of viscosity. A more convenient coefficient is $\frac{\mu}{\rho} = \nu$, called the kinematic coefficient of viscosity. On an element $dx\,dy\,dz$ of fluid the tractions in the direction of the x-axis are

$$\frac{\partial p_{xx}}{\partial x} + \frac{\partial p_{yx}}{\partial y} + \frac{\partial p_{zx}}{\partial z} = -\frac{\partial p}{\partial x} + \tfrac{1}{3}\mu \frac{\partial}{\partial x}\left(\frac{\partial u}{\partial x} + \frac{\partial v}{\partial y} + \frac{\partial w}{\partial z}\right) + \mu\nabla^2 u \quad \ldots\ldots(67).$$

Hence in equations (39), (40) and (41) above the contributions to X, Y, Z due to the viscous forces are

$$\left.\begin{aligned}
&\tfrac{1}{3}\nu \frac{\partial}{\partial x}\left(\frac{\partial u}{\partial x} + \frac{\partial v}{\partial y} + \frac{\partial w}{\partial z}\right) + \nu\nabla^2 u \\[2mm]
&\tfrac{1}{3}\nu \frac{\partial}{\partial y}\left(\frac{\partial u}{\partial x} + \frac{\partial v}{\partial y} + \frac{\partial w}{\partial z}\right) + \nu\nabla^2 v \\[2mm]
&\tfrac{1}{3}\nu \frac{\partial}{\partial z}\left(\frac{\partial u}{\partial x} + \frac{\partial v}{\partial y} + \frac{\partial w}{\partial z}\right) + \nu\nabla^2 w
\end{aligned}\right\} \quad \ldots\ldots(68).$$

When the fluid is incompressible the term in $\frac{\partial u}{\partial x} + \frac{\partial v}{\partial y} + \frac{\partial w}{\partial z}$ is zero. In the atmosphere, which is compressible, the effect of this term is considerably smaller than that of the final term in the above equation, and the viscous forces may be regarded as accounted for with all the accuracy which is required in meteorological discussion by the terms $\nu\nabla^2 u$, $\nu\nabla^2 v$, $\nu\nabla^2 w$, respectively.

The effect of the terms such as $\nu\nabla^2 u$ is analogous to the term in the equation of heat transfer by diffusion, and indicates that in a viscous fluid there is a diffusion of fluid momentum due to the viscosity, of amount proportional to the coefficient of viscosity.

Lamb shows (*loc. cit.* pp. 579 *et seq.*) that in a viscous fluid kinetic energy is dissipated at a rate

$$\mu\left\{2\left(\frac{\partial u}{\partial x}\right)^2 + 2\left(\frac{\partial v}{\partial y}\right)^2 + 2\left(\frac{\partial w}{\partial z}\right)^2 + \left(\frac{\partial w}{\partial y} + \frac{\partial v}{\partial z}\right)^2 + \left(\frac{\partial u}{\partial z} + \frac{\partial w}{\partial x}\right)^2 + \left(\frac{\partial v}{\partial x} + \frac{\partial u}{\partial y}\right)^2\right\} \ldots(69)$$

per unit volume of the fluid. To obtain the total dissipation through a finite volume of the fluid, this expression should be integrated through the volume in question.

Lamb shows (*loc. cit.* p. 578) that the equations for the change in the components of vorticity ξ, η, ζ in a viscous fluid are

$$\left.\begin{aligned}
\frac{d\xi}{dt} &= \xi\frac{\partial u}{\partial x} + \eta\frac{\partial u}{\partial y} + \zeta\frac{\partial u}{\partial z} + \nu\nabla^2\xi \\[2mm]
\frac{d\eta}{dt} &= \xi\frac{\partial v}{\partial x} + \eta\frac{\partial v}{\partial y} + \zeta\frac{\partial v}{\partial z} + \nu\nabla^2\eta \\[2mm]
\frac{d\zeta}{dt} &= \xi\frac{\partial w}{\partial x} + \eta\frac{\partial w}{\partial y} + \zeta\frac{\partial w}{\partial z} + \nu\nabla^2\zeta
\end{aligned}\right\} \quad \ldots\ldots(70).$$

The first three terms on the right-hand side of each of these equations represent the change of vorticity when the vortex lines move with the fluid, the strengths of the vortices remaining constant. The last term in each equation represents the diffusion of vorticity produced by viscosity, following a law similar to the conduction of heat. It is evident from this analogy that vorticity cannot originate in the body of a fluid, but must be diffused outward from solid boundaries.

The value of ν for different temperatures is shown in a table on p. 421 below. For the purpose of rough computation it will usually suffice to adopt the value $\nu = 0.15$, in C.G.S. units.

CHAPTER IX

MOTION UNDER BALANCED FORCES: THE GRADIENT WIND

§ 112. *Conditions for steady motion*

WHEN the motion of air is purely horizontal the equations of motion are

$$\frac{du}{dt} - 2\omega v \sin \phi = -\frac{1}{\rho}\frac{\partial p}{\partial x} + X \qquad \text{......(1)},$$

$$\frac{dv}{dt} + 2\omega u \sin \phi = -\frac{1}{\rho}\frac{\partial p}{\partial y} + Y \qquad \text{......(2)},$$

where X, Y are the components of forces other than those due to the pressure distribution. These equations state that the forces X, Y, together with the gradient of pressure and the deviating force due to the earth's rotation, balance the accelerations of the element of air in its path.

An important application of this is to the steady motion of air at heights sufficiently great to be removed from all effects due to friction and turbulence at the surface of the earth. The motion at any point is then independent of the time, or $\frac{\partial u}{\partial t} = \frac{\partial v}{\partial t} = 0$. Equations (1) and (2), in which we now put $X = Y = 0$, state that the acceleration of the air in its path will balance the combined effect of the pressure gradient and the deviating force, and we are at liberty to write the accelerations relative to the earth in any convenient form. The most convenient form is that which gives the components of acceleration along the tangent and the normal to the path, the latter being the centripetal acceleration. If V be the velocity at any point of the path where the radius of curvature is r, the total acceleration is made up of V^2/r directed towards the centre of curvature, and dV/dt or $V dV/ds$ along the tangent to the path. The algebraic equivalence of these two expressions to du/dt and dv/dt is readily seen by writing $u = V \cos \psi$, $v = V \sin \psi$, and differentiating. The result follows immediately when $d\psi/ds$ is replaced by $1/r$.

In considering horizontal motion under balanced forces, it is the custom of most writers to omit the tangential acceleration $V dV/ds$, but it is not obvious that this is justifiable *a priori*. We can, however, justify it by an appeal to observations. Shaw and Lempfert found, in their study of the *Life-History of Surface Air Currents*, that these currents frequently flow for thousands of miles with no appreciable change of velocity. Also Durward,[*] in a study of pilot balloon observations made behind the British lines in France during the war of 1914–18, found that the changes in velocity downwind were very slight. We shall therefore assume that the tangential acceleration is negligible, and that only the centripetal acceleration perpendicular to the path need be

---

[*] M.O., *Professional Notes*, No. 24.

considered. The centripetal acceleration, of magnitude V^2/r, is directed towards the centre of curvature. This is balanced by the deviating force and the pressure gradient. The deviating force is along the same line as the resultant acceleration, being at right angles to the path. If the motion is exactly balanced, the pressure is along the same direction, and must be normal to the path. This means that the path must be tangential to the isobars, and the motion is around the isobars.

The wind which blows around the isobars, and has such a magnitude that it produces the right centripetal acceleration and deviating force to balance exactly the pressure gradient, is called the *gradient wind*. The conditions that it fulfils, which have been stated above, assume that the pressure distribution is not changing. It is therefore not justifiable to use the conception of a gradient wind in conditions when the pressure distribution is changing, without some consideration of the rate at which the pressure distribution is changing.

§ 113. *The gradient wind equation*

In fig. 54 let P be a point on the path of an element of air, and let PP' be the circle of curvature of the path at P, PP' being a small circle on the earth's surface. The points P, P' are taken at opposite points of a diameter, and N the midpoint of PP' is the centre of the small circle. The angular radius of the circle, or the angle subtended by NP at the centre of the earth, is equal to c. If R be the radius of the earth, the radius of the small circle is $R \sin c$.

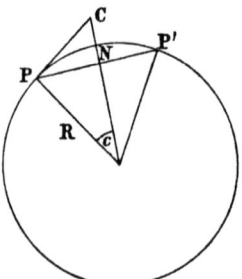

Fig. 54. Motion in a small circle.

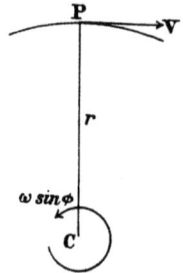

Fig. 55. Centrifugal force on an element of air.

The horizontal velocity of the element of air at P being V, the centripetal acceleration along PN is $V^2/R \sin c$, and the horizontal component of this along the line PC is $V^2/R \tan c$. Or, if r be the radius of curvature at P of the projection of the path of the air on the plane of the horizon at P, the horizontal acceleration along PC is V^2/r. The vertical component, directed downwards, is $V^2/R \sin c \times \sin c = V^2/R$.

The three terms in our equations of motion, which must be considered are

(*a*) the pressure gradient acting from high to low pressure, and at right angles to the isobars, (*b*) the deviating force acting at right angles to the direction of motion, and towards the right (in the Northern hemisphere), and

(*c*) the centripetal acceleration, acting at right angles to the direction of motion, towards the concave side of the isobars.

The condition for steady motion is that these three items shall balance, or that (*c*) shall be equal to the resultant of (*a*) and (*b*). It is convenient to distinguish two cases, according as (*a*) and (*c*) are in the same or in opposite directions, i.e. according as the isobars have anticyclonic or cyclonic curvature. In the first case

$$\text{(Anticyclone)} \qquad 2\omega V \sin\phi - V^2/r = \frac{\text{pressure gradient}}{\rho} \quad \ldots\ldots(3),$$

and in the second case

$$\text{(Cyclone)} \qquad 2\omega V \sin\phi + V^2/r = \frac{\text{pressure gradient}}{\rho} \quad \ldots\ldots(4).$$

Equation (4) may be regarded as included in (3), if we postulate that r shall be regarded as positive when it is directed to the right of the direction of motion, and negative when it is directed to the left of the direction of motion.

The solution of the appropriate one of equations (3) or (4) is called the *gradient wind*, and the equation is called the *gradient wind equation*. The gradient wind is such that the accelerations it calls forth are exactly balanced by the gradient of pressure. When the isobars are straight and parallel, the path of an element of air is along a great circle. The term in V^2/r in equations (3) and (4) then vanishes, and the value of V is defined by

$$2\omega V \sin\phi = \frac{\text{pressure gradient}}{\rho} \qquad \ldots\ldots(5).$$

When the isobars are only slightly curved, it will be found that the term in V^2/r may be neglected by comparison with $2\omega V \sin\phi$. The value of V which is derived when the term V^2/r is neglected is called the *geostrophic wind*. If we denote the geostrophic wind by G, then the

$$\text{gradient of pressure} = 2\omega\rho G \sin\phi \qquad \ldots\ldots(6).$$

We shall frequently find it convenient to express the gradient of pressure by the expression in (6). The geostrophic wind is thus the wind which, blowing around the isobars, would produce a deviating force of the right magnitude to balance exactly the pressure gradient. The use at any time of the symbol G, or references to the geostrophic wind, must not be taken to imply that the actual motion is along a great circle (or along straight isobars). The geostrophic wind is to be regarded as a first approximation to the gradient wind.

The gradient wind equation may be rewritten in the form

$$2\omega \sin\phi\,(V - G) = \pm\, V^2/r \qquad \ldots\ldots(7)$$

in which the upper or lower sign is used according as the original equation is (3) or (4); the upper sign is used for anticyclonic isobars, and the lower sign for cyclonic isobars. From equation (7) it follows that G is an underestimate of the gradient wind in an anticyclone, and an overestimate in a cyclone.

In equations (3) and (4) the first term on the left-hand side is called the geostrophic component of the pressure gradient, and the second is called the cyclostrophic component. It is frequently assumed that in middle and high latitudes the first of these outweighs the second, so that the geostrophic wind may be regarded as a close approximation to the true wind, provided the pressure distribution is not changing. The validity of this assumption is discussed later, in § 117. In low latitudes, on account of the smallness of $\sin \phi$, the cyclostrophic component is the more important, and the geostrophic component is then negligible. Equation (4) is then replaced by

$$\frac{V^2}{r} = \frac{\text{pressure gradient}}{\rho} = \frac{1}{\rho}\frac{\partial p}{\partial r} \qquad \ldots\ldots(8),$$

which gives the velocity required to produce a balance between the gradient of pressure and the centripetal acceleration. The left-hand side of this equation must always be positive, and steady cyclostrophic motion is only possible when $\partial p/\partial r$ is positive, or the pressure distribution is cyclonic. Equation (8) cannot be satisfied when $\partial p/\partial r$ is negative, or with an anticyclonic distribution of pressure. This conclusion is to be interpreted as indicating, *either* that closed anticyclonic isobars are not possible as stable systems in very low latitudes, *or* that when closed isobars occur in low latitudes the wind is directly across the isobars, and not around them.

The gradient wind is defined as the appropriate solution of equations (3) or (4), in which allowance is made for the curvature of the path of the air, the path being regarded as identical with the isobars. The name geostrophic wind is used to denote that wind which would call into being a deviating force which would exactly balance the acceleration due to the pressure gradient. Some writers use the term gradient wind to denote what is here called the geostrophic wind, but the distinction here drawn between the two appears to be desirable.

§ 114. *Comparison of the gradient wind with observed winds*

The first detailed comparison of the gradient wind with observed winds was made by E. Gold,* who found that at a height of 500 metres above the ground the direction of the wind was almost exactly along the mean sea level isobars, while the velocity was slightly below the computed gradient wind. The agreement was found to be sufficiently good to justify the adoption of the gradient wind as a close approximation to the actual wind, within the limits of the errors of observation. In Gold's investigation the term V^2/r was computed for each individual occasion, but r was taken as the curvature of the isobars, and not of the actual path of the air. The curvature of the path of the air is not the same as the curvature of the isobars, except when the conditions are unchanging. Gold found† that in a cyclone the radius of the curvature of the path exceeded that of the isobars, at points to the right-hand side of the path of the centre, and was less than that of the isobars at points to the left-hand side of the path of the centre; in an anticyclone the inequalities were reversed. But the main value of Gold's investigation is its confirmation that the direct use of the synoptic chart gives a reasonably close approximation to the observed wind. See also § 117 below.

* M.O. 190, *Barometric Gradient and Wind Force.*   † *Ibid.* p. 42.

The geostrophic wind is readily evaluated by the use of a scale graduated in miles per hour or metres per second, which is placed normally across the isobars at the place for which the geostrophic wind is required. The scale is graduated on a scale of reciprocals of the velocities, and the nearer the isobars the greater is the gradient of pressure, and the greater the geostrophic wind. The scale is graduated for a standard density, but the small correction required to allow for variations in density is readily made.

The computation of the gradient wind is more difficult, since it requires first an estimate of the curvature of the path of the air, and secondly the solution of a quadratic equation. The result of such a computation is some-what unreliable on account of the uncertainty of any estimate of the curvature of the path, which is not of necessity the same as the curvature of the isobars, and no serious attempt has been made to use the gradient wind in practical meteorology. It is partly for this reason that the geostrophic wind has been adopted as the most convenient approximation to the wind at heights of 500 or 1000 metres.

Attention is again directed to the fact that the gradient wind has no definite meaning when the pressure distribution is changing, and the motion is there-fore not under balanced forces. The principle of motion under balanced forces was first elaborated by Sir Napier Shaw, in the *Manual of Meteorology*, **4**, where it is maintained that the adjustment of the wind to the pressure gradient is automatic and complete. If this were so the motion would be under balanced forces, and the gradient wind equation would be applicable.

§ 115. *The solution of the gradient wind equation*

The gradient wind equation has been given in several forms in equations (3), (4) and (7) above. The equation is a quadratic, and thus has two solutions. When the isobars are straight the quadratic degenerates into a simple equation whose solution is $V = G$, and the wind becomes equal to the geostrophic wind. Any solution of the quadratic which has a real physical meaning must therefore give $V = G$ when the radius of curvature r is increased indefinitely.

We shall consider first the equation which is appropriate to anticyclonic isobars, which have high pressure on the concave side. The equation is

$$\frac{V^2}{r} - 2\omega \sin \phi \, (V - G) = 0 \qquad \qquad \ldots\ldots(9).$$

Solving this in the usual way, we find

$$V = r\omega \sin \phi \left\{ 1 \mp \sqrt{1 - \frac{2G}{r\omega \sin \phi}} \right\} \qquad \ldots\ldots(10).$$

Expanding the expression under the radical, we find

$$V = r\omega \sin \phi \left\{ 1 \mp \left(1 - \frac{G}{r\omega \sin \phi} - \frac{\frac{1}{2}G^2}{r^2\omega^2 \sin^2 \phi} - \ldots \right) \right\} \quad \ldots\ldots(11).$$

The upper sign gives $\quad V = G + \frac{1}{2}G^2 / r\omega \sin \phi + \ldots$ etc. $\qquad \ldots\ldots(12),$

which gives $V = G$ when $r = \infty$, and so yields a solution which is continuous near straight isobars. The lower sign gives

$$V = 2r\omega \sin \phi - G - \tfrac{1}{2}G^2/r\omega \sin \phi - \text{etc.} \qquad \ldots\ldots(13).$$

For indefinitely great radius of curvature r, equation (13) yields indefinitely great velocities, and therefore this solution of the gradient wind equation requires an infinite discontinuity near straight isobars, since in a system of straight isobars the wind is finite and equal to G. Such an arrangement would also require an infinite amount of energy to set it in motion, and this is not available in the earth's atmosphere.

Equation (13) requires that the velocity should decrease as the gradient increases, while equation (12) requires that the velocity and the gradient should increase together. When the gradient is zero, and G zero, equation (13) appears at first sight to demand a finite velocity. But if there is no gradient of pressure, no isobars can be drawn, and r in equation (13) cannot then be defined, so that the second solution then becomes indeterminate. The known facts of observation are in direct contradiction to the results here derived from a consideration of equation (13).

Thus the only solution of equation (9) which can extend out to a region of straight isobars is given by equation (10) with the negative sign before the radical. In an anticyclone, therefore, the only physically significant solution of the gradient wind equation is

$$V = r\omega \sin \phi \, \{1 - \sqrt{1 - 2G/r\omega \sin \phi}\} \qquad \ldots\ldots(14).$$

It should be noted that $V < r\omega \sin \phi$. Now the rate of rotation of the horizon about the vertical is $\omega \sin \phi$. If, in fig. 55, C is the centre of curvature of the isobar at the point P, and $CP = r$, the rate of rotation relative to the earth of the air at P about the point C is V/r, and thus the rate of rotation of the air in space is $(\omega \sin \phi - V/r)$ in a counter-clockwise direction. This is a positive quantity since $V < r\omega \sin \phi$. Thus we arrive at the result that the solution of the gradient wind equation which is significant represents a motion which is counter-clockwise in space. The anticyclone therefore rotates in space in the same direction as the earth, but as its rate of rotation is less than that of the earth, it appears to an observer on the earth to be a clockwise circulation. This fact will be reconsidered in some of its dynamical bearings later, in Chapter XVIII.

The solution which we rejected as physically inappropriate, having the lower sign in equation (10), gives $V > r\omega \sin \phi$, and consequently represents a circulation which rotates faster than the earth beneath it, so that it is a clockwise rotation in space, and therefore in the opposite sense to the earth's rotation. There is no mechanism in the atmosphere capable of producing large-scale rotations in the opposite sense to the earth's rotation, and the second solution should therefore be regarded as an algebraic accident which has no physical significance.

For the cyclone, the gradient wind equation is the same as that for the anticyclone, with the sign of r changed. The solution of the equation is therefore

$$V = r\omega \sin \phi \, \{\sqrt{1 + 2G/r\omega \sin \phi} - 1\} \qquad \ldots\ldots(15).$$

This solution is continuous near straight isobars, being in fact the solution given in (14) with the sign of r changed. Since the circulation represented by (15) is counter-clockwise relative to the earth it is also counter-clockwise in space, and thus has in space the same sense of rotation as the earth. The rejected solution for a cyclone would have a negative sign in front of the radical in (15), and would therefore represent a clockwise rotation relative to the earth, the velocity being then greater than $r\omega \sin \phi$ (arithmetically). This solution would thus represent a circulation which has in space the sense of rotation which is opposed to that of the earth.

Thus the rejected solution of the gradient wind equation for cyclone or anticyclone represents a circulation whose direction is opposite to that of the earth beneath it; and we may be satisfied that in rejecting it we are rejecting a hypothetical system which cannot come into existence on a large scale, and which has never been observed in the earth's atmosphere.

A point of some dynamic interest in connection with the solutions derived above is that for the anticyclone V is always less than $r\omega \sin \phi$ while for the cyclonic system there is no limit to the value of V so far as the algebraic form of the equation is concerned.

§ 116. *Direct derivation of the gradient wind equation*

In fig. 56 let C be the centre of curvature of the isobar at P. Let the velocity at P be V, in the clockwise direction, so that we restrict the discussion for the moment to the case of anticyclonic isobars. Let the angular velocity about C of the air at P be ζ, so that $V = r\zeta$. The angular velocity in space about C of the

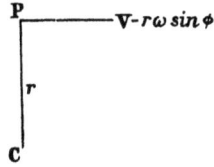

Fig. 56. Direct derivation of the gradient wind equation.

air at P is $(\omega \sin \phi - \zeta)$, while the angular velocity in space of the observer on the earth's surface at P about the point C is $\omega \sin \phi$. The condition from which we derive the equation of motion may be stated thus: "The gradient of pressure balances the acceleration of the air relative to the system in which the pressure is observed, or relative to the observer who measures the pressure".

Now the acceleration of the air at P relative to the observer at P on the earth's surface

$$= \text{Acceleration of the air at } P \text{ relative to } C$$
$$- \text{Acceleration of the observer at } P \text{ relative to } C \quad \ldots (16)$$
$$= r(\omega \sin \phi - \zeta)^2 - r\omega^2 \sin^2 \phi = -r\zeta(2\omega \sin \phi - \zeta) \quad \ldots (17).$$

The acceleration is along PC, and must be balanced by the pressure gradient.

Hence $\qquad \dfrac{1}{\rho}\dfrac{\partial p}{\partial r} = -r\zeta\,(2\omega\sin\phi - \zeta)$

$$= -2\omega\sin\phi\,.\,V + V^2/r$$

or $\qquad -\dfrac{1}{\rho}\dfrac{\partial p}{\partial r} = \dfrac{\text{pressure gradient}}{\rho} = 2\omega\sin\phi\,.\,V - V^2/r \qquad \ldots\ldots(18),$

which is a repetition of equation (3) above.

The cyclonic counterpart is derived by drawing PC in the opposite direction to that drawn in the figure, and this has only the effect of changing the sign of r in the whole of the analysis, leading to equation (4) instead of (3). This may be regarded as an independent derivation of the magnitude and direction of the so-called "deviating force".

There is one interesting detail in the above derivation of the gradient wind equation which is worthy of note. In (16) one might be tempted to overlook the second term, which is equivalent to overlooking the need to refer the accelerations to the same system of co-ordinates as that in which the pressure gradients are measured. This is perhaps the only point in meteorology at which relativity of motion demands close attention. The omission of the second term in (16) is equivalent to the omission of the second term in (17), of which the remaining term is always positive, and we appear to reach the conclusion that pressure will increase outwards from the centre in all circulations, whether clockwise or counter-clockwise, and that centres of high pressure are impossible.

§ 117. *The effect of changing pressure distribution*

In $2\omega\sin\phi = l$, the equations of motion for purely horizontal motion may be written

$$\frac{du}{dt} - lv = -\frac{1}{\rho}\frac{\partial p}{\partial x} \qquad\qquad \ldots\ldots(19),$$

$$\frac{dv}{dt} + lu = -\frac{1}{\rho}\frac{\partial p}{\partial y} \qquad\qquad \ldots\ldots(20).$$

Multiplying the second equation by $i = \sqrt{-1}$, and adding, we obtain a single equation

$$\frac{d}{dt}(u+iv) = -\frac{1}{\rho}\left(\frac{\partial p}{\partial x} + i\,\frac{\partial p}{\partial y}\right) - il\,(u+iv),$$

or $\qquad u+iv = \dfrac{i}{l}\dfrac{d}{dt}(u+iv) + \dfrac{i}{l\rho}\left(\dfrac{\partial p}{\partial x} + i\,\dfrac{\partial p}{\partial y}\right) \qquad \ldots\ldots(21).$

Differentiating equation (21) partially with respect to time, we find

$$\frac{\partial}{\partial t}(u+iv) = \frac{i}{l}\frac{\partial}{\partial t}\frac{d}{dt}(u+iv) + \frac{i}{l\rho}\left(\frac{\partial\dot{p}}{\partial x} + i\,\frac{\partial\dot{p}}{\partial y}\right) \qquad \ldots\ldots(22),$$

where ρ is assumed constant and $\dot{p} = \dfrac{\partial p}{\partial t}$.

Substituting for $\frac{\partial}{\partial t}(u+iv)$ in equation (21)

$$\left(u\frac{\partial}{\partial x}+v\frac{\partial}{\partial y}\right)(u+iv)+il(u+iv)$$

$$=-\frac{1}{\rho}\left(\frac{\partial p}{\partial x}+i\frac{\partial p}{\partial y}\right)-\frac{i}{l\rho}\left(\frac{\partial\dot{p}}{\partial x}+i\frac{\partial\dot{p}}{\partial y}\right)-\frac{i}{l}\frac{\partial}{\partial t}\frac{d}{dt}(u+iv)\ \ \ldots\ldots(23).$$

This equation was given by Brunt and Douglas,[*] who showed that in general the third term on the right-hand side of the equation may be neglected by comparison with the others. The first term on the left-hand side represents the ordinary accelerations of the air in its path, and if, following the same authors, we neglect these terms, we are left with

$$u+iv=\frac{i}{l\rho}\left(\frac{\partial p}{\partial x}+i\frac{\partial p}{\partial y}\right)-\frac{1}{l^2\rho}\left(\frac{\partial\dot{p}}{\partial x}+i\frac{\partial\dot{p}}{\partial y}\right)\ \ \ \ldots\ldots(24).$$

Here $\dot{p}$ is the rate of change of pressure, and to a rough degree of approximation the field of $\dot{p}$ may be taken as the field of isallobars obtained by plotting the barometric tendencies, which give the changes of pressure during three hours. The second term on the right-hand side of equation (24) thus represents a component of velocity of magnitude proportional to the gradient of the isallobars, and directed normally across the isallobars into the low values. The first term on the right-hand side is merely another way of expressing the geostrophic wind.

Thus we derive the result that the effect of changing pressure distribution can be allowed for by adding to the geostrophic wind a component blowing across the isallobars at right angles into the low values of isallobars, with a magnitude proportional to the gradient of the isallobars. A scale similar to a geostrophic wind scale can readily be made for measuring the magnitude of this component, or the ordinary geostrophic wind scale may be used for the purpose, the results directly read off being corrected by a factor appropriate to the scale of the charts, and the units in which they are plotted.

Brunt and Douglas found that the isallobaric component of wind frequently amounted to as much as 5 metres/sec, and they accounted for the rather unexpected course of development of the pressure distribution on a number of occasions by the effect of this component. Some further consideration will be given in a later chapter to the effect of this component on weather.

It was necessary to make a number of simplifying assumptions above in order to reduce the mathematical aspect of the question to a reasonably tractable form. If we refer back to equation (23) above, and only assume that the third term on the right-hand side may be neglected, while retaining the first term on the left-hand side, we may describe the resulting form of equation (23) as the standard form of the gradient wind equation, with the addition to the pressure gradient of a force directed around the isallobars, of a magnitude proportional to the gradient of the isallobars.

[*] *Mem. R. Met. Soc.* **3**, No. 22.

The idea that there is convergence of air into an isallobaric low, and divergence out of an isallobaric high, is borne out by the fact that an isallobaric low is usually a region of heavy cloud, while an isallobaric high is usually a region of clear sky. This appears to be the best confirmation of the general nature of the results derived above.

Möller and Sieber* attempted a direct comparison of the isallobaric components computed from the above equation, with the results of pilot balloon observations. They made no allowance for the curvature of the path of the air. It is probable that most of the pilot balloon ascents were made in anticyclonic conditions, so that, in accordance with equation (7) above, the geostrophic wind should on the whole be an underestimate of the true wind, as actually appeared from the results which they derived. In view of the uncertainties underlying such a comparison, it appears preferable to accept the phenomena of weather as described above as a better indication of the truth. The implications of the last term on the right-hand side of equation (23) above are obscure, and it is not to be expected that the whole truth is represented by equation (24).

§ 118. *An alternative form of the equations of motion in two dimensions*

Equations (19) and (20) above can also be written in a slightly different form, which is occasionally useful. Let u_G, v_G, be the components of the geostrophic wind along the axes of x and y respectively. Then the equations of motion can be written

$$\frac{du}{dt} = l\left(v - v_G\right) \qquad \qquad \text{......(25)}$$

$$\frac{dv}{dt} = -l\left(u - u_G\right) \qquad \qquad \text{......(26).}$$

This form of the equations shows that the acceleration of any element of air is at right angles to the vectorial difference between the actual wind and the geostrophic wind.

§ 119. *The variation of pressure gradient and geostrophic wind with height*

The discussion which follows relates only, so far as variation of wind with height is concerned, to those levels of the atmosphere which are beyond the reach of the effects of surface turbulence. If the distribution of pressure at any one level, and the distribution of temperature in the horizontal between this level and any other level, are known, the pressure gradient at this second level can be evaluated.

Let p, T, be the pressure, absolute temperature and density at a point (x, y, z); the axis of z is vertical, and the axes of x and y are any convenient horizontal axes. Let u and v be the components of the geostrophic wind along the axes of x and y respectively.

* *Ann. Hydr.* **7**, 1937, p. 312.

From the statical equation

$$\frac{\partial p}{\partial z} = -g\rho$$

it follows that

$$\frac{1}{p}\frac{\partial p}{\partial z} = -\frac{g}{RT} \qquad\qquad ......(27).$$

Differentiating this equation with respect to x, and changing the order of the differentiations with respect to x and z, we have

$$\frac{\partial}{\partial z}\left(\frac{1}{p}\frac{\partial p}{\partial x}\right) = \frac{\partial^2}{\partial z\,\partial x}\log p = \frac{\partial}{\partial x}\left(\frac{1}{p}\frac{\partial p}{\partial z}\right) = \frac{g}{RT^2}\frac{\partial T}{\partial x} \qquad ......(28).$$

This equation may be integrated, yielding

$$\left(\frac{1}{p}\frac{\partial p}{\partial x}\right)_1 - \left(\frac{1}{p}\frac{\partial p}{\partial x}\right)_2 = \frac{g}{R}\int_{z_2}^{z_1}\frac{1}{T^2}\frac{\partial T}{\partial x}\,dz \qquad ......(29).$$

This equation may be used to compute the pressure gradient at a height z_1, if the pressure gradient is known at a height z_2, and the horizontal temperature distribution is known between the levels z_1 and z_2. The only assumption involved in the derivation of equations (28) and (29) is the truth of the statical equation.

The components of the geostrophic wind are given by the equations

$$\left.\begin{aligned} 2\omega\sin\phi\,.\,\rho u &= -\frac{\partial p}{\partial y} \\[2mm] 2\omega\sin\phi\,.\,\rho v &= \frac{\partial p}{\partial x} \end{aligned}\right\} \qquad ......(30)$$

which may be written in the form

$$\left.\begin{aligned} \frac{2\omega\sin\phi}{R}\frac{u}{T} &= -\frac{1}{p}\frac{\partial p}{\partial y} \\[2mm] \frac{2\omega\sin\phi}{R}\frac{v}{T} &= \frac{1}{p}\frac{\partial p}{\partial x} \end{aligned}\right\} \qquad ......(31).$$

Substituting from the second of these equations for $\dfrac{1}{p}\dfrac{\partial p}{\partial x}$ in (28), we obtain

$$\frac{\partial}{\partial z}\left(\frac{v}{T}\right) = \frac{g}{2\omega\sin\phi}\frac{1}{T^2}\frac{\partial T}{\partial x} \qquad ......(32).$$

And similarly

$$\frac{\partial}{\partial z}\left(\frac{u}{T}\right) = -\frac{g}{2\omega\sin\phi}\frac{1}{T^2}\frac{\partial T}{\partial y} \qquad ......(33).$$

Equations (32), (33) may also be written in the form

$$\left.\begin{aligned} \frac{\partial u}{\partial z} &= \frac{u}{T}\frac{\partial T}{\partial z} - \frac{g}{2\omega\sin\phi}\frac{1}{T}\frac{\partial T}{\partial y} \\[2mm] \frac{\partial v}{\partial z} &= \frac{v}{T}\frac{\partial T}{\partial z} + \frac{g}{2\omega\sin\phi}\frac{1}{T}\frac{\partial T}{\partial x} \end{aligned}\right\} \qquad ......(34).$$

If for g we substitute $-\dfrac{\text{I}}{\rho}\dfrac{\partial p}{\partial z}$ in these equations, they can be readily reduced to the form

$$\left.\begin{aligned}\frac{\partial u}{\partial z} &= \frac{-\text{I}}{2\rho T\omega\sin\phi}\left\{\frac{\partial p}{\partial y}\frac{\partial T}{\partial z}-\frac{\partial p}{\partial z}\frac{\partial T}{\partial y}\right\}\\\frac{\partial v}{\partial z} &= \frac{\text{I}}{2\rho T\omega\sin\phi}\left\{\frac{\partial p}{\partial x}\frac{\partial T}{\partial z}-\frac{\partial p}{\partial z}\frac{\partial T}{\partial x}\right\}\end{aligned}\right\} \quad\ldots\ldots(35).$$

The condition that u and v shall not vary with height, or that

$$\frac{\partial u}{\partial z}=\frac{\partial v}{\partial z}=\text{o},$$

may be written

$$\frac{\partial p}{\partial x}:\frac{\partial p}{\partial y}:\frac{\partial p}{\partial z}=\frac{\partial T}{\partial x}:\frac{\partial T}{\partial y}:\frac{\partial T}{\partial z} \quad\ldots\ldots(36).$$

This is the condition that the tangent planes to the isobaric and isothermal surfaces through any point shall coincide; it is the condition that the isobaric surfaces shall also be isothermal surfaces. This condition was first enunciated by W. H. Dines.[*]

Equations (32), (33) can be integrated in the form

$$\left.\begin{aligned}\frac{u}{T} &= \frac{u_0}{T_0}-\frac{g}{2\omega\sin\phi}\int_{z_0}^{z}\frac{\text{I}}{T^2}\frac{\partial T}{\partial y}\,dz\\\frac{v}{T} &= \frac{v_0}{T_0}+\frac{g}{2\omega\sin\phi}\int_{z_0}^{z}\frac{\text{I}}{T^2}\frac{\partial T}{\partial x}\,dz\end{aligned}\right\} \quad\ldots\ldots(37),$$

where u_0, v_0 are the components of the geostrophic wind at height z_0, where the absolute temperature is T_0. Thus the geostrophic wind can be regarded as made up of two components, (a) a component equal to the geostrophic wind at level z_0 reduced in the ratio T/T_0, and (b) a thermal wind whose components are

$$-\frac{gT}{2\omega\sin\phi}\int_{z_0}^{z}\frac{\text{I}}{T^2}\frac{\partial T}{\partial y}\,dz \quad\text{and}\quad \frac{gT}{2\omega\sin\phi}\int_{z_0}^{z}\frac{\text{I}}{T^2}\frac{\partial T}{\partial x}\,dz.$$

The magnitude of the thermal wind will increase steadily with height in regions where the temperature gradient maintains its general direction unchanged. The signs of the components show that it blows around low temperature in the same sense that the geostrophic wind blows around low pressure, and that it keeps low temperature to its left.

§ 120.   *Some applications of the above formulae*

Some special cases are worth considering in further detail. (a) If low temperature is associated with low pressure, the wind will increase with height. (b) If low temperature is associated with high pressure, the wind will decrease with height and may even be reversed at moderate heights if the horizontal gradients of temperature be large. (c) If the geostrophic wind blows from high temperature to low, it will veer with height. (d) If the geostrophic wind blows from low temperature to high, it will back with height.

[*] *Nature*, **99**, 1917, p. 24.

These results are not easy to apply to individual cases, and they are perhaps more useful when inverted from the form given above. Thus (c) and (d) may be interpreted to indicate that a wind veering with height will bring up higher temperatures, and that a wind backing with height will bring up lower temperatures.

The same kind of reasoning can be applied to the larger scale movements of the atmosphere. Since the main latitude variation in the troposphere is an increase of temperature from pole to equator, the thermal wind in the troposphere is from West to East, so that the westerly component of wind should increase with height. This fact accounts for the tendency of westerly winds to increase with height, and of easterly winds to decrease with height. Within the stratosphere the temperature gradient is reversed, and the result is to cause a steady diminution with height of the westerly component of wind.

A useful summary of some investigations of wind in the troposphere and stratosphere will be found in a paper by Dobson.* On the whole the wind increases with height in the upper troposphere, and then falls off rapidly in the lower stratosphere.

By the use of equations (32) and (33) above it is readily possible to compute the horizontal temperature gradients for a given wind distribution, if the temperatures at different levels are known. For most cases it is sufficiently accurate to adopt mean values of the temperatures at different levels. The equations are written in the form

$$
\left.
\begin{aligned}
d\left(\frac{u}{T}\right) &= -\frac{g}{2\omega \sin\phi}\, \frac{1}{T^2}\, \frac{\partial T}{\partial y}\, dz \\[2mm]
d\left(\frac{v}{T}\right) &= \frac{g}{2\omega \sin\phi}\, \frac{1}{T^2}\, \frac{\partial T}{\partial x}\, dz
\end{aligned}
\right\} \qquad \ldots\ldots(38).
$$

The differences $d\left(\dfrac{u}{T}\right)$ and $d\left(\dfrac{v}{T}\right)$ are computed for any convenient steps of dz, say $dz = 1$ km, and the corresponding values of $\dfrac{\partial T}{\partial x}$, $\dfrac{\partial T}{\partial y}$ are then obtained by a simple computation. An example of such a computation is shown in the table below. Here the height interval dz is 1 km. The observed components of velocity to east and north, u and v, are given in metres per second in columns 3 and 4. The mean temperatures at different heights are given in

Height in km	T	u	v	$100u/T$	Diff.	$\frac{\partial T}{\partial y}$ 10^5	$100v/T$	Diff.	$\frac{\partial T}{\partial x}$ 10^5	$10^5 q$	Azim.
6	246	17·7	−10·3	7·19			−4·15				
	250				2·03	1·45		−1·17	0·84	1·67	30°
5	252	13·0	− 7·5	5·16			−2·98				
	255				1·92	1·44		−2·40	1·80	2·30	51°
4	259	8·4	− 1·5	3·24			−0·58				
	262				0·82	0·65		−0·81	0·64	0·91	44°
3	265	6·4	0·6	2·42			0·23				
	267				−0·36	−0·30		−0·77	0·63	0·70	116°
2	270	7·5	2·7	2·78			1·00				
	273				1·22	1·05		0·09	−0·08	1·05	356°
1	276	4·3	2·5	1·56			0·91				

* Q.J. Roy. Met. Soc. **46**, 1920, p. 54.

column 2. The remainder of the computation is carried out in accordance with the equations above, the last column but one giving $10^5 q$, which is the horizontal temperature gradient in degrees C per 100 km. The last column gives the azimuth of the temperature gradient.

The magnitude of the effect is readily seen from this example. A horizontal temperature gradient of 1° C per 100 km produces in a range of height of 1 km a thermal wind of about 3 m per second. If this gradient of temperature persisted through 2 km of height, the thermal wind would amount to 6 m per second, assuming a mean temperature of about 270° A. The amount of the vectorial wind change produced in a given range of height by a given horizontal temperature gradient is very nearly inversely proportional to the absolute temperature, so that with a constant horizontal temperature gradient the vectorial rate of change of wind with height will itself increase with height, on account of the decrease of temperature with height.

It will be recalled that reference was made above (Chapter 1) to the high correlation which W. H. Dines found to exist between pressure and temperature at different levels in the middle troposphere, yielding the result that on the average high temperature is associated with high pressure. This does not mean that the isothermal and isobaric surfaces are nearly coincident, so that the wind should change relatively little with height.* In the diagram which Dines gave of the mean distribution of upper air temperatures and pressures over anticyclones the average inclination of the isotherms is about three times as great as that of the isobars. The wind current which would correspond with this distribution would be a current increasing with height, while maintaining a steady direction. Direct observations show the frequent occurrence of large changes of wind with height at all levels, though the greatest changes are usually confined to the levels from 1 to 3 km.

An examination of the results derived by Cave† shows that solid currents, i.e. currents whose velocity and direction change little with height, are moderately frequent in the conditions in which observations are possible. Moreover, it is difficult to see how the depressions of middle latitudes could persist if the wind varied very greatly with height. Of Cave's ascents, 200 in number, excluding 10 rather indefinite cases,

(a) Thirty-two showed a solid current, the wind attaining the gradient value and not increasing beyond this limit.

(b) Forty-nine showed a considerable increase of velocity with height, surpassing the gradient value. These winds, in the typical cases, were associated with cyclonic systems of some intensity to West or North. The directions were mainly grouped about SE and WNW, and the surface temperatures were in most cases such as to account for a thermal wind in the right direction.

(c) Twenty-seven showed a marked decrease of velocity with height. These winds were predominantly easterly, and the fall off in velocity was explainable as the result of a decrease of temperature from South to North.

---

* See Douglas, *Mem. Roy. Met. Soc.* **3**, No. 29, especially pp. 160, 161.
† *The Structure of the Atmosphere in Clear Weather.*

(d) Thirty-four showed marked change of wind direction at different heights. It is difficult to summarise these in any manner.

(e) Thirty-seven showed an upper wind blowing out of the centre of low pressure, frequently with reversals in a lower layer.

(f) Eleven were followed well into the stratosphere, and most of these showed a definite fall of velocity in the stratosphere, without any great change of direction. The decrease of velocity in the stratosphere indicates a reversal of the horizontal temperature gradients in going into the stratosphere.

Richardson and Munday,* in an investigation of the possibility of treating the atmosphere as a single layer, came to the conclusion that barotropy is not even an approximation to the truth in the atmosphere. These authors examined the results of a large number of sounding balloons, and came to the conclusion that the equations which represent the motions of the atmosphere cannot be summarised in the form of the equation for a single layer. We need not therefore be surprised that the motion of the atmosphere should deviate widely from the motion of a barotropic fluid.

C. K. M. Douglas† compared the variations of temperature predicted by equations (35) above with the actual observed changes of temperature, and found that the correlation coefficient between the observed and predicted changes was a little less than 0·5, both for 6 hour and 24 hour changes, but was greater when only the occasions of large changes of temperature were considered. The relatively low value of the correlation coefficient may be due partly to the uncertainty of the determination of wind velocity by the pilot balloon method, partly to large vertical movements of air masses, and partly to the deviation of the winds from the geostrophic value which is assumed as the basis of the theory developed above. The uncertainty of the wind determination by the pilot balloon method is probably the most serious source of error. It should be noted that equations (30) above assume that the observed wind is equal to the geostrophic wind at the same level. In equation (7) it was shown that in steady conditions the geostrophic wind is an underestimate of the wind in an anticyclone and an overestimate in a cyclone, so that the assumption of geostrophic wind leads to an error which may be systematic if the observations are made in conditions which are predominantly cyclonic or anticyclonic, though the error is not likely to be great in average conditions.

Douglas assumed that the air masses which he investigated moved without change of temperature. It then follows that

$$\frac{dT}{dt} = \frac{\partial T}{\partial t} + u\,\frac{\partial T}{\partial x} + v\,\frac{\partial T}{\partial y} = 0.$$

$\partial T/\partial x$ and $\partial T/\partial y$ could be computed from the observed winds by the method outlined above, and $\partial T/\partial t$ computed from the equation

$$\frac{\partial T}{\partial t} = -u\,\frac{\partial T}{\partial x} - v\,\frac{\partial T}{\partial y}.$$

* Mem. Roy. Met. Soc. **1**, No. 2.
† Ibid. **1**, No. 7.

But from equation (34), multiplying by v and u and subtracting,

$$v\,\frac{\partial u}{\partial z} - u\,\frac{\partial v}{\partial z} = -\frac{g}{2\omega \sin \phi}\,\frac{1}{T}\left(u\,\frac{\partial T}{\partial x} + v\,\frac{\partial T}{\partial y}\right).$$

Hence

$$\frac{\partial T}{\partial t} = \frac{2\omega \sin \phi}{g}\,T\left(v\,\frac{\partial u}{\partial z} - u\,\frac{\partial v}{\partial z}\right) \qquad \ldots\ldots(39).$$

Over a finite interval of time Δt the change of temperature ΔT is given by
$$\frac{\Delta T}{\Delta t} = \frac{2\omega \sin \phi}{g}\,T \times \text{observed wind at height } z \times (\text{change of component at right}$$
angles to observed wind at height z, between the levels $z - \frac{1}{2}\Delta z$ and $z + \frac{1}{2}\Delta z) \div \Delta z$. When the relationship is used in this form, the amount of computation is reduced to a minimum and the predicted temperature is readily evaluated.

Douglas suggested that the low correlation coefficients which he obtained were in part due to the vertical motion of the masses. This motion would lead to another term in the equation for $\partial T/\partial t$, a term involving the vertical component of velocity. But it is possible that the low coefficients were also in part due to the assumption that the air moved without change of temperature. It would appear safer to assume that the air moved adiabatically, though it is clear that even that assumption is only a very crude approximation. For adiabatic motion

$$\frac{d}{dt}\left(\frac{p^{\gamma-1}}{T^{\gamma}}\right) = 0, \qquad \frac{\gamma}{\gamma-1}\,\frac{1}{T}\,\frac{dT}{dt} = \frac{1}{p}\,\frac{dp}{dt},$$

$$\frac{\partial T}{\partial t} = -u\,\frac{\partial T}{\partial x} - v\,\frac{\partial T}{\partial y} + \frac{\gamma-1}{\gamma}\,\frac{T}{p}\left(\frac{\partial p}{\partial t} + u\,\frac{\partial p}{\partial x} + v\,\frac{\partial p}{\partial y}\right) \qquad \ldots\ldots(40).$$

It is again assumed that the motion is purely horizontal, and the w term in the full equations is omitted. If the motion is along the isobars at each level

$$u\,\frac{\partial p}{\partial x} + v\,\frac{\partial p}{\partial y} = 0,$$

and the equation reduces to

$$\frac{\partial T}{\partial t} = -u\,\frac{\partial T}{\partial x} - v\,\frac{\partial T}{\partial y} + \frac{\gamma-1}{\gamma}\,\frac{T}{p}\,\frac{\partial p}{\partial t}$$

$$= \frac{2\omega \sin \phi}{g}\,T\left(v\,\frac{\partial u}{\partial z} - u\,\frac{\partial v}{\partial z}\right) + \frac{\gamma-1}{\gamma}\,\frac{T}{p}\,\frac{\partial p}{\partial t} \qquad \ldots\ldots(41).$$

This equation would make it possible to take account of the changes of pressure which may be brought about by advection in the upper air.

Errors will arise in the forecasts of temperature by the above method on account of neglected physical processes such as radiation, turbulent mixing, vertical motion, which may be accompanied by condensation of water-vapour or evaporation of water drops falling through the air. It is therefore not surprising that the coefficient of correlation between the predicted and observed temperatures should come out rather low.

CHAPTER X

SURFACES OF DISCONTINUITY

§ 121. *The slope of a surface of discontinuity with steady motion*

IT was first shown by Helmholtz[*] that currents of air at different temperatures, moving with different velocities, could remain in steady motion with a surface of discontinuity separating them, this surface being inclined to the horizontal at an angle whose magnitude depends upon the differences of temperature and velocity on the two sides of the surface. We shall first discuss briefly the relation of the discontinuities of temperature and velocity to the slope of the surface for the case of steady motion in the horizontal. This is admittedly a simple case, which may not be reproduced very frequently in practice, but its consideration will be of value as a preliminary to the discussion of more complicated cases. We shall, however, bear in mind that the analysis will not be strictly applicable to the more frequent phenomenon of warm air climbing up a sloping surface above cold air.

In fig. 57 let OX, assumed to be a straight line, be the intersection of the surface of discontinuity with the horizontal plane, and let OY be at right angles with OX in the horizontal plane, while the axis OZ is vertical. Let OS be the intersection of the surface of discontinuity with the plane YOZ, the angle YOS being α. Let ρ_1, u_1, and ρ_2, u_2, be the densities and velocities parallel to OX, of the air on the two sides of the surface of discontinuity. The velocities are taken to be both positive in the direction OX. Further let OX make an angle β with the East-West line, so that the components of velocity towards East and $u_1 \cos \beta$ and $u_2 \cos \beta$.

In the plane YOZ draw a small rectangle $ABCD$, whose sides are dy and dz. If p_a, p_b, p_c, p_d denote the pressures at A, B, C, D, respectively,

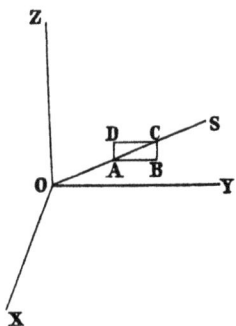

Fig. 57. Surface of discontinuity.

$$p_c - p_a = p_b - p_a + p_c - p_b = \left(\frac{\partial p}{\partial y}\right)_1 dy + \left(\frac{\partial p}{\partial z}\right)_1 dz,$$

$$= p_c - p_d + p_d - p_a = \left(\frac{\partial p}{\partial y}\right)_2 dy + \left(\frac{\partial p}{\partial z}\right)_2 dz,$$

since the pressure is continuous across the surface. Hence

$$\tan \alpha = \frac{dz}{dy} = -\frac{\left(\frac{\partial p}{\partial y}\right)_1 - \left(\frac{\partial p}{\partial y}\right)_2}{\left(\frac{\partial p}{\partial z}\right)_1 - \left(\frac{\partial p}{\partial z}\right)_2} \qquad \text{......(1).}$$

[*] *Wissenschaftliche Abhandlungen*, **3**.

Equations (39), (40) and (41) of p. 171 will apply separately to each of the air masses. It is further assumed that the motion is purely horizontal and parallel to the axis of x, and that frictional forces need not be considered, so that $w = v = 0$, and $X = Y = Z = 0$. For steady linear motion, in each air mass separately

$$2\omega u \sin \phi = -\frac{1}{\rho} \frac{\partial p}{\partial y} \qquad \ldots\ldots(2),$$

$$g - 2\omega u \cos \phi \cos \beta = -\frac{1}{\rho} \frac{\partial p}{\partial z} \qquad \ldots\ldots(3).$$

By substitution of $\partial p/\partial y$, $\partial p/\partial z$ from (2) and (3) in (1), we find

$$\tan \alpha = -\frac{2\omega \sin \phi \, (\rho_1 u_1 - \rho_2 u_2)}{g \, (\rho_1 - \rho_2) - 2\omega \cos \phi \cos \beta \, (\rho_1 u_1 - \rho_2 u_2)} \qquad \ldots\ldots(4).$$

When the vertical component of acceleration is neglected, the last equation reduces to

$$\tan \alpha' = -\frac{2\omega \sin \phi \, (\rho_1 u_1 - \rho_2 u_2)}{g \, (\rho_1 - \rho_2)} \qquad \ldots\ldots(5).$$

The angle α' may be regarded for the moment as an approximation to α. When $\beta = \pi/2$ and the discontinuity runs along the North-South line, α and α' are equal.

If it is assumed that the differences of density are due to differences of temperature alone, so that at any point on the surface of discontinuity $\rho_1 T_1 = \rho_2 T_2$, equation (4) becomes

$$\tan \alpha = -\frac{2\omega \sin \phi \, (u_1 T_2 - u_2 T_1)}{g \, (T_2 - T_1) - 2\omega \cos \phi \cos \beta \, (u_1 T_2 - u_2 T_1)} \qquad \ldots\ldots(6).$$

§ 122. *Discontinuity of velocity alone; density uniform*

Here $\rho_1 = \rho_2$, and equation (4) reduces to

$$\tan \alpha = \tan \phi \sec \beta \qquad \ldots\ldots(7).$$

When the intersection of the surface of discontinuity with the earth's surface runs West-East, and $\beta = 0$, $\tan \alpha = \tan \phi$, and the surface of discontinuity is parallel to the earth's axis. When the surface of discontinuity runs North-South, $\beta = \pi/2$, and $\tan \alpha$ is infinite, so that the surface of discontinuity is vertical. These are two special cases of a general rule that when the discontinuity is one of velocity alone, the surface of discontinuity is parallel to the earth's axis. This result is readily derived from a consideration of the direction cosines of the polar axis of the earth referred to the co-ordinate system used above. They are $\cos \phi \sin \beta$, $\cos \phi \cos \beta$ and $\sin \phi$. The direction cosines of the normal to the surface of discontinuity are 0, $-\sin \alpha$ and $\cos \alpha$. The two are therefore at right angles since

$$-\sin \alpha \cos \phi \cos \beta + \sin \phi \cos \alpha = 0$$

from equation (7).

§ 123. *Discontinuity of density alone*

Let $u_1 = u_2 = u$, and $\rho_1 \neq \rho_2$. Equation (4) now becomes

$$\tan \alpha = \frac{2\omega \sin \phi . u}{2\omega \cos \phi \cos \beta . u - g}$$

$$= \frac{\tan \phi \sec \beta}{1 - g/2\omega \cos \phi \cos \beta . u}.$$

The value of α depends on the value of $g/2\omega \cos \phi . u$. The values of u which occur in the atmosphere are of the order of 10 metres per second or 10^3 cm/sec. In C.G.S. units, $g = 1000$, $\omega = 7 \cdot 3 \times 10^{-5}$,

$$\frac{g}{2\omega u} = \frac{10^3 \times 10^5}{2 \times 7 \cdot 3 \times 10^3} = 7 \times 10^3 \text{ approximately.}$$

Hence $\tan \alpha$ is a very small fraction of $\sin \phi$, which means that the lighter fluid will float on the heavier with a practically horizontal surface of separation.

§ 124. *Approximate form of equation* (6)

The first point to be decided is the relative magnitude of the two terms in the denominator of equation (4) or (6). The difference between u_1 and u_2 is usually of the same order of magnitude as either of the two velocities, so that only a small error is involved in writing $T(u_1 - u_2)$ for $u_1 T_2 - u_2 T_1$, T being the mean of T_1 and T_2. The denominator then becomes

$$g(T_2 - T_1) - 2\omega \cos \phi \cos \beta . T(u_1 - u_2).$$

Let $\qquad\qquad u_1 - u_2 = 1 \text{ m/sec} = 10^2 \text{ cm/sec,}$

and let $\qquad T_2 - T_1 = 1° \text{ C,} \quad T = 275°, \quad \omega \cos \phi = 4 \times 10^{-5},$

$$\frac{g(T_2 - T_1)}{2\omega \cos \phi . T(u_1 - u_2) \cos \beta} = \frac{10^3}{2 \times 4 \times 10^{-5} \times 275 \times 10^2 \cos \beta} = 450 \sec \beta$$

approximately.

The ratio decreases as the temperature difference decreases, or as the velocity difference increases, but for all practical purposes, where the differences of temperature and velocity are appreciable, and of the orders of magnitude which actually occur in the atmosphere, the first term in the denominator will always far exceed the second term, and equation (6) may be written

$$\tan \alpha = - \frac{2\omega \sin \phi (u_1 T_2 - u_2 T_1)}{g(T_2 - T_1)} \qquad \ldots\ldots(8).$$

For most purposes it is legitimate to simplify it still further to

$$\tan \alpha = \frac{2\omega \sin \phi}{g} \frac{T(u_1 - u_2)}{T_1 - T_2} \qquad \ldots\ldots(9).$$

This equation shows that the slope of the surface of discontinuity increases in direct proportion to the difference of the wind velocities on the two sides, and in inverse proportion to the difference of temperatures.

To obtain an idea of the slope of the surface of discontinuity the values of T_1, T_2, u_1, u_2 used in the preceding computation may be inserted in equation (9). Then

$$\tan \alpha = 2 \times 7 \cdot 3 \times 10^{-5} \sin \phi \times 275 \times 10^2 \times 10^{-3}$$
$$= 4 \times 10^{-3} \sin \phi$$

and $\alpha = 10'$ approximately. If Δu and ΔT represent the differences of velocities in metres per second and in degrees C respectively,

$$\tan \alpha = 4 \times 10^{-3} \sin \phi \times \Delta u / \Delta T.$$

Δu is usually of the order of 10, and ΔT may be about $5°$ C, or may be as small as $\frac{1}{2}°$ C. With the differences which usually occur α is never likely to exceed a few degrees, even in extreme cases, and in practice it is usually found to be a small fraction of a degree; except that when $\tan \alpha$ is negative, it is $180° - \alpha$ which is a small angle. The inclination of the surface of discontinuity to the horizontal is however a small angle, if the difference of temperature is such as to be appreciable.

The approximation involved in the neglect of the second term in the denominator of equation (4) or (6) is only of importance when the difference of temperature is very slight. There is a possible small value of $T_1 - T_2$ which will give a vertical surface of separation between the two currents, and between this small difference of temperature and zero difference there is a wide range of possible slope of the surface.

§ 125. *General nature of the results derived above*

Apart from the slight approximation involved in equation (9), the result is readily stated verbally. Imagine the observer placed within the colder current looking towards the warmer current. If he is situated somewhere on OY in fig. 57, then $T_1 < T_2$. If the velocities are measured positive to his left, i.e. along the positive direction of OX, then the condition that $\tan \alpha$ shall be positive, and that the cold air shall lie as a wedge under the warm air, is that the warm air shall move more rapidly than the cold air towards the observer's left. If the cold air moves the more rapidly towards the observer's left, $\tan \alpha$ is negative and the warm air lies as a wedge below the cold air. (In the Southern hemisphere for "left" read "right" above.)

Thus at the surface of separation of a cold easterly current with a warm westerly current to the south of it, or of a cold northerly wind with a warm southerly wind to east of it, the cold air will lie as a wedge under the warm air. Also at the surface of separation of a cool westerly wind with a warm easterly wind south of it, or of a cold northerly wind with a warm southerly wind west of it, the warm air lies as a wedge below the cold air. If the cold air is to lie in the form of a wedge beneath the warm air, one of the following must hold:·

(*a*) If the cold air lies to the north of the warm air, it must have the greater velocity towards west.

(b) If the cold air lies south of the warm air, it must have the greater velocity towards east.

(c) If the cold air lies west of the warm air, it must have the greater velocity towards south.

(d) If the cold air lies east of the warm air, it must have the greater velocity towards north.

Thus it is possible to have a cone of cold air with a centre of high pressure and an anticyclonic circulation, surrounded by warm air on all sides if the anticyclonic circulation is stronger in the cold air than in the warm air around it. The same is obviously true of a long tongue of high pressure colder than its surroundings. The circulation in the cold air must be more vigorous than that in the surrounding warm air. Similarly a cone or tongue of cold air can be maintained with a sharp surface of discontinuity between it and the surrounding warm air when its circulation is cyclonic if the circulation in the cold air is less vigorous than that in the surrounding warm air.

An interesting result which appears from the above is that at the boundary between the westerly winds of middle latitudes and the trade winds the cooler westerly winds should flow over the warmer trade winds. There is also some interest in considering for this case the application of the accurate equation (6). Here $T_2 > T_1$, $u_1 - u_2 > 0$, and $u_1 T_2 - u_2 T_1 > 0$. Hence in equation (6) the two terms in the denominator are of opposite sign. For a given value of $u_1 - u_2$ there is a definite small value of $T_2 - T_1$ which gives an infinite value of $\tan \alpha$, or a vertical surface of separation. Thus as the difference of temperature decreases the surface of separation rises rapidly, passing through the vertical for a certain small value of $T_2 - T_1$, and then falling to a slope parallel to the earth's axis when the difference of temperature becomes zero. The boundary between a westerly current and an easterly current to the south of it is therefore liable to be very unstable when the difference of temperature is very small.

A further point to be noted from equation (9), which is sufficiently accurate for all practical purposes, is that the slope of the surface of discontinuity does not depend on the actual velocities u_1 and u_2. No appreciable change is produced by any changes in u_1 and u_2 provided the difference $u_1 - u_2$ remains constant. The winds on the two sides of the surface of discontinuity may therefore be in the same or in opposite directions.

§ 126. *Extension to motion in small circles, with a surface of discontinuity*

For steady motion with a curved surface of discontinuity the intersection of the surface of discontinuity with the horizontal plane must be an isobar. Let the radius of curvature be r, then the equation

$$2\omega \sin \phi \,.\, u = -\frac{1}{\rho}\frac{\partial p}{\partial y}$$

must be replaced by
$$2\omega \sin \phi \,.\, u \pm u^2/r = -\frac{1}{\rho}\frac{\partial p}{\partial r}.$$

The sign is + or − according as the curvature of the isobars is cyclonic or anticyclonic. The transformation of the equations in § 121 to meet the conditions now specified is readily made by adding a term $\pm u^2/r$ wherever the term $2\omega \sin \phi . u$ occurs. The result is

$$\tan \alpha = \frac{2\omega \sin \phi \, (T_2 u_1 - T_1 u_2) \pm \dfrac{1}{r} \, (T_2 u_1{}^2 - T_1 u_2{}^2)}{g \, (T_1 - T_2) + 2\omega \cos \phi \cos \beta \, (T_2 u_1 - T_1 u_2)} \quad ......(10).$$

The second term in the denominator is negligible by comparison with the first when the change of temperature in crossing the surface is greater than about $\frac{1}{2}°$ C, since in general the difference $T_2 - T_1$ is only a small fraction of T_1 or T_2, while the difference $u_1 - u_2$ is of the same order as either velocity. The numerator can therefore be written

$$T \, (u_1 - u_2) \left\{ 2\omega \sin \phi \pm \frac{u_1}{r} \mp \frac{u_2}{r} \right\} \quad ......(11).$$

From § 115 it follows that algebraically it is always true that $u_1/r < \omega \sin \phi$ and $u_2/r < \omega \sin \phi$, so that the quantity in the second bracket is always positive. Hence the imposition of curvature on the motion cannot change the sign of $\tan \alpha$, though it increases $\tan \alpha$ in cyclonic, and decreases it in anticyclonic curvature.

§ 127. *The form of the isobaric surfaces*

We might use the general equations of motion to obtain the form of the isobaric surfaces by writing

$$dp = \frac{\partial p}{\partial x} \, dx + \frac{\partial p}{\partial y} \, dy + \frac{\partial p}{\partial z} \, dz \quad ......(12),$$

and substituting for $\partial p/\partial x$, etc., then integrating the resulting equations in the general case. For steady motion this is unnecessary, and we may assume that the geostrophic relation is satisfied to a sufficient degree of approximation. The pressure increases from right to left across each current. Thus the surface of separation of two currents in opposite directions, with the cold current lying in a wedge under the warm current, will correspond to a trough of low pressure; while the surface of separation of two currents in opposite directions with the warm air lying as a wedge under the cold air will correspond to a ridge of high pressure. When the two currents of different temperatures are in the same direction but with different velocities the gradient of pressure will be in the same direction, but there will be a discontinuous change of the pressure gradient at the surface of separation.

§ 128. *The general equations for the slope of surfaces of discontinuity*

So far only stationary surfaces of discontinuity have been considered. When the assumption that the surfaces are stationary is abandoned it is necessary to use the full equations of motion (39), (40) and (41) of Chapter VIII, p. 171, substituting the values of $\partial p/\partial y$, $\partial p/\partial z$ from these equations in the equation

$$\tan \alpha = -\frac{\left(\dfrac{\partial p}{\partial y}\right)_1 - \left(\dfrac{\partial p}{\partial y}\right)_2}{\left(\dfrac{\partial p}{\partial z}\right)_1 - \left(\dfrac{\partial p}{\partial z}\right)_2} \qquad \ldots\ldots(13).$$

Putting $X = Y = Z = 0$, and so neglecting all forces other than the pressure forces and gravitation,

$$\tan \alpha =$$
$$-\frac{(\rho_1 \dot{v}_1 - \rho_2 \dot{v}_2) - 2\omega \cos\phi \sin\beta (\rho_1 w_1 - \rho_2 w_2) + 2\omega \sin\phi (\rho_1 u_1 - \rho_2 u_2)}{g(\rho_1 - \rho_2) + (\rho_1 \dot{w}_1 - \rho_2 \dot{w}_2) - 2\omega \cos\phi \cos\beta (\rho_1 u_1 - \rho_2 u_2) + 2\omega \cos\phi \sin\beta (\rho_1 v_1 - \rho_2 v_2)}$$
$$\ldots\ldots(14),$$

where the dots denote differentiation with respect to time. For unaccelerated horizontal motion and a stationary surface $w = v = 0$, and $\dot{w} = \dot{v} = 0$, and equation (14) reduces to the earlier form given in equation (4). If we assume the motion to be horizontal and unaccelerated, $w = 0$, $\dot{v} = \dot{w} = 0$, then in equation (14) the denominator reduces to its first term $g(\rho_1 - \rho_2)$, the third and fourth terms being negligible in all practical problems, as shown in § 124 above. Equation (14) then reduces to equation (5), showing that a horizontal velocity of translation of the surface of discontinuity parallel to itself makes no difference to the slope of the surface, as given by equation (5). From equation (39), p. 171, since the pressure is continuous at the surface of discontinuity, and therefore

$$\left(\frac{\partial p}{\partial x}\right)_1 = \left(\frac{\partial p}{\partial x}\right)_2,$$

$$\rho_1 \dot{u}_1 - \rho_2 \dot{u}_2 = -2\omega \cos\phi \cos\beta (\rho_1 w_1 - \rho_2 w_2) + 2\omega \sin\phi (\rho_1 v_1 - \rho_2 v_2) \ldots\ldots(15).$$

Equations (14) and (15) are extremely complicated and require for their strict application a complete knowledge of all three components of velocity and of acceleration. In practice considerable simplification is possible. In the denominator of (14) the term $g(\rho_1 - \rho_2)$ will by far exceed the other terms in magnitude, provided the difference in temperature of the two currents is not less than about $\frac{1}{2}°$ C. We shall therefore omit all the other terms in the denominator of equation (14), bearing in mind that by doing so we leave out of consideration all cases where the difference of temperature is very small. This is not a serious practical loss, as cases of small differences of temperature cannot be recognised on a synoptic chart. Also, as we only seek to obtain an idea of the general nature of the phenomena, we shall neglect the term in $\rho_1 w_1 - \rho_2 w_2$ in the numerator of (14) by comparison with $\rho_1 u_1 - \rho_2 u_2$. This is legitimate in view of the fact that the motion is very nearly horizontal. We then have
$$\rho_1 \dot{v}_1 - \rho_2 \dot{v}_2 = -2\omega \sin\phi (\rho_1 u_1 - \rho_2 u_2) - g \tan\alpha (\rho_1 - \rho_2) \quad \ldots\ldots(16).$$

From the same considerations equation (15) is reduced to

$$\rho_1 \dot{u}_1 - \rho_2 \dot{u}_2 = 2\omega \sin\phi (\rho_1 v_1 - \rho_2 v_2) \qquad \ldots\ldots(17).$$

There is a simple kinematical relationship between the values of v and w on the two sides of the surface of discontinuity, if it may be assumed that the surface moves forward without appreciable change of shape, and that the density of either fluid is not changing. Let v_f be the forward horizontal velocity of the surface. Then if in fig. 58 ABC be a small right-angled triangle

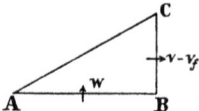

Fig. 58. Velocities at a surface of discontinuity.

whose sides dy, dz are respectively parallel to the axes of y and z, the net flow into a wedge of unit length parallel to the x-axis, of which the traingle ABC is the vertical section, is zero in the absence of change of density,

$$w \, dy = (v - v_f) \, dz,$$

$$v - v_f = w \cot \alpha, \quad \text{or} \quad w = (v - v_f) \tan \alpha \qquad \ldots\ldots(18).$$

From this it follows that

$$\rho_1 w_1 - \rho_2 w_2 = (\rho_1 v_1 - \rho_2 v_2) \tan \alpha - (\rho_1 - \rho_2) v_f \tan \alpha \qquad \ldots\ldots(19).$$

If the surface moves forward at a uniform velocity, we find by differentiation of (19)

$$\rho_1 \dot{w}_1 - \rho_2 \dot{w}_2 = (\rho_1 \dot{v}_1 - \rho_2 \dot{v}_2) \tan \alpha \qquad \ldots\ldots(20),$$

$$= -2\omega \sin \phi \, (\rho_1 u_1 - \rho_2 u_2) \tan \alpha - g \, (\rho_1 - \rho_2) \tan^2 \alpha \ldots\ldots(21).$$

Differentiating (16) and substituting from (17), we find

$$\rho_1 \ddot{v}_1 - \rho_2 \ddot{v}_2 = -2\omega \sin \phi \, (\rho_1 \dot{u}_1 - \rho_2 \dot{u}_2)$$

$$= -(2\omega \sin \phi)^2 \, (\rho_1 v_1 - \rho_2 v_2) \qquad \ldots\ldots(22),$$

which can be integrated giving

$$\rho_1 v_1 - \rho_2 v_2 = A \cos \{(2\omega \sin \phi) \, t - \epsilon\} \qquad \ldots\ldots(23).$$

By substituting in equations (17) and (19) from (23) it can be shown that $\rho_1 u_1 - \rho_2 u_2$ and $\rho_1 w_1 - \rho_2 w_2$ also have the same form, so that the surface of discontinuity is stable for the special type of disturbance which does not involve deformations of the surface of discontinuity itself.

If the surface is accelerated, so that v_f is not constant while the inclination α remains unaltered,

$$\rho_1 \dot{w}_1 - \rho_2 \dot{w}_2 = (\rho_1 \dot{v}_1 - \rho_2 \dot{v}_2) \tan \alpha - (\rho_1 - \rho_2) \dot{v}_f \tan \alpha \qquad \ldots\ldots(24).$$

At the typical warm front surface of a depression (see Chapter XVII) the motion of air is as shown in fig. 59 (a). Here

$$v_1 - v_f < 0, \quad w_1 < 0, \quad v_2 - v_f > 0, \quad w_2 > 0,$$

$$\rho_1 v_1 - \rho_2 v_2 < 0, \quad \rho_1 > \rho_2.$$

In equation (17) therefore

$$\rho_1 \dot{u}_1 - \rho_2 \dot{u}_2 < 0.$$

Since ρ_1/ρ_2 is very nearly unity we may infer that the system of accelerations is as shown in fig. 59 (b). Figs. 59 (c) and (d) show the reverse system of velocities and accelerations.*

J. Bjerknes† has shown that when u_1, $\dot{u}_1$, u_2, $\dot{u}_2$, etc. refer to points at an infinitesimal distance apart, as for example when the sharp surface of discontinuity is replaced by a finite layer of rapid transition, the differences in equations (16), (17), and (19) may be replaced by differentials,

$$\frac{\partial}{\partial y}(\rho\dot{v}) = -2\omega\sin\phi\,\frac{\partial}{\partial y}(\rho u) - g\tan\alpha\,\frac{\partial\rho}{\partial y},$$

$$\frac{\partial}{\partial y}(\rho\dot{u}) = 2\omega\sin\phi\,\frac{\partial}{\partial y}(\rho v),$$

$$\frac{\partial}{\partial y}(\rho\dot{w}) = \tan\alpha\,\frac{\partial}{\partial y}(\rho\dot{v}) + v_f\,\frac{\partial\rho}{\partial y}\tan\alpha.$$

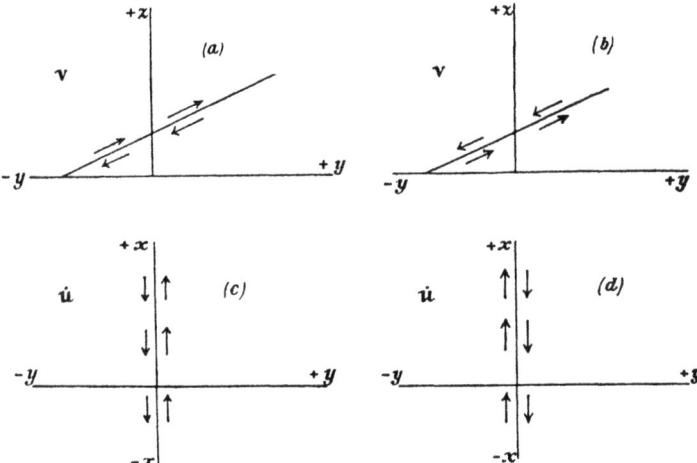

Fig. 59. Motion at a front.

If the surfaces are not accelerated, and the changes of density are negligible,

$$\frac{\partial}{\partial y}(\rho\dot{w}) = \tan\alpha\,\frac{\partial}{\partial y}(\rho\dot{v}).$$

Here $\frac{\partial v}{\partial y}$ is <0. Hence $\frac{\partial}{\partial y}(\rho\dot{u}) < 0$,

$$\frac{\partial}{\partial y}(\rho\dot{v}) \lessgtr 0 \quad \text{as} \quad \frac{\partial}{\partial y}(\rho u) \lessgtr -\frac{g}{2\omega\sin\phi}\frac{\partial\rho}{\partial y}\tan\alpha.$$

In the first case the warm air has an acceleration upward with reference to the cold wedge, whereas in the second case the warm air has a downward acceleration with reference to the cold wedge.

* From J. Bjerknes, *Geofys. Publ.* **3**, No. 6.

† *Geofys. Publ.* **3**, No. 6.

CHAPTER XI

THE GENERAL ASPECTS OF TURBULENCE

§ 129. *Stream-line and turbulent flow*

THE smooth motion of a fluid which can be represented by a stream-line function does not occur in nature except under certain definite conditions. When these conditions are not fulfilled the motion becomes irregular or "turbulent". We may define turbulence as an irregular motion which in general makes its appearance in fluids, gaseous or liquid, when they flow past solid surfaces, or even when neighbouring streams of the same fluid flow past or over one another. The existence of turbulence in the atmosphere is made visible by the trail of smoke from a chimney, or by the gusts and lulls in the trace of an anemometer. While the air in general has a "mean velocity" and a "mean direction", which are reasonably constant over intervals of from 10 to 100 minutes, the instantaneous velocity and direction may differ widely from the mean values. We have no clear idea why or how turbulence arises, or of the exact nature of the eddies which form in a fluid in turbulent flow, but something is known of the conditions which must be satisfied if a fluid is to flow without turbulence, or as "stream-line" motion; and some information has been accumulated concerning the magnitude of the eddy-velocities in certain cases.

The local variations of velocity in a fluid are associated with, and are presumably due to, the local variations of static pressure, which travel downstream with the mean velocity of the stream. The eddy which becomes visible as a dimple in the surface of a stream of water travels in this manner; at the same time it appears to be a simple rotational circulation, the motion being in its essentials two-dimensional. When a current of air moves over uneven ground we might be tempted to think of the eddies which form in it as cylindrical eddies, having their axes in the horizontal plane at right angles to the mean motion. This would require that the anemometer trace should show a ribbon of finite width for velocity, with a practically constant direction. Observation shows that this is not the case in practice, wide direction traces being associated with wide velocity traces.

If over a certain interval of time the mean velocity of flow remains constant, having components $\bar{u}$, $\bar{v}$, $\bar{w}$, the instantaneous deviations of the component velocities are known as the components of the eddy velocity. If the total velocity at any instant is represented by the components $\bar{u}+u'$, $\bar{v}+v'$, $\bar{w}+w'$, then u', v', w' are the components of the eddy velocity.

§ 130. *The Reynolds stresses*

The study of turbulence is largely the creation of Osborne Reynolds,[*] who carried out many beautiful experiments on turbulence in fluids, the motion being made visible by the introduction of colouring matter into the fluid. Reynolds found that turbulence occurred mainly when (*a*) the transverse variation of velocity exceeded a certain limit, and (*b*) the fluid was bounded by solid boundaries. The mathematical method outlined below follows Reynolds in assuming the fluid to be incompressible. This is not strictly true in the atmosphere, but it is probable that the turbulence in the atmosphere is in the main due to other causes than its compressibility, and that compressibility only modifies the motion to a slight degree.

The equations of motion of an incompressible viscous fluid, as shown in Chapter VIII, p. 185, are of the form

$$\rho \frac{du}{dt} = \rho X - \frac{\partial p}{\partial x} + \mu \nabla^2 u \qquad \ldots\ldots(1),$$

or

$$\rho \frac{du}{dt} = \rho X + \frac{\partial p_{xx}}{\partial x} + \frac{\partial p_{yx}}{\partial y} + \frac{\partial p_{zx}}{\partial z} \qquad \ldots\ldots(2),$$

which may be re-arranged in the form

$$\rho \frac{\partial u}{\partial t} = \rho X + \frac{\partial}{\partial x}(p_{xx} - \rho uu) + \frac{\partial}{\partial y}(p_{yx} - \rho uv) + \frac{\partial}{\partial z}(p_{zx} - \rho uw) \ \ldots\ldots(3).$$

Reynolds defines the mean velocities $\bar{u}, \bar{v}, \bar{w}$ as follows

$$\bar{u} = \frac{1}{\tau}\int_{t-\frac{1}{2}\tau}^{t+\frac{1}{2}\tau} u\, dt, \quad \bar{v} = \frac{1}{\tau}\int_{t-\frac{1}{2}\tau}^{t+\frac{1}{2}\tau} v\, dt, \quad \bar{w} = \frac{1}{\tau}\int_{t-\frac{1}{2}\tau}^{t+\frac{1}{2}\tau} w\, dt \qquad \ldots\ldots(4).$$

The mean values of u', v', w' are zero when the means are taken over an interval of time sufficiently long to permit of a large number of alternations of the velocities about their mean value. Then

$$\overline{uu} = \text{mean of } (\bar{u}\bar{u} + u'u' + 2\bar{u}u')$$
$$= \bar{u}\bar{u} + \overline{u'u'}.$$

Similarly
$$\overline{uv} = \bar{u}\bar{v} + \overline{u'v'},$$
$$\overline{uw} = \bar{u}\bar{w} + \overline{u'w'}.$$

Substituting these values, we find

$$\rho \frac{\partial \bar{u}}{\partial t} = \rho X + \frac{\partial}{\partial x}(\bar{p}_{xx} - \rho\bar{u}\bar{u} - \rho\overline{u'u'}) + \frac{\partial}{\partial y}(\bar{p}_{yx} - \rho\bar{u}\bar{v} - \rho\overline{u'v'})$$
$$+ \frac{\partial}{\partial z}(\bar{p}_{zx} - \rho\bar{u}\bar{w} - \rho\overline{u'w'}) \qquad \ldots\ldots(5),$$

with two similar equations. The equation of continuity gives

$$\frac{\partial \bar{u}}{\partial x} + \frac{\partial \bar{v}}{\partial y} + \frac{\partial \bar{w}}{\partial z} = 0 \qquad \ldots\ldots(6).$$

* *Collected Papers*, **2**, p. 51.

We thus find that the equations of mean motion are of the same form as the exact equations (3) above, provided we add the additional stresses

$$\widehat{xx}= -\rho \overline{u'u'}, \quad \widehat{yx}= -\rho \overline{u'v'}, \quad \widehat{zx}= -\rho \overline{u'w'} \left.\right\}$$
$$\widehat{yy}= -\rho \overline{v'v'}, \quad \widehat{yz}= -\rho \overline{v'w'}, \quad \widehat{zz}= -\rho \overline{w'w'} \quad \dots\dots(7).$$

These are the six eddy stresses of Osborne Reynolds. The notation is that of Karl Pearson, and the convention as to signs conforms to that of Love's *Theory of Elasticity*.

The equations derived above are not based on any assumptions as to the nature of turbulence. All that is necessary is that the interval τ over which the mean values $\bar{u}$, $\bar{v}$, $\bar{w}$ are taken should be sufficiently long to permit of a large number of oscillations of the wind components during that interval.

Note that
$$\widehat{xx} = -\rho\overline{u'u'} = -\rho\,\overline{(u-\bar{u})(u-\bar{w})} = -\rho\sigma_u{}^2 \left.\right\}$$
$$\widehat{xy} = -\rho\overline{u'v'} = -\rho\,\overline{(u-\bar{u})(v-\bar{w})} = -\rho r_{uv}\sigma_u\sigma_v \quad \dots\dots(8),$$
$$\widehat{xz} = -\rho\overline{u'w'} = -\rho\,\overline{(u-\bar{u})(w-\bar{w})} = -\rho r_{uw}\sigma_u\sigma_w.$$

etc., where σ_u, σ_v, σ_w are the standard deviations of the wind components, and r_{uv}, r_{uw}, etc. are the coefficients of correlation between the eddy components of velocity.

§ 131. *Dynamical similitude; the Reynolds number*

If the geometrical conditions of two separate motions are precisely similar, as for example when geometrically similar bodies are surrounded by fluid or immersed in fluid, we can find certain conditions which must be satisfied in order that the fluid motions may likewise be similar. The conditions in the one case may, be specified by the density ρ_1, the velocity v_1, the coefficient of kinematic viscosity ν_1, and a linear dimension l_1 of the body or bodies which bound the fluid in any way. The corresponding variables for the second fluid are ρ_2, v_2, ν_2, l_2. All characteristic lengths such as x will be measured in terms of the appropriate l, and the velocities being of the nature dx/dt, will also be proportional to l. We require to find the condition that the flow in the two cases shall be geometrically similar.

The ratios of the three forces due to the pressure gradient, friction and inertia must be the same at coresponding points. Since these forces are in equilibrium we need only consider two of them, say the friction and the inertia forces. One of the components of the inertia forces is $-\rho u\,\partial u/\partial x$. This must be proportional to $\rho v^2/l$, since the condition of geometrical similarity of flow demands that the u's should be proportional to the selected characteristic velocities v_1, v_2, etc., and the lengths x must be proportional to l. Again the frictional forces are of the nature $\mu\,\partial^2 u/\partial x^2$, which must be proportional to $\mu v/l^2$. The ratio of the inertia forces to the frictional forces must be the same for the two fluids, if the conditions of flow are to be geometrically similar, and

$$\frac{\rho v^2}{l} \div \frac{\mu v}{l^2} = \frac{\rho v l}{\mu} = \frac{v l}{\nu} \qquad \dots\dots(9)$$

must be the same for the two fluids. The two systems of flow will therefore be similar if

$$\frac{v_1 l_1}{\nu_1} = \frac{v_2 l_2}{\nu_2} = \text{const.} = R \qquad \ldots\ldots(10).$$

It is readily seen that this ratio has no dimensions. The ratio is known as the *Reynolds number*. When this number is small the viscous forces are great by comparison with the inertia forces, and turbulent motion is readily damped out. When the ratio is large the dynamic forces set up turbulence in spite of the viscous forces.

In his experiments on the flow of liquids in pipes Reynolds distinguished an "upper" and a "lower" Reynolds number. The lower gives the critical value above which initially turbulent flow entering the tube ceases to become laminar in the tube, while the upper gives the critical value at which initially laminar flow becomes turbulent.

It is found that the general mathematical treatment of turbulence is in practice so difficult that only special cases are amenable to discussion, in particular the cases where the Reynolds number is either very small or very large. A few cases of motion of solids in a viscous fluid have been worked out in detail when the shape of the solid is simple. One of the best known results is that derived by Stokes for the resistance of a viscous fluid to the motion of a sphere, which is found to be

$$6\pi\mu v r,$$

r being the radius of the sphere and v its velocity. When the sphere falls through the air, and v is its terminal velocity, the resistance is balanced by the buoyancy forces, $\frac{4}{3}\pi r^3 g (\rho_1 - \rho_2)$, where ρ_1 and ρ_2 are the densities of sphere and air, respectively. Neglecting ρ_2 by comparison with ρ_1, and putting $\rho_1 = 1$, we find

$$6\pi\mu v r = \frac{4}{3}\pi r^3 g, \quad \text{or} \quad v = \frac{2}{9}\frac{r^2 g}{\mu}.$$

This formula only holds for values of R small by comparison with unity. If r is in cm, $v = 1\cdot3 \times 10^6 r^2$, and the formula only holds for droplets with radius $0\cdot001$ mm or less.

Reynolds, in his experiments on the flow of liquids, found that turbulence set in when the Reynolds number reached 6400, but later experimenters have been able by careful adjustment to obtain non-turbulent flow with values of vl/ν as high as 25,000.

In the atmosphere, for motions reaching up to the tropopause, l is presumably to be taken as equal to the height of the tropopause, or perhaps the height of the homogeneous atmosphere. With velocities of 1 metre per second, vl/ν is then of the order of 10^9, and motion on such a scale must always be turbulent. But atmospheric motions on a much smaller scale than this must be turbulent, on account of the low value of the coefficient of viscosity of air.

For the consideration of motion in the atmosphere it is possible to give a slightly different form to Reynolds number. Let l be the linear dimension of a disturbance in the flow, and $\partial\bar{v}/\partial z$ the vertical gradient of velocity. Then instead of

$$R = \frac{lv}{\nu}$$

we may write

$$\frac{l^2}{\nu}\frac{\partial\bar{v}}{\partial z} = f(R).$$

This equation indicates that the linear dimension of any disturbance in the flow is given by

$$l^2 = \frac{f(R)\,\nu}{\partial\bar{v}/\partial z}.$$

If $f(R)$ were known, we could estimate the linear dimension of the disturbance corresponding to any particular value of $\partial\bar{v}/\partial z$. If $\partial\bar{v}/\partial z$ be small then l is great, and the disturbances are large in extent, but a longer time will be required to set them up.

It is not possible to apply the Reynolds number in a straightforward manner to motions in the atmosphere. The analysis used by Reynolds to derive his criterion R presupposes that in the different states of flow compared the only differences are those of size of the boundary and the rate and scale of the flow, the nature of the fluid being unchanged. In the atmosphere there is an added complexity due to the lapse-rate of temperature. The existence of a large lapse-rate facilitates the formation of turbulence, while the existence of an inversion checks the formation of turbulence, thus suggesting that the value of Reynolds number alone is not sufficient to determine whether an atmospheric motion is turbulent or not. Practically all wind-tunnel experiments refer to conditions in which the lapse-rate is zero, and are therefore strictly comparable with each other, but not with atmospheric motions.

§ 132. *The partition of eddy energy*

Taylor[*] has given a simple proof that the eddying velocity transverse to the mean wind is on the average about equal to the eddying velocity along the direction of the mean wind. Let OA (fig. 60) represent the mean wind in velocity and direction, and let OC and OD represent the extremes in the gusts and lulls. With A as centre and AC as radius, draw the circle $CEDF$. Then if the wind is always represented by a vector OP with P always within the circle and capable of taking all positions within it, we have the relation

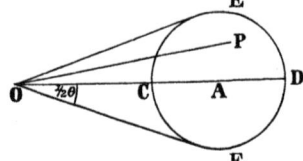

Fig. 60. Mean and eddy winds.

$$\frac{V_{max} - V_{min}}{V_{max} + V_{min}} = \sin\tfrac{1}{2}\theta,$$

θ being the angular width of the trace. Taylor found that this relationship was on the whole very closely confirmed in practice. We therefore conclude that

---

[*] *Aero. Res. Committee*, R. and M., No. 345.

if u', v', w' be the components of eddy velocity about the mean value $\bar{u}$ ($\bar{v}=\bar{w}=0$),

$$\overline{u'^2}=\overline{v'^2}.$$

Taylor also compared the magnitudes of v' and w' by observing the oscillations of a tethered balloon from some distance to leeward of its point of attachment, following the balloon with a pointer to which was attached a pencil whose motion was recorded on a piece of smoked glass. Taylor's records show complex curves which have no obvious axis of symmetry, and which could be roughly enclosed within a circle, thus showing on the average equal values of $\overline{v'^2}$ and $\overline{w'^2}$. This establishes a complete equipartition of eddying energy,

$$\overline{u'^2}=\overline{v'^2}=\overline{w'^2}.$$

Taylor's observations were made on a balloon tethered at 20 feet above the ground. Scrase* carried out somewhat similar observations by means of Taylor's bi-directional vane, the observations giving values of v' and w'. His observations were limited to occasions when the lapse-rate was small (a difference of $-0.1°$ F to $0.5°$ F in 17 metres) and the wind was from a direction between 270° and 310°, or between 160° and 200°, measured from North. The conditions in his observations were therefore reasonably comparable. The ratio of the v' component to the w' component was found to be 1·59 at 2 metres, diminishing to 1·20 at 18 metres. Scrase states that on the average at 2 metres the ratio of the lateral to the vertical diameter of the trace was about 1·5, indicating that $\overline{v'^2}$ is more than twice as great as $\overline{w'^2}$. He also found that while the diameters of the traces increased rapidly during the first minute, the subsequent increase was slow. From this we infer that most of the eddies have periods of, at most, a few seconds, the later slower growth of the diameter of the trace being due to the passing of large individual eddies. Further, when the duration of the record was kept constant (1 minute), and the instrument was maintained at a height of 3·3 metres, the lateral width of the trace showed no variation with wind velocity, while the vertical diameter increased slowly with wind velocity. The mean ratio of the lateral to the vertical diameter was 1·46. It is to be noted that the observations made with the bi-directional vane are observations of extremes of wind in each direction. From this it may be inferred that the extremes of the eddy winds are closely proportional to the wind velocity. In a further series of experiments Scrase found that the lateral and vertical diameters of the trace increased with height above ground up to a height of about 1·4 metres, and then decreased slowly, the ratio of the horizontal to the vertical diameter decreasing from 1·63 at 0·5 metres above the ground to 1·2 at 18 metres. The departure from isotropic distribution of the eddy components observed at heights of a metre or so thus appears to diminish with increasing height, leading to a nearer approach to equipartition in the v, w components at heights of the order of about 18 metres. It should be noted that the analysis by which Taylor established the relation $\overline{u'^2}=\overline{v'^2}$ was applied to observations at 40 metres above the ground.

* M.O., *Geophys. Mem.* No. 52.

The results derived by Scrase are in reasonable agreement with some observations described by G. I. Taylor in a lecture before the Royal Meteorological Society,* which show that at 2 feet $v/w = 3$, and at 8 feet $v/w = 1\cdot4$. At 25 feet over grassland Taylor found a rough equality of v and w, but Scrase's lengthier series of measurements indicate that the departure from complete isotropy of v and w extends to a height of at least 20 metres.

Scrase (*loc. cit.*) has also described a series of observations of a light bi-directional vane photographed kinematographically, 16 photographs being taken per second. These records thus give the true mean value of the wind components, and not the extremes as yielding by the pen traces. The reproductions of readings taken from the photographs at intervals of $1/5$ second, $3/5$ second and 5 seconds respectively indicate that by far the greater part of the variability is associated with eddies of periods of about 1 second or less. The mean deviation from zero of the v component decreased from $0\cdot24$ for intervals of $1/5$ second to $0\cdot14$ for intervals of 5 seconds. This is equivalent to a decrease of the eddying energy to $\left(\dfrac{0\cdot14}{0\cdot24}\right)^2$ or $1/3$. Thus at least two-thirds of the eddying energy is associated with eddies of period of less than 5 seconds. Reducing the interval from $1/5$ to $1/16$ second increased the mean deviation from $0\cdot115$ to $0\cdot132$ for one short record analysed, equivalent to an increase of the eddying energy by one-third. The kinematograph records confirmed the results previously derived that the horizontal transverse variations exceeded the vertical variations. They also showed that there is a tendency for positive u' to be associated with negative w', so that the slower moving air goes upward, while the faster moving air comes downward, but the correlation between the signs of v' and w', and between the signs of v' and u' was not clearly marked. Other conditions being equal (contour and lapse-rate), the mean magnitudes of u', v' and w' at a given height are proportional to the mean velocity of the wind.

Möller† has also considered Scrase's results, using mean values of the eddy components over intervals of 6 seconds, and found a coefficient of correlation of $-0\cdot8$ between u' and w', again indicating a tendency for fast-moving air to fall, and for slow-moving air to rise. These results are strongly suggestive of a tendency for the eddying motion to be in planes strongly inclined to the horizontal.

Further observational details of wind structure will be found in *Geophysical Memoir*, No. 54, by the late M. A. Giblett and others. The wind structure near the ground has also been examined in detail by A. C. Best,‡ using a hot-wire anemometer and a bi-directional vane. Best found that the eddy velocities are distributed according to the Maxwell law (the normal error law), and his records with the bi-directional vane show clearly the effect of the temperature gradient on the magnitude of these oscillations. The general tendency is for the lateral component to decrease with height, whilst the vertical component increases with height, isotropy being reached at a height estimated (over short cropped grass) at 25 m, after which it is presumably maintained.

* *Q.J. Roy. Met. Soc.* **53**, 1927, p. 201.   † *Beitr. Phys. fr. Atmos.* **20**, 1933, p. 79.
‡ M.O., *Geophysical Memoir*, No. 65.

Scrase showed that there are usually present large numbers of small scale eddies whose periods are of the order of one second, which give at a height of 1·5 metres mean components in the u, v, w directions respectively in the ratios 1·0 : 1·16 : 0·75, and at a height of 19 metres in the ratios 1·0 : 0·73 : 0·56. He found that the mean u' component showed very little variation from 1·5 to 19 metres, while the v' and w' components diminished in the same range to about two-thirds of their value at 1·5 metres. Scrase only summarises the v', w' components for eddies of periods of the order of a few minutes, but he states that he found the v' component greater than the w' component. Each of these increased from the ground up to a height of about 1–2 metres, and then decreased, the values at 18 metres being about two-thirds the values at 1·5 metres. Scrase found that mean values of the eddy velocity over intervals of an hour were proportional to the mean wind velocity, and that at a height of 13 metres the mean u', v' and w' components were in the ratios 1·0:0·76:0·39.

§ 133. The nature of eddies

So far no effort has been made to define clearly what is meant by an eddy, nor is it considered likely that any definition could be given which would be universally acceptable. The eddies which form at the edge of a stream flowing into a millpond are of the nature of vortices with vertical axes, but the name "eddy" as used in discussing motion in the atmosphere is not restricted to circular motions. We can only define an eddy as a physical entity which disturbs the uniform flow of air, and this definition will include rotating eddies, convection currents, and any other type of disturbance. There is a type of eddy which can be produced and made visible in the laboratory, that has become known largely through the work of H. Bénard, for which the name of "convection cell" is suggested as appropriate. Bénard* showed that when a shallow layer of fluid containing volatile constituents is cooled at the upper surface by evaporation, the whole mass breaks up into a number of separate cells in each of which there is an upward motion at the centre, diverging motion at the top, and descending motion in the outer regions (see fig. 61). The diameter of the cells is from three to four times the depth of the fluid, and in very steady conditions the cells become hexagonal.

The motion in the Bénard cells was investigated mathematically by Rayleigh,† who showed that it is possible for a fluid to remain in equilibrium with the density greater above than below, so long as the excess of density does not exceed the limit given by the following inequality

$$\frac{\rho_1 - \rho_0}{\rho_0} < \frac{27\pi^4 \kappa \nu}{4gh^3},$$

where κ is the coefficient of molecular conduction of heat, and ν the kinematic coefficient of viscosity, ρ_1 the density of the fluid at the top, ρ_0 the density of the fluid at the bottom of the layer whose thickness is h. This result is confirmed when the cells are formed in a dish which is slightly tilted so that the

* Ann. Phys. Chem., Paris, 23, 1901.　　　† Phil. Mag., London, 32, 1916, p. 529.

fluid becomes shallow at one edge. The cells become smaller as the fluid becomes shallower, but a narrow belt of very shallow fluid remains free of cells.

The mathematical treatment of the problem was extended and clarified by Jeffreys,* who corrected certain of the boundary conditions used by Rayleigh, and further showed that when the fluid has a motion of steady translation the cells are drawn out into long strips, as had been suggested earlier by A. R. Low.†

The simple Bénard cell is analogous to the ring vortex or the smoke ring sometimes produced by a locomotive. It is an interesting case of the application of the considerations of § 109 above. In the liquid we may regard potential and absolute temperature as identical for all practical purposes, so that stability requires that the density and pressure gradients should be parallel. In the unstable state the density is greatest at the top, and any disturbance will tend to produce circulations in vertical planes. As the flow of heat tends to

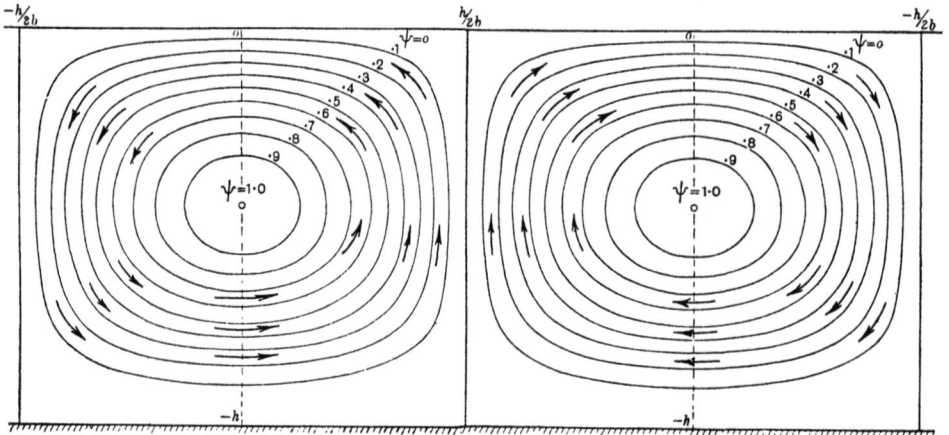

Fig. 61. Circulation in a Bénard cell.

maintain the unstable condition, the circulation persists. The Bénard cell is perhaps the clearest illustration of the possibility of generating vorticity in a fluid in which pressure is not a function of density alone. Frictional forces have little or nothing to do with the genesis of the circulation, which in fact dies away rapidly through the action of friction as soon as the transfer of heat dies away. If, for example, the dish in which the cells are formed by the evaporation of volatile constituents of a liquid is covered by a glass plate, the evaporation ceases and the circulation dies away rapidly.

The attention of English meteorologists was first directed to the work of Bénard by Low and Brunt,‡ who suggested that the existence and persistence of super-adiabatic lapse-rates in the atmosphere might be explained by the joint effects of radiation and turbulence taking the place of molecular conduction and viscosity in Rayleigh's inequality above. Low§ also showed that it was possible for the unstable layer of fluid to be filled by several layers of

* *Proc. Roy. Soc.* A, **118**, 1928, p. 195.   † *Nature*, **115**, 1925, p. 299.
‡ *Nature*, **115**, 1925, pp. 299–301.   § *3rd Int. Congress App. Mech.*, Stockholm, 1930.

cells, though it is not clear how far this occurs in practice. In the atmosphere the multiple cells, if they occur at all, are probably transitory.

Later developments of these ideas by Idrac,* Mal,† and by Phillips and Walker‡ have shown that a wide variety of forms of cellular structure can be obtained by suitable variation of the rate of shear in the original motion. Idrac showed that when the fluid is sheared a series of vortices were formed having their axes parallel to the direction of the shear. Mal investigated the detailed application of these ideas to explain cloud forms, and showed that, in some of the cases he investigated, particular cloud forms were associated with instability and wind-shear. Phillips and Walker extended the work of Mal, and described a number of laboratory experiments which showed that by suitable adjustment of the variation of velocity with height it was possible to obtain, in unstable air in a wind channel, polygons, transverse vortices, crossed vortices, and longitudinal vortices. The resemblance to cloud forms shown by the details of the motions found in the photographs reproduced in their paper is very remarkable. The work of Mal and of Phillips and Walker renders it very probable that a wide variety of cloud forms can be explained as the effect of instability combined with shear. In particular it should be remarked that long rolls of cloud may thus be explained, the direction of the roll being in the direction of the shear. The earlier view that such clouds were to be explained as Helmholtz gravitational waves would require the direction of the roll to be perpendicular to the direction of shear, and thus the test of observation of wind at different heights can distinguish between the two theories.

The present writer§ has described a remarkable hailstorm which occurred in France in 1865, in which the cloud was described as resembling a huge net, and the hailstones produced damage in a series of areas which formed an irregular network at distances of 50 to 100 metres apart. This appears to suggest a more or less cellular distribution of convection currents, involving a shallow layer of violent instability.

If the phenomena observed in the laboratory are reproduced with any fidelity in the atmosphere, in which the instability set up as the result of the heating of the earth's surface by solar radiation is limited to the lowest 2000 feet or so, the convection currents induced should be at distances apart of the order of 5000 to 8000 feet. The fact that the typical cumulus forms in long rows immediately suggests an analogy with the forms depicted in the photographs of Mal and Phillips.

§ 134. *The nature of convection*

The comparison of the phenomena observed in viscous fluids in the laboratory with the forms of clouds suggests that what is loosely termed "thermal convection" is to be explained largely by means of the circulation shown in fig. 61. The horizontal extent of the convection cell may therefore be expected to be of the order of three times the depth of the unstable layer. The idea thus given

---

* *Comptes Rendus*, 1920, July–Dec. p. 42.   † *Beitr. Phys. fr. Atmos.* **17**, 1930, p. 40.
‡ *Q.J. Roy. Met. Soc.* **58**, 1932, p. 23; see also Brunt, *Q.J. Roy. Met. Soc.* **63**, 1937, p. 277.
§ *Met. Mag.* **63**, 1928, p. 14.

of convection differs materially from the ascent of isolated bubbles of warm air which penetrate through the environment much as an ascending balloon does, the place of the ascending air being filled by a general drift inward of air from all sides. The most recent work on the form of clouds, such as that of G. T. Walker and his pupils, very definitely supports the idea that thermal convection is to be represented by the circulation in fig. 61 rather than by the ascending bubble. The acceptance of such a view of the mechanism of convection makes it practically impossible to specify a "temperature of the environment" through which the ascending air passes, such as is represented by T' in § 21 above. True polygonal cells in the laboratory have a circulation which can be represented accurately by fig. 61. Large cells in the atmosphere appear to have some rotation about a vertical axis superposed on the motion shown in fig. 61. Thus atmospheric convection cells are of rather more complex structure than the cells observed in the laboratory.

When the convective process is set up in still air the convection cell will be roughly polygonal in shape, and fig. 61 will represent any axial cross-section of the cell. When on the other hand the air has a general motion in a particular direction, the cells will be long strips, and fig. 61 will then represent the circulation in a vertical section at right angles to the direction of flow. The clouds then formed will be typical cumulus in long parallel strips.

On this view the typical cumulus cloud is at the top of an ascending current, and the temperature within the cloud may be lower than that in the surrounding air, the cloud being partly supported by the ascending current. It is then not necessary to suppose the density within the cloud to be equal to the density outside at the same level, as was done by Kopp;* and Kopp's argument in favour of very great supersaturation within the cloud loses its validity.

§ 135. *Stability of motion*

Numerous writers have attacked directly the problem of determining the stability of laminar motion, and though the analysis used in some cases has been questioned by later writers it is generally regarded as probable that, for infinitely small disturbances, smooth flow is stable, and that a finite disturbance is required to produce turbulence in initially steady stream-line flow. In the atmosphere there is no difficulty in imagining the occurrence of finite disturbances, on account of the magnitude of the impediment to smooth flow produced by obstacles at the ground, and the effect of widely varying types of surface upon the flow of air.

When a given state of motion becomes unstable two possible courses arise. The motion may become turbulent, so that its details become unpredictable, or it may change to another type of steady motion. The Bénard convection cell to which reference has been made above is a special case of the latter. In the initial state of unstable equilibrium of a liquid cooled from above or heated from below there is no motion, but when the equilibrium breaks down a steady motion sets in which is not turbulent. In his mathematical investiga-

* *Beitr. Phys. fr. Atmos.* **16**, 1930, p. 173.

tion of the conditions in the convection cell, Rayleigh assumed that the component velocities u, v, w, and the departure of the temperature from that corresponding to stable equilibrium, were all sufficiently small to permit of his neglecting all squares and products by comparison with terms of the first degree. He further assumed that these small quantities varied as $e^{ilx}e^{imy}e^{int}$. If in is real and positive the corresponding disturbance will increase exponentially with time. Rayleigh's treatment was essentially based on the assumption that the disturbance which had the largest rate of growth would eventually predominate. Taylor,[*] in his investigation of the motion of fluid between two rotating cylinders, followed substantially the same method, and found close agreement with his experimental investigations, in which the uniform circular motion of rotation about an axis broke down, and was replaced by toroidal ring flow combined with rotation about the common axis of the cylinders. In the Bénard-Rayleigh problem it is the density gradient which attains a critical value, but in Taylor's experiments it is the gradient of velocity whose critical value determines the change in the type of motion.

For the details of these investigations reference should be made to the original papers. They are mentioned here in order to emphasise the difference between a change to a new type of steady motion and a breakdown of laminar flow leading to turbulent flow. In Bénard's experiments the motion in the cells is to be regarded as a steady motion, and in no sense as turbulent motion, and we should perhaps make it clear that in including a description of these phenomena in the present chapter we regard the cellular structure which may occur in the atmosphere as continually breaking down and re-forming.

In the next chapter will be found some further discussion of stability in relation to vertical distribution of density.

---

[*] *Phil. Trans. Roy. Soc.* A, **223**, 1922, p. 289.

CHAPTER XII

TURBULENCE IN THE ATMOSPHERE: THE EDDIES AS DIFFUSING AGENCIES

§ 136. *Diffusion by eddies and molecular diffusion*

It is readily understood that in a convection current the air which rises from the ground on account of its excess of heat content will carry some of its excess heat into any higher layer with which it partially mixes. It will also carry some of its excess (or defect) of momentum in the same way. An eddy which moves from one place to another may therefore be regarded as an agency in a process of diffusion of heat, momentum, content of water-vapour or carbon dioxide, or other properties of the fluid.

The molecular diffusion of momentum in a given azimuth by viscosity, the molecular conduction of heat, and gaseous diffusion such as the diffusion of water-vapour into air, can all be represented by an equation of the form

$$\frac{dV}{dt} = k\,\frac{\partial^2 V}{\partial z^2} \qquad\qquad \ldots\ldots(1),$$

where V represents component velocity, temperature, or water content per unit mass. For diffusion of momentum k has the value ν, the kinematic coefficient of viscosity; for conduction of heat, k has the value κ; and for the diffusion of water-vapour into air k has the value represented usually by D. The values of these coefficients are tabulated on p. 421.

§ 137. *The vertical transfer of heat by turbulence*

We shall consider the transfer of heat by eddies moving upward or downward across a horizontal surface at level z, where p, T, and ρ denote the pressure, absolute temperature and density, respectively. Let c_p be the specific heat of air at constant pressure. Then the quantity of heat required to raise the temperature of 1 gramme of air by an amount dT is $c_p dT$.

An eddy which crosses the horizontal surface z will be regarded as having originated at height $z-l$, where it was originally a normal sample of the environment. The lapse-rate $\dfrac{\partial T}{\partial z}$ will be treated as constant within the range $z-l$. Initially, when at level $z-l$, the eddy has the temperature $T-l\dfrac{\partial T}{\partial z}$. In moving adiabatically from $z-l$ to z the temperature is decreased by Γl. The temperature of the moving eddy when crossing the level z is therefore $T-l\left(\dfrac{\partial T}{\partial z}+\Gamma\right)$. If the vertical component of velocity at any point of the horizontal surface is w, the rate of vertical transfer of air per unit area and per unit time is ρw, and the total heat content of the air so transferred is

$$\rho w c_p \left\{ T - l \left(\frac{\partial T}{\partial z} + \Gamma \right) \right\} \qquad \ldots \ldots (2).$$

The total net transfer of mass must be zero, and hence $\overline{\rho w}$ is zero when averaged over an area and a time both sufficiently large to allow for the passage of a large number of eddies in each direction. Hence we find, by averaging (2) above over a large area and over a long interval of time, that the net upward transfer of heat per unit area per unit time is

$$- \rho \overline{w l} c_p \left(\frac{\partial T}{\partial z} + \Gamma \right) \qquad \ldots \ldots (3).$$

It should be noted that the product wl is always positive, w and l being both positive for upward moving eddies, and negative for downward moving eddies. Strictly speaking, the assumption that there is no net transfer of mass across the horizontal surface implies that the surface should be an isobaric surface, and (3) above is better defined as the net transfer across unit area of the isobaric surface of pressure p.

Following Taylor,* we represent $\overline{wl}$ by the symbol K, the eddy-diffusivity. Then the net upward flux of heat across the isobaric surface p is

$$- K \rho c_p \left(\frac{\partial T}{\partial z} + \Gamma \right) \qquad \ldots \ldots (4).$$

The net gain of heat in the layer between the surfaces p and $p + dp$ is

$$\frac{\partial}{\partial p} \left\{ K \rho c_p \left(\frac{\partial T}{\partial z} + \Gamma \right) \right\} dp = c_p \frac{\partial}{\partial z} \left\{ K \rho \left(\frac{\partial T}{\partial z} + \Gamma \right) \right\} dz \qquad \ldots \ldots (5).$$

This must be equal to $\rho \, dz \, c_p \dfrac{dT}{dt}$. Hence

$$\rho \, c_p \frac{dT}{dt} = c_p \frac{\partial}{\partial z} \left\{ K \rho \left(\frac{\partial T}{\partial z} + \Gamma \right) \right\} \qquad \ldots \ldots (6).$$

On the left-hand side dT/dt is to be interpreted as implying differentiation following the isobaric surface p, which is equivalent to differentiation following the mean motion. The left-hand side may without serious error be reduced to $\rho \, \partial T / \partial t$, so that equation (6) becomes

$$\rho \frac{\partial T}{\partial t} = \frac{\partial}{\partial z} \left\{ K \rho \left(\frac{\partial T}{\partial z} + \Gamma \right) \right\} \qquad \ldots \ldots (7).$$

The variations of ρ with height being relatively slow, we may without serious error write the equation in the form

$$\frac{\partial T}{\partial t} = \frac{\partial}{\partial z} K \left(\frac{\partial T}{\partial z} + \Gamma \right) \qquad \ldots \ldots (8).$$

* *Phil. Trans. Roy. Soc.* A, **215**, 1915, p. 1.

If further we neglect the variation of K with height, the equation is further simplified to the form

$$\frac{\partial T}{\partial t} = K\frac{\partial^2 T}{\partial z^2}$$ ......(9).

Equations (7), (8) and (9) are to be regarded as approximations to equation (6), which is the equation directly derived by Taylor's analysis.

Equation (9) shows that there is an analogy between the diffusion of heat by eddies, and the ordinary diffusion of heat by molecular conduction, the coefficient K taking the place of the coefficient of thermal diffusivity in the equation. For this reason the quantity K is called the *eddy diffusivity*. It will be shown later that K is very much greater than the corresponding coefficient for molecular diffusion, and usually greater than the coefficient of radiative diffusivity (see § 85).

One other feature of the equations of eddy transfer of heat requires to be emphasised. The net upward flux of heat across unit horizontal surface is

$$-K\rho c_p\left(\frac{\partial T}{\partial z}+\Gamma\right),$$

where Γ is the dry adiabatic lapse-rate. If the lapse-rate $-\partial T/\partial z$ is less than Γ, or if the atmosphere is stable, this expression is negative, and the flow of heat produced by the eddies is downward. The direction in which heat is transferred by eddies is downward or upward according as the atmosphere is stable or unstable. If the atmosphere is stable, churning it up will cause the bottom to grow warm and the top to grow cold, until the lapse-rate becomes equal to the dry adiabatic, after which further churning can produce no result.

§ 138. *Richardson's treatment of diffusion by eddies*

Let Z stand for the property whose diffusion we wish to study, and let χ be the amount of Z per unit mass of air. χ will be a function of the height z, and of the time t. In a layer of thickness dz the amount of Z is $\chi\rho\,dz$ per unit of horizontal area. The upward flux of Z is the amount of Z which flows across unit horizontal surface. The rate of increase of Z in the layer dz is

$$-\frac{\partial}{\partial z}\,(\text{upward flux})=\frac{\partial}{\partial t}\,(\rho\chi)$$ ......(10).

When an eddy moves from one level to another carrying its original quantity χ of Z with it, the new level gains an amount proportional to the difference of χ at the two levels, and if χ is uniform at all levels, there can be no gain or loss by transport. It thus appears reasonable to assume

$$\text{upward flux}=-c\,\frac{\partial\chi}{\partial z}$$ ......(11),

where c will depend in part on the amount of air crossing the surface across which the flux is measured. It may also be a function of z, of χ, and of $\partial\chi/\partial z$, but it must remain finite when $\partial\chi/\partial z = 0$, since by definition the flux must then be zero. Substituting in (10), we find

$$\frac{\partial}{\partial t}(\rho\chi) = \rho\,\frac{\partial\chi}{\partial t} = \frac{\partial}{\partial z}\left(c\,\frac{\partial\chi}{\partial z}\right) \qquad \ldots\ldots(12).$$

The limitations imposed upon χ are of importance. We must not use χ to represent properties whose measures are changed by delay or by transportation to a new level, and presumably χ must be such that its upward flux has a physical meaning. Properties which appear to satisfy these conditions are content of water-vapour, dust or carbon dioxide, etc., momentum in a fixed azimuth, deviation of temperature from an adiabatic distribution.

Richardson calls the quantity c the *eddy conductivity*. Its dimensions are $ML^{-1}T^{-1}$. Since $dp = -g\rho\,dz$, we might write (12) in the form

$$\rho\,\frac{\partial\chi}{\partial t} = g\rho\,\frac{\partial}{\partial p}\left(g\rho c\,\frac{\partial\chi}{\partial p}\right) = \rho\,\frac{\partial}{\partial p}\left(\xi\,\frac{\partial\chi}{\partial p}\right) \qquad \ldots\ldots(13),$$

where $\xi = g^2\rho c$.

Richardson calls ξ the *turbulivity*.

§ 139. *Application of Taylor's results to the atmosphere*

The equation of eddy transfer of heat will be used in the form (9) of p. 226, it being assumed that K does not vary with height. We then have

$$\frac{\partial T}{\partial t} = K\,\frac{\partial^2 T}{\partial z^2} \qquad \ldots\ldots(14).$$

Equation (14) can be solved for a number of special cases in which the boundary conditions are specified. One of the most interesting of Taylor's applications of his formula was to discuss the changes in the distribution of temperature within a current of air which, after being heated during its passage over warm land, passed over a cold sea. Suppose the air on reaching the coast had a constant lapse-rate β, its surface temperature being T_0. Let the surface temperature of the sea be T_1. Then the air immediately in contact with the sea has its temperature suddenly lowered to T_1. The problem is then specified, so far as the boundary conditions are concerned. The solution of the equation which is appropriate is given in any textbook on the theory of the conduction of heat, and may be written

$$T = T_0 - \beta z + (T_1 - T_0)\left\{1 - \frac{2}{\sqrt{\pi}}\int_0^{z/\sqrt{4Kt}} e^{-\mu^2}\,d\mu\right\} \qquad \ldots\ldots(15).$$

The term which multiplies $(T_1 - T_0)$ in this equation is unity at $z = 0$, and falls to 0·1 at a height given by $z/\sqrt{4Kt} = 1\cdot2$. Taylor assumes that for all practical purposes this term has no effect beyond $z/\sqrt{4Kt} = 1$. Thus in a time t the height to which the effect of the surface changes of temperature extends is given by

$$z/\sqrt{4Kt} = 1, \quad \text{or} \quad z^2 = 4Kt \qquad \ldots\ldots(16).$$

The observations which Taylor made of the vertical distribution of temperature above the Great Banks of Newfoundland showed a marked inversion in the lower layers, with a lapse-rate approaching the dry adiabatic at higher levels. From the height at which the inversion ceased, combined with an estimate of the time which the air had been flowing above the cold sea, Taylor deduced values of K of the order of 10^3 in C.G.S. units. He found K to be 1×10^3 with winds of force 2, and 3×10^3 with winds of force 3. It must be borne in mind that in inversions the atmosphere is very stable, and that the upward movement of eddies is thereby checked. Thus we should expect to find with large lapse-rates greater values of K than were found by Taylor during inversions over the Great Banks.

The relation
$$z^2 = 4Kt$$

for the height to which turbulence is effective in a time interval t was derived on the supposition that the surface temperature underwent a sudden change. But the result is only slightly modified when the rate of change of temperature at the surface is uniform. Let the surface temperature diminish $n°$ per unit time, so that the surface temperature is

$$T_0 - nt.$$

Then at time t the temperature at height z is

$$T_0 - \beta z - nt \left\{ \left(1 + \frac{z^2}{2Kt}\right) \left(1 - \frac{2}{\sqrt{\pi}} \int_0^{z/\sqrt{4Kt}} e^{-\mu^2} d\mu \right) - \frac{2}{\sqrt{\pi}} \frac{z}{\sqrt{4Kt}} e^{-z^2/4Kt} \right\} \quad \dots (17).$$

The term which multiplies nt in this expression is unity at the surface, and 0.1 at $z/\sqrt{4Kt} = 0.8$. The height attained in time t is therefore rather less than in the case of a sudden fall of the surface temperature, but we may still take as an approximation

$$z^2 = 4Kt.$$

This method gives only the order of magnitude of K, but at this stage of the development of the subject an estimate of the order of magnitude of the effect is of value, in that it shows that turbulence must be a much more effective agent than molecular diffusion in the transport of heat in the vertical. Even in quiet conditions, such as those over the Great Banks of Newfoundland investigated by Taylor, the mean value of K from the surface to 1000 feet is of the order of 10^3, and therefore at least as great as K_R, the coefficient of radiative diffusivity. In more normal conditions K for the same levels is of the order of 10^5, and the effect of eddies on the transfer of heat in the vertical direction is then enormously greater than that of radiation, which is itself enormously more effective than molecular conduction.

At quite low levels, a few inches from the ground, K is of the order of 10^{-1}. It increases with height at first, probably up to about 300 or 500 metres above the ground, after which there is a slow decrease.

A more elaborate discussion of the equations of heat transfer with the boundary conditions assumed above will be found in Riemann-Weber, *Die Differential- und Integral-Gleichungen der Mechanik und Physik*, **2**, p. 220.

§ 140.   *The effect of turbulence on the diurnal variation of temperature*

If we assume that the transfer of heat in the vertical is brought about entirely by eddies, we can readily find the change with height in the form of the curve of diurnal variation of temperature, for the case when K is constant with height. Let the temperature at the ground be given by

$$T = T_0 + A \sin qt \qquad \dots\dots(18).$$

The solution of the equation $\quad \dfrac{\partial T}{\partial t} = K \dfrac{\partial^2 T}{\partial z^2} \qquad \dots\dots(19),$

which has this boundary condition at $z = 0$, is

$$T = T_0 - \beta z + A e^{-bz} \sin (qt - bz) \qquad \dots\dots(20),$$

where b is a constant defined by $b^2 = q/2K$. The term $-\beta z$ is included on the right-hand side of the equation to allow for the mean lapse-rate during the period.

Now the diurnal variation of temperature at the ground can be represented with reasonable accuracy by a single sine-term of period 24 hours. Thus

$$q = 2\pi/24 \times 60 \times 60 = 7 \cdot 3 \times 10^{-5}.$$

Then the diurnal variation at any height z is in accordance with equation (20) above. The amplitude $A e^{-bz}$ falls off exponentially, the ratio of the amplitudes at any two heights z_1 and z_2 being $e^{-b\,(z_1 - z_2)}$. The lag in the occurrence of maximum temperature from z_1 to z_2 is $\dfrac{b}{q} (z_2 - z_1)$, and from this, or from the ratio of amplitudes, the value of b can be readily derived. Taylor compared this theory with observations made at Parc St Maur and on the Eiffel Tower at heights of $1 \cdot 8$, 123, 197 and 302 metres above the ground. By evaluating K for a number of stages from the ground upward, he found a definite tendency for K to increase with height in summer, and to decrease with height in winter, the mean value of K for the whole year, deduced from a comparison of the amplitudes at $1 \cdot 8$ and 302 metres, being 10^5. It is therefore seen that the mean value of K for the Eiffel Tower is much in excess of the values which Taylor derived from observations of inversions over the sea. The highest values of K at the Eiffel Tower occur in summer, when large lapse-rates are most frequent. In winter the lapse-rates are much smaller, and turbulence is much less active. Presumably the eddies in winter are on a smaller scale, and do not disturb the atmosphere to such heights as in the summer.

Theoretically it should be possible to compute K also from the lag in the time of maximum between any two levels, but in practice this method is unreliable on account of the difficulty in estimating accurately the extent of the lag. It is also open to doubt whether the afternoon temperatures at the top of the tower are altogether reliable, on account of the effect of the solarisation of the tower itself. It is not likely that the total range of temperature is very seriously affected by this, but it is probable that the time of maximum is appreciably affected.

Let us see how the diurnal variation of temperature should diminish with height if we adopt the value $K = 10^5$. We then have $b^2 = \pi/10^5 \times 24 \times 60 \times 60$ and $b = 2.10^{-5}$ approximately. The diurnal variation at a height z bears to the diurnal variation at the ground the ratio e^{-bz}. If we measure z in metres this becomes $e^{-2.10^{-5}z}$. The diurnal variation thus falls to $1/e$ (or 0·37) of the surface value at $z = 500$ metres, and to $1/e^3$ (or 0·05) at 1500 metres. This is in fairly close agreement with observation. The Lindenberg observations give for the amplitude of the 24 hour variation of temperature the following values:

Height in km	0	0·5	1·0	1·5	2·0
Amplitude °C	3·0	1·1	0·7	0·5	0·2

These figures agree fairly well with our conclusions based on a value $K = 10^5$, up to 500 metres, but beyond this level the agreement is less accurate, and appears to indicate an increase in the value of K at heights above 500 metres. The comparison must not be pushed too far, however, as the data are a little uncertain above 500 metres, and the effect of the condensation of water-vapour begins to be important at heights of about 1 km and above.

Similar arguments to those used above might be applied to the second, third, and higher harmonics of the diurnal variation of temperature, the only difference in the analysis being that the value assigned to q will now be 2, 3, etc. times the value previously used. Since b is proportional to $\sqrt{q}$, its value will increase with the order of the harmonic, and thus e^{-bz} will diminish with increasing order of the harmonic. Thus the diurnal variation of temperature, if transferred upward purely by the effects of turbulence, should approach more and more closely with increasing height to a sine-curve of period 24 hours.

Haurwitz[*] has shown that if K be a linear function of height the amplitude of the diurnal wave falls off more rapidly with height at first, while the time of maximum is more retarded, as compared with the case of K having a constant value equal to the mean value of the linearly varying K up to heights of say 200 m.

The assumption of $K = $ constant implies that the phase varies linearly with height. Best[†] found that the lag of time of maximum temperature from 1 inch to 56 feet varied fairly accurately as $z^{0.19}$, so that the use of an eddy conductivity independent of height gives results which are approximate only.

§ 141. *Taylor's discussion of the eddy transfer of momentum contrasted with that of Schmidt and Prandtl*

If the wind varies with height it is natural to suppose that an eddy originating at one height and travelling to another in which the velocity is different will transfer horizontal momentum to its new level. Let U_z, V_z be the average components of velocity at height z parallel to rectangular axes x, y in a horizontal plane, and let u', v', w' be the three components of the eddy velocity,

---

[*] *Trans. R.S. Canada*, Sect. III, 3rd Series, **30**, 1936, p. 1.
[†] M.O., *Geophysical Memoir*, No. 65.

so that the three components of the total velocity of the air at x, y, z are

$$U_z+u', \quad V_z+v', \quad w'. \qquad \ldots\ldots(21).$$

The rate of transfer of x-momentum upward across a horizontal surface is

$$\iint\rho\,(U_z+u')\,w'\,dx\,dy = \iint\rho u'w'\,dx\,dy \qquad \ldots\ldots(22),$$

and of y-momentum $\iint\rho\,(V_z+v')\,w'\,dx\,dy = \iint\rho v'w'\,dx\,dy \qquad \ldots\ldots(23),$

each integral being taken over the area in question. The terms $\iint\rho U_z w'\,dx\,dy$, $\iint\rho V_z w'\,dx\,dy$ vanish, since it is assumed that there is no net transfer of mass across the surface. Relations (22) and (23) should be compared with equations (5) and (7) of Chapter XI and it is then seen that what Reynolds calls the eddy stresses are the rates of transport into unit volume of momenta parallel to the co-ordinate axes.

The immediate problem, which is in fact the central problem to be faced in the discussion of turbulence, is that of putting expressions (22) and (23) into a form which involves only the mean motions U_z, V_z, W_z, and their derivatives relative to x, y and z. Schmidt and Prandtl assume that each eddy conserves the momentum of the layer in which it originates, so that

$$U_z+u'=U_{z_0}, \quad V_z+v'=V_{z_0},$$

where z_0 is the height at which the eddy originated. These equations may be written

$$u'=U_{z_0}-U_z=\frac{\partial U}{\partial z}\,(z_0-z)+\frac{1}{2}\frac{\partial^2 U}{\partial z^2}\,(z_0-z)^2+\ldots \text{ etc.}$$

If it is assumed that z_0-z is small so that the first term of the infinite series predominates,

$$u'=\frac{\partial U}{\partial z}\,(z_0-z).$$

The integral for the eddy transfer of momentum upward then becomes

$$U\iint\rho w'\,dx\,dy+\iint\rho\,(z_0-z)\,w'\,\frac{\partial U}{\partial z}\,dx\,dy.$$

Since there is no net transfer of mass across the horizontal plane, the first term is zero, and the second term may be written

$$\iint\rho\,(z_0-z)\,w'\,\frac{\partial U}{\partial z}\,dx\,dy = -K\rho\,\frac{\partial U}{\partial z} \text{ per unit surface } \ldots\ldots(24),$$

where K is readily seen to have the same meaning as was previously given to it in discussing the transfer of heat in § 137.

The net rate of gain of momentum by unit volume at height z is

$$\frac{\partial}{\partial z}\left(K\rho\,\frac{\partial U}{\partial z}\right) \qquad \ldots\ldots(25).$$

This is the method which has been followed by Schmidt and other writers. The coefficient which Schmidt* calls the *Austausch* is equal to $K\rho$ in our notation. Prandtl follows substantially the same method, and defines the mean value of $z-z_0$ as the *Mischungsweg* or "path of mixing".

* W. Schmidt, *Der Massenaustausch in freier Luft und verwandte Erscheinungen*. Hamburg, 1925.

Taylor starts from a different standpoint. He makes it clear that the irregularities of motion in a turbulent fluid are associated with irregularities of distribution of static pressure, and emphasises the possibility of these pressure differences influencing the horizontal momentum of the moving eddies. In his paper of 1915 he restricts the motion to two-dimensional flow, the mean motion being parallel to the axis of x, and the turbulent flow being restricted to the xz plane. When the motion is thus restricted to two dimensions the *vorticity* of any element of the fluid is not affected by the local variations of pressure, and it is the constancy of vorticity, and not the constancy of momentum, of a moving element which we must apply in order to evaluate the net loss of momentum per unit volume of the turbulent fluid.

From (22) above it follows that the net gain of momentum per unit volume

$$= -\frac{\partial}{\partial z} \iint \rho u' w' \, dx \, dy = I, \text{ say.}$$

We neglect the variations of density ρ, and treat the fluid as incompressible. Then

$$-I = \rho \iint \left(u' \frac{\partial w'}{\partial z} + w' \frac{\partial u'}{\partial z} \right) dx \, dy \qquad \ldots \ldots (26).$$

The equation of continuity is $\quad \dfrac{\partial u'}{\partial x} + \dfrac{\partial w'}{\partial z} = 0 \qquad \ldots \ldots (27).$

The condition that the moving element retains its original vorticity may be written

$$\frac{\partial}{\partial z} (U_z + u') - \frac{\partial w'}{\partial x} = \frac{\partial U_0}{\partial z} \qquad \ldots \ldots (28).$$

Substituting in (26) the values of $\partial w'/\partial z$ and $\partial u'/\partial z$ given by equations (27) and (28), we find

$$-I = \rho \iint \left\{ -u' \frac{\partial u'}{\partial x} + w' \frac{\partial w'}{\partial x} + w' \left[\left(\frac{\partial U}{\partial z} \right)_{z_0} - \left(\frac{\partial U}{\partial z} \right)_z \right] \right\} dx \, dy$$

$$= \tfrac{1}{2} \rho \iint \frac{\partial}{\partial x} (w'^2 - u'^2) \, dx \, dy + \rho \iint w' \left[\left(\frac{\partial U}{\partial z} \right)_{z_0} - \left(\frac{\partial U}{\partial z} \right)_z \right] dx \, dy.$$

The first term integrates out and vanishes, since it may be assumed that $\overline{u'^2}$ and $\overline{w'^2}$ do not vary over the horizontal area. It follows that

$$-I = \rho \iint w' \left\{ (z_0 - z) \frac{\partial^2 U}{\partial z^2} + \tfrac{1}{2} (z_0 - z)^2 \frac{\partial^3 U}{\partial z^3} + \ldots \right\} dx \, dy \ldots \ldots (29).$$

This equation is rigidly true for all disturbances, but if $z_0 - z$ is sufficiently small so that within this limit the changes in $\dfrac{\partial^2 U}{\partial z^2}$ are small by comparison with itself, we may write down the first term only of the series

$$I = -\rho \iint w' (z_0 - z) \frac{\partial^2 U}{\partial z^2} \, dx \, dy = \rho \overline{w' (z - z_0)} \frac{\partial^2 U}{\partial z^2} \qquad \ldots \ldots (30),$$

where $\overline{w'\,(z-z_0)}$ is the mean value of $w'\,(z-z_0)$ taken over the horizontal area. Or, returning to the notation of equation (24),

$$I = K\rho\,\frac{\partial^2 U}{\partial z^2} \qquad \ldots\ldots(31),$$

I represents the rate of loss of momentum by eddy transfer from unit volume of the turbulent medium.

It is of interest to compare equation (31) with equation (25). The latter may be written

$$I = \frac{\partial}{\partial z}\left(K\rho\,\frac{\partial U}{\partial z}\right) = K\rho\,\frac{\partial^2 U}{\partial z^2} + \rho\,\frac{\partial K}{\partial z}\,\frac{\partial U}{\partial z} \qquad \ldots\ldots(32),$$

neglecting the variations of ρ with height. Equation (31) only contains the first term on the right-hand side of the above equation. Since the terms on the right-hand side of equation (32) represent the net difference between the transfers into and out of unit volume, whereas the first term, according to Taylor's analysis, represents the net gain or loss of momentum by unit volume, it follows that the second term on the right-hand side of (32) must represent the rate at which momentum is destroyed by the action of the local differences of pressure.

Taylor's proof of equation (31), which has been reproduced above from his original paper, makes no assumption that K is constant at all heights, and equation (29) can therefore be used for K varying with height. Taylor's discussion of the nature of the distribution of temperature and wind-velocity with height, based on the assumption that K is constant, shows that the theory predicts results in reasonable agreement with observation. The assumption of constant K simplifies the mathematical treatment, and is in any case the obvious assumption to make in a first attempt to compare the theory with observations. Some further details of the variation of wind with height with different assumptions as to the nature of the variation of K with height will be found in § 150 below.

In view of the importance of the result shown in equation (31) it is of interest to consider a restatement of this proof given by Taylor in a recent paper.* The motion is again restricted to two dimensions, x and z. If we write η for the vorticity $\frac{1}{2}\left(\frac{\partial u}{\partial z}-\frac{\partial w}{\partial x}\right)$, we may write the equation of motion in the x-direction

$$-\frac{\partial}{\partial x}\left(\frac{p}{\rho}+\tfrac{1}{2}u^2+\tfrac{1}{2}w^2\right) = \frac{\partial u}{\partial t} + 2w'\eta \qquad \ldots\ldots(33).$$

This equation is rigidly true for each element of mass. Again neglecting variations of ρ, we may write it

$$\frac{\partial u}{\partial t} = -\frac{1}{\rho}\frac{\partial p}{\partial x} - \frac{\partial}{\partial x}\left(\tfrac{1}{2}u^2+\tfrac{1}{2}w^2\right) - 2w'\eta \qquad \ldots\ldots(34).$$

If we suppose the eddying motion to be on the average uniform in the direction of x, the second term on the right-hand side vanishes when we take mean values

$$\frac{\partial U}{\partial t} = -\frac{1}{\rho}\frac{\partial \bar{p}}{\partial x} - \overline{2w'\eta} \qquad \ldots\ldots(35).$$

* *Proc. Roy. Soc.* A, **135**, 1932, p. 685.

Thus the term $-\overline{2w'\eta}$ represents the rate of increase of mean velocity as a result of the eddying motion. If the eddy has retained the vorticity which it has as a result of the mean motion when at height z_0, then at height z it still has a vorticity $(z_0 - z)\dfrac{\partial}{\partial z}\left(\dfrac{1}{2}\dfrac{\partial U}{\partial z}\right)$ in excess of the normal vorticity η at that level

$$2w'\eta = 2w'\left\{\eta_0 + \tfrac{1}{2}(z_0 - z)\frac{\partial^2 U}{\partial z^2}\right\},$$

and taking mean values over a large horizontal area

$$\overline{2w'\eta} = \overline{w'(z_0 - z)}\frac{\partial^2 U}{\partial z^2} \qquad\qquad \ldots\ldots(36).$$

Substituting in equation (35) we find

$$\frac{\partial U}{\partial t} = -\frac{1}{\rho}\frac{\partial \bar{p}}{\partial x} - \overline{w'(z_0 - z)}\frac{\partial^2 U}{\partial z^2} = -\frac{1}{\rho}\frac{\partial \bar{p}}{\partial x} + \overline{w'(z - z_0)}\frac{\partial^2 U}{\partial z^2},$$

$$\frac{\partial U}{\partial t} = -\frac{1}{\rho}\frac{\partial \bar{p}}{\partial x} + K\frac{\partial^2 U}{\partial z^2} \qquad\qquad \ldots\ldots(37).$$

Hence the rate of gain of momentum due to the eddying motion is

$$K\rho\,\frac{\partial^2 U}{\partial z^2} \text{ per unit volume.}$$

§ 142. *Extension to three dimensions*

In the latest paper to which reference was made above Taylor has extended the vorticity-transport theory to three dimensions, but as Taylor remarks the results which he derives are so complicated as to be of little practical use. The lengthy expressions which he derives for the eddy transport of momentum reduce to the result of the earlier discussion (equation (31) above), in the special case of two-dimensional flow parallel to the xz plane. When the turbulent flow is parallel to the plane of yz the results reduce to Prandtl's form, shown in equation (32) above.

In three-dimensional flow we can no longer assume that the moving eddy retains its original components of vorticity, since the vorticity will be affected by the local variations of pressure. The mathematical conditions become far more complex, and there is no simple condition which can take the place of the constancy of vorticity which was the basis of the two-dimensional treatment.

The momentum-transport theory, as developed by Prandtl, Schmidt, and others, is directly applicable to three-dimensional motion, since the u and v components of velocity are treated separately and independently, and the expressions $\dfrac{\partial}{\partial z}\left(K\rho\dfrac{\partial U}{\partial z}\right)$ and $\dfrac{\partial}{\partial z}\left(K\rho\dfrac{\partial V}{\partial z}\right)$ are derived together by the same argument.

In applications to the study of turbulent motions in the atmosphere we are faced with the difficulty that observations show that the turbulent motion is in three dimensions, and that near the ground the cross-wind or v component is greater than either of the other two components. The use of Taylor's expression $K\rho \dfrac{\partial^2 U}{\partial z^2}$, or of Prandtl's expression $\dfrac{\partial}{\partial z}\left(K\rho \dfrac{\partial U}{\partial z}\right)$, cannot be held to be a complete representation of the conditions which we know to exist.

Many writers on this subject have been attracted by the analogy between eddy viscosity and ordinary viscosity. The analogy appears to be a true one for two-dimensional motion, but is no longer strictly true for three-dimensional motion, and we are left in some doubt as to how close this analogy really is. There is in fact no mathematical theory of turbulence in three dimensions which is in a form applicable to motion in the atmosphere. The serious obstacle to the dynamical study of turbulence is the difficulty of visualising the nature of a single eddy. It is, moreover, probable that eddies should be divided into a number of classes, each of which might behave differently with regard to the transfer of mass, momentum, vorticity, etc. It is possible to form a mental picture of the working of a simple convection bubble, and to visualise it as an agent in the transfer of heat and momentum in the vertical direction. But a vortex ring which mixes with its surroundings can impart no vorticity to those surroundings, since any two diametrically opposite arcs have vorticity of opposite signs, and will cancel each other when complete mixing takes place.

Prandtl's form $\dfrac{\partial}{\partial z}\left(K\rho \dfrac{\partial U}{\partial z}\right)$ is applicable to motion in a tube in which the turbulence is in a plane at right angles to the direction of the mean motion, or the vortices have axes parallel to the mean motion. In some ways the atmosphere appears more nearly analogous to the motion in a tube than to the two-dimensional motion discussed by Taylor, but the conditions in the atmosphere are complicated by the change in the direction of the mean wind with height. Taylor's equations for the transport of vorticity in three dimensions are so complicated that one is forced to conclude that the atmospheric problem is not capable of dynamical solution, and that the line of approach suggested by Taylor in another paper, which starts from a more purely statistical basis, is the most hopeful. This method is discussed in §§ 157–160 below.

It should be noted that a frictional term such as $K\partial^2 U/\partial z^2$ tends to annihilate existing differences of velocity, no matter how K varies with height; while a frictional term $\dfrac{\partial}{\partial z}\left(K \dfrac{\partial U}{\partial z}\right)$ will not of necessity do so, but may in fact lead to an accentuation of the existing differences of velocity.

§ 143. Comparison of the momentum-transport and the vorticity-transport theories

In the paper referred to above Taylor compares the results derived for the transport of heat and momentum on the momentum-transport theory and on the vorticity-transport theory respectively, by considering the distribution of temperature and velocity in the wake behind a cylindrical obstacle. Taylor shows that on the Prandtl theory the distribution of temperature and of velocity across the wake should follow the same law, as is indeed obvious *a priori*, whereas on the vorticity-transport theory these distributions should differ.

If η_0 is the value of the vorticity η at the edge of the wake, and $\xi = \eta/\eta_0$, then the distribution of velocity on either theory should be given by

$$\frac{u}{u_0} = (1 - \xi^{\frac{3}{2}})^2 \qquad \qquad \ldots\ldots(38).$$

The distribution of temperature should be given by

$$\frac{\theta}{\theta_0} = (1 - \xi^{\frac{3}{2}})^2 \qquad \qquad \ldots\ldots(39),$$

on the momentum-transport theory, and by

$$\frac{\theta}{\theta_0} = 1 - \xi^{\frac{3}{2}} \qquad \qquad \ldots\ldots(40)$$

on the vorticity-transport theory. Measurements of the distribution of temperature and velocity in the wake behind a heated obstacle carried out at the National Physical Laboratory showed that the distribution of velocity agreed very closely with equation (38) above, and that the distribution of temperature agreed closely with equation (40) above, but failed to agree even approximately with equation (39). Taylor claims that it is thus established that, at least in these conditions, the vorticity-transport theory is a better representation of the facts than the momentum-transport theory.

The argument is not completely convincing. The velocity u is represented by

$$\frac{u}{u_0} = \phi(x) f(\eta)$$

and $\phi(x)$ is taken as $x^{\frac{1}{2}}$. Taking

$$f(\eta) = e^{-a\eta^2} \qquad \qquad \ldots\ldots(41)$$

we shall have a solution of the equation which fits the observational data with slightly greater accuracy than equation (40). The adoption of equation (41) requires that K should be constant, on either the momentum-transport or the vorticity-transport theory. The temperature observations can be fitted by a similar equation with a constant K, which is, however, different from the K for momentum. It must be remembered that the corresponding constants for molecular transport are different, being κ the coefficient of thermal diffusivity, and ν the kinematic coefficient of viscosity. (See also § 144 below.)

§ 144. *Eddy diffusivity and Austausch for different properties*

The eddy diffusivity and the *Austausch* coefficient can be defined separately for momentum, temperature, etc. In § 141 above K was defined as the mean value of $w'l'$, where l' is the vertical distance traversed by the eddy since it was last a normal specimen of its immediate environment, for the property which we are considering. It is clear that if we approach the subject from this point of view we may regard l' as differing for momentum, heat, etc. In the first place consider the diffusion of momentum.

Taylor has shown that $\overline{w'l'}$ may be written in several other forms, and that l' may be replaced by a distance l which the observed element has moved since some initial time t_0, provided t_0 is sufficiently early to ensure that each observed element has at least once since t_0 been a normal portion of its immediate surroundings. Then

$$\overline{w'l'} = \overline{w'l} - \overline{w'\,(l - l')} = \overline{w'l} \qquad \ldots\ldots(42)$$

since there is no correlation between w' and $l - l'$; also

$$\overline{w'l'} = \overline{w'l} = \frac{\overline{dl}}{dt}\,l = \frac{1}{2}\frac{d}{dt}\,\overline{l^2} \qquad \ldots\ldots(43).$$

For the diffusion of heat let the corresponding values of l be represented by l''. Then

$$\overline{w'l''} = \overline{w'l'} + \overline{w'\,(l'' - l')} \qquad \ldots\ldots(44).$$

If inequalities of heat were diffused more slowly than inequalities of momentum, then l'' would generally be measured from an earlier instant than l', and there would be no correlation between w' and $l'' - l'$. In this case

$$\overline{w'l''} = \overline{w'l'} = \overline{w'l} \qquad \ldots\ldots(45).$$

If, however, the inequalities of heat were diffused more rapidly than inequalities of momentum, then l'' would generally be measured from a later instant than l', and $l' - l''$ would in general have the same sign as w'. Hence

$$\overline{w'l''} = \overline{w'l'} - \overline{w'\,(l' - l'')},$$

$$\overline{w'l''} = \overline{w'l'} - \text{a positive quantity},$$

and

$$\overline{w'l''} < \overline{w'l'} \qquad \ldots\ldots(46).$$

Thus the value of K appropriate to momentum should be at least as great as that appropriate to heat or any other diffusing property, and may conceivably be very must greater. Taylor[*] has quoted some figures showing that in the sea K may be at least nineteen times as great for momentum as for salinity.

§ 145. *Stability and the criterion of turbulence*

Richardson[†] has given a criterion to determine whether turbulence will increase or decrease. The principle on which the criterion is based is that the flow will remain turbulent if the rate of supply of energy by the Reynolds stresses is at least as great as the work which has to be done to maintain the turbulence against gravity.

At a level z, where pressure, density and temperature are p, ρ, T respectively, let an eddy arrive from level $z - l'$, at which it had the normal temperature at that level, $T - l'\frac{\partial T}{\partial z}$. At its new level its temperature will be

---

[*] *Brit. Ass. Rep.* London, 1931.
[†] *Proc. Roy. Soc.* A, **97**, 1920, p. 354.

$T - l'\left(\frac{\partial T}{\partial z} + \Gamma\right)$, where Γ is the adiabatic lapse-rate. Its excess of density over the normal environment at that level will be

$$\frac{l'\rho}{T}\left(\frac{\partial T}{\partial z} + \Gamma\right),$$

and the downward force upon it due to this excess of density will be

$$\frac{gl'\rho}{T}\left(\frac{\partial T}{\partial z} + \Gamma\right) \text{ per unit volume,}$$

or

$$\frac{gl'}{T}\left(\frac{\partial T}{\partial z} + \Gamma\right) \text{ per unit mass.}$$

The rate of upward flow of fluid per unit horizontal area per unit time is w', and hence the amount of work done against gravity per unit volume per unit time is

$$\frac{g\rho\overline{w'l'}}{T}\left(\frac{\partial T}{\partial z} + \Gamma\right),$$

or

$$\frac{gK_T\rho}{T}\left(\frac{\partial T}{\partial z} + \Gamma\right),$$

where K_T is the appropriate constant for diffusion of heat, defined by

$$K_T = \overline{w'l'}.$$

The work done by the eddy stresses per unit volume is

$$\widehat{xz}\,\frac{\partial \overline{u}}{\partial z} + \widehat{yz}\,\frac{\partial \overline{v}}{\partial z} \qquad \ldots\ldots(47),$$

or

$$\rho\left\{K_{mx}\left(\frac{\partial \overline{u}}{\partial z}\right)^2 + K_{my}\left(\frac{\partial \overline{v}}{\partial z}\right)^2\right\} \qquad \ldots\ldots(48),$$

where K_{mx}, K_{my} are the appropriate coefficients of eddy diffusivity for momentum along the x and y axes. K_{mx} and K_{my} need not of necessity be equal, as was pointed out by Richardson.

The condition that turbulence should increase may therefore be written

$$K_{mx}\left(\frac{\partial \overline{u}}{\partial z}\right)^2 + K_{my}\left(\frac{\partial \overline{v}}{\partial z}\right)^2 > \frac{gK_T}{T}\left(\frac{\partial T}{\partial z} + \Gamma\right) \qquad \ldots\ldots(49).$$

If $K_{mx} = K_{my} = K_m$, the condition becomes

$$\left(\frac{\partial \overline{u}}{\partial z}\right)^2 + \left(\frac{\partial \overline{v}}{\partial z}\right)^2 > g\,\frac{K_T}{K_m}\,\frac{1}{T}\left(\frac{\partial T}{\partial z} + \Gamma\right) \qquad \ldots\ldots(50).$$

As we have seen above, the factor K_T/K_m is at most unity and may be considerably less than unity. In the atmosphere it does not appear to differ widely from unity, if we judge by the reasonably close agreement of the values of K derived from Eiffel Tower observations of wind and temperature.

Fig. 62 reproduces an application of this criterion to the breakdown of an inversion, taken from a paper by C. S. Durst.[*] The lower diagram gives

[*] *Q.J. Roy. Met. Soc.* **59**, 1933, p. 131.

$\left(\dfrac{\partial \bar{u}}{\partial z}\right)^2 + \left(\dfrac{\partial \bar{v}}{\partial z}\right)^2$, here represented as $\left(\dfrac{\partial v}{\partial h}\right)^2$, and it is seen that the motion becomes turbulent when this quantity exceeds $\dfrac{g}{T}\left(\dfrac{\partial T}{\partial z}+\Gamma\right)$.

Richardson's criterion agrees with that given by Prandtl except for a factor $\frac{1}{2}$ which appears on the right-hand side of the inequality as derived by Prandtl. The difference between the treatments of Richardson and Prandtl is that the latter treats the element of air as starting from rest and attaining a velocity w after a vertical distance l, whereas Richardson effectively assumes that it has the velocity w during the whole of the distance l. It is readily seen that this accounts for the factor $\frac{1}{2}$ in Prandtl's inequality. The example quoted above from Durst's paper, and other examples of the application of this criterion given by Richardson in various papers, appear to indicate the validity of Richardson's form of the criterion.

It should be noted that if the atmosphere is vertically stable, so that $(\partial T/\partial z)+\Gamma$ is positive, turbulence can only be maintained if $\partial \bar{u}/\partial z$ exceeds a certain definite limit. If $(\partial T/\partial z)+\Gamma$ is negative, any motion will become turbulent if slightly disturbed. In the atmosphere large gradients of velocity occur mainly in the immediate neighbourhood of the ground, as an effect of surface friction, and it is here that $(\partial T/\partial z)+\Gamma$ is most liable to assume negative values. Thus the layer of air in contact with the ground is the one in which turbulence will most readily occur. Richardson has pointed out (loc. cit.) that vertical gradients of wind of very considerable magnitude may occur near the upper boundary of the troposphere, and that turbulence may also arise in that region. In the neighbourhood of a surface of discontinuity between two streams of air moving with different velocities large gradients of velocity occur, and eddies form which tend to smooth out the surface of discontinuity.

A criterion such as is given above affords an explanation of the smoothing out of turbulence at the ground after sunset, when the ground cools by radiation to the sky. The growth of an inversion is accompanied by the suppression of turbulence, though there may be a large increase of wind with height.

G. I. Taylor* has investigated the effect of variation in density on the stability of superposed streams of fluid, and has shown that for three streams, of which the intermediate one is of small thickness h, instability of the wave-motion first arises when

$$\left(\frac{\partial \bar{u}}{\partial z}\right)^2 = -2g\,\frac{1}{\rho}\frac{\partial \rho}{\partial z} \qquad \ldots\ldots(51).$$

Taylor found that the first type of instability which occurs when $\left(\dfrac{\partial \bar{u}}{\partial z}\right)^2$ reaches the limiting value $-\dfrac{2g}{\rho}\dfrac{\partial \rho}{\partial z}$ involves a tendency for the intermediate fluid to collect into lumps which form separate eddies and are projected alternatively upward and downward into the neighbouring fluid.

* *Proc. Roy. Soc.* A, **132**, 1931, p. 499.

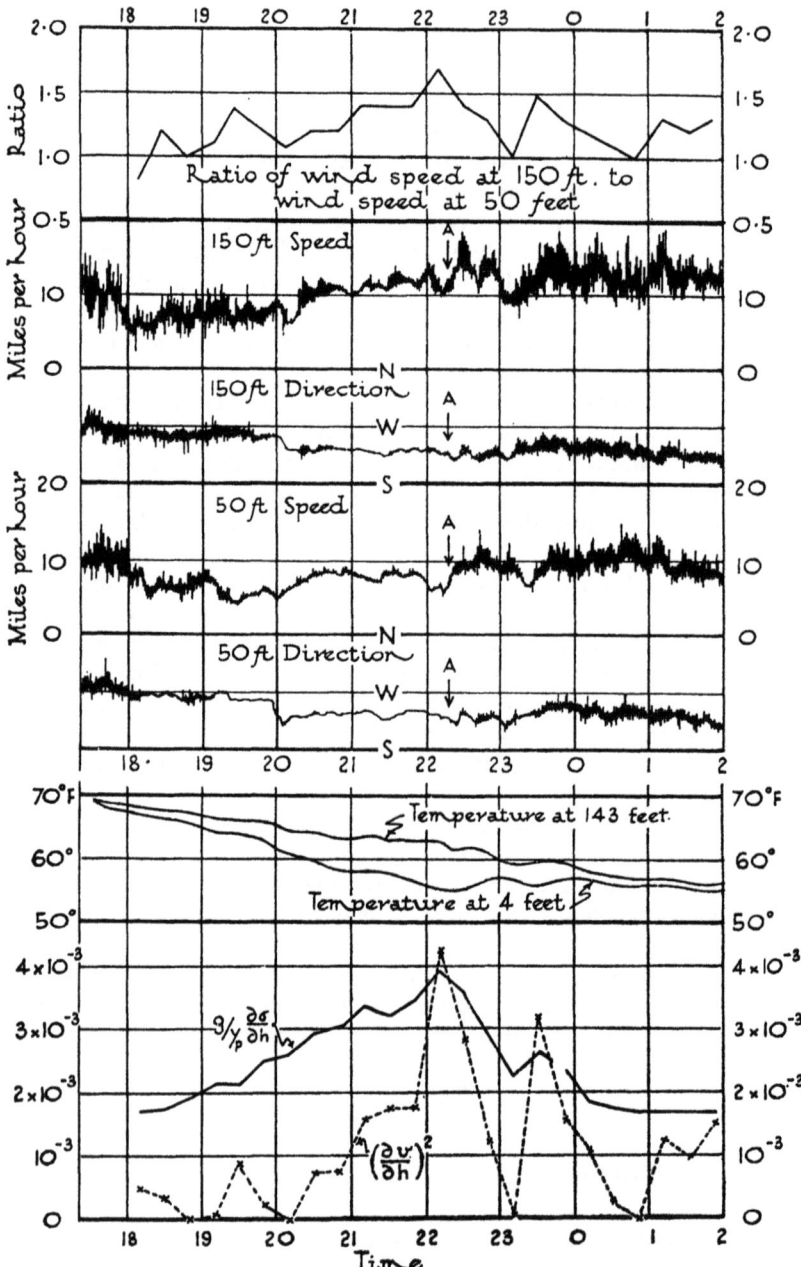

Fig. 62. Durst's application of Richardson's criterion.

Taylor further showed that in a fluid in which $\frac{\partial \overline{u}}{\partial z}$ and $\frac{1}{\rho}\frac{\partial \rho}{\partial z}$ are not functions of the height z, all waves are stable if

$$\left(\frac{\partial \overline{u}}{\partial z}\right)^2 < -4g\frac{1}{\rho}\frac{\partial \rho}{\partial z} \qquad \ldots\ldots(52),$$

but that no waves are possible, stable or unstable, if

$$\left(\frac{\partial \overline{u}}{\partial z}\right)^2 > -4g\frac{1}{\rho}\frac{\partial \rho}{\partial z} \qquad \ldots\ldots(53).$$

These inequalities apply to continuous distributions of density, and should be contrasted with the equality (51) above which applies to stability with one intermediate layer. In (51) the coefficient 2 becomes 2·11 for two intermediate layers, and Taylor has suggested that it may become 4 when the number of layers becomes infinite. Taylor's discussion applies, however, only to infinitesimal displacements, and is not strictly applicable when the motion

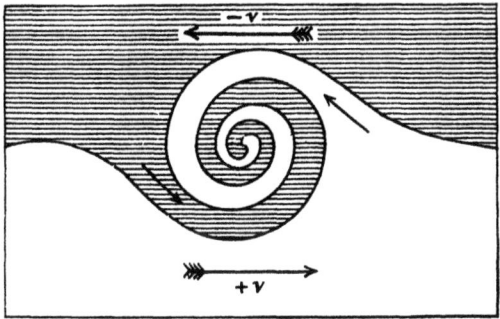

Fig. 63. Mallock's representation of the eddy.

becomes finite. The work of Rosenhead[*] is probably more closely akin to the actual physical phenomena. The idea underlying Rosenhead's work is to be found in a paper by Mallock.[†] Fig. 63, which shows the nature of an eddy as pictured by Mallock, gives a picture of the process of mixing which is readily understood.

Rosenhead investigates the flow of a stream of density ρ and velocity U in the direction of the axis of x, above a stream of the same density, having velocity U in the opposite direction. The motion is treated as two-dimensional, the axis of x being the undisturbed surface. The flow is continuous and irrotational on both sides of the surface of separation. Initially the surface of separation is of the form of a sine-curve of small amplitude. The solution of the equations of motion by the method of small oscillations contains a first and a second order term. The first order term is a sine-term of increasing amplitude. The second order term, which eventually dominates, introduces a term which is antisymmetrical with respect to a crest, and hence the disturbance does not grow symmetrically.

[*] *Proc. Roy. Soc.* A, **134**, 1932, p. 170.   [†] *Aero. Res. Committee, R. and M.*, No. 314.

In Rosenhead's analysis the surface of discontinuity, which is a vortex sheet, is replaced by a distribution of finite elemental vortices along its trace, and the paths of the vortices are determined by a numerical step by step method. It appears that the effect of instability on a surface of discontinuity of sine form is to produce concentrations of vorticity at equal intervals along the surface, and that the surface of discontinuity tends to roll up round these points of concentration. The process of development is shown in fig. 64, which is a rough reproduction of Rosenhead's fig. 4 (*loc. cit.*), and the rolling up of the surface of separation is seen to be similar to that represented by Mallock, as shown in fig. 63 above. An interesting feature of the phenomenon is that while the initial conditions may be represented by two irrotational motions, the conditions represented by the final stage of fig. 64 indicate that in portions of the fluid a finite distribution of vorticity may become observable.*

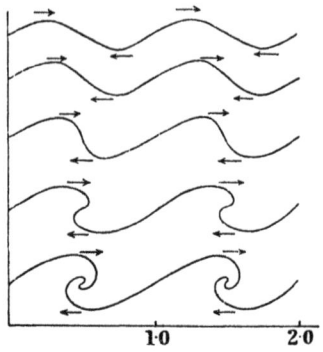

Fig. 64.
Rosenhead's representation of the development of the eddy.

Prandtl† summarises the question of the criterion of turbulence as follows: In large volumes stability depends on the dimensionless number

$$-\frac{g}{\rho}\frac{\partial\rho}{\partial z}\bigg/\left(\frac{\partial u}{\partial z}\right)^2,$$

which in the atmosphere is replaced by

$$\frac{g}{T}\left(\frac{\partial T}{\partial z}+\Gamma\right)\bigg/\left(\frac{\partial u}{\partial z}\right)^2 \qquad \ldots\ldots(54).$$

The critical value of this number, above which turbulence tends to die out, and below which turbulence sets in, is estimated at unity by Richardson, at $\frac{1}{2}$ by Prandtl, and $\frac{1}{4}$ by Taylor and Goldstein (from small oscillations). Prandtl states (*loc. cit.*) that preliminary experiments indicated a limiting value of $\frac{1}{4}$ to $\frac{1}{2}$. The agreement with Richardson's criterion shown in fig. 62 appears, however, to indicate that in the atmosphere the limiting value of expression (54) above is unity.

Douglas‡ has used Richardson's criterion to show that there is an upper limit to the slope of a surface of discontinuity beyond which even an inversion or an isothermal patch will become turbulent, and consequently will be smoothed out.

The above discussion has no application to purely dynamical stability. Rayleigh§ considered the stability of laminated steady motion in which the undisturbed velocity U is parallel to the axis of x, and is a function of z, with the vorticity changing suddenly from one layer to another. He discussed the effect of superposing a small disturbing velocity of a simple harmonic type, and found that the two-dimensional motion should be stable if $\partial^2 U/\partial z^2$ is of

---

* See also A. R. Low, *Nature*, **121**, 1928, p. 576.
† *Beitr. Phys. fr. Atmos.* **19**, 1932, p. 188.
‡ *Q.J. Roy. Met. Soc.* **50**, 1924, Appendix, p. 359.
§ *Collected Papers*, **3**, p. 375.

the same sign throughout the fluid. Taylor* derived the same result for a perfectly generalised disturbance, and gave a physical explanation of the origin of the instability in cases where $\partial^2 U/\partial z^2$ changes sign, in terms of the eddy transport of momentum.

Early investigations into the stability of laminar flow did not take into account friction, or the curvature of the velocity profile. Later investigators, who took partial account of friction, but not of the curvature of the velocity profile, were equally unsuccessful in finding a stability limit. Thus Prandtl† and Tietjens,‡ taking account of the larger frictional terms and employing a velocity profile composed of straight line segments, found that instability could arise, but could not establish a stability limit. When Tollmien§ took into consideration the curvature of the velocity profile, he obtained a stability limit in the form of a critical Reynolds number. Schlichting‖ found much lower values for the non-dimensional criterion than those quoted above. Squire¶ obtained the interesting result that if a plane current is unstable for a determined Reynolds number for three-dimensional disturbances, it is unstable for two-dimensional disturbances for a lower Reynolds number; thus two-dimensional disturbances are more "dangerous" than three-dimensional disturbances.

We cannot here enter into the details of these researches, but the following table given by Schlichting as a summary of the work which has been done in the subject will indicate the nature of the field which has been covered.

Scheme of researches carried out on stability and the initiation of turbulence

Velocity distribution	Homogeneous fluids		Non-homogeneous fluids (with density stratification)	
	Without friction	With friction	Without friction	With friction
Linear profile	Rayleigh	Sommerfeld, v. Mises, Hopf, Prandtl, Tietjens, etc.	Taylor, Goldstein	—
Curved profile	—	Tollmien	—	Schlichting

§ 146. The maintenance of super-adiabatic lapse-rates in the atmosphere

Reference was made earlier, p. 219, to the result derived by Rayleigh that a fluid can remain stable with density increasing upward, if

$$\frac{\rho_1 - \rho_0}{\rho_0} < \frac{27\pi^4 \kappa \nu}{4gh^3} \qquad \ldots\ldots(55),$$

or, if $\partial\rho/\partial z$ be the upward gradient of density, if

$$\frac{h}{\rho}\frac{\partial\rho}{\partial z} < \frac{27\pi^4 \kappa \nu}{4gh^3} \qquad \ldots\ldots(56).$$

* *Phil. Trans. Roy. Soc.* A, **215**, 1915, p. 1.
† *Zeit. f. angew. Math. u. Mech.* **1**, 1921, p. 431.   ‡ *Ibid.* **5**, 1925, p. 200.
§ *Nachr. Ges. Wiss. Göttingen, Math. Phys. Kl.* 1929, p. 21.
‖ *Zeit. f. angew. Math. u. Mech.* **15**, 1935, p. 313.
¶ *Proc. Roy. Soc.* A, **142**, 1933, p. 621.

In the atmosphere the transfer of heat and momentum is not brought about by the molecular processes of viscosity and conduction, but by eddy diffusion or eddy diffusion and radiative diffusion combined. Jeffreys[*] has shown that it is permissible in a compressible fluid to use the potential temperature in lieu of absolute temperature, or, to a sufficient degree of approximation, to use the deviations of temperature from an adiabatic distribution. We then replace $\frac{1}{\rho}\frac{\partial \rho}{\partial z}$ by $-\frac{1}{T}\left(\frac{\partial T}{\partial z}+\Gamma\right)$ for compressible fluid, and the inequality then becomes

$$-\frac{\partial T}{\partial z}-\Gamma < \frac{27\pi^4 \kappa \nu T}{4gh^4} \qquad\qquad ......(57).$$

In a turbulent atmosphere the transport of heat and momentum by turbulence greatly exceeds that by molecular diffusion, and κ and ν should be replaced by K. If K is of the order of 10^3, the maximum lapse-rate in a depth h is given by

$$-\frac{\partial T}{\partial z}-\Gamma = \frac{27\pi^4 10^6}{4gh^4}T = \frac{2\cdot1\,.\,10^8}{h^4} \text{ approximately.}$$

If $h = 10$ metres $= 10^3$ cm,

$$\left(-\frac{\partial T}{\partial z}-\Gamma\right)_{\text{max}} = 2\cdot1 \times 10^{-4}\,^{\circ}\text{C}/\text{cm} = 2\cdot1^{\circ}\text{ C}/100\text{ m.}$$

If $h = 2$ metres, the maximum lapse-rate becomes about $1\cdot3^{\circ}$ C per metre, and for $h = 1$ metre, the maximum becomes 21° C per metre. For other values of K the maximum possible value of $-(\partial T/\partial z)-\Gamma$ which can persist will be proportional to K^2.

In quieter conditions when turbulence is less active the transfer of heat by radiation may exceed that by eddy diffusion, and it is then necessary to replace κ in the inequality (57) above by the radiative diffusivity K_R, and to replace ν by K. The resulting maximum lapse-rate is smaller than was derived on the supposition that κ and ν should be replaced by K, whose value was taken as 10^5, but the values so derived are still much in excess of the adiabatic lapse-rate, for very shallow layers. The point which is of interest is that the occurrence of very large lapse-rates in shallow layers of air could have been predicted by theory, and that such lapse-rates actually occur and are readily observed. Best[*] found that in England, on sunny days, the lapse-rate in the layer from 1 inch to 1 foot above grassland can reach a value about 2000 times the dry adiabatic lapse-rate. Further, even when K is assumed to have a value as great as 10^6, the maximum value of $-(\partial T/\partial z)-\Gamma$ which can persist in deep layers is only a very minute fraction of Γ. It follows that, except in very shallow layers, the limiting condition for stability is fixed by the adiabatic lapse-rate as derived in § 21 above.

§ 147. *Prandtl's theory of the Mischungsweg*

Reynolds showed that the effect of turbulent motion might be taken into account by the addition to the equations of motion of certain shearing stresses of the type

$$\widehat{xz} = -\rho\overline{u'w'},$$

* *Proc. Camb. Phil. Soc.* **26**, pt 2, 1930.   † M.O., *Geophys. Mem.* No. 65.

where u' and w' are the components of eddy velocity, or the deviations of the instantaneous velocities u, w from their true mean values $\bar{u}$, $\bar{w}$. The problem consists essentially in finding a method of expressing $\overline{u'w'}$ in terms of measurable quantities. Prandtl* makes the same assumption as Taylor that an eddy consists of a mass of fluid originally a normal sample of the fluid at level z, which moves to a new level $z + l$, and there mixes with its environment. Its eddy velocity just before mixing is $\bar{u}(z) - \bar{u}(z + l)$, which to a first approximation is $-l\dfrac{\partial \bar{u}}{\partial z}$. The parameter l is called the *Mischungsweg*, or mixing length. It is to be noted that the theory assumes that mixing is a discontinuous process.

Prandtl further assumes that w' must be of the same order of magnitude as u' and proportional to it, so that

$$\widehat{xz} = \pm \rho l^2 \left(\frac{\partial \bar{u}}{\partial z}\right)^2 \qquad \ldots\ldots(58)$$

except for a possible factor of proportionality on the right-hand side. This factor may be taken as unity, thereby adding somewhat to the indefinite nature of the parameter l. In order to determine which sign is appropriate, take a case of mean velocity increasing with height. Then upward motion, with w' positive, involves negative u', and *vice versa*, and the equation must be written

$$\widehat{xz} = + \rho l^2 \left|\frac{\partial u}{\partial z}\right| \frac{\partial u}{\partial z} \qquad \ldots\ldots(59).$$

Schmidt's *Austausch* coefficient A may then be written

$$A = K\rho = \rho l^2 \left|\frac{\partial u}{\partial z}\right| \qquad \ldots\ldots(60).$$

It is advantageous, at this stage, to consider the application of the above ideas to pipe flow. For a circular pipe of radius r_0, an elementary discussion shows that the shearing stress τ at any point in the fluid is related to the shearing stress between the wall and the fluid, τ_0, by

$$\widehat{xz} \equiv \tau = \tau_0 \left(1 - z/r_0\right) \qquad \ldots\ldots(61)$$

where z is the distance from the wall.† This law of linear distribution of shearing stress is quite general, holding for all types of motion, and constitutes an overriding condition to which all forms of velocity distribution across the pipe must adjust themselves. We have, then, from (58)

$$l\frac{\partial \bar{u}}{\partial z} = \sqrt{\frac{\tau_0}{\rho}} \sqrt{1 - z/r_0} \qquad \ldots\ldots(62)$$

as the equation which fixes the distribution of mixing length with distance from the boundary.

In order to apply this basic equation, von Kármán appeals to the experimental fact, established by Nikuradse,‡ that for the same value of τ_0, the

---

* *Vide* Prandtl, *The Physics of Solids and Fluids* (Blackie), pp. 277–83; or Prandtl, *Abriss der Strömungslehre* (Vieweg).

† Bakhmeteff, *The Mechanics of Turbulent Flow*, 1936, p. 3.

‡ *Proc. 3rd Int. Congress Tech. Mech.* Stockholm, 1930.

quantity $\bar{u}_{max} - \bar{u}$ is a function of z/r_0 only; that is, although such factors as the radius of the pipe, the average velocity, etc., be varied, the shape of the curve of velocity remains invariant so long as the local shearing stress is unaltered. Von Kármán's investigation therefore aims at expressing a relationship, governing the shape of the velocity profile, which does not involve such "bulk" factors as the average velocity or the radius of the pipe. He makes the assumption that the local factors which determine the stresses at two points act in a dynamically similar fashion, and therefore, when referred to axes moving with the mean velocity, differ only in their scales of length and time. Taking l as the characteristic length, the time scale being implied in the velocity derivatives, it is possible to express similarity of pattern by a series of proportionalities between the factors affecting the shape of the velocity profile. Von Kármán's result is that

$$l = k_0 \frac{\partial u}{\partial z} \Big/ \frac{\partial^2 u}{\partial z^2} \qquad \qquad \ldots \ldots (63)$$

where k_0 is a dimensionless constant.

Numerous proofs of this formula have been given, but the result is immediately obvious from a consideration of dimensions, without any appeal to the idea of similarity, if we assume only that l depends on the form of the velocity profile in the immediate neighbourhood. The mean velocity is removed by taking axes moving with the fluid, and hence l must depend on $\frac{\partial u}{\partial z}, \frac{\partial^2 u}{\partial z^2}$, etc. If only the immediate neighbourhood of the point be considered, the higher derivatives can be neglected, and since the only combination of $\frac{\partial u}{\partial z}$ and $\frac{\partial^2 u}{\partial z^2}$ which will yield a length is their quotient, von Kármán's formula follows at once.

We thus have, from (58) and (63),

$$\frac{\rho k_0^2 \left(\frac{\partial u}{\partial z}\right)^4}{\left(\frac{\partial^2 u}{\partial z^2}\right)^2} = \tau_0 \left(1 - z/r_0\right) \qquad \qquad \ldots \ldots (64)$$

which, with the condition $\frac{\partial \bar{u}}{\partial z} \to \infty$ as $z \to 0$, yields von Kármán's logarithmic law for the velocity distribution in a circular pipe

$$\bar{u}_{max} - \bar{u} = \frac{-1}{k} \sqrt{\frac{\tau_0}{\rho}} \log \{(1 - \sqrt{1 - z/r_0}) - \sqrt{(1 - z/r_0)}\}.$$

A simpler form results from the assumption that the change of shearing stress with distance may be neglected near the wall, and taking $\tau = \tau_0 = $ constant, we find

$$\left(\frac{\partial u}{\partial z}\right)^2 \Big/ \frac{\partial^2 u}{\partial z^2} = -\frac{1}{k} \sqrt{\frac{\tau_0}{\rho}}$$

which yields on integration,

$$u = \frac{1}{k} \sqrt{\frac{\tau_0}{\rho}} \log z + C \qquad \qquad \ldots \ldots (65).$$

Prandtl* suggested that this equation could be applied to the whole region of turbulent flow, and not merely to the layer near the wall, and a comparison with Nikuradse's measurements shows that the simpler equation is, if anything, a better fit than von Kármán's more elaborate expression. This is all the more remarkable since the equation is based on the assumption that τ/ρ is constant. Thorade suggests that the variations in τ/ρ are compensated by variations in l, which keep the formula correct within the limits of experimental error. The most remarkable feature of the logarithmic law is that it holds for widely varying Reynolds numbers, and brings within the ambit of one formula all the experimental data available.

§ 148. *Development of Prandtl's theory for the surface layers*

Prandtl has applied the above ideas to a discussion of the vertical distribution of the wind in the layer of air next to the ground, this layer being regarded as sufficiently thin to permit of our regarding the motion as being controlled entirely by the viscous drag on its two horizontal boundaries, the effects of the horizontal pressure gradient and of the deviating acceleration due to the earth's rotation being neglected. The velocity will then be constant in direction through the whole layer, its direction being that of the viscous drag at the ground.

Following Rossby and Montgomery,† we assume that the mixing length l is proportional to the height z above the lower boundary. At the ground, l has a finite value which is assumed to be proportional to the average height, ϵ, of the surface roughnesses. Then l may be expressed by .

$$l = k_0 \left(z + z_0\right) \qquad \ldots\ldots(66)$$

where k_0 is a non-dimensional constant whose value, according to Prandtl and von Kármán, is 0·38, and z_0 is equal to $s\epsilon$, s being a numerical factor for which Prandtl suggests a tentative value 1/30. This value of s is based on experiments in a wind tunnel, in which the size and packing of the roughness elements were closely related. This value of s is not of necessity appropriate to atmospheric flow.

Equation (58) above then gives for the eddy stress

$$\widehat{xz} = \rho l^2 \left(\frac{\partial u}{\partial z}\right)^2 = \rho k_0^2 \left(z + z_0\right)^2 \left(\frac{\partial u}{\partial z}\right)^2 \qquad \ldots\ldots(67).$$

Taking the square root of this expression and integrating, we find

$$u = \frac{1}{k_0} \sqrt{\frac{\widehat{xz}}{\rho}} \log_e \frac{z + z_0}{z_0} \qquad \ldots\ldots(68)$$

$$K = l^2 \left|\frac{\partial u}{\partial z}\right| = k_0^2 \left(z + z_0\right) \frac{U_a}{\log_e \frac{z_a + z_0}{z_0}} \qquad \ldots\ldots(69)$$

where U_a is the mean wind velocity measured by anemometer at height z_a.

* *Z.V.D.I.* No. 5, 1933.
† Papers in *Phys. Oceanography and Meteorology, Mass. Inst. Tech.* **3**, No. 3.

Equation (68) can also be written in the form

$$\widehat{xz} = \rho k_0^2 u^2 \Big/ \Big(\log_e \frac{z + z_0}{z_0} \Big)^2 \qquad \qquad \dots\dots(70).$$

A logarithmic law for the variation of wind with height has been given by various writers. E. H. Chapman[*] has shown that a number of sets of observations can be represented by the formula

$$\frac{V}{V_{10}} = a \log z + b \qquad \qquad \dots\dots(71)$$

and Hellmann[†] has suggested a slightly different formula

$$V = a \log (z + c) + b \qquad \qquad \dots\dots(72).$$

This law acquires a special interest in view of the wide generality of equation (65) above for the flow in pipes.

Rossby and Montgomery have shown that the logarithmic law of variation of wind with height is in good agreement with observations, when the atmosphere is adiabatic or unstable. They give as examples of the measure of z_0, the roughness parameter,

Over short cropped grass	$z_0 = 0.54$ cm,
Over open grass land	$z_0 = 3.2$ cm,
Over sea with small swell	$z_0 = 3.9$ cm.

Rossby[‡] has shown that certain determinations of K made by Mildner from hodographs of vertical wind distribution at Leipzig give excellent agreement with the values derived from equation (69), up to heights of 250 metres, though it is obvious that the neglect of pressure and deviating force up to such a height cannot be justified theoretically.

The theory as developed above assumes that the atmosphere is adiabatic, that l is given by equation (66) above, and that the shearing stress $\widehat{xz}$ is constant. As in the case of flow in pipes, referred to in § 147 above, there is no doubt that $\widehat{xz}$ does vary widely. Scrase[§] has shown, from the analysis of cinematographic records of wind at 19 m and 1.5 m, that the shearing stress $\widehat{xz}$ varied from 3.6 gm cm² sec⁻² at 19 m, to rather less than a quarter of this value at 1.6 m. There is therefore no doubt that $\widehat{xz}$ does vary with height, but for some reason which has not yet been explained, the assumption that $\widehat{xz}$ does not vary with height yields a law of variation of wind with height which agrees with observations.

Equation (70) above shows that $\widehat{xz}$ is proportional to u^2, and that the velocity should increase in the same proportion at all heights. Experiments by Nikuradse[||] and Dryden[¶] in smooth tubes and over flat plates have shown

[*] M.O., *Professional Notes*, No. 6.   [†] *Preuss. Akad. Wiss. Berlin*, **10**, 1917, p. 174.
[‡] Papers in *Phys. Oceanography and Meteorology, Mass. Inst. Tech.* **1**, No. 4.
[§] M.O., *Geophys. Mem.* No. 52, 1930, p. 14.
[||] *Forschungsheft* 356 *des Vereins Deutsch. Ingen.*, 1932.
[¶] *N.A.C.A. Rep.* No. 562, 1936.

that equation (68) is satisfactory for values of $\sqrt{\dfrac{\overline{xz}}{\rho}}$ greater than about 30, but that for lower values the equation is unsatisfactory.

Sutton[*] has shown that the logarithmic law, adjusted so as to make u vanish at the ground, and take an assigned value u_1 at a fixed height z_1, has the form

$$\frac{u}{u_1} = \log\left(\alpha z/z_1 + 1\right)/\log\left(\alpha + 1\right) \qquad \ldots\ldots(73)$$

$$\sim 1 + \frac{1}{\log \alpha} \log z/z_1 + o\left(\frac{1}{\alpha}\right) \qquad \ldots\ldots(74)$$

for large values of α. The quantity α is a non-dimensional parameter which varies considerably with the lapse-rate of temperature, being about 3×10^2 for isothermal conditions. The second form of the equation has the advantage of making it immediately obvious how to evaluate α. Sverdrup[†] criticised Sutton's use of the logarithmic equation for other than adiabatic atmospheres, but Sutton[‡] later pointed out that the very careful series of observations made by Best[§] fitted the equation with satisfactory accuracy. Sutton suggested that the variations in α indicate that the influence of surface roughness is small in large lapse-rates, and large during inversions.

§ 149. *Variation of wind with height in the surface layers; Rossby and Montgomery's extension to stable atmospheres*

It has been shown by Richardson and Prandtl that the influence of stability must depend on the dimensionless ratio

$$\frac{-\dfrac{g}{T}\left(\dfrac{\partial T}{\partial z} + \Gamma\right)}{\left(\dfrac{\partial u}{\partial z}\right)^2}.$$

Stable stratification tends to damp out the turbulent vertical motion. At a height z above the ground the mixing length will now have a value l_s, the rate of shear being C_s, or $\dfrac{\partial u_s}{\partial z}$. The turbulent kinetic energy $\frac{1}{2}\left(\overline{u'^2} + \overline{v'^2} + \overline{w'^2}\right)$ must be proportional to $l_s^2 C_s^2$. If the atmosphere were stable, but had the same rate of shear, the mixing length would be

$$l = k_0\left(z + z_0\right)$$

and the turbulent kinetic energy would be $l^2 C_s^2$. The difference between $l^2 C_s^2$ and $l_s^2 C_s^2$ must be represented by the changes in potential energy due to the turbulent elements having everywhere a different density from their immediate environment. It is easily seen (see § 145) that, on this basis

$$l^2 C_s^2 = l_s^2 C_s^2 + \frac{\beta g}{T}\left(\frac{\partial T}{\partial z} + \Gamma\right) l_s^2 \qquad \ldots\ldots(75)$$

[*] *Q.J. Roy. Met. Soc.* **62**, 1936, p. 124. [†] *Ibid.* p. 461.
[‡] *Ibid.* **63**, 1937, p. 105. [§] M.O., *Geophys. Mem.* No. 65, 1935, p. 37.

where β is a proportionality factor as yet undetermined, though a plausible value is unity.

Rossby and Montgomery proceed to compare a homogeneous (labile), and a stratified medium, moving under the same shearing stress $\widehat{xz}$, the rates of shearing being C and C_s respectively.

From (75), if
$$f(z) = \frac{\beta g}{T}\left(\frac{\partial T}{\partial z} + \Gamma\right) \qquad \ldots\ldots(76),$$

$$l_s = \frac{1}{\sqrt{1 + \dfrac{f(z)}{C_s^2}}} \qquad \ldots\ldots(77),$$

$$\frac{\widehat{xz}}{\rho} = l^2\left(\frac{\partial u}{\partial z}\right)^2 = l^2 C^2.$$

Rossby and Montgomery assume that $\widehat{xz}/\rho = l_s C_s$, though this appears to be a doubtful assumption. They take $f(z)$ to be equal to κ^2, a constant. It then follows that
$$C_s = C\sqrt{1 + \frac{\kappa^2}{C_s^2}},$$

or
$$\frac{C_s}{C} = \sqrt{\tfrac{1}{2} + \tfrac{1}{2}\sqrt{1 + \frac{4\kappa^2}{C^2}}}.$$

But
$$C = \frac{1}{k_0(z + z_0)}\sqrt{\frac{\widehat{xz}}{\rho}} \qquad \ldots\ldots(78)$$

and so
$$\frac{\partial u_s}{\partial z} = \frac{1}{k_0(z + z_0)}\sqrt{\frac{\widehat{xz}}{\rho}}\sqrt{\tfrac{1}{2} + \tfrac{1}{2}\sqrt{1 + \frac{4\kappa^2}{C^2}}} \qquad \ldots\ldots(79).$$

The assumption of a constant κ^2 involves the assumption that the lapse-rate is constant, and this is not usually so near the ground.

Sverdrup[*] assumed instead of this that near the ground the variations of velocity and temperature follow the same law, so that
$$\frac{\partial u}{\partial z} = mf(z).$$

From this and from equation (79) above it follows that
$$\left(\frac{\partial u_s}{\partial z}\right)^3 - \left(\frac{1}{k_0(z + z_0)}\sqrt{\frac{\widehat{xz}}{\rho}}\right)^2\left(\frac{\partial u_s}{\partial z} + \frac{1}{m}\right) = 0 \qquad \ldots\ldots(80).$$

This equation is not capable of general solution. It should be noted that in (79) above, if $\frac{\kappa^2}{C^2}$ is very great, that is, if the stability is great, the equation reduces to
$$\frac{\partial u_s}{\partial z} \propto (z + z_0)^{-\frac{2}{3}} \qquad \ldots\ldots(81)$$

* *Geofysiske Publikationer*, **11**, No. 7, 1936; *Met. Zeit.* **53**, 1936, p. 10.

while for indifferent equilibrium

$$\frac{\partial u_s}{\partial z} \propto (z+z_0)^{-1} \qquad \ldots\ldots(82).$$

Sverdrup next introduces the assumption

$$\frac{\partial u_s}{\partial z} \propto (z+z_0)^{\frac{1-n}{n}} \text{ where } 3 < n < \infty \qquad \ldots\ldots(83)$$

and derives the series of relationships which he compares with actual observations over a snow surface, obtaining surprisingly good agreement. For details of this work reference should be made to the original paper by Sverdrup.

The physical bases of the equations derived above are by no means clear, and some of the assumptions made appear to be questionable, but many of the relationships derived are in striking agreement with the results of observation. The great merit of this line of approach is that it makes it possible to separate the effects of stability and surface roughness.

§ 150. *The variation of wind with height*

(*a*) THE LAYER NEAR THE GROUND (0–10 METRES)

Observations of the flow of liquids in pipes have shown that near the boundary the mean velocity in turbulent flow can be represented by a fractional power law of the distance from the boundary, i.e.

$$u \propto z^{\frac{1}{q}} \qquad \ldots\ldots(84).$$

The constant q is equal to 7 for a wide range of conditions (say up to values of 50,000 for Reynolds number), but increases as the Reynolds number increases, being approximately 10 when Reynolds number is 10^6. A full account of the researches into these questions will be found in Wien-Harms, *Handbuch der Experimental Physik*, **4**, pt 4, Chapter IV. It may be also noted in passing that Stanton's[*] observations of velocity distribution in air have been shown by von Kármán[†] to fit a 1/7th power law with great accuracy.

In the layer of the atmosphere near the ground the conditions approach those near the boundary of the tube containing the fluid, the deviating force being negligible by comparison with the turbulent stresses and the pressure gradient. The distribution of velocity with height should therefore approximate to a fractional power law.

Sutton[‡] has computed the value of $1/q$ for observations at Leafield, and has found that it varies in summer from about 1/6 at midnight to about 1/14 in the afternoon, and in winter from 1/8 at night to about 1/12 in the afternoon. The lapse-rate appears to have a very marked effect on the value of $1/q$ the largest values appearing in inversions and the smallest values in large lapse-rates. A similar investigation by Barkat Ali[§] of the winds at Agra for the range 6 feet to 72 feet shows still larger diurnal and seasonal variations of $1/q$, values as high as 0·9 occurring at Agra in the small hours of the morning.

* *Proc. Roy. Soc.* A, **85**, 1911, p. 355.   † *Zeit. f. angew. Math. u. Mech.* **1**, 1921, p. 239.
‡ *Q.J. Roy. Met. Soc.* **58**, 1932, p. 74.   § *Ibid.* p. 285.

The table reproduced below (from *Geoph. Mem.* No. 54, Table XXI, p. 65) gives the ratio of the wind speed at 150 feet to that at 50 feet, for different values of the wind speeds at 150 feet and of the lapse-rate. It is seen that the ratio is nearly unity in large lapses, but increases to nearly 2 in very large inversions. For high wind velocities the ratio never differs very much from 1·2. The values of $1/q$ corresponding to the ratios 2 and 1·2 are respectively 0·63 and 0·166. The results shown in this table are of practical importance in observational meteorology, and show that it is not possible to draw up a table of wind velocities for different heights for the Beaufort scale numbers. A different table would be required for each value of the lapse-rate, in order to yield comparable values of the Beaufort numbers.

Ratio of wind speed at 150 feet to that at 50 feet in relation to wind speed at 150 feet and vertical temperature gradient

Wind speed at 150 feet m.p.h.	Vertical temperature difference 143 feet–4 feet (°F)												
	−5·0	−4·0	−3·0	−2·0	−1·0	0·0	1·0	2·0	3·0	4·0	5·0	6·0	7·0
10–14	1·01	1·02	1·03	1·04	1·07	1·15	1·23	1·35	1·50	1·60	1·73	1·85	1·98
15–19	—	1·05	1·07	1·09	1·12	1·17	1·23	1·32	1·40	1·49	1·58	—	—
20–24	—	—	1·16	1·16	1·18	1·20	1·20	1·21	—	—	—	—	—
25–29	—	—	1·17	1·15	1·14	1·16	1·18	1·20	—	—	—	—	—

For a discussion of the representation of the variation of wind with height by a logarithmic formula, see § 148.

(b) The layers from 10 to 1000 metres, K being assumed constant and the motion steady

Except in the narrow region in the immediate neighbourhood of the ground the deviating force due to the earth's rotation has to be taken into consideration. The variation of wind with height for these conditions has been investigated by Taylor, whose analysis is reproduced below.

It has been shown that if K is constant the rate of eddy transfer of x-momentum to unit volume is $K\rho \dfrac{\partial^2 u}{\partial z^2}$. Equations (1) and (2) of § 112 then become

$$\left. \begin{aligned} \frac{\partial u}{\partial t} - 2\omega \sin\phi \cdot v &= -\frac{1}{\rho}\frac{\partial p}{\partial x} + K\frac{\partial^2 u}{\partial z^2} \\ \frac{\partial v}{\partial t} + 2\omega \sin\phi \cdot u &= -\frac{1}{\rho}\frac{\partial p}{\partial y} + K\frac{\partial^2 v}{\partial z^2} \end{aligned} \right\} \qquad \ldots\ldots(85).$$

The motion being steady $\partial u/\partial t$ and $\partial v/\partial t$ are zero. Multiplying the second of these equations by $i\,(\sqrt{-1})$, and adding to the first, we find

$$K\frac{\partial^2}{\partial z^2}(u+iv) = \frac{1}{\rho}\left(\frac{\partial p}{\partial x} + i\frac{\partial p}{\partial y}\right) + 2i\omega \sin\phi\,(u+iv) \qquad \ldots\ldots(86).$$

Now let $2\omega \sin\phi \cdot Gi$ represent the pressure gradient, G being the geostrophic wind. For convenience we take the axis of x to be tangential to the isobar, so that G is along the axis of x. Further let $V = u + iv$. Then equation (86) may be written

$$\frac{d^2V}{dz^2} - (1+i)^2 B^2 (V - G) = 0 \qquad \ldots\ldots(87),$$

where
$$B^2 = \omega \sin \phi / K \qquad \ldots\ldots(88).$$

Within the limits of height with which we are concerned G may be treated as a constant. The solution of equation (87) may be written

$$V - G = C_1 e^{(1+i)\,Bz} + C_2 e^{-(1+i)\,Bz} \qquad \ldots\ldots(89).$$

Since the velocity must not become infinite at great heights, $C_1 = 0$. Equation (89) then reduces to the second term, which we now write

$$V - G = C e^{-(1+i)\,Bz + i\gamma} \qquad \ldots\ldots(90),$$

where C and γ are both real constants, whose values are to be determined from the boundary conditions.

In fig. 65 let O be the origin, OL the axis of x, and let the length Og represent the geostrophic wind G. OP is drawn to represent the wind at height z. Then $gP = Ce^{-Bz}$, and $\angle LgP = \gamma - Bz$. If OS is the surface wind (at $z = 0$), $gS = C$, and $\angle SgL = \gamma$.

We assume with Taylor that the slip is in the direction of strain, i.e. that V and $\dfrac{\partial V}{\partial z}$ are parallel at $z = 0$. Differentiating (90) we find

$$\frac{\partial V}{\partial z} = -C(1+i)Be^{-(1+i)\,Bz + i\gamma} = -\sqrt{2}\,Be^{i\pi/4}(G - V) \quad \ldots\ldots(91)$$

Thus $\dfrac{\partial V}{\partial z}$ always makes an angle $\pi/4$ with Pg, in the counter-clockwise direction from Pg. Since at the surface $\dfrac{\partial V}{\partial z}$ is coincident with OS in direction, the exterior angle formed by OS continued with Sg is $\pi/4$, and $\angle OSg = 3\pi/4$. Hence

$$\frac{C}{\sin \alpha} = \frac{G}{\sin 3\pi/4}, \quad \text{or} \quad C = \sqrt{2}\,G \sin \alpha \qquad \ldots\ldots(92)$$

and
$$\gamma = 3\pi/4 + \alpha. \qquad \ldots\ldots(93).$$

Hence the complete solution of the equation (87) which is appropriate to the conditions stipulated is

$$V - G = \sqrt{2}\,G \sin \alpha \,.\, e^{-Bz + i\,(\alpha + 3\pi/4 - Bz)} \qquad \ldots\ldots(94)$$

or $u + iv - G = \sqrt{2}\,G \sin \alpha \,.\, e^{-Bz} \left\{ \cos\left(\alpha + \dfrac{3\pi}{4} - Bz\right) + i \sin\left(\alpha + \dfrac{3\pi}{4} - Bz\right) \right\}.$

Hence
$$u = G - \sqrt{2}\,G \sin \alpha \,.\, e^{-Bz} \cos\left(Bz + \frac{\pi}{4} - \alpha\right) \Bigg|$$
$$\qquad\qquad\qquad\qquad\qquad\qquad\qquad \ldots\ldots(95).$$
$$v = \sqrt{2}\,G \sin \alpha \,.\, e^{-Bz} \sin\left(Bz + \frac{\pi}{4} - \alpha\right) \Bigg|$$

If we plot the wind at all heights on a plane diagram, as in fig. 65, where O is the origin, OL is the axis of x, and $Og = G$, then

$$gP = \sqrt{2}\,G \sin \alpha \cdot e^{-Bz}, \quad \text{and} \quad \angle PgL = \alpha + \frac{3\pi}{4} - Bz.$$

If OS is the surface wind,

$$\angle SOg = \alpha, \quad \angle SgL = \alpha + \frac{3\pi}{4}, \quad \angle SgO = \frac{\pi}{4} - \alpha \quad \text{and} \quad \angle OSg = \frac{3\pi}{4}.$$

The wind at a height z is made up of the geostrophic wind G together with an added component whose magnitude is $\sqrt{2}\,G \sin \alpha \cdot e^{-Bz}$, acting in a direction making an angle $\alpha + 3\pi/4 - Bz$ with the geostrophic wind. The point P therefore sweeps out an equiangular spiral of angle $\pi/4$.

At the ground $(z = 0)$ the velocity is $G (\cos \alpha - \sin \alpha)$. If this is to be positive α must be less than $\pi/4$, and when the surface wind is nearly zero α must be very nearly equal to $\pi/4$. Normal daytime values of α are more nearly

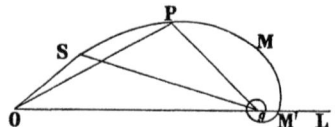

Fig. 65. The variation of wind with height; the equiangular spiral.

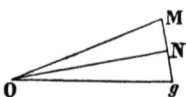

Fig. 66. The height at which the geostrophic wind is attained.

$\pi/8$, the wind blowing in across the isobars at about this angle in steady conditions. From fig. 65 we should expect that with increasing height the wind would veer, slowly at first, and then more rapidly, the velocity increasing and attaining the geostrophic value at a height represented by M, where $OM = Og$. In fig. 66 let N be the midpoint of Mg. Then since ONg is a right angle,

$$Mg = 2Ng = 2G \cos NgO = -2G \cos MgL = -2G \cos\left(\alpha - \frac{\pi}{4} - BH\right),$$

where H is the height corresponding to M. Then

$$\sqrt{2}\,G \sin \alpha \cdot e^{-BH} = -2G \cos\left(\alpha - \frac{\pi}{4} - BH\right)$$

$$\sin \alpha \cdot e^{-BH} = -\sqrt{2} \cos\left(\alpha - \frac{\pi}{4} - BH\right) \qquad \ldots\ldots(96).$$

Hence if α be known, and H can be obtained from observations, B can be evaluated by the use of equation (96). At a greater height the wind attains the gradient direction, corresponding to the point M' in fig. 66. The height H' corresponding to this point is given by

$$\alpha + \frac{3\pi}{4} - BH' = 0 \qquad \ldots\ldots(97).$$

Beyond this height the wind veers still further, and its velocity diminishes.

Observations made by Dobson[*] at Upavon enabled Taylor to compute values of H and H', the mean values being 300 metres and 800 metres respectively. From these the values of K were deduced, yielding 6×10^4, 5×10^4 and 3×10^4 for strong, moderate and light winds respectively, the values deduced from H and H' agreeing very closely. These values of K are greater than those obtained by Taylor from the observations in inversions over the sea, referred to on p. 228 above, but are in fair agreement with the values deduced by Taylor from the observations of temperature on the Eiffel Tower. Equation (86), or its equivalent (87), was derived without any assumption as to the constancy of K with height. It can be solved in finite terms in a few special cases of K varying with height. Brunt[†] gave a solution for the cases of K varying directly as z^2 or z^4, and Apte[‡] showed that a solution is possible when

$$K = K_0 (1 + cz)^{2 - \frac{2}{2n+1}}.$$ These solutions are of little more than academic interest. When K is assumed to be proportional to z, the solution is obtained in terms of the ber and bei and ker and kei functions, which are Bessel functions of a complex variable.

§ 151. *The internal friction due to turbulence*

The effect of turbulence on the mean motion is allowed for in equation (85) by the addition of terms $K\rho \partial^2 u/\partial z^2$, $K\rho \partial^2 v/\partial z^2$. These terms may be regarded as representing a virtual frictional force, and they may be combined into one term $K\rho \partial^2 V/\partial z^2$. Differentiating equation (94) twice, we find that this term amounts to

$$2\sqrt{2} KB^2 \rho G \sin \alpha \cdot e^{-Bz + i(\alpha + 5\pi/4 - Bz)} = 2\sqrt{2}\rho\omega \sin \phi \cdot G \sin \alpha \cdot e^{-Bz + i(\alpha + 5\pi/4 - Bz)}$$
$$\ldots\ldots(98).$$

A comparison of this equation with equation (85) shows that the virtual friction at the ground makes an angle of $\pi/4$ with the direction of the wind reversed. This should be contrasted with Guldberg and Mohn's assumption[§] that the frictional effect is opposite to the wind direction. The magnitude of the frictional effect decreases with height in proportion to e^{-Bz}. Since the term $K\rho \partial^2 V/\partial z^2$ balances the pressure gradient and the deviating force, it is possible to evaluate it from observations, if a synoptic chart is available. A series of values of the virtual frictional force at the top and bottom of the Eiffel Tower was evaluated by Akerblom. From these values, or rather from the ratios of the values at the top and bottom of the tower, it is possible to evaluate B directly, and from this to deduce the value of K. The mean values of K so derived were $6 \cdot 8 \times 10^4$ in winter, and $9 \cdot 3 \times 10^4$ in summer.[||]

[*] *Q.J. Roy. Met. Soc.* **40**, 1914, p. 123.   [†] *Ibid.* **46**, 1920, p. 175.
[‡] *Journ. Indian Math. Soc.* **15**, 1924, p. 183.
[§] See Abbe, *Mechanics of the Earth's Atmosphere*, Smithsonian Misc. Coll., 1910.
[||] For details of the computation see Brunt, *Q.J. Roy. Met. Soc.* **46**, 1920, p. 175.

§ 152. *The height to which the effect of surface turbulence extends*

We can obtain some idea of the height to which the effects of surface turbulence are appreciable by considering equations (95) and (20) above. The former shows that the vectorial effect of surface turbulence upon the wind is measured by

$$\sqrt{2}\, G \sin \alpha \,.\, e^{-Bz}.$$

We assume $K = 10^5$, so that $B = 2\cdot 4 \times 10^{-5}$, and $\alpha = 22\frac{1}{2}°$. The height at which e^{-Bz} is approximately $0\cdot 05$ is given by

$$z = 3/B \text{ cm} = 4 \times 10^4 \times 3 \text{ cm} = 1200 \text{ m}.$$

Thus the effect of turbulence on the wind is practically inappreciable beyond a height of a little more than 1 km. The height thus deduced is inversely proportional to B, and therefore directly proportional to $\sqrt{K}$.

Again from equation (20) the effect of surface turbulence on the temperature at different heights is given by a term e^{-bz}, and if we again assume the effect to become negligible when the exponential term reaches the same limit $0\cdot 05$, we find that the corresponding height bears to that deduced for the wind the ratio $\sqrt{2} \sin \phi : 1$. Thus below latitude $30°$ the effect upon the wind is appreciable to greater heights than the effect upon temperature, while above latitude $30°$ the effect upon temperature is appreciable to the greater height. For a given value of K the height to which the wind is affected by surface turbulence increases as $1/\sqrt{\sin \phi}$, and it appears from the equation that with approach to the equator the effects of surface turbulence should become appreciable at ever increasing heights.

§ 153. *Total flow along and across the isobars*

By the use of equations (94) or (95) it is readily possible to compute the flow of air along and across the isobars up to any given height, provided the geostrophic wind G does not change with height. The total flow across a vertical surface of height h and unit width, set up at right angles to the isobar, is

$$\int_0^h \rho u \, dz = D_x, \text{ say} \qquad \qquad \text{......(99)},$$

and the total flow across a vertical surface of height h and unit width, set up parallel to the isobar, is

$$\int_0^h \rho v \, dz = D_v, \text{ say} \qquad \qquad \text{......(100)}.$$

Neglecting variations of ρ,

$$D_x + i D_v = \rho \int_0^h (u + iv) \, dz$$

$$= \rho G h + \sqrt{2} \rho G \sin \alpha \int_0^h e^{-Bz + i(\alpha + \frac{3}{4}\pi - Bz)} \, dz$$

$$= \rho G h - \frac{\rho G \sin \alpha}{B} \left[e^{-Bz + i(\alpha + \pi/2 - Bz)} \right]_0^h$$

$$= \rho G h + \frac{\rho G \sin \alpha}{B} \{ -\sin \alpha + i \cos \alpha - e^{-Bh} [-\sin (\alpha - Bh) + i \cos (\alpha - Bh)] \},$$

and
$$D_x = \rho G h - \frac{\rho G \sin \alpha}{B} \{\sin \alpha - e^{-Bh} \sin (\alpha - Bh)\}$$
$$D_y = \frac{\rho G \sin \alpha}{B} \{\cos \alpha - e^{-Bh} \cos (\alpha - Bh)\}$$
......(101).

When h is great the factor e^{-Bh} becomes very small and may be neglected. The total flow across the isobar into low pressure is then $\rho G \sin \alpha \cos \alpha / B$ per unit length of the isobar, while the total flow across the surface of unit width at right angles to the isobars is $\rho G h - \rho G \sin^2 \alpha / B$. As shown in § 152, e^{-Bh} is in practice negligible when h is 1 km, and the flow across the isobars above this level is negligible.

When $K = 10^5$, $\alpha = 22\frac{1}{2}°$, $B = 2 \cdot 4 \times 10^{-5}$, $h = 1$ km $= 10^5$ cm,

$G = 10$ m/sec $= 1000$ cm/sec, $\rho = 1 \cdot 25 \times 10^{-3}$ gm/cm³,

the values of D_x and D_y are
$$D_x = 1 \cdot 25 \times 10^5 \text{ gm/sec} = 125 \text{ kg/sec}$$
$$= 100 \text{ (m)}^3 \text{ of air per sec,}$$
$$D_y = 0 \cdot 2 \times 10^5 \text{ gm/sec} = 20 \text{ kg/sec}$$
$$= 16 \text{ (m)}^3 \text{ of air per sec.}$$

Here D_x and D_y are the flows across surfaces whose width is 1 cm. In the lowest kilometre the flow across the isobars is comparable with the flow along the isobars. The drift of air across the isobars may be of importance in connection with the formation of rain in depressions, since the air carries with it a supply of moisture. Some further consideration is given to this point in a later section in connection with the discussion of rainfall in depressions (see § 198). It is probable that the compensating outflow across the isobars is at higher levels, where the moisture content is small, and the contribution of the inward drift of moisture in the lowest layers to the rainfall in the depression may be considerable, particularly at a front.

The frictional flow of air across a front inclined at an angle β to the isobars is deduced from the component normal to the front of the deviation from the geostrophic wind. The total flow up to great heights is readily deduced from equations (101) above, leaving out the geostrophic term which is independent of α, and neglecting e^{-Bh}. The total flow is

$$\frac{\rho G \sin \alpha}{B} \cos (\alpha - \beta) = D_y \sec \alpha \cos (\alpha - \beta) \qquad ......(102),$$

where D_y is as defined above, and expressed in equations (101). This is greatest when $\alpha = \beta$, and decreases as β increases beyond this value.

§ 154. *The boundary layer*

When a fluid flows over an aerodynamically "smooth" surface, the drag exerted by it on the surface is purely viscous. This drag is usually known as *skin friction*. In the immediate neighbourhood of the surface the flow is purely laminar, but usually in the greater portion of the fluid the flow is turbulent. The flow in the turbulent fluid above the laminar sub-layer, as discussed by Prandtl and von Kármán, on the assumption that the shearing stress does not vary with height, is given in § 148 above.

The earth's surface is not normally aerodynamically smooth, but with light winds the sea surface may be sufficiently smooth for a layer in laminar flow to form in immediate contact with the surface. Over most of the earth's surface, however, the surface is not sufficiently smooth to permit of the formation of a laminar layer, nor is it sufficiently smooth to permit of our ascribing to it any definite measure of its roughness. If the variation of wind with height be represented by a lower law $z^{1/q}$, it is found that $1/q$ is increased by an increase in either roughness* or stability. The fact that the value of $1/q$ depends on both these factors has the consequence that $1/q$ is not in itself a complete indicator of the turbulent state of the atmosphere.

The laminar sub-layer or "boundary layer" is a time-mean phenomenon, and it is not to be supposed that this layer is always constituted of the same fluid. The turbulence which may prevail at some distance from the boundary will from time to time break through the layer, carrying away portions of the fluid which instantaneously constitute it, but as soon as the individual eddy has removed a portion of the boundary layer normal processes will tend to build it up again.

Reference will be made in § 155 below to Stanton's measurements of flow in pipes, according to which the velocity at the outer limit of the boundary layer is 0·6 times the velocity in the middle. In equation (103) is represented Taylor's estimate of the skin friction at the ground as $0 \cdot 004 \rho V^2$ per unit area, where the coefficient 0·004 is a constant deduced from the flow in tubes, and V is the velocity "near" the surface. It is assumed that V may be taken at the outer limit of the boundary layer.

From the above considerations it is possible to obtain an estimate of the thickness of the boundary layer. Call this thickness δ. Then the shearing force per unit area of the surface is $\mu V/\delta$. This is equal to the skin friction, and hence

$$\mu V/\delta = 0 \cdot 004 \rho V^2 \qquad \ldots\ldots(103),$$

or
$$\delta = \frac{\mu}{0 \cdot 004 \rho V} = \frac{\nu}{0 \cdot 004 V}.$$

Putting $\nu = 0 \cdot 15$, this reduces to $\quad \delta = 37/V.$

If V_0 is the velocity at a great distance from the boundary, $V = 0 \cdot 6 V_0$,

$$\delta = \frac{37}{V} = \frac{60}{V_0} \qquad \ldots\ldots(104).$$

The thickness δ is given in cm and the velocity V in cm/sec. For such velocities as are important in the atmosphere, V is of the order of at least 400 cm/sec, and δ is then of the order of 1 mm. Equation (104) above is in reasonable agreement with such aerodynamical observations as are available. For a discussion of the boundary layer from the aerodynamical standpoint the

* See Paeschke, Beitr. Phys. fr. Atmos. **24**, h. 3, 1937, p. 163.

reader is referred to Prandtl-Tietjens, *Hydro- und Aero-Mechanik* (Berlin), Chapter v. Reference should also be made to the observations of Stanton in the paper cited in § 155 below.

The existence of the boundary layer is a fact of considerable importance in meteorology, particularly in connection with problems involving evaporation. Within this layer the transport of heat or of water-vapour is brought about entirely by molecular diffusion, while outside it, in the turbulent region, the transfer is brought about by eddies. It has already been emphasised that the boundary layer is a time-mean phenomenon, and that there is a slow exchange of fluid between the layer and the outer region of turbulent motion. Thus the water-vapour produced by evaporation at a liquid surface spreads outward through the laminar layer by molecular diffusion, while the air within that layer is from time to time exchanged by the action of eddies which penetrate it.

The concept of the boundary layer is useful in discussing the action of the wet-bulb thermometer, as well as the phenomena of evaporation from sheets of water. Some brief consideration of these two cases will be found in §§ 162, 163 below.

§ 155. *Skin friction at the ground*

For fluid flowing through a pipe Stanton[*] showed that with high values of lV/ν the velocity V of the fluid near the wall of the pipe is about 0·6 of the velocity in the middle, and that the mean velocity V_0 is about 0·85 of the velocity in the middle, so that

$$V = 0 \cdot 7 V_0.$$

The skin friction for the highest values of lV/ν which Stanton was able to obtain could be expressed by the formula

$$F = 0 \cdot 002 \rho V_0{}^2,$$

which is equivalent to

$$F = 0 \cdot 004 \rho V^2 \qquad \qquad \dots\dots(105).$$

(l represents the linear dimension of the system, and ν is the coefficient of kinematic viscosity.)

Taylor[†] has suggested that as the friction between the wind and the ground is due to projections such as trees and houses, the square law may also hold in such a case, the resistance being represented by

$$F = \kappa \rho V^2,$$

where V is the velocity at the ground, and κ is therefore comparable with 0·004.

If F_x, F_y be the components of the wind friction acting on unit area of the ground

$$F_x = \int_0^\infty K\rho \frac{d^2u}{dz^2}\, dz, \quad F_y = \int_0^\infty K\rho \frac{d^2v}{dz^2}\, dz,$$

---

* *Coll. Res. N.P.L.* **9**, 1913, plate I, p. 6.
† *Proc. Roy. Soc.* A, **92**, 1916, p. 196.

$$F_x + iF_y = \int_0^\infty K\rho \frac{d^2}{dz^2}(u+iv)\,dz$$

$$= K\rho \frac{d}{dz}(u+iv), \quad \text{at } z=0$$

$$= KB\rho\sqrt{2}G \sin\alpha\,(1+i)\,e^{-Bz+i\,(\alpha+3\pi/4-Bz)}, \quad \text{at } z=0$$

$$= 2KB\rho G \sin\alpha.e^{-Bz+i\,(\alpha+\pi-Bz)}, \quad \text{at } z=0,$$

$$F = \sqrt{F_x{}^2 + F_y{}^2} = 2KB\rho G \sin\alpha \qquad \ldots\ldots(106),$$

and the direction is exactly opposed to the surface wind as shown by the inclination to the isobars, $\alpha+\pi$.

From (105) and (106) and the relation

$$V_s = G\,(\cos\alpha - \sin\alpha)$$

it follows that
$$\kappa = \frac{2KBG \sin\alpha}{G^2\,(\cos\alpha - \sin\alpha)^2} = \frac{2KB \sin\alpha}{G\,(\cos\alpha - \sin\alpha)^2} \qquad \ldots\ldots(107),$$

or
$$\kappa = 2K \sin\alpha\,\frac{\left(\dfrac{3\pi}{4}+\alpha\right)G}{H'V_s{}^2} \qquad \ldots\ldots(108),$$

where H' is the height at which the wind attains the gradient direction, as defined by equation (97). From Dobson's observations over Salisbury Plain Taylor deduced the following results:

	K	α	H'	G	V	κ
Light winds	$2\cdot8 \times 10^4$	$13°$	600 m	460 cm/s	330 cm/s	0·0023
Moderate winds	$5\cdot0 \times 10^4$	$21\frac{1}{2}$	800	910	590	0·0032
Strong winds	$6\cdot2 \times 10^4$	20	900	1560	950	0·0022

The coefficient κ does not increase with wind velocity, and the use of equation (105) is thus justified. The mean value of κ may be taken as 0·0025, as compared with 0·004 in a pipe. The ratio of the values of lV/ν in the two cases is more than 10^5, and the same law thus appears to hold over this very wide range.

It would appear as a corollary that the surface wind should be about 0·7 times the geostrophic value for light winds, and 0·6 times the geostrophic value for strong winds.

The wind at the ground, V, is the wind derived from the anemometer, and so it is obvious that κ must depend on the height of the anemometer. Some light is thrown on the nature of this dependence by the theory of § 148. Equation (105) above becomes identical with the value derived from equation (70) when V is the wind at anemometer height z_a, i.e.

$$F = \widehat{zx} = \frac{\rho k_0{}^2 V^2}{\left(\log_e \dfrac{z_a + z_0}{z_0}\right)^2} \qquad \ldots\ldots(109)$$

if
$$\kappa = \frac{k_0{}^2}{\left(\log_e \dfrac{z_a + z_0}{z_0}\right)^2} \qquad \ldots\ldots(110).$$

In Taylor's experiments, $z_a = 30$ metres. Sutcliffe* derived rather higher values of κ, averaging about 0·006 over land, using winds at about 10 metres. It would be anticipated from equation (110) that the use of winds at a lower level would yield higher values of κ, but without some estimate of the appropriate values of z_0 it is not possible to make a direct comparison of Sutcliffe's results with those of Taylor.

If we accept Rossby and Montgomery's suggestion of $z_0 = 3\cdot2$ cm for open grass land as applying to Dobson's observations over Salisbury Plain, which formed the basis of Taylor's computations, we find that equation (110) above gives for κ the value 3.10^{-3}, in close agreement with the upper limit obtained by Taylor. Values of about 2.10^{-3} are obtained with $z_0 = 0\cdot5$ cm, the value appropriate to a smooth field or lawn. Again, if we assume $z_a = 10$ m, $z_0 = 3\cdot2$ cm, the value of κ is $4\cdot3 . 10^{-3}$, which is rather lower than the values derived by Sutcliffe from a number of observations over land.

§ 156. *The diurnal variation of wind at different heights*

Equation (107) may also be written in the form

$$\kappa = \frac{2\omega \sin \phi}{BG} \frac{\sin \alpha}{(\cos \alpha - \sin \alpha)^2},$$

or

$$\frac{1}{BG} = \frac{\kappa}{2\omega \sin \phi} \frac{(\cos \alpha - \sin \alpha)^2}{\sin \alpha} = \frac{20\cdot4}{\sin \alpha} (\cos \alpha - \sin \alpha)^2 \quad \ldots\ldots(111).$$

Equation (111) gives a relationship between B, G and α, and since

$$1/B^2 G^2 = K/\omega \sin \phi \ G^2,$$

it follows that K/G^2 is a function of α. Taylor has given a table of the corresponding values of K/G^2 and α, from which the following is an extract:

α in °	4	6	8	10	12	14	16	18	20	24	30	36
K/G^2	3·54	1·35	0·635	0·338	0·192	0·116	0·069	0·042	0·027	0·0094	0·0017	0·00016

Taylor also gave a series of curves showing the variation of wind with height for different values of α, which he then transformed into a series of curves showing the variation of V/G for given values of z/G, as α varies from 6° to 30°. From these he deduced curves for the diurnal variation of wind at different heights assuming α to vary from 10° at midday to 30° at midnight, and obtained the result that from the ground up to the height for which $z/G = 1$ the maximum wind velocity should occur at about midday, but that at heights above that at which $z/G = 15$ there should be a minimum at midday and a maximum at midnight, while at intermediate heights there should be two maxima.

These results are in good agreement with observation. At the top of the Eiffel Tower the diurnal variation of wind velocity is the reverse of the surface variation, the maximum occurring in the night and the minimum in the middle

* *Q.J. Roy. Met. Soc.* **62**, 1936, p. 63.

of the day. Hellmann* set up three anemometers at heights of 2, 16 and 32 metres above the ground in a flat meadow at Nauen, with a view to investigating the height at which the change in the nature of the diurnal variation took place. His results may be briefly summarised as follows: With light winds the anemometer at 2 metres showed a maximum in the early afternoon and a minimum at night, while those at 16 and 32 metres showed two maxima, one about midday and the other about midnight, with minima in the early morning and afternoon. At 16 metres, in winter the two maxima were about equal, while in summer the night maximum tended to exceed the day maximum; but at 32 metres the night maximum was greater both in winter and in summer. Thus with light winds the height at which the night maximum is equal to the day maximum is below 16 metres in winter, and between 16 and 32 metres in summer. With strong winds the day maximum was the stronger up to 32 metres. Hellmann confirmed these results by an examination of the records at Potsdam, where the anemometer is fixed at a height of 41 metres above the ground. With strong winds the maximum always occurs in the middle of the day, while with light winds there is in winter a minimum in the middle of the day and a maximum in the middle of the night; with light winds in summer there is a weak maximum in the middle of the day and a stronger maximum in the middle of the night. Thus the height at which the reversal in type of the diurnal variation takes place is greater than 41 metres for strong winds, but less than 41 metres for light winds.

The analysis of the wind velocities at 13 metres and at 95 metres at Leafield, given by Heywood,† confirms these results. Heywood classifies the winds into three groups, light winds (< 6 m/s), moderate winds (6–9 m/s), and strong winds ($\geqslant 9$ m/s), the classification being based on the mean wind for 24 hours at 95 metres above the ground. In summer with strong winds there is only one maximum, shortly after midday, even at 95 metres, while in winter the night maximum is appreciable at 13 metres, and is still a little lower than the day maximum at 95 metres. With moderate winds, in summer the night maximum is absent at 13 metres, and is slightly lower than the day maximum at 95 metres; while in winter the night maximum is just appreciable at 13 metres, and at 95 metres there is a maximum in the middle of the night and a minimum in the middle of the day. With light winds in summer there is a faint maximum in the middle of the night at 13 metres, and a night maximum and a day maximum at 95 metres, while in winter the day maximum is weaker than the night maximum at 13 metres, and is replaced by a day minimum at 95 metres. It should be noted that Heywood's strong winds are stronger than Hellmann's "strong winds", and so the height of transition from day maximum to night maximum which he deduces is higher than the corresponding estimate given by Hellmann and quoted by Taylor.

From the above brief description of the observations available it will be seen that the changes with height of the diurnal variations of wind are in reasonable agreement with the theory deduced by Taylor, based on the variations of α during the day.

---

* *Met. Zeit.* Jan. 1915.   † *Q.J. Roy. Met. Soc.* **57**, 1931, p. 433.

§ 157. *Diffusion by continuous movement*

The analysis of § 142 above leads to a conviction that any effort at discussion of atmospheric turbulence from a purely dynamical standpoint is likely to be fruitless, on account of the impossibility of visualising the individual eddy, and the dynamical processes involved in its movement from place to place. A totally different line of approach, also due to G. I. Taylor,* is available, and will be briefly described here.

Consider a condition in which the turbulence in a fluid is uniformly distributed, so that the average conditions are the same at every point. Let u be the velocity of any particle parallel to the axis of x. Let R_ξ be the coefficient of correlation between the velocities of the particles considered at time t, and at an interval of time ξ later. Then by definition

$$[u_t u_{t+\xi}] = [u_t^2] R_\xi,$$

where the square brackets denote mean values taken over a long period. Integrating with respect to ξ we find

$$[u^2] \int_0^t R_\xi d\xi = \int_0^t u_t u_{t+\xi} d\xi = u_t \int_0^t u_{t+\xi} d\xi = [u_t X] \quad \ldots\ldots(112),$$

where X is the distance travelled by the particle in time t. Thus

$$[u^2] \int_0^t R_\xi d\xi = [u_t X] = \frac{1}{2} \frac{d}{dt} [X^2], \quad \ldots\ldots(113)$$

and
$$[X^2] = 2 [u^2] \int_0^t \int_0^t R_\xi d\xi dt \quad \ldots\ldots(114).$$

This remarkable equation reduces the problem of diffusion to that of the determination of the mean square velocity, and of the correlation between the velocity of a particle at one instant and at a time ξ later.

When t is so small that there is not sufficient time for the coefficient of correlation to fall appreciably from unity the double integral becomes $\frac{1}{2}t^2$, so that the equation (114) reduces to

$$[X^2] = [u^2] t^2 \quad \text{or} \quad \sqrt{[X^2]} = \sqrt{[u^2]} t \quad \ldots\ldots(115),$$

or, in other words, the deviation of the particle from its initial position is proportional to the time.

We should anticipate that in a turbulent fluid R_ξ should fall off to zero for large values of ξ. It might remain positive, or might oscillate between positive and negative values, but in either case there should be a finite interval of time T_1 such that at the end of this interval there should be no correlation with the velocity at the beginning. In this case suppose that $\lim_{t\to\infty} \int_0^t R_\xi d\xi$ is finite and equal to I. Then at any time $T (> T_1)$ after the beginning of the motion equation (81) becomes

$$\frac{d}{dt} [X^2] = 2 [u^2] I, \quad \text{or} \quad [X^2] = 2 [u^2] IT \quad \ldots\ldots(116),$$

* *Proc. Lond. Math. Soc.* **20** 1922, p. 196.

so that $[X^2]$ increases at a uniform rate. Thus a continuous eddying motion may be expected to have the property that the standard deviation of X is proportional to the square root of the time.

It will also be noted from equation (112) that for $T > T_1$

$$[Xu] = [u^2] I = \text{const.} \qquad \qquad \text{......(117)}.$$

Thus though X increases with the time, the product $[Xu]$ is constant. It follows that X must be positively correlated with u, but the correlation coefficient decreases with increasing X. Let this correlation coefficient be r_{Xu}. Then

$$r_{Xu} = \frac{[Xu]}{\sqrt{[X^2][u^2]}} = \frac{I\sqrt{[u^2]}}{\sqrt{[X^2]}} = \frac{I}{\sqrt{2IT}} = \sqrt{\frac{I}{2T}} \qquad \text{......(118)}$$

for large values of T.

Consider the application of the above results to the diffusion of smoke emitted at a point, and carried away downwind. At short distances from the point of emission, at which the correlation coefficient R_ξ is still nearly unity, X is proportional to the time, and therefore the outline of the smoke should be a cone. At greater distances we should use the result shown in equation (116) according to which the deviation is proportional to the square root of the time, and the outline is therefore a paraboloid. These anticipations are in accordance with observation, and appear to bear out Taylor's deduction that $[Xu]$ becomes constant after a certain interval of time, this interval being long enough to allow $\int_0^T R_\xi d\xi$ to become practically constant.

Richardson* has also derived an equation similar to equation (116) above. He showed that the increase in the standard deviation of a set of particles in time T was equal to $2KT$, where K is the coefficient of eddy diffusivity as defined by Taylor. Comparing this with equation (116) above, we find

$$K = [u^2] I = [u^2] \int_0^\infty R_\xi d\xi \qquad \qquad \text{......(119)}.$$

Equation (119) was also derived independently by Taylor. In terms of the vertical component of the eddy velocity, we have

$$R_\xi = \frac{\overline{w'(t)\, w'(t+\xi)}}{\overline{w'^2}} \qquad \qquad \text{......(120)}$$

or

$$w'(t)\, w'(t+\xi) = \overline{w'^2} R_\xi \qquad \qquad \text{......(121)}.$$

This equation may be integrated with respect to ξ from o to t_0,

$$\overline{w'^2} \int_0^{t_0} R_\xi d\xi = \overline{w'(t)} \int_0^{t_0} w'(t+\xi)\, d\xi = \overline{w'l} = K \qquad \text{......(122)}.$$

The conditions to be observed are clearly that

   (a) $R_\xi = 1$ for $\xi = 0$,

   (b) R_ξ decreases as ξ increases,

   (c) R_ξ is negligibly small for $\xi = t_0$.

* *Phil. Trans. Roy. Soc.* A, **221**, 1920, p. 1.

§ 158. *The form of R_ξ*

The direct detailed investigation of R_ξ is not in general a feasible proposition, since it would require a series of anemometers strung in a line downwind. If such a line of anemometers could be set up it would only be utilisable when the wind was directed along that line. Some general results can, however, be derived from simple considerations. If the turbulence were of the nature of indestructible whirls moving downwind, the correlation between the velocity of a selected element at a time t and at an interval ξ later would be perfect for all values of ξ, though the correlation would not of necessity be linear. This supposition is far from the truth, which must be more closely represented by the statement that eddies are continually being destroyed by mixing, others being created to take their place. Let us assume for the moment that all the eddies are of the same size, and let the correlation between the velocity of a selected element at the beginning and at the end of an interval ξ be R_ξ. The part of $u^2_{t+\xi}$ which is due to correlation with u_t is $R_\xi^2 u_t^2$, and the part of $u^2_{t+\xi}$ which is independent of u_t is $(1 - R_\xi^2) u_t^2$. In other words the part of the total eddying energy, at the end of the interval, which is due to the eddies which have persisted throughout the interval is $\frac{1}{2}R_\xi^2 u_t^2$, and the remaining portion $\frac{1}{2}(1 - R_\xi^2) u_t^2$ is due to eddies which have come into existence during the interval. Considering the same process as taking place during a further interval ξ subsequent to the first, we arrive at the result that of the eddying energy at the end of the double interval 2ξ a portion $\frac{1}{2}R_\xi^4 u_t^2$ is due to eddies originally in existence. Thus

$$R_{2\xi} = R_\xi^2 \quad \text{or} \quad R_\xi = e^{-a\xi}.$$

This result depends on the assumption that all the eddies are of the same size, with a uniform rate of creation and destruction, and is only true with that assumption. The constant a in the last equation must obviously depend on the size of the eddy, and is no longer true when there are present eddies of varying sizes.

Returning to equation (114) and substituting $R_\xi = e^{-a\xi}$, we find

$$[X^2] = 2\,[u^2] \int_0^T \int_0^t e^{-a\xi}\,d\xi\,dt$$

$$= 2\,[u^2] \int_0^T \frac{1}{a}(1 - e^{-at})\,dt$$

$$= 2\,[u^2] \left\{ \frac{T}{a} - \frac{1}{a^2}(1 - e^{-aT}) \right\} \qquad \ldots\ldots(123).$$

Since T is sufficiently great to permit of our neglecting $\int_T^\infty R_\xi\,d\xi$, aT must be large by comparison with unity, and e^{-aT} is negligible by comparison with unity, and $1/a^2$ by comparison with T/a, so that the equation reduces to

$$[X^2] = 2\,[u^2]\,T/a \qquad \ldots\ldots(124).$$

It will be noted that Taylor's form of the equation requires that $\int_0^\infty R_\xi d\xi$ shall be constant. This is clearly satisfied by that treatment of turbulence which regards the eddies as analogous to molecules, and treats the dynamics of turbulence essentially as "collision dynamics". These theories regard the eddy as preserving its identity and its motion until it reaches a certain point at time t_0, when it suddenly mixes with its new environment. Implicitly it is assumed that

$$R_\xi = 1 \quad \text{for} \quad \xi \leqslant t_0,$$
$$R_\xi = 0 \quad \text{for} \quad \xi > t_0.$$

Such assumptions make $\int_0^t R_\xi d\xi = t_0$ for all values of $t \leqslant t_0$, and $= t$ for all values of $t < t_0$.

The outstanding feature of Taylor's treatment by correlation methods is that it permits of our regarding mixing as a process which is continuous, and not intermittently explosive.

§ 159.　Sutton's extension of Taylor's theory

It was suggested by O. G. Sutton[*] that R_ξ might be of the form

$$R_\xi = \left(\frac{a}{\bar{u}\xi} \right)^n \qquad \ldots\ldots(125),$$

where a is a constant length, and $\bar{u}$ the mean wind speed, n being a real quantity whose exact value is defined below. In a second paper[†] Sutton defined R_ξ as

$$R_\xi = \left(\frac{\nu}{\nu + \overline{w'^2}\xi} \right)^n \qquad \ldots\ldots(126),$$

where ν, the kinematic coefficient of viscosity, is small by comparison with $\overline{w'^2}t_0$; here t_0 is the time taken to move through the distance l. Both Taylor[‡] and Scrase[§] have shown that in the lower atmosphere $\overline{w'^2}$ is proportional to $(\bar{u})^2$, so that R_ξ behaves like $((u)^2\xi)^{-n}$. The experiments of Relf and Lavender[||] in a wind tunnel, and Scrase's[¶] observations in the lower atmosphere, show that t_0 is of the order of 10^{-4} sec, $\overline{w'^2}$ is of the order of 10^4 cm^2 sec^{-2}, and hence $\overline{w'^2}t_0$ is of the order of at least 10^3 cm^2 sec^{-4}. Since ν is of the order of 10^{-1} cm^2 sec^{-1}, it follows that for such values of n as hold in the atmosphere ($n = 1/4$ will be found to fit a wide variety of conditions), the correlation at time $\xi = t_0$ is about 0·1, and is therefore negligible.

Substituting for R_ξ in equation (112) and integrating, we find for σ, the standard deviation of the particle from its mean position, after time T, the following relation,

$$\sigma^2 = \frac{2\nu^n}{(1-n)(2-n)\overline{w'^2}} (\overline{w'^2}T)^{2-n} \qquad \ldots\ldots(127).$$

* Proc. Roy. Soc. A, **135**, 1932, p. 143.　　† Ibid. **146**, 1934, p. 701.
‡ Q.J. Roy. Met. Soc. **53**, 1927, p. 201.　　§ M.O., Geophys. Mem. No. 52, 1930.
|| Advisory Committee Aero., R. and M., No. 597, 1918.　　¶ Loc. cit. p. 15.

If the velocities are now assumed to follow the Maxwellian law of distribution[*] there is a simple relationship between $\overline{w'^2}$ and $(\overline{w'})^2$.

$$\overline{w'^2} = \tfrac{1}{2}\pi\,(\overline{w'})^2 \qquad \qquad \ldots\ldots(128).$$

Following Prandtl, we now assume

$$\overline{w'} = l\frac{\partial u}{\partial z},$$

where the quantity u now denotes the mean velocity at level z. Then

$$\overline{w'^2} = \tfrac{1}{2}\pi\,(\overline{w'})^2 = \tfrac{1}{2}\pi l^2 \left(\frac{\partial u}{\partial z}\right)^2 \qquad \qquad \ldots\ldots(129).$$

We can now estimate the time t_0,

$$t_0 = \int_z^{z+l} \frac{dz}{\overline{w'}} = \frac{l}{\overline{w'}} = \left(\frac{\partial u}{\partial z}\right)^{-1} \qquad \qquad \ldots\ldots(130)$$

approximately. Equation (127) may be re-written

$$\sigma^2 = \frac{2\nu^n\,(\tfrac{1}{2}\pi)^{1-n}}{(1-n)\,(2-n)} \frac{\left\{l^2\left(\dfrac{\partial u}{\partial z}\right)^2 T\right\}^{2-n}}{l^2\left(\dfrac{\partial u}{\partial z}\right)^2} \qquad \qquad \ldots\ldots(131).$$

From equation (122) above

$$\overline{w'l} = \overline{w'^2} \int_0^{t_0} \left(\frac{\nu}{\nu + \overline{w'^2}\xi}\right)^n d\xi$$

$$= \frac{\nu^n}{1-n}\left\{(\nu + \overline{w'^2}t_0)^{1-n} - \nu^{1-n}\right\}.$$

Since ν is assumed small by comparison with $\overline{w'^2}t_0$, and since it is found that n is less than unity, it follows that this may be written

$$K = \overline{w'l} = \frac{\nu^n}{1-n}\,(\overline{w'^2}t_0)^{1-n} \qquad \qquad \ldots\ldots(132).$$

The approximation amounts to the neglect of the molecular forces by comparison with the eddy forces. Also

$$\overline{w'^2}t_0 = \tfrac{1}{2}\pi l^2 \left(\frac{\partial u}{\partial z}\right) \qquad \qquad \ldots\ldots(133).$$

If we assume with von Kármán that

$$l = k_0 \frac{\partial u}{\partial z} \Big/ \frac{\partial^2 u}{\partial z^2},$$

where $k_0 = 0.38$,

$$\overline{w'^2}t_0 = 0.08\pi \left(\frac{\partial u}{\partial z}\right)^3 \left(\frac{\partial^2 u}{\partial z^2}\right)^{-2} \qquad \qquad \ldots\ldots(134)$$

and finally, from (132)

$$K = f(n)\,\nu^n \left\{\left(\frac{\partial u}{\partial z}\right)^3 \left(\frac{\partial^2 u}{\partial z^2}\right)^{-2}\right\}^{1-n} \qquad \qquad \ldots\ldots(135),$$

[*] *Vide* Hesselberg and Bjorkdal, *Beitr. Phys. fr. Atmos.* **15**, 1929, p. 121.

where $f(n)$ depends on n and k_0 only. For pipe flow, Sutton employed von Kármán's value $k_0 = 0\cdot38$, and calculated the surface traction, finding a result which is numerically in close agreement with observation. It does not follow, however, that this value of $f(n)$ is appropriate to flow over the surface of the earth.

§ 160. *The variation of wind with height on Sutton's theory*

The assumption of a power law for the correlation coefficient R_ξ leads, as might have been anticipated, to a power law for the variation of wind with height. If we assume the eddy stress does not vary with height,

$$K \frac{\partial u}{\partial z} = \text{const.} \qquad \qquad \dots\dots(136).$$

Substituting for K from (135), we find

$$\left(\frac{\partial^2 u}{\partial z^2}\right)^{2-2n} \bigg/ \left(\frac{\partial u}{\partial z}\right)^{4-3n} = \text{const.}$$

or

$$\frac{\partial^2 u}{\partial z^2} \bigg/ \left(\frac{\partial u}{\partial z}\right)^{\frac{4-3n}{2-2n}} = \text{const.}$$

By integration we find

$$u \big/ z^{\frac{n}{2-n}} = \text{const.}$$

or

$$\frac{u}{u_1} = \left(\frac{z}{z_1}\right)^{\frac{n}{2-n}} \qquad \qquad \dots\dots(137),$$

where u_1 is the velocity at a standard height z_1. To ensure physical reality, it may be shown that $0 < n < 1$, and hence it follows that the velocity in the surface layers cannot increase more rapidly than the height. This deduction from the theory has been verified by observations at Cardington,[*] and by Best's measurements.[†]

The parameter n is readily determined if the form of the curve of variation of wind with height is known. If the index of z in (137) is written as $1/q$, it is readily seen from (136) that K is proportional to $z^{1-1/q}$, or to $uz\left(\dfrac{\nu}{uz}\right)^n$. Equation (137) above becomes

$$K = f(n)\, uz \left(\frac{\nu}{uz}\right)^n \qquad \qquad \dots\dots(138).$$

The complete representation of K is obtained when we substitute for u and its derivatives from (137) in (135) above. We then find

$$K(z) = \frac{(0\cdot251)^{1-n}\,(2-n)^{1-n}\,n^{1-n}\,\nu^n}{(1-n)\,(2n-2)^{2(1-n)}}\,(u_1 z_1)^{1-n}\left(\frac{z}{z_1}\right)^{\frac{2(1-n)}{2-n}} \qquad \dots\dots(139)$$

$$= a'\,(u_1)^{1-n}\, z^{\frac{2(1-n)}{2-n}} \text{ say,} \qquad \qquad \dots\dots(140),$$

where a' is a quantity which depends only on n and the physical constants of

---

* M.O., *Geophys. Mem.* No. 54.   † M.O., *Geophys. Mem.* No. 65.

the atmosphere, and is independent of the mean velocity u_1 measured at a fixed height. The result is only valid within the layer in which equation (136) is applicable, in which the shearing stress may be treated as constant.

§ 161. *Recent developments in the study of turbulence*

In recent years, the statistical theory of turbulence has been greatly developed by Taylor and von Kármán. It is impossible to give an adequate summary of this large body of work here, but in general terms it may be described as an attempt to develop a theory of turbulence without employing the idea of a mixing length. So far, attention has been concentrated on the phenomenon of the downstream decay of isotropic turbulence set up by the introduction of a rectangular mesh in a wind tunnel, and very striking agreement has been obtained between theory and observation. While there is, at the moment, no direct application of this work to meteorological problems, there is no doubt that these developments constitute a very valuable contribution to the general problem of turbulent motion.

§ 162. *The derivation of the hygrometer equation*

In deriving the August-Apjohn equation in § 51 above it was assumed that a portion of the normal air of the environment of the wet bulb takes up the required amount of water from the wet bulb to become saturated at the temperature of the wet bulb. The weakness of this theory is that it is not possible in practice to separate out the air into two streams, one of which becomes saturated while the other is unaffected by the presence of the wet bulb.

An alternative theory due to G. I. Taylor was first published by Skinner, in an article in the *Dictionary of Applied Physics*, **3**. It was reproduced by Whipple in a recent paper,[*] and is summarised below.

Let T and e be the absolute temperature and vapour-pressure in the free air at some distance from the wet bulb, T' and e' the values at the surface of the wet bulb, and T'' and e'' the values at the boundary between the laminar layer and the turbulent layer. Let κ be the molecular diffusivity (for heat) of air, and let D be the coefficient of diffusion of water-vapour into air. Then if δ be the thickness of the boundary layer, the rate at which heat is conducted to the wet bulb is, per unit area,

$$\kappa \rho c_p \, (T'' - T')/\delta.$$

The rate at which water-vapour diffuses outward from density $\epsilon \rho e'/p$ to the density $\epsilon \rho e''/p$ is

$$D \epsilon \rho \, (e' - e'')/p\delta.$$

The incoming heat provides the latent heat of evaporation of the outward diffusing water-vapour. Hence

$$\kappa \rho c_p \, (T'' - T') = L D \epsilon \rho \, (e' - e'')/p \qquad \ldots\ldots(141).$$

Let E be the rate at which the air within the boundary layer is being replaced by normal air from the turbulent region. E will be reckoned as volume per

[*] *Proc. Phys. Soc. London*, **45**, pt 2, 1933, p. 307.

unit area of the surface of the outer boundary of the laminar layer, per unit time. The physical significance of E is not very clear, but this is of no great importance if we may assume that E has the same significance in the transfer of heat and of water-vapour.

The rate at which heat is carried to the boundary layer is

$$E\left(T-T''\right)\rho c_p.$$

The rate at which water-vapour is carried outward is

$$E\epsilon\rho\left(e''-e\right)/p.$$

Since these two must balance

$$\left(T-T''\right)c_p = L\epsilon\left(e''-e\right)/p \qquad \ldots\ldots(142).$$

The two equations (141) and (142) involve T'' and e'', which are not measured quantities. These two variables can be eliminated together only if $\kappa = D$. For then by addition of the two equations we find

$$
\left.
\begin{aligned}
T-T' &= \frac{L\epsilon}{c_p}\left(e'-e\right)/p \\[2mm]
e'-e &= \frac{pc_p}{L\epsilon}\left(T-T'\right)
\end{aligned}
\right\} \qquad \ldots\ldots(143).
$$

or

This is the hygrometer equation already derived in § 51. Taylor justified the above derivation of equation (143) by the use of a value of D of 0·198, a value appropriate to a temperature of 0° C, which should be compared with the corresponding value of κ, 0·17. There is some uncertainty as to the most reliable value of D. The estimate of 0·198, due to Winkelmann, was published in 1884. More recent estimates of D are nearer to 0·24. If this value is accepted, the validity of the derivation of the hygrometer equation given above is considerably impaired. (Some measured values of D are given in Table IX, p. 421.)

§ 163. *Evaporation from the surface of water*

(a) EVAPORATION INTO STILL AIR[*]

When the evaporation takes place into still air, the water-vapour is diffused by molecular motion. Let D be the coefficient of diffusion of water-vapour into air. Then if V is the mass of water-vapour per unit mass of air, expressed as a function of three co-ordinates x, y, z, of which z is vertical, the equation of diffusion is

$$\frac{dV}{dt} = \frac{\partial V}{\partial t} = D\left(\frac{\partial^2 V}{\partial x^2} + \frac{\partial^2 V}{\partial y^2} + \frac{\partial^2 V}{\partial z^2}\right) \qquad \ldots\ldots(144).$$

Let V_0 be the difference between V at $z=0$ and at a great distance. In the steady state $\partial V/\partial t = 0$, and the equation reduces to

$$\nabla^2 V = 0 \qquad \ldots\ldots(145).$$

---

[*] This treatment follows a paper by Jeffreys, *Phil. Mag.* **35**, 1918, p. 270.

The rate of transfer of water-vapour upward from the boundary $(z=0)$

$$= -\iint D\rho \frac{\partial V}{\partial z} \, dS \qquad \ldots\ldots(146),$$

the integral being taken over the whole surface of the water. But (145) is the equation which must be satisfied by the potential in electrostatic problems, and in our case the analogous electrostatic problem is that of a charged plate coinciding with the water surface. Let σ be the density of electrostatic charge on the surface. Then

$$\partial V/\partial z = -4\pi\sigma$$

and hence

$$-\iint \frac{\partial V}{\partial z} \, dS = 4\pi \iint \sigma \, dS = 4\pi C V_0 \qquad \ldots\ldots(147),$$

where C is the electrostatic capacity of the plate. Hence the rate of transference outwards is

$$4\pi D\rho C V_0 \qquad \ldots\ldots(148).$$

This determines the total amount of evaporation from the water surface. Since C is proportional to the linear dimension of the conductor for similar plates, it follows that the evaporation is proportional to the linear dimension of the surface. It is also proportional to the difference in V at the water surface, where the air is saturated, and at a great distance, where the value of V corresponds to the state of the air before it has drifted over the water surface.

(b) Evaporation into a steady wind

Jeffreys (loc. cit.) discussed the evaporation into a steady wind, taking account of the effects of turbulence. The following is a brief abstract of his treatment. Let u be the velocity of the wind, which will be assumed to blow along the x-axis. Then, since there is no wind along the y- or z-axis, equation (144) may be written

$$\frac{dV}{dt} = \frac{\partial V}{\partial t} + u \frac{\partial V}{\partial x} = K\nabla^2 V \qquad \ldots\ldots(149),$$

where K is the eddy diffusivity, assumed not to vary with height. Outside the boundary layer of rapid shearing the equation may be written

$$u \frac{\partial V}{\partial x} = K \frac{\partial^2 V}{\partial z^2} \qquad \ldots\ldots(150).$$

Let $h^2 = K/u$. Then a solution of (150) is

$$\left.\begin{array}{l} V = V_0 \left(1 - \mathrm{Erf}\, \dfrac{zx^{-\frac{1}{2}}}{2h}\right) \text{ when } x \text{ is positive} \\[2mm] V = 0 \text{ when } x \text{ is negative} \end{array}\right\} \qquad \ldots\ldots(151).$$

This makes $\partial V/\partial z = -V_0/h\sqrt{\pi x}$ over the wet surface. The rate of evaporation is therefore

$$-K\rho \frac{\partial V}{\partial z} = V_0 \rho \sqrt{\frac{Ku}{\pi x}} \qquad \ldots\ldots(152)$$

per unit area. The amount evaporated between 0 and x over a strip dy in width is by integration $\qquad 2\rho V_0 \,(Kux/\pi)^{\frac{1}{2}} \, dy.$

If the length of the strip from one margin to the other is l, thus neglecting the diffusion sideways at the edges, the total evaporation is

$$2\rho V_0 \left(Ku/\pi\right)^{\frac{1}{2}} \int l^{\frac{1}{2}} \, dy \qquad \qquad \ldots\ldots(153),$$

the integration being taken over the whole area. For areas of the same shape, in which a is a linear dimension, the evaporation is proportional to $a^{\frac{3}{2}}$, and in particular for a circle of radius a the total evaporation is

$$3\cdot95\rho V_0 \left(Kua^3\right)^{\frac{1}{2}} \qquad \qquad \ldots\ldots(154).$$

In equations (151) to (154) V_0 is to be taken as the difference of the humidity mixing ratio at the water surface, where the air is saturated, and at a great distance. Jeffreys also considered the limitations of his results, and showed that they apply in the open air to sheets of water for which a is between say 10 cm and 250 metres.

(c) GIBLETT'S INVESTIGATION

Giblett[*] solved the diffusion equation, using as his boundary condition an empirical formula for the total evaporation from unit surface.

$$\text{Evaporation} = A \left(e_s - e\right) \left(1 + cu\right) \qquad \qquad \ldots\ldots(155),$$

where A and c are constants, e_s is the saturation vapour-pressure at the surface temperature, and e is the vapour-pressure in the air "near" the surface (where "near" presumably means just outside the boundary layer), and u is the horizontal velocity of the wind "near" the surface. Giblett regarded the eddy diffusivity K as not varying with height. In view of the assumptions made the results are of limited application.

(d) SUTTON'S TREATMENT OF EVAPORATION

In his treatment of the problem of evaporation, Sutton makes the assumption that the turbulent transfer of matter is governed by the same laws as the transfer of momentum.

The evaporating surface is taken to be level with the earth, its dimensions being such that it does not interfere with the normal wind structure. It is also assumed that the increase in the vapour content of the air as it flows over the surface is small enough to allow variations in ρ, the air density, to be neglected.

Let $\chi\,(x, y, z)$ be the portion of the air mass per unit volume of the atmosphere due to evaporated water-vapour. χ is zero at all points to windward of the surface, and vanishes at great heights. The axes of reference are, x downwind from the windward edge of the evaporating surface, y across wind, and z vertical. The equation for the diffusion of water-vapour may then be written

$$\rho \frac{d\chi}{dt} = \frac{\partial}{\partial x} \left\{ K_x \rho \frac{\partial \chi}{\partial x} \right\} + \frac{\partial}{\partial y} \left\{ K_y \rho \frac{\partial \chi}{\partial y} \right\} + \frac{\partial}{\partial z} \left\{ K_z \rho \frac{\partial \chi}{\partial z} \right\} \qquad \ldots\ldots(156),$$

where K_x, K_y and K_z are eddy diffusivities in the directions x, y and z respectively. Sutton limits the problem to two dimensions, and retains only the last term on the right-hand side of the above equation. The axis of x being

Proc. Roy. Soc. A, **99**, 1921, p. 472.

in the direction of the mean wind, $\bar{v} = \bar{w} = 0$. Then in the steady state, with $\frac{\partial \chi}{\partial t} = 0$, the equation becomes

$$\bar{u} \frac{\partial \chi}{\partial x} = \frac{1}{\rho} \frac{\partial}{\partial z} \left\{ K_z \rho \frac{\partial \chi}{\partial z} \right\} \qquad \ldots\ldots(157).$$

Substituting for K_z its value for the transfer of momentum from (140) above, and giving $\bar{u}$ its appropriate form as a power of z, from (137) above, we have, for ρ independent of z,

$$\frac{\bar{u}_1^n}{a} z^{\frac{n}{2-n}} \frac{\partial \chi}{\partial x} = \frac{\partial}{\partial z} \left\{ z^{\frac{2(1-n)}{2-n}} \frac{\partial \chi}{\partial z} \right\} \qquad \ldots\ldots(158)$$

with the boundary conditions

    (a) $\lim\limits_{z \to 0} \chi(x, z) = \chi_0$ (a constant, which may be the saturation value),

    (b) $\lim\limits_{z \to \infty} \chi(x, z) = 0$ $(0 \leqslant x \leqslant x_0)$,

    (c) $\lim\limits_{x \to 0} \chi(x, z) = 0$ $(0 < z)$,

where x_0 is the length of the surface downwind. The vapour density, χ, is the solution of the equation, with its boundary conditions, valid for $0 \leqslant x \leqslant x_0$, $0 < z$.

The solution of the problem may be obtained, as Sutton shows, involving Bessel functions, but it is possible to find the variation of the rate of evaporation with the wind velocity, the length of the surface downwind, and the degree of turbulence present, by an artifice which avoids the mathematical difficulties inherent in solving the differential equation. The mean quantity of vapour passing in unit time through unit surface perpendicular to the direction of the mean wind is $\chi\bar{u}$, and the total mass of vapour removed in unit time is found by integrating this quantity over the whole atmosphere. Hence, if $\int_0^\infty \chi\bar{u}\,dz$ can be calculated for $x = x_0$, the problem is solved. This is carried out as follows.

    Write $\qquad\qquad \xi = \frac{x}{x_0}, \qquad \zeta = \left(\frac{u_1^n}{ax_0} \right)^{\frac{1}{2}} z^{m+\frac{1}{2}} \qquad \ldots\ldots(159),$

where $\qquad\qquad\qquad m = n/(2-n).$

Equation (158) now becomes

$$\frac{4}{(2m+1)^2} \frac{\partial \chi}{\partial \xi} = \frac{\partial^2 \chi}{\partial \zeta^2} + \frac{1}{(2m+1)\zeta} \frac{\partial \chi}{\partial \zeta} \qquad \ldots\ldots(160)$$

with the boundary conditions

    (a) $\lim\limits_{\zeta \to 0} \chi(\xi, \zeta) = \chi_0$ $(0 < \xi \leqslant 1)$,

    (b) $\lim\limits_{\zeta \to \infty} \chi(\xi, \zeta) = 0$ $(0 \leqslant \xi \leqslant 1)$,

    (c) $\lim\limits_{\xi \to 0} \chi(\xi, \zeta) = 0$ $(0 < \zeta)$.

The evaporating surface, it will be noticed, is now defined by $0 \leqslant \xi \leqslant 1$, $\zeta = 0$, and the transformed equation and its boundary conditions do not involve a, $\bar{u}_1$, and x_0 explicitly. Hence, if $F(\xi, \zeta)$ be the solution of this latter problem, the solution of the original problem must be

$$\chi(x, z) = F\left\{\frac{x}{x_0}, \left(\frac{u_1^n}{ax_0}\right)^{\frac{1}{2}} z^{m+\frac{1}{2}}\right\} \qquad \ldots\ldots(161)$$

and the rate of evaporation, E, is given by

$$E = \int_0^\infty \bar{u}_1 z^m F\left\{1, \left(\frac{u_1^n}{ax_0}\right)^{\frac{1}{2}} z^{m+\frac{1}{2}}\right\} dz$$

$$= \int_0^\infty \bar{u}_1 \left(\frac{u_1^n}{ax_0}\right)^{-\frac{1+m}{2m+1}} F(1, \zeta) \frac{d\zeta}{\zeta^{\frac{2(m+1)}{2m+1}}}$$

$$= B\bar{u}_1^{\frac{2-n}{2+n}} a^{\frac{2}{2+n}} x_0^{\frac{2}{2+n}} \qquad \ldots\ldots(162),$$

where

$$B = \int_0^\infty F(1, \zeta) \frac{d\zeta}{\zeta^{\frac{2(m+1)}{2m+1}}}$$

is a function of n only, and thus a constant for any given wind structure. It follows immediately that the evaporation for a strip of length x_0 and width y_0 (assuming no edge losses due to lateral diffusion) is

$$E = B\bar{u}_1^{\frac{2-n}{2+n}} a^{\frac{2}{2+n}} x_0^{\frac{2}{2+n}} y_0 \qquad \ldots\ldots(163).$$

For an elliptic area of semi-axes r_1 (downwind) and r_2 (across wind)

$$E = B'\bar{u}_1^{\frac{2-n}{2+n}} a^{\frac{2}{2+n}} r_1^{\frac{2}{2+n}} r_2 \qquad \ldots\ldots(164)$$

so that for a circular area

$$E = B'\bar{u}_1^{\frac{2-n}{2+n}} a^{\frac{2}{2+n}} r^{\frac{4+n}{2+n}} \qquad \ldots\ldots(165).$$

For turbulent flow in a wind tunnel, it is known that the velocity distribution over a smooth plane surface is given very accurately, for a wide range of Reynolds numbers, by the formula $\bar{u} = \bar{u}_1 z^{\frac{1}{7}}$, which requires $n = \frac{1}{4}$ on the above theory. Hence on the basis of the above theory of evaporation, it would be expected that for rectangular and circular areas the rate of evaporation should vary as $\bar{u}_1^{0.78}$, while the variation with length of surface downwind is given by $x_0^{0.89}$. Recent work on evaporation makes it possible to test these expressions.

As regards the variation with velocity, Lettau* quotes the results of Dorffel, who found the value of the index of u to be 0·71, and that of Mrose, who found an index of 0·75 with similar apparatus. Himus† gave 0·77, and Sutton (*loc. cit.*) analysed Hine's results with liquids other than water, showing that there was good agreement with the index 0·78. Powell and Griffiths,‡ using a

---

* *Ann. Hydr. u. Mar. Met.* **65**, h. 4, 1937.
† *Inst. of Chem. Eng.*, Conference on Vapour Absorption and Adsorption (1929).
‡ *Trans. Inst. Chem. Eng.* **13**, 1935, p. 175.

somewhat large value for the rate of evaporation in a calm, gave the value 0·85 for the index, but taking the evaporation to be negligibly small in the absence of wind, their results yield the theoretical value 0·78. Powell and Griffiths also state that Thiesenhausen's[*] experiments give the value 0·78 for the index.

In dealing with results from evaporimeters, it must be remembered that the evaporation will be influenced chiefly by eddies which are not greater in size than the dishes themselves, and consequently the turbulence set up by the dishes themselves will tend to be the controlling factor. Results from small pans, therefore, will not in general reflect the major diurnal changes in wind structure, but will tend to approximate more to the wind tunnel type of evaporation. This is clearly shown by the results of Millar,[†] who found the laws $\bar{u}_1^{0·77}$ and $\bar{u}_1^{0·725}$ for circular dishes.

As regards the variation of evaporation with the dimensions of the surface, Lettau and Dorffel[‡] found the law $x_0^{0·873}$. Powell and Griffiths found the index to vary with the cross-wind dimension, indicating that in their experiments there was appreciable lateral diffusion.

The above results show that, on the whole, the theory is in good agreement with experiment, justifying the assumption that the transfer of matter and momentum in the vertical are identical.

(e) Transfer of heat by eddies

It is clear that the above analysis may be applied to the problem of the loss of heat from a plane surface by introducing the specific heat, and letting χ indicate temperature. Direct proof of the identity of the transfer of heat with that of matter and momentum is afforded by the results of Elias,[§] who found for the loss of heat from a rectangular plate the law $x_0^{0·89} \bar{u}_1^{0·8}$, which is in practically perfect agreement with the theory. Hence we may conclude that heat, matter, and momentum obey the same law of transfer in the vertical, a result suggested by G. I. Taylor in 1915.

[*] *Gesundheits Ingenieur*, **53**, 1930, p. 113.
[†] *Can. Met. Mem.* **1**, No. 2.
[‡] *Ann. Hydr. u. Mar. Met.* **64**, 1936, p. 342.
[§] *Zeit. f. angew. Math. u. Mech.* **9** and **10**, 1929–30.

CHAPTER XIII

THE CLASSIFICATION OF WINDS

§ 164. *The terms in the equations of motion*

THE equations of motion of air over the earth's surface contain a statement of the forces acting upon the air, and of the accelerations which the air undergoes. The forces acting upon the air are (*a*) gravity, (*b*) hydrostatic pressure, and (*c*) friction. In addition there is the so-called "deviating force" due to the earth's rotation. The equations of motion of air moving over a rotating earth referred to axes drawn to East, North and vertical respectively, may be written

$$\frac{du}{dt} + 2\omega\,(w\cos\phi - v\sin\phi) = -\frac{1}{\rho}\frac{\partial p}{\partial x} + K\frac{\partial^2 u}{\partial z^2} \qquad \ldots\ldots(1),$$

$$\frac{dv}{dt} + 2\omega u\sin\phi \qquad\qquad = -\frac{1}{\rho}\frac{\partial p}{\partial y} + K\frac{\partial^2 v}{\partial z^2} \qquad \ldots\ldots(2),$$

$$\frac{dw}{dt} - 2\omega u\cos\phi \qquad\qquad = -\frac{1}{\rho}\frac{\partial p}{\partial z} - g \qquad \ldots\ldots(3).$$

These equations are sufficiently exact for all practical purposes except possibly for the terms $K\partial^2 u/\partial z^2$, $K\partial^2 v/\partial z^2$ in the first and second equations. These terms represent the frictional effects, whose nature is not known with complete certainty, as was shown in the preceding chapter. These equations were used by Jeffreys* as a basis for a classification of winds. Most of the present chapter is based on Jeffreys' paper.

In the first place, the pressure terms in the equations are always important, otherwise each portion of the fluid would pursue its path independently, without being appreciably interfered with by the impacts of surrounding portions. The latter is far from being true in the atmosphere, the mean free path being very small by comparison with the horizontal displacements with which we are concerned in atmospheric motions. In any type of wind the pressure terms are therefore of fundamental importance.

The terms on the left-hand side of equation (3) are usually negligible, and are in practically all circumstances small by comparison with gravity. The easterly wind velocity u is at most 100 metres per second, and ω is 7×10^{-5} sec^{-1}. Thus $2\omega u\cos\phi$ is at most about $1\cdot5$ cm/sec^2, which is small by comparison with g. The term dw/dt can be estimated roughly. The fall of a hailstone requires that the frictional resistance of the upward current shall be less than the weight of the hailstone. The resistance to the motion of a sphere is of the order of $0\cdot1\pi\rho a^2 w^2$, and weight is $\frac{4}{3}\pi da^3 g$, where a is the radius of the

---

* *Q.J. Roy. Met. Soc.* **48**, 1922, p. 29.

stone and d its density. Hence the velocity required in an ascending current to maintain the hailstone without acceleration is given by

$$0 \cdot 1 \pi \rho a^2 w^2 = \tfrac{4}{3} \pi d a^3 g \quad \text{or} \quad w^2 = 10^7 a.$$

If $a = 3$ cm for an unusually large hailstone, the vertical velocity is about 55 m/sec. If the vertical acceleration is f, and h is the height to which the air has ascended,

$$w^2 = 2fh.$$

Let $h = 3$ km $= 3 \times 10^5$ cm, then f is 50 cm/sec². This value of the vertical acceleration is small by comparison with gravity, and it is therefore legitimate to omit the term dw/dt from the left-hand side of equation (3) which we can now write

$$\frac{1}{\rho} \frac{\partial p}{\partial z} + g = 0, \quad \text{or} \quad \frac{1}{\rho} \frac{\partial p}{\partial z} = -g \qquad \ldots \ldots (4).$$

In general vertical motion of any appreciable magnitude only takes place in eddies, and for most purposes it is legitimate to neglect w in equation (1). The equations of mean horizontal motion are then

$$\frac{du}{dt} - 2\omega v \sin \phi = -\frac{1}{\rho} \frac{\partial p}{\partial x} + K \frac{\partial^2 u}{\partial z^2} \qquad \ldots \ldots (5),$$

$$\frac{dv}{dt} + 2\omega u \sin \phi = -\frac{1}{\rho} \frac{\partial p}{\partial y} + K \frac{\partial^2 v}{\partial z^2} \qquad \ldots \ldots (6),$$

$$0 = -\frac{1}{\rho} \frac{\partial p}{\partial z} - g \qquad \ldots \ldots (7).$$

Since the pressure term is always important, it follows that at least one of the other terms is comparable in magnitude with it. Winds may be classified according to the identity of the term or terms in question.

Case 1. Rotational and frictional terms negligible. Here

$$\frac{du}{dt} = -\frac{1}{\rho} \frac{\partial p}{\partial x} \qquad \ldots \ldots (8),$$

$$\frac{dv}{dt} = -\frac{1}{\rho} \frac{\partial p}{\partial y} \qquad \ldots \ldots (9).$$

The acceleration is measured by the pressure gradient. Jeffreys calls this class of wind "Eulerian".

Case 2. Rotational terms far in excess of both accelerational and frictional terms. The equations now reduce to

$$-2\omega v \sin \phi = -\frac{1}{\rho} \frac{\partial p}{\partial x} \qquad \ldots \ldots (10),$$

$$2\omega u \sin \phi = -\frac{1}{\rho} \frac{\partial p}{\partial y} \qquad \ldots \ldots (11).$$

The gradient of pressure is here balanced by the "deviating force". Winds of such a type are called "geostrophic".

Case 3. Frictional terms exceeding the rotational and accelerational terms. The equations of motion reduce to

$$K\frac{\partial^2 u}{\partial z^2} = \frac{1}{\rho}\frac{\partial p}{\partial x} \qquad \qquad ......(12),$$

$$K\frac{\partial^2 v}{\partial z^2} = \frac{1}{\rho}\frac{\partial p}{\partial y} \qquad \qquad ......(13).$$

The wind will in general blow along the pressure gradient, friction preventing its velocity from increasing steadily. For such winds Jeffreys suggests the name "antitriptic".

§ 165. *The application of the classification*

It is convenient, following Jeffreys, to classify the winds of the globe according to their horizontal extent, as follows:

(*a*) World-wide phenomena, including the general circulation and its seasonal variation.

(*b*) Phenomena on a continental scale, including monsoon winds and associated changes of pressure.

(*c*) Phenomena on a scale comparable with the British Isles. In view of the fact that even so large an island as Australia only produces modifications of secondary importance in the general circulation, it seems likely that moderate-sized and small islands do not to any great extent modify the general and continental circulations in their neighbourhood.

(*d*) Small scale phenomena, including all atmospheric disturbances whose horizontal dimensions in at least one direction are of the order of kilometres or tens of kilometres. This category includes tropical cyclones, tornadoes, land and sea breezes, mountain and valley winds, and winds of the Föhn type.

It is advantageous to begin by deciding in which of these cases the accelerational term is greater than the rotational term. A strong wind of any of the first three classes has a velocity of about 20 m/sec. The acceleration du/dt is comparable with the ratio u^2/l, where l is the linear dimension of the disturbance. This will be small compared with $u/2\omega \sin\phi$. If u is 20 m/sec, then at the poles $u/2\omega \sin\phi$ is 140 km, and in latitude 20° its value is 800 km. When l has this value, du/dt and the rotational term are about equal. For greater velocities the horizontal extent is increased in proportion. It is clear, however, that in cases *a*, *b* and *c* above the rotational terms exceed the accelerational term, so that none of the winds in these classes can be Eulerian, but are either geostrophic or antitriptic. In the typical tropical revolving storm, the velocities frequently reach 70 m/sec, while the dimensions in typical cases are of the order of 50–100 km. The dimensions are therefore such that the accelerational term exceeds considerably the rotational term. The tropical cyclone is therefore not geostrophic.

Observations show that winds of type (*a*), (*b*) and (*c*) deviate from the direction of the isobars by not more than 2 to 4 points at the surface, and by much less at heights above 0·5 km. These winds are therefore mainly geostrophic in character. In the tropical storm, a particle performs one revolution about the centre in a few hours, while the life of the storm is several days. The frictional term must therefore be much smaller than the accelerational term, and the motion in these storms is therefore Eulerian. The same result has been stated by Sir Napier Shaw, who pointed out that the winds are "cyclostrophic" in the tropical storm and the tornado.

§ 166. *Geostrophic winds*

The result stated above, that large scale phenomena in the atmosphere are essentially geostrophic in character, is in accordance with the results already stated in Chapter IX, where it was shown that at levels removed from the effects of surface friction the wind could be treated as closely approximating to the gradient wind, provided no rapid change of pressure were taking place. Further, the comparison of observed winds with the gradient winds computed from synoptic charts, carried out by Gold (see p. 190 above), also shows that the winds are in the main geostrophic. The only point which is left obscure is the closeness of the geostrophic approximation in general. The comparison of observed and computed winds, being based on pilot balloon observations in clear weather, is thereby restricted to occasions when the pressure distribution is normally not changing rapidly, and does not apply to the central regions of depressions, in which cloud and rain make pilot balloon observations impracticable.

The closeness of the geostrophic approximation can be readily deduced by the use of equation (4) of Chapter IX, if the pressure distribution is not changing. For simplicity take the symmetrical depression, so that the pressure is a function only of distance r from the centre. Let the velocity at distance r be v, and let $v = r\zeta$, so that ζ is the angular velocity about the centre. The equation of motion for cyclonic curvature is

$$\frac{1}{\rho}\frac{dp}{dr} = r\zeta\,(2\omega\sin\phi + \zeta).$$

The gradient wind velocity $r\zeta$ is obtained by solving this equation for ζ, taking the appropriate solution as shown in Chapter IX. The geostrophic wind is

$$\frac{1}{2\omega\sin\phi}\frac{1}{\rho}\frac{dp}{dr} = \frac{1}{2\omega\sin\phi}\,r\zeta\,(2\omega\sin\phi + \zeta) = r\zeta\,(1 + \zeta/2\omega\sin\phi).$$

Thus the ratio of the geostrophic to the true wind in such a system is $1 + \dfrac{\zeta}{2\omega\sin\phi}$. If therefore the angular velocity of the air about the centre is comparable with $2\omega\sin\phi$, the difference between the geostrophic wind and the true wind will be comparable in magnitude with these two winds. In some intense depressions ζ approaches $2\omega\sin\phi$. A rough computation indicated that in the depression of December 16, 1917, ζ was about $1\cdot6\omega\sin\phi$. In such a case therefore the geostrophic wind is nearly double the true wind. In an anticyclone of November 16, 1922, ζ was about $\tfrac{1}{6}\omega\sin\phi$, and the geostrophic wind was therefore about five-sixths of the true wind.

The statement in the preceding paragraph can be extended to any non-symmetrical system, by defining r as the radius of curvature of the path of the air at any point, so that ζ is the angular velocity about the centre of curvature. When the curvature is small, the difference between the geostrophic and the true wind is usually relatively small. But it is clear that it is not justifiable in general to assume the geostrophic wind to be a close approximation to the true wind without further consideration. In particular near the central regions of anticyclones, and still more near the centres of depressions, the approximation breaks down. This adds considerably to the difficulties of theoretical meteorology, in which the equations only become reasonably tractable when the winds are assumed to be geostrophic. Very brief consideration shows that the winds of the globe cannot be accurately geostrophic. For geostrophic winds would blow accurately round the isobars, and transfer of air into or out of the isobar would be impossible. The fundamental problems of meteorology are, however, associated with the transfer of air across the isobars, and it is clear that any satisfactory theory must take into consideration the accelerational and frictional terms in the equations of motion.

The frictional terms are about one-third the magnitude of the rotational terms at the ground, since the air at the ground flows across the isobars at an angle of about $\tan^{-1}\frac{1}{3}$. At greater heights the influence of friction rapidly decreases and at heights above about 1 km it appears to be negligible. It can be said that, in general, friction exerts a modifying rather than a controlling influence in atmospheric phenomena. Friction enters only when the motion has been initiated by other agencies.

There are two aspects of the problem of comparing the actual wind with the geostrophic wind. In the first place we may perhaps require an estimate of the upper wind at some given level for some such practical purpose as air navigation, and it is then legitimate to regard the geostrophic wind as an approximation to the actual wind, though the use of the approximation will demand a reasonable exercise of judgment. In the second place we may require to consider the physical changes taking place in a mass of air, and while the geostrophic wind may again be regarded as an approximation to the actual wind it is the deviation from geostrophic motion which is of importance in determining changes of pressure. Here the accelerational terms become important. Thus while a good instantaneous picture of the winds is derived by the geostrophic approximation, the changes which are taking place are to be measured by deviations from the geostrophic wind, and involve the accelerational terms in the equations of motion.

§ 167. *Antitriptic winds*

Jeffreys has discussed (*loc. cit.*) the nature of the land and sea breeze and of mountain and valley winds, and has shown from mathematical arguments that these winds are mainly antitriptic. The sea breeze at Aberdeen has been found to set in suddenly as a breeze from the sea, and to veer in the course of the day until it blows nearly parallel to the coast. This suggests that the rotational terms are of importance in the motion of these winds. They have, however, the essential characteristics of all antitriptic winds, in that they blow roughly at right angles to the isotherms, and extend to only a small altitude.

No satisfactory detailed theory of land and sea breezes, or of katabatic winds, has been worked out hitherto. Some idea of the volume of papers on mountain and valley winds can be gathered from a paper by Wagner,[*] but it is clear that a number of different phenomena has been grouped under the name of "Mountain and Valley winds". A discussion of some detailed observations of katabatic winds in a valley of the Cotswold Hills has been given by Heywood,[†] who showed that on occasions these winds are extremely shallow. The land and sea breezes are also very shallow, extending normally to a height of about 200 feet only. These breezes are produced by the difference of temperature over land and sea.

[*] *Met. Zeit.* **49**, 1932, p. 329.
[†] *Q.J. Roy. Met. Soc.* **59**, 1933, p. 47.

CHAPTER XIV

THE TRANSFORMATIONS OF ENERGY
IN THE ATMOSPHERE

§ 168. *Incoming radiant energy*

THE direct source of energy in the atmosphere is the incoming solar beam, whose amount can be measured by the "solar constant". The phenomena which are directly associated with the disposal of the incoming radiant energy have already been discussed in Chapter VI. For our present purpose we may regard this energy as in four categories, which are:

(*a*) Reflected from clouds, and in varying degrees from the atmosphere itself, and portions of the earth's surface: this radiation is ineffective in producing thermodynamical changes in the atmosphere and may be left out of further consideration.

(*b*) Absorbed in the atmosphere.

(*c*) Used up in evaporation of water from the earth's surface.

(*d*) Used in heating the earth's surface.

The energy in class (*c*) above become latent heat of evaporation, and is again liberated when the water-vapour condenses, mainly in middle levels of the atmosphere. The energy absorbed by the earth's surface is radiated again outwards, and, possibly after repeated absorption and re-radiation, passes out again through the atmosphere into space. The energy in all four categories eventually passes out into space as radiation, usually with much longer wavelengths than occur in the incoming solar beam, but a portion at least of this energy goes through varied transformations before it is finally passed out to space.

One of the effects of the heating of the atmosphere is to produce an increase of gravitational potential energy, which is convertible into kinetic energy, and the main problem to be considered in the present chapter is the maintenance of the kinetic energy of the atmosphere substantially unchanged from year to year.

§ 169. *The classification of energy*

The complete statement of the types of energy in the atmosphere should include electrical and magnetic energy. We restrict our attention, however, to four types, which appear to be those of meteorological importance:

(1) Kinetic energy of the general circulation, and of the currents in depressions and anticyclones.

(2) Turbulent energy, or the energy associated with eddies in the main currents.

(3) Potential energy (gravitational).

(4) Thermal energy.

It will be noted that energy of radiation is not included in the above classification. Meteorologically this energy is ineffective until it has been absorbed by the atmosphere when it is manifested by a change in items 3 and 4 of the classification.

If no condensation or evaporation is taking place, changes in items 3 and 4 are connected by a simple relationship. Any element of mass in the atmosphere which is heated or cooled behaves as though its specific heat were c_p, the specific heat at constant pressure. For a change ΔT in the temperature of unit mass, the loss of heat is $c_p \Delta T$. Of this a portion $c_v \Delta T$ represents the change of thermal energy, and the remainder $(c_p - c_v) \Delta T$ the work done against the pressure of the environment.

In a column of unit horizontal cross-section, extending to the top of the atmosphere assumed of uniform constitution, there is a simple relationship between the total gravitational potential energy and the total internal energy $c_v \int_0^\infty \rho T dz$. The total potential energy

$$= \int_0^\infty g\rho z\, dz = - \int_{p_0}^0 z\, dp = - \left[pz \right]_0^\infty + \int_0^\infty p\, dz$$

$$= R \int_0^\infty \rho T dz$$

$$= \frac{AR}{c_v} \times \text{total internal energy in heat units.}$$

The coefficient AR/c_v may be written $(c_p - c_v)/c_v$; or $\gamma - 1$. Hence the potential energy is $\gamma - 1$ times the internal energy, and if any change of temperature takes place within the column, the resulting change of potential energy is $\gamma - 1$ times the resulting change of internal energy. To put this in another way, when a given quantity of heat is added to any column of the atmosphere, a fraction $1/\gamma$ of it is used in increasing the internal energy of the air, and the remaining fraction $(\gamma - 1)/\gamma$ is used in increasing the potential energy. Changes in internal and potential energy are thus closely bound together.

Various writers, including Margules, include in their statements of classes of energy the energy associated with distribution of pressure. To add the effect of pressure distribution to the four types named above would be to include some types of energy twice over, since this effect is accounted for under 3 and 4 above.

§ 170. *Transformations of energy in the atmosphere*

Since the great atmospheric motions are without exception turbulent, there is a continual transformation of their kinetic energy into turbulent kinetic energy. This transformation is irreversible, and the energy of the great currents cannot be reinforced by eddy motions. There is a continual degradation of the kinetic energy of eddies into heat, as the effect of viscosity. This process also acts in a lesser degree upon the kinetic energy of the great currents, and this kinetic energy is therefore being continually degraded to heat energy, either directly, or through an intermediate stage as turbulent energy.

There is in addition an interchange between the turbulent energy and the combined potential and thermal energies, in a direction which depends on the degree of stability of the vertical distribution of temperature. Richardson[*] has shown that when the rate of variation of wind with height exceeds a critical value fixed by the existing lapse-rate there is a transformation of thermal and potential energy into turbulent energy, and that when the variation of wind with height is less than the critical value the transformation is in the opposite direction (see § 145, p. 237).

This statement cannot include all the possible transformations in the atmosphere, since it only provides for a steady degradation of the kinetic energy of the great currents of the atmosphere. There must therefore be some other transformation which provides for the reinforcement of the kinetic energy of these currents, and this transformation can only be from the combined potential and thermal energies.

The possible transformations of energy are therefore the following:

(a) $1 \big\langle {\begin{smallmatrix} {}^{\nearrow 2} \to (4+3) \\ {}_{\searrow} (4+3) \end{smallmatrix}}$

(b) $2 \to (4+3)$ by viscosity

(c) $2 \to (4+3)$ if $\left(\dfrac{\partial u}{\partial z}\right)^2 + \left(\dfrac{\partial v}{\partial z}\right)^2 < \dfrac{g}{T}\left(\dfrac{\partial T}{\partial z}+\Gamma\right)$

$\qquad (4+3) \to 2$ if $\left(\dfrac{\partial u}{\partial z}\right)^2 + \left(\dfrac{\partial v}{\partial z}\right)^2 > \dfrac{g}{T}\left(\dfrac{\partial T}{\partial z}+\Gamma\right)$

(d) $(4+3) \to 1$.

Items 3 and 4 occur together in this list, since changes of internal energy and of gravitational potential energy are always associated as shown in § 169 above.

The first three classes of transformation have been discussed in an earlier chapter. The fourth class of transformation, from the thermal and gravitational potential energy to kinetic energy, which remains to be considered, is the central problem of meteorology. It includes (1) the maintenance of the general circulation of the atmosphere against turbulence and friction, and (2) the origin of depressions and anticyclones of middle latitudes. It is shown above, and is indeed obvious *a priori*, that the kinetic energy of the large scale currents can only be produced and maintained by the consumption of potential and thermal energy. The real problem is to explain the precise mechanism by which the transformation of energy is brought about, and to picture the method by which the kinetic energy is organised, to produce and maintain the currents observed in the atmosphere.

In the first place we shall endeavour to assess the rate at which the kinetic energy of the main currents is destroyed by turbulence, in order to have a working estimate of the rate at which the kinetic energy is being renewed.

---

[*] *Proc. Roy. Soc.* A, **97**, 1920, p. 354.

§ 171. *The kinetic energy of the atmosphere, and its dissipation by turbulence*

If we take the parallels of 30° North and South as separating the main easterly and westerly circulations of the atmosphere, the easterly belt, covering a range of 60° in latitude, will contain half of the mass of the atmosphere. Assuming the average pressure to be one atmosphere, we deduce the approximate mass of air in this belt to be $2 \cdot 7 \times 10^{21}$ grammes or $2 \cdot 7 \times 10^{15}$ metric tons. Taking the mean velocity to be 10 m/sec, a value which is in agreement with the distribution of pressure, we find the kinetic energy of the equatorial belt to be $\frac{1}{2} \times 2 \cdot 7 \times 10^{21} \times 10^{6}$ ergs $= 1 \cdot 35 \times 10^{27}$ ergs.

The combined masses of the polar caps will not differ appreciably from the mass of the equatorial belt, and the amount of momentum of the two circulations should balance, since they are produced by internal reactions within the earth's atmosphere. The equality of moments of momentum demands that the westerly velocity should be the greater, and the energy of the combined polar caps should exceed somewhat that of the equatorial belt.

It appears safe to assume that the total kinetic energy of the earth's atmosphere is of the order of 3×10^{27} ergs.

This estimate does not specifically include the energy of the circulations around cyclones and anticyclones. An estimate has been given by Sir Napier Shaw[*] of the kinetic energy of a cyclone which formed over the lower part of the North Sea between July 27 and August 3, 1917. Its diameter was about 1400 km and the depth at the centre was 10 millibars. The kinetic energy developed was $1 \cdot 5 \times 10^{24}$ ergs.

If the kinetic energy of the earth's atmosphere, roughly computed above, were spread uniformly over the earth's surface, the amount contained in the portion of the atmosphere over a circle of diameter 1400 km would be 10^{24} ergs, so that in the case considered the kinetic energy of the cyclone was about 50 per cent higher than the average kinetic energy of the general circulation.

It has been shown in Chapter XII, p. 255, that the effect of turbulence may be represented by an internal frictional force R, given in equation (98), p. 255. The rate at which work is done by this force is equal to the product of R by the component of the velocity along the direction of R. The total rate of loss of kinetic energy of the main current from the ground up to height z is readily obtained by integration. Its amount is[†]

$$2\omega\rho \sin \alpha \sin \phi \frac{G^{2}}{B} \{\cos \alpha - e^{-Bz} \cos (\alpha - Bz)\} \qquad \ldots\ldots(1).$$

If the same simple law of distribution of wind with height given in equations (95), p. 253, persisted up to the top of the atmosphere, then for a column extending from the ground to the top of the atmosphere the rate of dissipation of energy would amount to

$$2\omega\rho \sin \phi \sin \alpha \cos \alpha \frac{G^{2}}{B} \qquad \ldots\ldots(2).$$

[*] *Dict. App. Phys.* **3**, p. 84.　　　　　[†] Brunt, *Phil. Mag.* Feb. 1926.

With the assumed law of distribution of wind with height this dissipation is effectively brought about in the lowest kilometre of the atmosphere.

The rate of dissipation of energy per unit volume is

$$K\rho \left(u \frac{\partial^2 u}{\partial z^2} + v \frac{\partial^2 v}{\partial z^2} \right) \qquad \ldots\ldots(3).$$

A rough estimate by Brunt (*loc. cit.*) shows that the rate of dissipation of energy in the column above 1 m² of the earth's surface is as follows:

From the ground to 1 km	3×10^{-3} kilowatts
From 1 km to 10 km	2×10^{-3} ,,
Total	5×10^{-3} ,,

This total is the rate at which the kinetic energy of the earth's atmosphere is being dissipated by turbulence. It is probably an overestimate for those regions of the earth's atmosphere in which the winds are normally light at all heights, but so far as order of magnitude is concerned it may be accepted with a fair degree of faith in spite of the numerous approximations made. On the whole it appears likely to be an overestimate rather than an underestimate.

The total energy above 1 m² of the earth's surface, on the assumption of uniform velocity of 10 m/sec, is 5×10^{12} ergs. The rate of dissipation by turbulence is 5×10^7 ergs per m² per sec. Hence if the same rate of dissipation were maintained for 10^5 seconds, or $1\frac{1}{6}$ days, the whole kinetic energy would be destroyed in that time. If the rate of dissipation is assumed to be proportional at each instant to the total kinetic energy, then the total kinetic energy is reduced to one-hundredth of its original value in six days. As no sensible change takes place in the kinetic energy of the general circulation of the atmosphere, we conclude that the loss by turbulence is being continually made up by the conversion of solar energy into kinetic energy.

§ 172. *Comparison of the eddy dissipation of kinetic energy with the radiation coming in from the sun*

The value of the solar constant being taken as 2 gramme-calories per cm² per minute, then the amount of radiation coming from the sun into the earth's atmosphere

$$= 2 \text{ gm-cal per cm}^2 \text{ per min}$$
$$= 8\cdot36 \text{ joules per cm}^2 \text{ per min}$$
$$= 83600 \text{ joules per m}^2 \text{ per min}$$
$$= 1400 \text{ joules per m}^2 \text{ per sec}$$
$$= 1\cdot4 \text{ kilowatts per m}^2.$$

But the radiation intercepted by the earth is the portion of the beam of solar radiation intercepted by an area πr^2, where r is the radius of the earth. This is spread over the whole area of the earth, $4\pi r^2$, and hence the effective incoming radiation per m² is one-fourth of the figure given above, and amounts to $0\cdot35$ kilowatt per m².

Also Aldrich estimates that the effect of reflexion from the earth's surface and from clouds is to reduce by 37 per cent the amount of solar radiation available for absorption and subsequent conversion into kinetic energy. The effective incoming solar radiation is therefore 0·22 kilowatt per m² when averaged over the whole of the earth's surface.

The rate of dissipation by turbulence was given above as 5×10^{-3} kilowatt per m², which is only a little over 2 per cent of the effective incoming solar radiation.

Thus the conversion of a little over 2 per cent of the incoming solar radiation into kinetic energy will suffice to make up for the dissipation by turbulence.

§ 173. *The development of circulation between sources of heat and cold*

Sandström* and Bjerknes† have given a theorem which appears at first sight to be of great importance in meteorology, that permanent circulation is only possible if the source of heat is lower than the source of cold. Wenger‡ gave a slightly different form of the same theorem, that the source of heat must be at higher pressure than the source of cold.

It is necessary to be clear as to the meaning of the terms warm and cold sources. The definition given by Bjerknes was as follows: "If a fluid during its motion is gaining heat, i.e. its potential temperature is rising, it is said to pass a warm source, if it is losing heat, it is passing a cold source."

The general aspects of these theorems have been discussed by H. Jeffreys§ in a paper on fluid motions produced by differences of temperature and humidity. Jeffreys shows that if a fluid is in equilibrium, every level surface within it, or in contact with it, must be isothermal, and if any solid is wholly surrounded by fluid, the total rate of inflow of heat from the solid to the fluid is zero. For physical applications he states the theorem in another form as follows: "if a difference of temperature is maintained over any level surface within or in contact with a fluid, or if heat is supplied to, or withdrawn from any region within the fluid, the fluid will move, and will continue to move until such difference of temperatures or supply or removal of heat ceases". Jeffreys' theorem also holds with regard to the supply of material constituents as well as heat, and permanent equilibrium will only be possible if the theorem stated above holds separately for temperature, and for every material constituent.

Sandström's argument is that the supply of energy required in order to maintain a steady motion of fluid against viscosity must arise from the work done on the fluid. If the fluid is in a steady state its centre of mass has no vertical motion, and no work is done by gravity. The energy must therefore arise from work done against fluid pressure in expansion and contraction, and if there is to be a net gain of work, the expansion must be done under greater pressure than the contraction. The portion of their paths in which elements of fluid become hotter must lie in regions of higher pressure than the places

---

* Goteborg, *Vet. Handl.* **17** (4), 1916.
† Leipzig, *Abh. Ges. Wiss.* **35**, 1916, p. 29; also *Physikalische Hydrodynamik*, para. 52.
‡ *Phys. Zeit.* **17**, 1916.
§ *Q.J. Roy. Met. Soc.* **51**, 1925, p. 347.

where they become colder. But as Jeffreys states, the correct interpretation of the last statement is that, as the fluid is contracting on its way from the warm to the cold source and expanding on its way from the cold to the warm source, the part of the path from cold to warm source must lie below the part of the path from warm to cold.

It is in any case dangerous to proceed by analogy with the results derived in the experiments of Sandström and Bjerknes in closed tubes to interpret phenomena in the atmosphere. In the latter heat is redistributed to only a very slight extent by molecular conduction, and turbulence and radiation are of far greater importance, as was shown earlier in Chapter XII. The thermal phenomena in the atmosphere cannot be related to strictly localised sources of heat and cold. Cooling or heating by radiation may take place from a very large volume of air. Further, a given portion of the earth's surface will be a source of heat for polar currents and a source of cold for equatorial currents. Any interpretation of meteorological phenomena as due to the action of simple sources of heat and cold must therefore proceed with caution, taking account of radiation and turbulence. If the lower atmosphere is heated, or a region in the upper atmosphere is cooled, say by radiation, there will be a tendency to set up instability, leading to turbulence and possible large scale convection. Thus the instability which sets in during the afternoon in summer is mainly due to the diurnal heating of the surface levels while the upper levels maintain their initial temperature. Reference should also be made to later notes on the same subject by Godske[*] and Jeffreys.[†]

§ 174. *Equations of energy*

If equations (39), (40) and (41) of p. 171 are multiplied by u, v, w respectively and added, the result may be written

$$\frac{1}{2}\frac{d}{dt}(u^2+v^2+w^2) = -gw - \frac{1}{\rho}\left(u\frac{\partial p}{\partial x}+v\frac{\partial p}{\partial y}+w\frac{\partial p}{\partial z}\right) + (uX+vY+wZ) \quad ......(4),$$

or

$$\frac{d}{dt}\{\tfrac{1}{2}(u^2+v^2+w^2)+gz\} = -\frac{1}{\rho}\left(\frac{dp}{dt}-\frac{\partial p}{\partial t}\right) + (uX+vY+wZ) \quad ......(5).$$

The quantity in the curly bracket on the left-hand side of the last equation is the sum of the kinetic and potential energies of unit mass of fluid. The equation states that the rate of increase of the sum of the kinetic and potential energies of unit mass of fluid is equal to the rate at which work is done upon it by the joint effect of the pressure forces and the frictional forces.

The thermal equation of energy as given in equation (23), p. 37, may be written

$$J\frac{dQ}{dt} = Jc_v\frac{dT}{dt}+p\frac{d}{dt}\frac{1}{\rho} \quad ......(6),$$

or

$$J\frac{dQ}{dt} = Jc_p\frac{dT}{dt}-\frac{1}{\rho}\frac{dp}{dt} \quad ......(7),$$

* *Q.J. Roy. Met. Soc.* **62**, 1936, p. 446.   † *Ibid.* p. 449.

where $\mathcal{J}$ is the mechanical equivalent of heat. The second of these equations may be used to eliminate dp/dt from equation (5) above, giving

$$\mathcal{J}\frac{dQ}{dt} = \mathcal{J}c_p\frac{dT}{dt} + \frac{d}{dt}\{\tfrac{1}{2}(u^2+v^2+w^2)+gz\} - \frac{1}{\rho}\frac{\partial p}{\partial t} - (uX+vY+wZ) \quad \ldots\ldots(8).$$

If the pressure distribution is not changing $\partial p/\partial t$ is zero, and this equation reduces to

$$\mathcal{J}\frac{dQ}{dt} = \frac{d}{dt}\{\mathcal{J}c_p T + \tfrac{1}{2}(u^2+v^2+w^2)+gz\} - (uX+vY+wZ) \quad \ldots\ldots(9).$$

This equation is useful in some applications, but another form of equation is more useful in the first consideration of questions of energy. Equation (6) may be written

$$\mathcal{J}\frac{dQ}{dt} = \mathcal{J}c_v\frac{dT}{dt} - \frac{p}{\rho^2}\frac{d\rho}{dt} = \mathcal{J}c_v\frac{dT}{dt} + \frac{p}{\rho}\left(\frac{\partial u}{\partial x}+\frac{\partial v}{\partial y}+\frac{\partial w}{\partial z}\right) \quad \ldots\ldots(10),$$

from equation (30) of p. 168. From equations (4) and (10), by addition, we find

$$\mathcal{J}\frac{dQ}{dt} = \frac{d}{dt}\{\mathcal{J}c_v T + \tfrac{1}{2}(u^2+v^2+w^2)+gz\}$$

$$- (uX+vY+wZ) + \frac{1}{\rho}\left(u\frac{\partial p}{\partial x}+v\frac{\partial p}{\partial y}+w\frac{\partial p}{\partial z}\right) + \frac{p}{\rho}\left(\frac{\partial u}{\partial x}+\frac{\partial v}{\partial y}+\frac{\partial w}{\partial z}\right) \quad \ldots\ldots(11).$$

In this equation Q is the amount of heat added to unit mass of fluid, and the quantity following the operator d/dt on the right-hand side is the sum of the internal, kinetic and potential energies of unit mass of fluid. To obtain the corresponding relation for any mass of fluid, the equation is multiplied by $\rho\,dx\,dy\,dz$ and integrated through the mass under consideration.

Let
$$\begin{aligned} I &= \mathcal{J}c_v \textstyle\int\rho T\,dx\,dy\,dz \\ K &= \tfrac{1}{2}\textstyle\int\rho\,(u^2+v^2+w^2)\,dx\,dy\,dz \\ P &= \textstyle\int g\rho z\,dx\,dy\,dz \end{aligned} \qquad \ldots\ldots(12),$$

so that I, K and P are the total internal, molar-kinetic and potential energies of the mass in question. Then from equation (11) above

$$\mathcal{J}\int\rho\frac{dQ}{dt}\,dx\,dy\,dz = \text{total rate of addition of heat}$$

$$= \frac{\partial}{\partial t}(I+K+P) - \iiint\rho\,(uX+vY+wZ)\,dx\,dy\,dz$$

$$+ \iiint\left\{\left(u\frac{\partial p}{\partial x}+p\frac{\partial u}{\partial x}\right)+\left(v\frac{\partial p}{\partial y}+p\frac{\partial v}{\partial y}\right)+\left(w\frac{\partial p}{\partial z}+p\frac{\partial w}{\partial z}\right)\right\}dx\,dy\,dz\ldots\ldots (13).$$

The last integral in this equation is readily shown to be equal to

$$\int pV_n\,dS$$

taken over the whole boundary of the mass of fluid considered, where dS is an element of surface of the boundary, and V_n is the component of velocity normal to the boundary. If the fluid is limited by solid boundaries, or by a system of stream lines, V_n is everywhere zero, and the integral vanishes. For the moment we retain the term in the equation, which may now be written

Total rate of addition of heat

$$=\frac{\partial}{\partial t}(I+K+P)-\iiint \rho\,(uX+vY+wZ)\,dx\,dy\,dz+\int pV_n\,dS \quad\ldots\ldots(14).$$

The equation expresses the fact that the energy added to the fluid in the form of heat is used partly in overcoming frictional forces, partly in increasing the sum of the internal, kinetic and potential energies of the fluid, and partly in expansion of the volume occupied by the fluid against the pressure of its environment. This equation is a complete statement of the principle of conservation of energy in a fluid, and if we had initially assumed the principle of conservation of energy, the equation could have been written down immediately.

Margules proceeded somewhat differently. In equation (10), he considered

$$-\iiint p\,\frac{d}{dt}\left(\frac{1}{\rho}\right)\,dx\,dy\,dz,\quad\text{or}\quad\iiint \frac{p}{\rho^2}\frac{d\rho}{dt}\,dx\,dy\,dz,$$

which represents the total work of expansion against pressure, and equated it to $\partial A/\partial t$, so that A is in some respects of the nature of a potential. The thermal equation (10) may then be written

$$\text{Total rate of addition of heat}=\frac{\partial}{\partial t}(I-A) \quad\ldots\ldots(15).$$

Also equation (14) may be written

$$\frac{\partial}{\partial t}(K+P+A)=\iiint \rho\,(uX+vY+wZ)\,dx\,dy\,dz-\int pV_n\,dS \quad\ldots\ldots(16).$$

It is of interest to consider the magnitude of the effect of the frictional term. The rate of dissipation of energy per unit volume of fluid is given by expression (69), of § 111, p. 185 above, where $\mu=\nu\rho$. Let the change of velocity be of the order of 10 m/sec per km, a rate which is large in the free air. Then all the differential coefficients in the equation are of the order of 10^{-2} sec^{-1}, and their squares of the order 10^{-4} sec^{-2}. The rate of dissipation in 1 cm^3 is therefore (assuming $\mu=1\cdot5\times10^{-4}$)

$$1\cdot5\times10^{-4}\times10^{-4}\times20=3\times10^{-7}.$$

The amount of kinetic energy present in 1 cm^3, assuming a velocity of the order of 10 metres sec^{-1} or 10^3 cm sec^{-1}, is $5\times10^5\rho$. Thus the amount dissipated per second is only about 10^{-9} of the whole, so that in one day 10^{-4} of the kinetic energy would be dissipated by viscous forces in smooth currents having the gradient of velocity of the order of 10 metres sec^{-1} per kilometre. Thus if the flow of air were smooth or laminar flow, the rate of dissipation of the kinetic energy would be negligible.

Atmospheric motion as we observe it is very seldom laminar, being in almost all cases turbulent, and the forces X, Y, Z must contain the effects of turbulence. The energy degraded through the effect of turbulence finally into thermal energy is in fact converted into thermal energy through the action of viscous forces. The eddies produce large gradients of velocity within restricted regions, and these gradients are so great that the dissipation far exceeds that evaluated on the supposition of laminar flow.

The magnitude of the dissipation by turbulence of the kinetic energy of the large scale motions of the atmosphere has been estimated in § 171 above. It is always in the same direction, tending to destroy the large-scale motions. In discussing the generation of the motions of the atmosphere which are associated with depressions or anticyclones we may therefore in the first place leave the effects of viscosity and turbulence out of consideration, with the understanding that their neglect will lead to overestimates of the motions produced by any conversion of potential into kinetic energy. The same result is achieved if we let Q represent the sum of the heat added from external sources, and the heat produced by the viscous degradation of molar-kinetic energy.

It has been shown in § 169 above that in a clear atmosphere, in which there is no evaporation or condensation, there is a simple relationship between the gravitational potential and the heat content of a column of air extending from the ground to the upper limit of the atmosphere. Let P' and I' be the gravitational potential and the total heat content of a column standing on unit cross-section. Then

$$P' = \frac{c_p - c_v}{c_v} I' = (\gamma - 1) I' \qquad \ldots \ldots (17).$$

If the mass of fluid to be considered is bounded laterally by vertical solid boundaries, then it is readily seen from equation (17) that

$$P = \frac{c_p - c_v}{c_v} I,$$

where P and I now represent the total potential and internal energies. It is also seen that in any changes which take place

$$\frac{\partial P}{\partial t} = (\gamma - 1) \frac{\partial I}{\partial t} \qquad \ldots \ldots (18),$$

and that of any heat energy added to the mass a fraction $1/\gamma$ is used in increasing the internal energy, and a fraction $(\gamma - 1)/\gamma$ in increasing the potential energy. Any loss or gain of potential energy is accompanied by a proportionate loss of internal energy.

In these circumstances we may take

$$I + P = \gamma I = \mathfrak{J} c_p \iiint \rho T \, dx \, dy \, dz \qquad \ldots \ldots (19).$$

§ 175. *Energy liberated when vertical interchange of masses occurs*

The region over which the vertical interchange takes place is regarded as a closed region, having a boundary across which no transfer of mass takes place. It is assumed that friction is negligible and that no heat is given to or extracted from the fluid affected, so that the total

$$K + P + I \quad \text{or} \quad K + \gamma I$$

is unaltered. The air being initially at rest, the final amount of kinetic energy is

$$\gamma \times \text{change in } I \qquad \ldots \ldots (20).$$

For a layer from the ground to height z,

$$I = c_v \int_0^z T\rho\, dz = \frac{c_v}{R}\int_0^z p\, dz.$$

Fig. 67 shows a case considered in some detail by Margules. The initial state (a) shows a heavy cold mass of air 1, above a potentially warmer mass 2. The two masses are interchanged, and their final state is shown in (b).

The pressure at the ground is p_0, at the boundary of the two masses p_1, and at the upper boundary of the mass 1, p_2. In the final state p_0 and p_2 are unchanged, but the pressure at the middle boundary is now p_1'. For the sake of simplification in the computation we assume each mass to have a uniform adiabatic lapse-rate, so that the temperatures are completely specified by the potential temperatures θ_1 and θ_2 in conjunction with the pressures.

Let T_0, T_1 be the initial temperatures at the surface and at the lower limit of the second mass, and let T_0', T_1' be the final temperatures. The temperature at a height z in the lower mass is $T_0 - \Gamma z$ and the pressure is

$$p = p_0\left(\frac{T_0 - \Gamma z}{T_0}\right)^{\gamma/(\gamma-1)} \qquad \ldots\ldots(21),$$

where Γ is the dry adiabatic lapse-rate. Hence

$$\int_0^z p\, dz = \left[\frac{p_0 T_0}{\Gamma}\frac{\gamma-1}{2\gamma-1}\left(\frac{T_0-\Gamma z}{T_0}\right)^{(2\gamma-1)/(\gamma-1)}\right]_z^0,$$

$$\frac{1}{R}\int_0^z p\, dz = \frac{1}{g}\frac{\gamma}{2\gamma-1}(p_0 T_0 - p_z T_z) \qquad \ldots\ldots(22)$$

since

$$R\Gamma = g\frac{AR}{c_p} = g\frac{c_p - c_v}{c_p} = g\frac{\gamma-1}{\gamma}.$$

We now require to evaluate the expression (22) above for the initial and final conditions represented in fig. 67 (a) and (b). The initial value of $\frac{1}{R}\int_0^z p\, dz$ through the two masses is

$$\frac{1}{g}\frac{\gamma}{2\gamma-1}\{p_0 T_0 - p_1(T_0 - \Gamma h_1) + p_1 T_1 - p_2(T_1 - \Gamma h_2)\} \qquad \ldots\ldots(23).$$

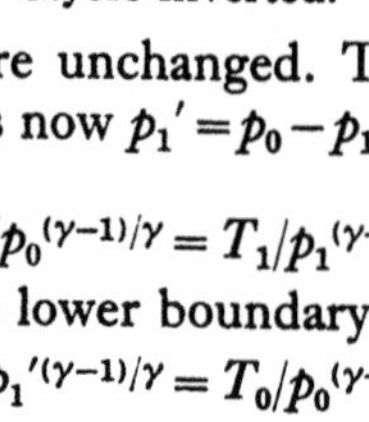

Fig. 67. Margules' diagram showing warm and cold
layers inverted.

In the final state p_0 and p_2 are unchanged. The pressure at the surface of separation of the two masses is now $p_1' = p_0 - p_1 + p_2$. The surface temperature T_0' is given by

$$T_0'/p_0^{(\gamma-1)/\gamma} = T_1/p_1^{(\gamma-1)/\gamma}$$

and the temperature T_1' at the lower boundary of the upper mass is given by

$$T_1'/p_1'^{(\gamma-1)/\gamma} = T_0/p_0^{(\gamma-1)/\gamma}.$$

Margules carried out a detailed computation for the case where initially the depth of each layer was 2000 metres, and the drop in temperature at the boundary was 3° C. Evaluating the change in I from equation (23) above, and equating the kinetic energy developed to the loss of potential and internal energy, he found that the velocity which would be acquired by the moving masses would be nearly 15 m/sec.

In a second example Margules took as the initial state two masses lying side by side, each of depth 3000 metres, with a difference of 5° C in temperature, and having equal horizontal extent. In the final state the warmer mass has ascended and lies above the colder, which has now spread laterally so as to cover the whole of the surface area previously covered by the combined masses. The details of the computation are similar to those of the first example, and in the final result it is shown that the amount of potential and internal energy given up is sufficient to produce a mean velocity of 12·2 m/sec, while if the temperature difference is 10° C, the mean velocity is 17·3 m/sec.

Margules' computations were carried through with great accuracy, since the final net gain of kinetic energy is the relatively small difference of two large quantities. We do not here propose to dwell upon the details of the computations. The important fact is that the re-adjustment of such unstable situations as Margules presupposes liberates sufficient energy to account for very considerable velocities.

Margules carried out his computations on the basis that all the potential energy given up during the readjustment was converted into kinetic energy. In practice this could never be strictly true. Some of the energy is converted into energy of eddies, and the values for kinetic energy deduced by Margules are upper limits to the kinetic energy developed.

It has been rasised as an objection to the work of Margules that large differences of temperature such as 10° C are not observed as discontinuous differences, and that differences of temperature in the atmosphere are in practice found to be continuous. But it is not unusual to find differences of the order of 10° C over a very small horizontal range of distance, and this objection need not be treated too seriously. The fundamental difficulty involved in the adoption of Margules' work as an explanation of the genesis of motions in the atmosphere lies rather in explaining how the energy liberated in the way he discusses is able to organise itself into the horizontal circulations which we observe in the atmosphere.

It should perhaps be added that the kinetic energy of the winds in a depression, and the formation of rain in a depression, are both in some measure to be ascribed to the horizontal (dynamical) instability associated with the occurrence of air currents of different density flowing side by side.

§ 176. *Single layer in unstable equilibrium*

The analogous problem of the energy liberated when a layer in which the lapse-rate exceeds the adiabatic can be computed by similar methods, though the solution is not capable of simple analytical expression.

Let fig. 68 (*a*) represent the distribution of temperature with height, the lapse-rate β exceeding the dry adiabatic. When the unstable layer is inverted, each portion of the layer retains its original potential temperature, and the potential temperature will now increase with height. Let the initial surface temperature be T_0.

The thin layer initially at P, at pressure p, with difference of pressure dp, will now be at P' (fig. 68 (b)) with pressure

$$p_0 + p_1 - p = p', \text{ say} \qquad \ldots\ldots(24).$$

Its initial temperature was $T = T_0 - \beta z$. This is related to the pressure by equation (19) of p. 35

$$\frac{T}{T_0} = \frac{T_0 - \beta z}{T_0} = \left(\frac{p}{p_0}\right)^{R\beta/g} = \left(\frac{p}{p_0}\right)^{(\gamma-1)/\gamma \cdot (\beta/\Gamma)} \qquad \ldots\ldots(25).$$

In the new position at P' let its temperature be T'. Then

$$\frac{T'}{T} = \left(\frac{p'}{p}\right)^{(\gamma-1)/\gamma} = \left(\frac{p_0 + p_1 - p}{p}\right)^{(\gamma-1)/\gamma} \qquad \ldots\ldots(26).$$

The total internal and potential energy of the layer in its initial state

$$= \frac{\mathfrak{J}c_p}{R} \int_0^h p\, dz = \frac{\mathfrak{J}c_p p_0}{R} \int_0^h \left(\frac{T_0 - \beta z}{T_0}\right)^{\gamma/(\gamma-1) \cdot (\Gamma/\beta)} dz = \frac{\mathfrak{J}c_p}{g + R\beta}(p_0 T_0 - p_1 T_h)$$
$$\ldots\ldots(27).$$

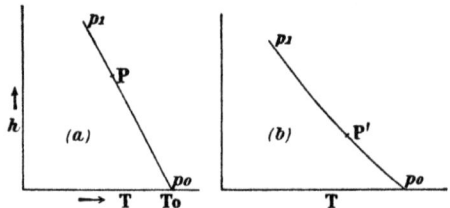

Fig. 68. An unstable layer inverted.

The total internal and potential energy of the layer in its final state

$$= \mathfrak{J}c_p \int \rho' T'\, dz' = -\frac{\mathfrak{J}c_p}{g} \int T'\, dp' = -\frac{\mathfrak{J}c_p}{g} \int T'\, dp \qquad \ldots\ldots(28).$$

From (25) and (26)

$$\frac{T'}{T_0} = \left(\frac{p_0 + p_1 - p}{p}\right)^{(\gamma-1)/\gamma} \left(\frac{p}{p_0}\right)^{(\gamma-1)/\gamma \cdot (\beta/\Gamma)}$$

$$= (p_0 + p_1 - p)^{(\gamma-1)/\gamma}\, p^{(\gamma-1)/\gamma \cdot (\beta-\Gamma)/\Gamma}\, p_0^{-(\gamma-1)/\gamma \cdot (\beta/\Gamma)}.$$

Hence the total internal and potential energy of the layer

$$= \frac{\mathfrak{J}c_p}{g} T_0 p_0^{-(\gamma-1)/\gamma \cdot (\beta/\Gamma)} \int_{p_1}^{p_0} (p_0 + p_1 - p)^{(\gamma-1)/\gamma}\, p^{(\gamma-1)/\gamma \cdot (\beta-\Gamma)/\Gamma}\, dp \ldots\ldots(29).$$

This integral is not capable of simple evaluation in a finite form. The exponent $(\gamma - 1)/\gamma$ is equal to 0·29, and if $(\beta - \Gamma)/\Gamma$ is much less than unity, as in practice it must be when the layer we are considering is deep, then $(\gamma - 1)/\gamma \cdot (\beta - \Gamma)/\Gamma$ is very small. If for example the lapse-rate is 10 per cent greater than the dry adiabatic $(\gamma-1)/\gamma.(\beta-\Gamma)/\Gamma=0\cdot029$, and the factor $p^{(\gamma-1)/\gamma.(\beta-\Gamma)/\Gamma}$ will only vary very slowly with p, and we may without serious error put it equal to

$$\left(\frac{p_0+p_1}{2}\right)^{(\gamma-1)/\gamma.(\beta-\Gamma)/\Gamma},$$

and take it outside the integral sign. The total internal and potential energy of the layer may then be written

$$\frac{Jc_p}{g}\,T_0\left(\frac{p_0+p_1}{2p_0}\right)^{(\gamma-1)/\gamma.(\beta-\Gamma)/\Gamma}\,p_0^{-(\gamma-1)\,\gamma.(\beta/\Gamma)}\int_{p_1}^{p_0}(p_0+p_1-p)^{(\gamma-1)/\gamma}\,dp$$

$$=\frac{Jc_p}{g}\,T_0\,\frac{\gamma}{2\gamma-1}\left(\frac{p_0+p_1}{2p_0}\right)^{(\gamma-1)/\gamma.(\beta-\Gamma)/\Gamma}\,p_0^{-(\gamma-1)\,\gamma.(\beta/\Gamma)}\,(p_0^{(2\gamma-1)/\gamma}-p_1^{(2\gamma-1)/\gamma})$$

$$\dots\dots(30).$$

The total kinetic energy generated is the difference between expressions (27) and (30). The result cannot be stated in a simple form, even with the approximations made above. The mean velocity generated in the particular case where $T_0=300°$, $p_0=1000$ mb, $p_1=700$ mb has been evaluated by Littwin, and the results are given by Koschmieder in his *Dynamische Meteorologie*, p. 337, from which we extract the results that for $\beta=1\cdot05\Gamma$, $\beta=1\cdot1\Gamma$, $\beta=1\cdot25\Gamma$ the mean velocity generated by the re-adjustment of the unstable layer is 8·7 m/sec, 10·8 m/sec, and 16·6 m/sec, respectively. Thus the amount of kinetic energy generated is sufficient to produce velocities of the order of magnitude of those which are observed in the atmosphere. But here again we are faced with the same difficulty of seeing how these velocities can organise themselves into the horizontal circulations which we observe in the atmosphere.

Normand* has discussed the readjustment of a layer in unstable equilibrium having a lapse-rate β times the dry adiabatic, with pressures p_h and p_0 at top and bottom, and mean pressure $\bar{p}$, $\bar{T}$ being the mean temperature at $\bar{p}$. If the layer is completely inverted, all the loss of potential energy reappearing as kinetic energy, the average kinetic energy of each particle of air concerned will correspond to a linear velocity v, where

$$v=10\cdot5\,\frac{p_0-p_h}{p_0+p_h}\,\sqrt{(\beta-1)\,\bar{T}}\,\text{m/sec.}$$

This formula gives good agreement with the results derived by Littwin by the use of the rigorous analysis of Margules.

§ 177. *Effect of the presence of water-vapour*

The examples quoted from Margules refer to dry air. Margules also discussed in detail the effect of condensation upon the liberation of energy, and concluded that the presence of water-vapour had little effect upon the amount of energy liberated. This is undoubtedly true when we consider the liberation of energy by the readjustment of masses of air such as are represented in figs. 67 and 68, when the depth of the disturbed layers is prescribed beforehand. The effect of water-vapour is to modify greatly the depths of the layers which are

* *Q.J. Roy. Met. Soc.* **64**, 1938, p. 71.

effective, and in this manner it exerts a predominant influence on the phenomena produced. Moreover, its presence leads to instability setting in with lapse-rates which would be stable in dry air.

In practice, in dealing with layers of damp air which become saturated during the vertical motions produced when an unstable arrangement breaks down, it is usually most convenient to make the computations of potential energy liberated by using the tephigram or other diagram. Reference should also be made to the thermodynamical discussion of § 47 above.

The resulting amount of energy computed represents the maximum amount of kinetic energy which can be produced as energy of *mean motion*. In practice some of this energy will become turbulent energy, which in turn will be dissipated into heat by the viscous forces, reappearing as thermal and potential energy. It is probable that in some cases at least the potential energy liberated by the readjustment of an unstable layer is entirely used up in producing energy of turbulent motion, which in turn is dissipated into thermal and potential energy. We then have the series of transformations,

thermal + potential energy → turbulent energy → thermal + potential energy.

But the final state will usually be a stable one.

§ 178. *Maintenance of a difference of pressure by addition of heat*

(a) DRY AIR

Margules has considered the maintenance of a circulation in which air moves from high pressure to low and tends to annihilate the difference of pressure and consequently the wind circulation associated with it. In fig. 69, the circulation is around $ABCD$ in the sense $A \to B \to C \to D \to A$. The pressure P_1 at A is higher than the pressure P_2 at B, and the pressure p_2 at C is higher than the pressure p_1 at D.

At the surface the air has initially the temperature T_1 at A. In moving to B its temperature falls adiabatically to T_1', but at B heat is added and its temperature raised to T_2. The air ascends from B to C adiabatically so that the temperature T_3 at C is approximately $T_2 - \Gamma h$, h being the height BC, and Γ the dry adiabatic lapse-rate. The air moves adiabatically from C to D, arriving at D with a temperature T_3'. At D heat is abstracted, so that the temperature falls to T_4, and the air descends along DA, arriving at A with the original temperature T_1. Hence $T_4 = T_1 - \Gamma h$.

$$\text{Heat added at } B \quad = c_p (T_2 - T_1'),$$
$$\text{Heat subtracted at } D = c_p (T_3' - T_4).$$

Since $T_4 = T_1 - \Gamma h$, $T_3 = T_2 - \Gamma h$, the quantity of heat converted into work

$$= c_p \{(T_2 - T_1') - (T_3' - T_4)\}$$
$$= c_p (T_1 - T_1' + T_3 - T_3')$$
$$= c_p T_1 \left\{ 1 - \left(\frac{P_2}{P_1}\right)^{(\gamma-1)/\gamma} \right\} + c_p T_3 \left\{ 1 - \left(\frac{p_1}{p_2}\right)^{(\gamma-1)/\gamma} \right\} \qquad \dots\dots(31).$$

The evaluation of the amount of energy converted into work in any definite cycle in the atmosphere is most readily carried out by means of the tephigram.

Let
$$P_1 = 1020 \text{ mb,} \qquad P_2 = 1000 \text{ mb,}$$
$$p_1 = 700 \text{ mb,} \qquad p_2 = 720 \text{ mb,}$$
$$T_1 = 273° \text{ A,} \qquad T_2 = 288° \text{ A.}$$

Any unit mass which goes around the circuit $ABCD$ goes through the cycle of changes $AB'BDD'A$ in fig. 70, where B' and D' represent the condition of the mass before the addition or subtraction of heat at B and D respectively. The amount of heat converted into work is represented by the area $B'BDD'$,

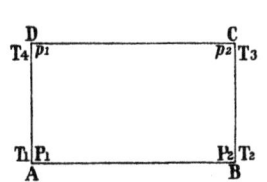

Fig. 69. Margules' diagram for a cycle in a vertical plane.

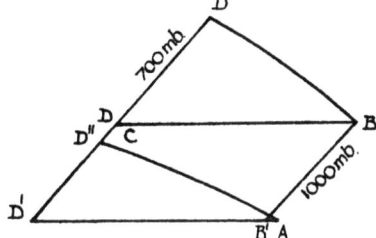

Fig. 70. The same cycle represented in a tephigram.

which computation in an actual tephigram showed to amount to $1·6 \times 10^7$ ergs, or $0·375$ calorie, per gramme of air. This energy, if completely converted into kinetic energy, would suffice to give the air a velocity of 60 m/s.

(b) SATURATED AIR

In saturated air the phenomena are complicated by the latent heat associated with condensation and evaporation. If unit mass of air is to perform a true cycle and to reach A with the same water content as it had initially, the changes of state must be represented by some such cycle as $AB'BD'''D''A$, whose area is clearly only very little different from that of $AB'BDD'A$. It is supposed that water is added to the air descending along $D''A$ to maintain it saturated at all points.

§ 179. *A cycle of changes in the atmosphere*

A possible cycle in the atmosphere has been discussed by Sir Napier Shaw,[*] in which air ascends at the equator and travels at high levels into latitude 60°, where it descends to the sea surface, after which it returns along the surface towards the equator. The cycle can be represented readily in a tephigram. In fig. 71 E represents the state of the air at the equator before it starts its ascent. It ascends at first along the dry adiabatic EA, attaining saturation at A, at a pressure of 900 mb. Its further ascent is along the saturated adiabatic AB, B having a pressure of 100 mb. Its journey northward is accompanied by a

---

[*] *Dict. App. Phys.* **3**, p. 82.

steady loss of heat by radiation, which enables it to descend to lower levels as it travels poleward, retaining its vapour content practically unchanged.

This part of its path is represented by *BC*. At *C* it is in a condition to descend adiabatically to the surface, which it reaches in a condition specified by the point *D*, at a temperature of 275° A. Its further history, during the return

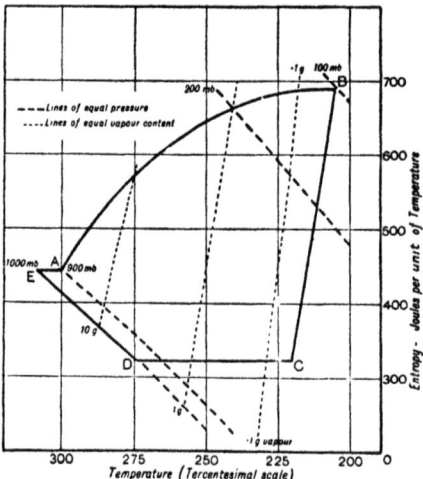

Fig. 71. An atmospheric cycle shown in the tephigram.

journey to the equator, is represented by *DE* in the diagram. In this stage of its path it acquires heat and moisture from the surface of the earth.

The amount of heat converted into work is represented by the closed area *EABCDE*, which in the example shown amounts of $2\cdot4 \times 10^8$ ergs per gramme. If this amount of energy were used in setting the gramme of air in motion, its velocity would be $\sqrt{4\cdot8 \times 10^8}$ cm/sec, or about 220 m/sec. Naturally we do not seriously suggest that all the energy in the cycle is devoted to setting in motion the limited mass of air which performs the cycle. It is actually devoted to setting in motion much larger masses of air which surround the moving mass.

CHAPTER XV

THE GROWTH OF CYCLIC CIRCULATIONS AND OF PRESSURE INEQUALITIES IN THE ATMOSPHERE

§ 180. *The growth of cyclic circulations in the atmosphere*

IT has been shown earlier, in § 109, p. 176, that, apart from the retarding effect of friction, the rate of growth of a cyclic circulation (taken positive in the cyclonic or counter-clockwise sense), can be represented by $-2\omega \dfrac{dF}{dt}$, where ω is the angular velocity of rotation of the earth, and F is the area of the projection on the plane of the equator of the circuit of particles around which the circulation is taken. If the circuit is horizontal, in latitude ϕ, and the area within it at any instant is E, then $F = E \sin \phi$, and the rate of growth of the cyclic circulation is $-2\omega \sin \phi \dfrac{dE}{dt}$, if we neglect the effect of the variation of latitude over the circuit.

If, in a given time, the area within the circuit changes from E_1 to E_2, and the circulation from C_1 to C_2, then

$$C_1 - C_2 = -2\omega \sin \phi \, (E_1 - E_2) \qquad \ldots\ldots(1).$$

If the circuit is a circle, whose radius changes from r_1 to r_2, while the tangential velocity, regarded as uniform around the circuit, changes from v_1 to v_2, equation (1) becomes

$$2\pi (r_1 v_1 - r_2 v_2) = -2\omega \sin \phi \, . \, \pi \, (r_1{}^2 - r_2{}^2)$$

or

$$r_2 v_2 = r_1 v_1 + \omega \sin \phi \, (r_1{}^2 - r_2{}^2) \qquad \ldots\ldots(2).$$

If a cyclonic circulation is to be produced, $\dfrac{dF}{dt}$ must be negative, or the circuit must contract (i.e. $r_2 < r_1$), the fluid converging inward across the instantaneous boundary, and being removed vertically from some area within the circuit. Conversely, the growth of an anticyclonic circulation requires the expansion of the circuit (i.e. $r_2 > r_1$), or the divergence of the fluid from some central area. Equation (2) covers both cyclonic and anticyclonic motion, except that in the anticyclonic case it applies only to the fluid originally in the plane of motion considered.

§ 181. *The formation of revolving fluid in the atmosphere*

The results above derived are regarded as being of sufficient importance to warrant their discussion from first principles. The motion will be treated as two-dimensional, and symmetrical about a centre. The results derived are only applicable outside the region in which the vertical removal of fluid takes place. This region will require some special consideration. The results obtained are those which are sometimes referred to as "revolving fluid".

In the first place we shall consider the type of motion which is produced in a revolving horizontal disc of incompressible fluid, when fluid is drawn off from the centre of the disc. This problem was first discussed by Rayleigh,[*] and the analysis which he gave was extended by Brunt[†] to take account of the rotation of the earth.

The motion is assumed to be symmetrical about the centre of the disc, which has unit vertical thickness. Except at the centre the motion will be assumed to be everywhere horizontal. The disc has initially an external radius R_0 and rotates with angular velocity ζ in a counter-clockwise direction, relative to an axis Ox fixed in the earth (fig. 72 (a)). Fluid is drawn off at O, and the primary result is to produce a convergence of the fluid in the disc inward towards O. At a later stage let the external radius of the disc be R, and let an element of air which was initially at P', distant r' from O, be now situated at P, distant r from O. Let the velocity at P be v transverse to OP relative to

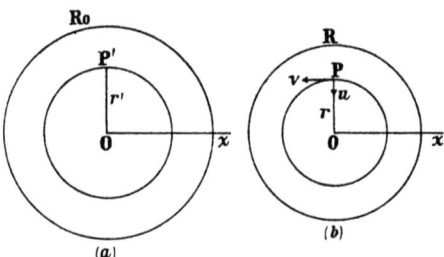

Fig. 72. The development of revolving fluid in the atmosphere.

the earth, and u along PO, measured positively inwards. The ring of particles of radius r' in fig. 72 (a) has become the ring of radius r in fig. 72 (b), and since the annulus originally between r' and R_0 is now between r and R,

$$\pi (R_0{}^2 - r'^2) = \pi (R^2 - r^2)$$

or
$$R_0{}^2 - r'^2 = R^2 - r^2 \qquad \text{......(3)}.$$

The rate of total inward flow across the cylinder of radius r is independent of r since it is equal to the rate of removal of fluid at O. Hence

$$ru = \text{const.} = B \qquad \text{......(4)}.$$

The horizontal plane relative to which the motion is measured itself rotates with the earth, with an angular velocity $\omega \sin \phi$ about the vertical. Hence the transverse motion in space of the fluid at P relative to O is $v + r\omega \sin \phi$. The only force in the horizontal plane acting on the air at P is the pressure gradient, which from considerations of symmetry must act along PO. Hence the angular momentum about O remains constant during the motion and the angular momentum about O of the air at P' in fig. 72 (a) and at P in fig. 72 (b) will be the same; i.e.

$$r'^2 (\zeta + \omega \sin \phi) = rv + r^2 \omega \sin \phi$$

or
$$rv = (r'^2 - r^2) \omega \sin \phi + r'^2 \zeta \qquad \text{......(5)}.$$

* *Proc. Roy. Soc.* A, **93**, 1916, p. 148.   † *Ibid.* **99**, 1921, p. 397.

Substituting for r'^2 from equation (3), we find

$$rv = (R_0^2 - R^2)\, \omega \sin \phi + (R_0^2 - R^2 + r^2)\, \zeta,$$

or
$$v = r\zeta + \frac{(R_0^2 - R^2)\,(\omega \sin \phi + \zeta)}{r} \qquad \ldots\ldots(6).$$

Equation (6) shows that the effect of the removal of fluid from the centre of a disc rotating with uniform angular velocity ζ is to superpose upon the uniform angular velocity a distribution of velocity

$$v = A/r,$$

where A is a constant. This added distribution is equivalent to a simple vortex. Its intensity is proportional to $\pi\,(R_0^2 - R^2)$, or to the amount of fluid which has been removed from a disc of unit thickness.

When the term $\omega \sin \phi$ is omitted from equation (6), the result is identical with that derived by Rayleigh. Equation (6) is also applicable to the case where $\zeta = 0$, corresponding to an initial state of rest. The distribution of velocity at any subsequent time is then given by

$$v = \frac{(R_0^2 - R^2)\, \omega \sin \phi}{r} \qquad \ldots\ldots(7).$$

There need be no definite outer boundary in this case, and $R_0^2 - R^2$ is to be interpreted as $1/\pi$ times the total amount of fluid removed since the beginning of the motion.

The analysis shows that if fluid be removed from the centre of a disc of incompressible fluid originally having uniform angular velocity, or in other words, rotating as a solid, the consequent convergence towards the centre has the effect of superposing a simple vortex

$$vr = \text{const.}$$

upon the solid rotation
$$v = \zeta r.$$

If the fluid outside the disc be initially at rest, it will converge inwards to fill up the space left otherwise vacant by the shrinkage of the disc, and will take up the velocity distribution given by equation (7) in the region outside the radius R of the disc.

The converse case of addition of fluid at the centre of a disc originally revolving as a solid with angular velocity ζ is equally readily derived by similar considerations. The added fluid spreads out horizontally, causing the disc to increase in size, so that the external radius R at the later stage is greater than the original radius R_0. Suppose at any instant t the added fluid forms a disc of radius R', while the outer radius of the original fluid is R. The fluid which originally constituted the disc of radius R_0 is now contained between $r = R'$ and $r = R$. Hence

$$\pi\,(R^2 - R'^2) = \pi R_0^2$$

or
$$R^2 - R_0^2 = R'^2 \qquad \ldots\ldots(8).$$

Equation (6) still holds for the original fluid, i.e. for $R' < r < R$. It may be written in the form

$$v = \zeta r - \frac{(\zeta + \omega \sin \phi)\, R'^2}{r} \qquad \ldots\ldots(9).$$

Thus the effect of the addition of fluid at the centre is to superpose upon the original solid rotation a simple vortex in a clockwise direction. This is true whether ζ is positive or negative, whether the original motion is counter-clockwise or clockwise.

The motion of the added fluid requires separate consideration. If the fluid enters at O with no horizontal velocity, it must spread out horizontally so as to have no rotation in space about the vertical axis through O. It therefore rotates clockwise relative to the earth with an angular velocity $\omega \sin \phi$. The complete motion thus requires three specifications:

$$\left. \begin{aligned} \text{For} \qquad r < R' \qquad & v = -r\omega \sin \phi \\ R' < r < R \qquad & v = \zeta r - \frac{(\zeta + \omega \sin \phi)\, R'^2}{r} \\ \text{and for} \qquad r > R \qquad & v = -\frac{(\omega \sin \phi)\, R'^2}{r} \end{aligned} \right\} \quad \dots\dots(10),$$

it being assumed that the fluid outside the disc was originally at rest. Here v is positive for counter-clockwise rotation about the centre.

If the initial motion is zero, so that fluid is added at a point to a medium at rest, the equations reduce to two:

$$\left. \begin{aligned} \text{For} \qquad r \leqslant R' \qquad & v = -r\omega \sin \phi \\ r \geqslant R' \qquad & v = -\frac{\omega \sin \phi \,.\, R'^2}{r} \end{aligned} \right\} \quad \dots\dots(11).$$

Thus the effect of the addition of fluid is to give the fluid originally present the motion of a simple vortex, revolving clockwise, the added fluid forming a central disc having clockwise rotation as a solid.

It is not here suggested that the results derived above afford complete explanations of the depression or anticyclone, but the results will have to be borne in mind in discussing these pressure distributions, in whose formation convergence and divergence have to be considered.

The results given above could also have been derived from equation (64) of Chapter VIII. Let ψ be the stream-line function, then if ζ is the vorticity, since the motion is symmetrical,

$$\frac{\partial^2 \psi}{\partial \theta^2} = 0 \quad \text{and} \quad \frac{1}{r}\frac{\partial}{\partial r}\left(r\frac{\partial \psi}{\partial r} \right) = 2\zeta \qquad \dots\dots(12),$$

which on integration yields

$$r\frac{\partial \psi}{\partial r} = 2\int \zeta r\, dr = \zeta r^2 + A, \text{ if } \zeta \text{ is constant,}$$

$$\frac{\partial \psi}{\partial r} = \zeta r + \frac{A}{r} \qquad \dots\dots(13).$$

The effect of the convergence or divergence, therefore, is to impose on the original rotation as a solid a vr-vortex of intensity A. By the adjustment of the appropriate value of the constant A in the last equation, we can as before derive all the results previously obtained. This is left to the reader as an exercise in analysis. Attention is drawn to the method here, in order to emphasise the utility of the equations of § 110.

The same results can also be derived by the use of equation (2) above, by substituting $v_1 = r_1 \xi$. The derivation given in the present section perhaps gives a better idea of the physical factors involved in the processes of convergence and divergence, showing that convergence is associated with the growth of cyclonic circulations, and divergence with the growth of anticyclonic circulations.

§ 182. *The tropical cyclone and the tornado as revolving fluid*

Let us consider how we may apply the considerations of revolving fluid developed above to explain the tropical cyclone or the tornado.

The most obvious method of visualising the removal of air is by means of a convection current. If convection on a large scale acts for a sufficiently long period over a restricted area, the effect on the air in the surrounding region is to superpose a simple vortex ($vr = $ const.) upon the previously existing motion. In the special case where the air is originally at rest, the final motion consists only of the simple vortex. We need not regard the external boundary R_0 as having any objective existence, and we may interpret $(R_0{}^2 - R^2)$ in equation (6) above as $1/\pi$ times the amount of air removed from a layer of unit thickness.

We may think of the convection current as produced by local inequalities of temperature or of water-vapour content, or by the effect of a surface of discontinuity in causing the warmer current to flow up over the colder. If a large mass of air is set in upward motion, the turbulence at the boundary of the rising air will cause a partial mixing of the rising air with the environment, some of the excess of temperature or moisture of the rising air being shared with the environment. The mixture formed will therefore also be lighter than the normal environment, and will also tend to rise. We may therefore regard the ascending current as having a scouring effect upon the environment; the process is known as "eviction of air".

The effect of the condensation of water-vapour is probably an important factor in maintaining convection on a large scale. In any layer in which the lapse-rate is intermediate between the dry adiabatic and the saturated adiabatic any air which attains saturation in the course of its upward motion will thereafter become increasingly warmer than its environment, except in so far as mixing at the boundary restricts the difference. If therefore we can visualise warm damp air at the surface being set in upward motion by its buoyancy, there is no difficulty in understanding the continuance of the upward motion as an effect of the condensation of water-vapour.

The main difficulty which confronts this theory at this stage is that of accounting for a diminution of pressure at the centre. At first sight it appears that the convergence of air towards the centre, combined with ascent in the vertical of air at the centre itself, should cause an increase of pressure at the centre, yielding a pressure gradient opposing the converging motion. The motion cannot continue for any length of time unless there is present some mechanism for removing the air which has ascended. The simplest mechanism which we can postulate is an upper current whose direction differs from that of the current in the lower troposphere. It is possible that the outward motion of cirrus from the centre of cyclonic systems is to be taken as evidence of the existence of such a current. In the absence of some means of removal of the evicted air the development of the system we have described is impossible and a thunderstorm is a more likely occurrence than a cyclonic system.

Observation has shown that convection is not by itself likely to be sufficient to produce a cyclonic system. Tropical cyclones originate in regions where the air is damp and the surface temperature is high, but if these conditions alone were sufficient to produce cyclonic systems, it is difficult to see why they are not far more plentiful than they actually are. Unfortunately little is known of the general wind structure associated with the formation of tropical cyclones, and it is left to future observation to decide the question.

Two main difficulties still confront the theory, and call for further explanation: (a) Why is the cyclone not torn to pieces by the variation of wind with height? (b) What protects the cyclone from filling in at the top? We shall consider these points briefly in turn:

(a) The developing cyclone does not, in its initial stages, extend to great heights, and so is not affected by the marked shear of wind which is frequently observed between 4 and 9 km.

(b) The top of the cyclone is protected by the gradual diminution of the horizontal pressure gradients and of the wind velocity, so that no tendency to fill in at the top should occur.

The central portion of the revolving storm should initially be a region of cloud and heavy rain. At a later stage, when the convection has ceased, the inner portion is protected from inflow of the surrounding air by the ring of high velocity outside the core. When this stage has been reached, the cyclone has placed itself on a dynamical rather than a thermal footing, and it continues in existence until its energy is dissipated by friction and turbulence. In this later stage, the central portion of the storm is not of necessity rainy or even cloudy.

If a cyclone could be brought into existence in the manner we have outlined above, it should, in its later stages, possess great stability, and should be capable of acting as a "centre of attraction" for currents of air originally outside its sphere of action. Such currents would be accelerated when approaching the cyclone, and slowed down when receding from it, eventually passing away with approximately the same speed with which they approached the system. Shaw and Lempfert, in their examination of surface air trajectories,* found many examples of such motion.

There is a growing conviction in the minds of those who are most intimately concerned with the forecasting of the formation of tropical storms, that these cyclones originate at the rather diffuse boundary between two air masses of widely different origin, one being of oceanic, and the other generally of continental origin.† This brings the tropical cyclone closely into line with the depression of middle latitudes, discussed below in Chapter XVII.

* "Life-history of Surface Air Currents", M.O. 174.
† See Normand, *Gerlands Beitr.* **34**, 1931, p. 233.

Before leaving this part of the subject, we would emphasise one point: that a diminution of pressure and a counter-clockwise wind circulation can only be brought about by vertical motion and convergence of the air to take the place of that removed by convection. It is not possible to form such a system by horizontal divergence from a point. For divergence from a centre must produce a clockwise circulation, as the diverging air is deviated round by the effect of the earth's rotation, and a balance between the wind and pressure distribution would then be impossible.

There is a feature of the revolving fluid theory to which no reference has yet been made, and which is usually left out of consideration. The velocity distribution which is produced by the effect of convergence is

$$vr = \text{const.}$$

This requires an infinite velocity at $r = 0$. In practice this is impossible. The air is removed not at a point, but over a finite area, and the revolving field is set up outside this area of convection.

In the formation of the tornado, the instability of the air near the ground is an essential feature of the initial conditions, and it is probable that it is the surface layer which is drawn in to form the core of the vortex. The surface air has its motion checked by friction with the ground, and it thus reaches the inner region with relatively small transverse velocity. Only by this supposition does it appear possible to account for the non-existence of the infinite velocities at the centre demanded by the formal theory.

It is to be noted that if a depression could be formed in still air, then apart from the drift of air across the isobars in the lowest layers, due to friction with the ground, the depression would always contain the same air, since the motion would be circular about the centre of lowest pressure.

Durst and Sutcliffe* have discussed the genesis of the tropical cyclone along lines somewhat similar to the above, using the equations of motion in three dimensions. They attribute the fall of pressure at the centre of the cyclone to the effect of the fast rotating converging air being forced to ascend into regions where its tangential velocity is too great for the gradient of pressure, so that it diverges outward. The central core, or eye of the storm, is regarded as a region of descending air, thus accounting for the clear skies observed in this region. The theory is in many ways attractive, and is novel in that it takes into account the vertical component of the velocity.

The difficulty which is usually expressed by most writers on this subject is that of accounting for the calm central area. It appears, however, that if a system of winds such as those represented by equations (4) and (9) above were to come into existence, the viscous dissipation of energy (see equation (69) § 111, p. 185) would contain terms in $1/r^4$, and these terms would represent a rate of dissipation capable of annihilating the velocity within a central region. We might then expect the tangential velocity to increase inward to a maximum, and rapidly fall off to very light values within an inner region.

Grimes† has associated the formation of tropical cyclones near the equator

* *Q.J. Roy. Met. Soc.* **64**, 1938, p. 75.
† *Mem. Malayan Met. Service*, No. 2 1937.

in the southern hemisphere to the transport of air with anticyclonic spin from the northern hemisphere. Such a transport represents a growth of $\dfrac{dF}{dt}$ of § 109 above, in the right sense.

§ 183.　*The relation of wind and isobars in a moving cyclone*

It is customary to regard a moving cyclone as capable of being analysed into a system of isobars which would represent a cyclone in a stationary state, combined with a uniform pressure gradient of the right magnitude to produce a geostrophic wind equal to the rate of translation of the cyclone. It will be found that this method of analysis of a cyclone has certain defects, but it is treated briefly below in order to demonstrate the difficulties of explaining the mechanism of the moving cyclone.

Following the lines of a discussion by C. K. M. Douglas,* we shall consider a moving circular depression, regarded as made up of a disc of fluid kept in motion by a suitable gradient of pressure at right angles to the direction of motion. It is assumed that if the system were not travelling the velocity would everywhere be equal to the gradient wind, and that when the system is given a velocity of translation U, the motion of any element is compounded of these two velocities.

Let O (fig. 73) be the centre of the revolving fluid, and let the system be

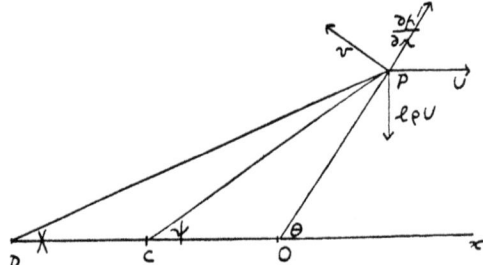

Fig. 73. Revolving fluid in any cyclone.

travelling with velocity U parallel to Ox. Let the field of pressure be regarded as made up of p', the pressure in the system when not moving ($p' = f(r)$), and $-2\omega \sin \phi . \rho U y$ which allows for travel. The pressure gradient at any point P is made up of $\partial p'/\partial r$ along OP and $-2\omega \sin \phi . \rho U$ parallel to the axis of y. Let v be the velocity due to the rotation about O alone. Then if PD and PC be respectively perpendicular to the wind direction and the isobars of the combined field, the following relations hold

$$\tan \chi = \frac{v \sin \theta - U}{v \cos \theta} = \tan \theta - \frac{2\omega \sin \phi . \rho U}{2\omega \sin \phi . \rho v \cos \theta} \quad \ldots\ldots(14),$$

$$\tan \psi = \tan \theta - \frac{2\omega \sin \phi . \rho U}{\dfrac{\partial p'}{\partial r} \cos \theta} \quad \ldots\ldots(15),$$

* *Q.J. Roy. Met. Soc.* **55**, pp. 123–51, 1929, Appendix, p. 146.

and
$$\frac{\partial p'}{\partial r} = 2\omega \sin \phi \cdot \rho v + \frac{\rho v^2}{r} \qquad \ldots\ldots(16).$$

whence
$$\frac{\partial p'}{\partial r} > 2\omega \sin \phi \cdot \rho v \qquad \ldots\ldots(17).$$

Equation (15) expresses the condition that the isobar is perpendicular to PC. In the last of these equations v is always positive. From equations (14), (15) and (17), if $\cos \theta > 0$, $\tan \psi > \tan \chi$ and $\psi > \chi$, while if $\cos \theta < 0$, $\psi < \chi$. Thus the air flows in across the isobars in the rear half of the depression and out across the isobars in the front half of the depression. But this forward flow across the instantaneous position of the isobars is less than the forward velocity of the isobars themselves, and so that the actual flow of air is into the depression in the front half, and outward in the rear half.

For let the total velocity at P be made up of V along the isobar at P of the total field (perpendicular to CP) and V' forward. Then

$$V \sin \psi - V' = v \sin \theta - U \qquad \ldots\ldots(18),$$

$$V \cos \psi = v \cos \theta \qquad \ldots\ldots(19).$$

By division of (18) by (19)

$$\tan \psi - \frac{V'}{V \cos \psi} = \tan \theta - \frac{U}{v \cos \theta} = \tan \psi - \frac{V'}{v \cos \theta} \qquad \ldots\ldots(20),$$

$$= \tan \theta - \frac{2\omega \sin \phi \cdot \rho U}{\frac{\partial p'}{\partial r} \cos \theta} - \frac{V'}{v \cos \theta}, \text{ from (15)}.$$

Hence
$$\frac{U - V'}{v \cos \theta} = \frac{2\omega \sin \phi \cdot \rho U}{\frac{\partial p'}{\partial r} \cos \theta} \qquad \ldots\ldots(21),$$

and
$$V' = \frac{U}{\frac{1}{\rho}\frac{\partial p'}{\partial r}} \left\{ \frac{1}{\rho}\frac{\partial p'}{\partial r} - 2\omega \sin \phi \cdot v \right\} = U \frac{v^2}{r} \Big/ \frac{1}{\rho}\frac{\partial p'}{\partial r} \qquad \ldots\ldots(22).$$

Equations (21) and (22) show that while the winds at any point in a cyclone are forward across the instantaneous position of the isobars, the motion relative to the moving isobars is backward across the isobars. This does not of necessity involve any real change in the identity of the air within the depression. It may be merely a consequence of the use of the isobars to define the depression, instead of the actual flow of the air.

This result can be generalised for any system of isobars, circular or otherwise, moving unchanged with a velocity of translation U. We now define O in fig. 73 as the centre of curvature of the isobar through P due to p' alone, and r as the radius of curvature. Equation (22) is then derived as before. This appears to indicate that the wind has a component across the isobars in their instantaneous position, parallel to the direction of motion, and of magnitude $U \frac{v^2}{r} \Big/ \frac{1}{\rho}\frac{\partial p'}{\partial r}$. This component is zero for straight isobars, so that in this case the actual wind blows along the instantaneous position of the isobars.

The above discussion is based on the assumption that the winds in the cyclone when considered stationary are strictly gradient winds. The result shows that the combination of these winds with the velocity of translation U yields velocities which are not gradient winds as computed from the instantaneous position of the isobars, since it was found that these winds will not blow round the isobars. This was shown by the fact that PD and PC, the normals to the wind direction and the isobar respectively, do not coincide, except in the special case of straight isobars. Nor do the winds blow in such a way as to ensure that any selected element of air shall always remain on the same isobar. Thus the addition of a pressure field of the right magnitude to give as a geostrophic wind the required velocity of translation U of the cyclone leads to a system of winds which no longer blow around the isobars. It is not clear that the pressure distribution will have any velocity of translation, or that if it has such a velocity, that it will be equal to U, in view of the result expressed in equation (22) above. There is an obvious weakness in the above discussion, in that it assumes a constant density within the cyclone, both when stationary and when moving. It is obviously impossible to impose a linear pressure field, and yet to retain the same constant density unchanged. So far no satisfactory explanation of the transmission of pressure changes in the atmosphere has been given.

It may be added that most of the results which Shaw* derived for the "normal" or "cartwheel" cyclone can be derived directly from the equations given above, by the substitution of $r\zeta$ for v.

§ 184. *The genesis of pressure inequalities*

The statical pressure p at any level z is given by

$$p = \int_z^\infty g\rho \, dz = g \int_z^\infty \rho \, dz \qquad \ldots\ldots(23)$$

neglecting variations of gravity. Differentiating with respect to time,

$$\frac{\partial p}{\partial t} = g \int_z^\infty \frac{\partial \rho}{\partial t} \, dz$$

$$\frac{\partial p}{\partial t} = -g \int_z^\infty \left\{ \frac{\partial}{\partial x}(\rho u) + \frac{\partial}{\partial y}(\rho v) + \frac{\partial}{\partial z}(\rho w) \right\} dz \qquad \ldots\ldots(24)$$

from the equation of continuity, equation (29), p. 168. The last term in equation (24) can be integrated, so that

$$\frac{\partial p}{\partial t} = -g \int_z^\infty \left\{ \frac{\partial}{\partial x}(\rho u) + \frac{\partial}{\partial y}(\rho v) \right\} dz - g(\rho w)_\infty g(\rho w)_0.$$

At $z = \infty$, $\rho = 0$, and hence

$$\frac{\partial p}{\partial t} = -g \int_z^\infty \left\{ \frac{\partial}{\partial x}(\rho u) + \frac{\partial}{\partial y}(\rho v) \right\} dz + g\rho w \qquad \ldots\ldots(25).$$

* *Manual of Meteorology*, 4, Chapter IX.

The last term on the right-hand side of equation (25) gives the effect on pressure changes of the vertical expansion or contraction of the air. At $z=0$, $w=0$, and the equation for the changes of surface pressure becomes

$$\frac{\partial p_s}{\partial t} = -g \int_0^\infty \left\{ \frac{\partial}{\partial x}(\rho u) + \frac{\partial}{\partial y}(\rho v) \right\} dz \qquad \ldots\ldots(26),$$

p_s denoting the surface pressure.

If the motion is strictly geostrophic, the wind components u, v, are given by

$$\left. \begin{aligned} 2\omega \sin\phi\,\rho u &= -\frac{\partial p}{\partial y} \\ 2\omega \sin\phi\,\rho v &= \frac{\partial p}{\partial x} \end{aligned} \right\} \qquad \ldots\ldots(27).$$

Differentiating these two equations with respect to x and y respectively, and adding, we find that the right-hand side of equation (26) vanishes, so that $\frac{\partial p_s}{\partial t} = 0$, and no change of pressure is then possible anywhere. The fact that in

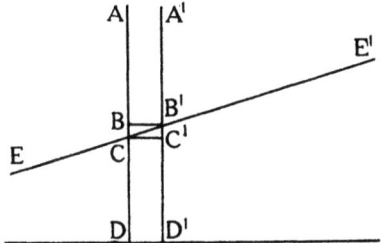

Fig. 74. Changes of pressure at a moving sharp surface of discontinuity.

a system of purely geostrophic winds no changes of pressure are possible was first clearly enunciated by H. Jeffreys.[*]

The analysis given above shows that changes of pressure must be associated with deviations from geostrophic winds. An expression for the rate of change of pressure in the general case might be derived from equation (26) above, by using equations (35) and (36) of Chapter VIII, with the forces X, Y, equated to zero, instead of using the approximate equations (27) above. The expression so derived is complicated, and cannot be interpreted in any simple manner. It is better to proceed from the original equation for the rate of change of pressure given in equation (26) above.

§ 185. *The effect on pressure changes of the presence of a surface of discontinuity, or of a zone of transition*

The possible limitation imposed upon the step at which equation (27) was differentiated now requires consideration. In fig. 74 let EE' be a surface of discontinuity. In accordance with the notation of Chapter X, we take the axis of x parallel to the intersection of the surface of discontinuity with the horizontal plane, the axis of y also horizontal, and the axis of z vertical.

* Phil. Mag. **36**, 1919, p. 1.

Let u_1, v_1, u_2, v_2, be the components of velocity within the two air masses, ρ_1 and ρ_2 their densities, suffix 1 applying to the upper mass, and suffix 2 to the lower; further let $ABCD$, $A'B'C'D'$ be two adjacent verticals, the plane through which is parallel to the yz plane. BB', CC' are horizontal, and $BC = B'C' = dz$; $BB' = CC' = dy$; $\angle B'CC' = \alpha$.

Everywhere below CC' and above BB', the motion is horizontal and geostrophic, and the relation

$$\frac{\partial}{\partial x}(\rho u) + \frac{\partial}{\partial y}(\rho v) = 0$$

is satisfied. Consider the flow of air into the parallelepiped of which the vertical section is $BB'C'C$, and of which the length in the direction parallel to the axis of x is unity. The inflow per unit time across BC is $\rho_1 v_1 dz$, and the outflow per unit time across BC is $\rho_2 v_2 dz$. The net gain of mass within the parallelepiped is

$$(\rho_1 v_1 - \rho_2 v_2)\, dz = (\rho_1 v_1 - \rho_2 v_2)\, dx \tan \alpha.$$

Hence the rate of increase of pressure over DD' is

$$g\,(\rho_1 v_1 - \rho_2 v_2) \tan \alpha = \frac{g}{2\omega \sin \phi}\left\{\left(\frac{\partial p}{\partial x}\right)_1 - \left(\frac{\partial p}{\partial x}\right)_2\right\} = 0 \qquad \ldots\ldots(28).$$

Thus when the winds are truly geostrophic, the presence of a sharp surface of discontinuity does not affect the conclusion drawn in § 184.

Ertel,* in a number of papers on singular advection, has claimed that the result derived above for sharp surfaces of discontinuity is no longer true when this is replaced by a zone of transition. He regards the zone of transition as the main cause of the pressure changes which accompany the change in air-mass over a station, while accepting the geostrophic wind as a sufficient approximation to the true wind everywhere else within the body of each airmass. But there is definite evidence that the geostrophic wind is not a close approximation to the true wind, and it appears preferable to accept the general deviation of the winds from the geostrophic value as a likelier cause of the pressure changes. The marked changes of curvature of the isobars at the fronts shown in figs. 83, 86 and 87 below must be accompanied by marked deviations from the geostrophic wind.

§ 186. *Gradient winds in a stationary pressure distribution*

The most obvious second approximation to actual conditions to consider after the geostrophic wind is the gradient wind, in a stationary system of isobars. Cyclonic curvature of the isobars will first be assumed.

The gradient wind equation can be written

$$V - G = -\frac{1}{2\omega \sin \phi}\frac{V^2}{r}$$

where V is the gradient wind, G the geostrophic wind, and r the radius of curvature of the isobars.

* *Met. Zeit.* **53**, 1936, p. 280; *Sitz. Preuss. Akad. Wiss.* 1936, p. 257. See also van Mieghem *Mem. Inst. Roy. Met. de Belgique*, **8**, 1938.

In fig. 75 let AB, CD be two adjacent isobars, whose radii of curvature are r_1 and r_2 at AC and BD respectively. The isobars are taken sufficiently near to each other to justify our assuming the curvature of the two to be effectively equal. Further let G_1, G_2 be the geostrophic winds at AC and BD respectively, V_1, V_2, the gradient winds, and ρ_1, ρ_2 the corresponding densities. Let $AC = d_1$, $BD = d_2$, both drawn normal to the isobars. Then the rate of flow of air into the area represented in fig. 75 is

$$\rho_1 V_1 d_1 \text{ inward across } AC,$$
$$\rho_2 V_2 d_2 \text{ outward across } BD.$$

The net rate of gain of mass within the area is

$$\rho_1 V_1 d_1 - \rho_2 V_2 d_2 = \rho_1 G_1 d_1 - \rho_2 G_2 d_2 - \frac{1}{2\omega \sin \phi} \left(\frac{\rho_1 d_1 V_1^2}{r_1} - \frac{\rho_2 d_2 V_2^2}{r_2} \right)$$

$$= \frac{-1}{2\omega \sin \phi} \left(\frac{\rho_1 d_1 V_1^2}{r_1} - \frac{\rho_2 d_2 V_2^2}{r_2} \right) \qquad \ldots\ldots(29).$$

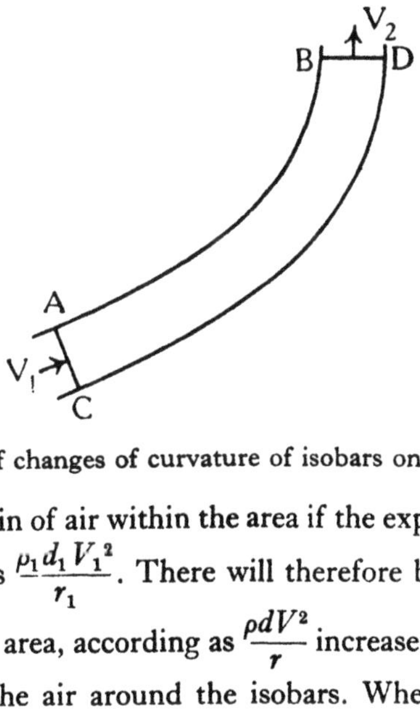

Fig. 75. Effect of changes of curvature of isobars on pressure changes.

There will be a net gain of air within the area if the expression (29) is positive, i.e., if $\frac{\rho_2 d_2 V_2^2}{r_2}$ exceeds $\frac{\rho_1 d_1 V_1^2}{r_1}$. There will therefore be convergence into, or divergence out of, the area, according as $\frac{\rho d V^2}{r}$ increases or decreases along the direction of flow of the air around the isobars. When the isobars are anti-cyclonic, the sign of r is changed in the expression above, and there will be convergence or divergence according as $\frac{\rho d V^2}{r}$ decreases or increases along the direction of flow of the air around the isobars.

The variable which is liable to vary most rapidly is the radius of curvature, as is shown in fig. 76, taken from a paper by Boyden.* It will be seen that the regions of convergence and divergence are arranged in accordance with the changes of curvature along the path of the air. Where the isobars change

* Q.J. Roy. Met. Soc. **64**, 1938, p. 85.

from anticyclonic curvature to nearly straight isobars, as at the lower left-hand side of the first section of fig. 76, or the upper right-hand side of the second section of fig. 76, there will be convergence. Where cyclonic isobars straighten out, or anticyclonic isobars become more curved, there will be divergence, in accordance with the rules derived above.

Several interesting conclusions can be drawn from the principle deduced above. In a system of gradient winds flowing around concentric isobars, no convergence or divergence can occur, so that no development or translation of the pressure system is possible. A depression in which the isobars are at any stage symmetrical about two rectangular axes will have the same pressure changes occurring at the two ends of any diameter, and the original centre

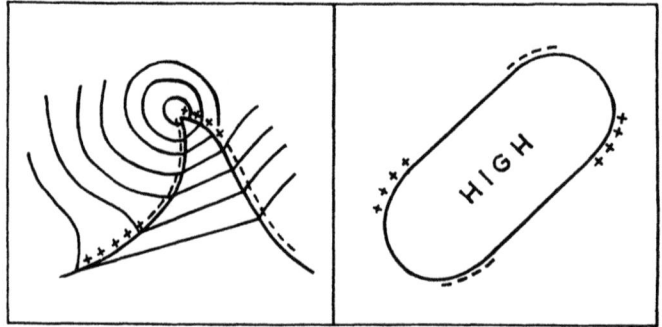

Fig. 76. Regions of convergence (+ + +) and divergence (– – –) resulting from isobaric curvature in a cyclone and an anticyclone.

will always remain a centre such that the points equidistant from it, along any diameter, will undergo the same pressure changes. Thus, though the isobars will suffer distortion, subject to the condition stated, there can be no translation of the system as a whole. This result should be compared with that derived by H. Jeffreys,* who found that a depression regarded as analogous to the disturbances produced by the variations in the height of an incompressible fluid could not have any velocity of translation as a whole if it had symmetry about two rectangular axes.

* *Phil. Mag.* **37**, 1919, p. 1.

CHAPTER XVI

THE IDEA OF AIR MASSES

§ 187. *The life-history of surface air currents*

THE earlier study of depressions of middle latitudes concentrated attention on the forms of the isobars, and related the weather to the isobars. The first clear effort to consider the physical processes in a depression from the point of view of the differences in the air masses which form the depression was that described by Shaw and Lempfert in the "Life-history of Surface Air Currents".* In this paper, whose value can scarcely be over-estimated, the authors traced the motions of air currents by means of hourly or two-hourly charts, and trajectories of surface air are reproduced in the memoir for a number of depressions, both fast- and slow-moving. The reader is referred to the original memoir for the fuller details of the results obtained, and a brief résumé only is given here of the main conclusions:

1. In travelling storms, in the front portion the motion of air is from higher to lower pressure, and is associated with falling temperature and the gradual formation of cloud; while in the rear portion, sometimes from quite near the centre, there is motion from low pressure to higher pressure, with rising temperature and improving weather.

2. A fast-travelling storm draws in air into the central region from the front right-hand side of the path, and throws out an equal amount on the same side in the rear. The trajectories which represent this exchange form loops enclosing the centre of the depression. In addition one of the two fast-travelling storms studied in detail showed a well-marked broad westerly current south of the centre, moving with a speed slightly greater than that of the centre of the depression. The heavy rainfall was limited to a narrow band to the left-hand side of the path of the centre.

3. A slow-moving depression takes in air from both sides of its path. In the example studied in detail, that of November 11–13, 1901, there were two distinct classes of trajectories. The front right-hand quadrant of the depression was filled by a warm southerly current which flowed practically in a straight line towards the centre, where it was apparently lost by ascent over the colder air of the front left-hand quadrant. This current yielded a series of practically straight trajectories. The other type of trajectory originated to the left-hand side of the path of the centre, and swept round the rear of the depression, finally taking a generally west-east direction, though some of these trajectories in their later stages approached the centre from a south-westerly direction. The current represented by these trajectories was distinctly cooler than the southerly current in the front right-hand quadrant, and so it formed a barrier over which the latter could rise on reaching the line of the path of the centre. In the depression of November 11–13, 1901, the rainfall was far more

* M.O. 174.

intense than in the fast-moving storms, though the distribution was very similar, the intensest rain occurring in a narrow band to the left of the path of the centre.

4. The distribution of rainfall in all the cases considered could be explained as the effect of (a) convergence within a warm damp current, (b) the ascent of warm damp air over a colder current, (c) the displacement of warm damp air by a colder current. These three factors were found to be of varying importance in the different synoptic situations, studied, but in the main they afforded a satisfactory explanation of the distribution of rainfall with the depressions. In particular, the heaviest rainfall, which occurred in a relatively narrow band to the left-hand side of the path of the centre, is readily explained by the ascent of the warm southerly or south-westerly current of the front right-hand quadrant over the colder easterly current of the front left-hand quadrant. Shaw summarised these results by the diagram of fig. 77. Of this diagram Shaw wrote "The north-west quadrant is not filled up. It is the region where the air sometimes bends round from the east to the north-west and west, but

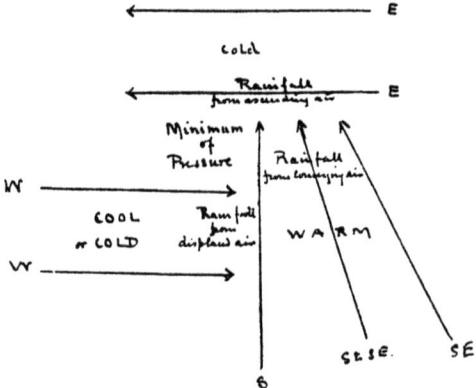

Fig. 77. Shaw's representation of the air currents in a cyclone.

the air supply for the westerly current is not always derived in that way." The representation in fig. 77 is a simplified one which must not be taken as fitting in detail all cases which can occur, but it is perhaps no more exaggerated than any other single diagram which has been put forward to represent the complexities of the depression of middle latitudes.

5. The difference between a fast- and a slow-travelling storm cannot be accounted for by regarding the storm as a vortex travelling in a main current. In fact, the study of the trajectories drawn for different depressions leave the reader with a very pronounced impression that the travel of a depression is not the travel of a definite mass of air, but the travel of a state of disturbance; it gives, however, no hint of the precise nature or cause of the disturbance.

6. The transition from one persistent wind direction to another is relatively sudden.

7. Over the eastern North Atlantic currents from the South soon disappear in the centre of depressions. Only in fast-travelling storms does the air curl round the centre of depressions, forming a loop, but it is possible that the

portion of such a trajectory from the central region outward is followed by a different air supply.

(This appears to be rather a dangerous generalisation from an isolated case. One would more naturally expect looped trajectories in a slow-moving rather than in a fast-moving depression. It is found in practice that when trajectories are drawn by the use of gradient winds, loops are frequent in "occluded" depressions, the air in some cases going round the centre two or three times.)

8. Air currents not from the South are of longer duration, and may persist and reach the trade winds, or turn round the rear of a depression and approach the centre from southward.

9. The flow of air along the southern side of the great Atlantic area of low pressure, which is associated with a series of approximately parallel isobars, may consist of a combination or alternation of currents of different direction, force, and temperature, with marked meteorological changes attending the sudden transition from one current to another.

10. When air moves over the sea the temperature of the air is governed by, and rapidly approaches, the temperature of the sea. The passage of warm air over cold sea generally leads to the formation of mist or fog.

11. The centres of well-marked anticyclonic areas could not be identified as regions of origin of surface air currents. These appeared rather to originate in the shoulders or protuberances of anticyclones, in ridges of high pressure, or along the trough lines of V-shaped depressions and parts of the central areas of travelling storms.

12. Well-defined anticyclones which persist for many days are for the most part inert and comparatively isolated masses of air, taking little part in the circulation which goes on around them.

The twelve paragraphs above give only a bare outline of the total results derived by Shaw and Lempfert. They suffice, however, to indicate the development of the conception of air masses as the features to be studied in relation to the phenomena of weather. These ideas have been much more fully developed by the Norwegian school of meteorologists in connexion with the "polar front" methods of analysis of synoptic charts, and the discussion of the depression of middle latitudes in greater detail will be taken up in the next chapter in relation to the Norwegian methods.

§ 188. *The classification of air masses*

An examination of synoptic charts shows the existence in the atmosphere of extensive air masses possessing a homogeneous or quasi-homogeneous character. Within such air masses there will not usually be found large local differences of temperature, wet-bulb temperature, or visibility. The boundary separating two air masses of different history is in general recognisable by the abrupt local change in motion and in other properties of the air.

The earlier writers on the polar front methods (see next chapter) distinguished two main classes of air only, polar and tropical, whose properties were sharply defined.

Polar air is defined as air which originates in high latitudes, and travels southward over a surface of sea or land whose temperature increases southward. The heating of the lower layers of the air by contact with the earth's surface produces a tendency to instability in the lower layers, with consequent turbulent mixing. The main features of polar air are therefore low temperature, low absolute humidity, good visibility, and instability in the lower layers.

Tropical air is defined as air which originates in low latitudes and travels poleward over a surface of land or sea whose temperature decreases with increasing latitudes. The cooling of the lower layers by the surface produces stable stratification in the surface layers, and the main features of tropical air are therefore relatively high temperature and humidity, and stable stratification in the lower layers. It is frequently stated that in tropical air the visibility is low, and it is clear that in tropical air which has passed over desert regions there might be present a considerable quantity of dust which in the absence of turbulence would be retained in the lower layers, so giving poor surface visibility. Pick* has shown that the visibility criterion cannot be used with any degree of safety, "exceptional" visibility being sometimes found in tropical air. Pick's data suggest that the recent history of the air mass is more important than its remote history, in determining the visibility within it.

The features described in the last two paragraphs will, in theory, draw a clear distinction between polar and tropical air. The phenomena of condensation also differ in the two classes on account of the differences in relation to stability. In polar air the clouds tend to take the form of cumulus or cumulonimbus, and any rain which occurs is in showers with squally winds. In tropical air condensation tends to take the form of stratus clouds at different heights, and sea fogs are also frequent.

The nomenclature has a special sense which is frequently overlooked. The names "polar air" and "tropical air" are not applied to air masses while in the polar and tropical regions respectively, but to air masses which originate in these regions and move to middle latitudes. It is therefore incorrect to say that tropical air which has spent a long time in polar regions becomes polar air.

Later writers† on polar front analysis have distinguished four classes of air— equatorial, sub-tropical, sub-polar, and arctic. Bergeron further divides each of these into two sub-classes according as they are of maritime or continental origin. Other writers adopt somewhat different criteria, and it is perhaps fair to say that at present there is a striking absence of uniformity of practice. This is in part due to the difference of emphasis laid upon geographical origin and subsequent history.

The essential fact to be remembered is that air masses tend to retain at high levels their original characteristics, for long periods. The origin of an air mass can therefore usually be detected by means of upper air observations more readily than by any other means. Very strong evidence in support of this view has been given by Douglas.‡ In the lower layers the characteristics of an air mass may be very markedly affected by its recent history, in par-

* *Q.J. Roy. Met. Soc.* **55**, 1929, p. 81; *ibid.* p. 195.

† See Bergeron, *Met. Zeit.* **65**, 1930, p. 246. For a detailed discussion of the problem of air masses see also Bergeron, "Über die drei-dimensionale Wetter-Analyse", *Geof. Publ.* **5**, No. 6.

‡ *Q.J. Roy. Met. Soc.* **51**, 1925, p. 229.

ticular the nature of the surface over which it has passed, and the time it has spent in the same latitude.

Polar air which has passed over a long stretch of sea is usually described as "maritime polar air". Good examples of this are the outbursts in winter of cold air between the west coast of Greenland and the American continent, the air moving South, then East, and frequently approaching the British Isles as a south-westerly current. In such currents there is frequently a high lapse-rate in the surface layers, associated with the original polar conditions at high levels.

In modern Norwegian practice "tropical air" is not very common, the name being rightly restricted to air originating in really low latitudes, say south of 40° N. Other warm masses may be called "maritime polar air", "returning maritime polar air", and in summer "continental air". There is a tendency to use the same "maritime polar air" too frequently, even for air which has spent a long time in summer between latitudes 50° and 62°, which might more appropriately be called "maritime air".

The number of possible classes of air masses is almost infinite, and in practice the forecaster has to make a compromise between the excessive complexity of the atmosphere and undue simplicity of classification. There is thus an arbitrary element in the classification of air masses and the drawing of fronts separating them which makes it unprofitable in this place to devote much time to a discussion of details. We shall return to the subject in the next chapter, in discussing the drawing of fronts between air masses.

There is no direct evidence that above 1 or 2 km the mean lapse-rate differs appreciably in polar and equatorial air. True instability for dry air rarely extends beyond 1 km, but a lapse-rate greater than the saturated adiabatic, and slightly greater than the average, is often found in polar air which has been subjected to prolonged heating over the sea. This may lead to the development of winter thunderstorms, heavy local rain, and in some cases to the formation of depressions. Various statistical studies of polar air masses have included cases where there had been no prolonged heating, and also cases where subsidence had occurred, and the importance of the effects produced by prolonged surface heating has not always been brought out in such studies.

CHAPTER XVII

THE POLAR FRONT AND ITS RELATION TO THE DEVELOPMENT OF CYCLONES

§ 189. *The depression as a wave disturbance in a surface of discontinuity*

HELMHOLTZ* many years ago drew attention to the possibility of two currents flowing side by side, having different temperatures and different velocities, and separated by a surface of discontinuity, the currents flowing either in the same or in opposite directions. The dynamical conditions for equilibrium yield a relatively simple equation for the inclination of the surface of separation (*vide* Chapter x).

A number of writers have suggested that cyclones tend to form at such surfaces of separation of cold and warm air. Thus Bigelow† in 1902 pointed out that on the whole cyclones could not be regarded as having warm centres or cold centres, but that the centres of cyclones were found on lines separating warm and cold currents. The evidence adduced by Bigelow in favour of this view related to North America, but similar results were confirmed for the depressions of North-western Europe by Hanzlik, and for Asiatic depressions by von Ficker. Further Shaw and Lempfert, in the "Life-history of Surface Air Currents" (*vide* Chapter XVI), showed that the phenomena of weather in depressions were to be explained by the interaction of air currents of different origin and of different temperatures and humidities.

The relation of the depression to the lines of discontinuity between different air masses has been developed in recent years by the Norwegian school of meteorologists from two different points of view, the first, developed by Professor V. Bjerknes‡ from a mathematical standpoint, regarding the depression as a wave in the surface of separation, the second developed by J. Bjerknes,§ H. Solberg, and others, concerning itself rather with the discussion of the physical processes at fronts, and the evolution of methods of application of physical principles to the analysis of synoptic charts and the practical problems of forecasting. The surface of separation between the warm and cold currents in the depression of middle latitudes has been named the "polar front", and the Norwegian methods of analysis are usually referred to as the polar front methods, or frontal methods.

V. Bjerknes visualises the cyclone as a wave in the *polar front*, the surface of discontinuity between the mild westerly currents of middle latitudes and the cold currents of high latitudes.

* H. Helmholtz, "Über Atmosphärische Bewegungen", *Ges. Abh.* **2**, p. 289.
† *Monthly Weather Review*, 1902, p. 251.
‡ "On the dynamics of the circular vortex", *Geof. Publ.* **3**, No. 4.
§ J. Bjerknes and H. Solberg, *Geof. Publ.* **2**, No. 3, **3**, No. 1; also J. Bjerknes, *ibid.* **1**, No. 2.

Later developments of the wave theory of the cyclone have been described in detail by V. Bjerknes and others.* Some further reference to the theory will be made in § 200 below, after a discussion of the observed phenomena in cyclones.

§ 190. *The polar front methods of analysis of charts*

The original picture of the phenomena associated with the formation of a depression, as put forward by the Norwegian meteorologists, visualised the

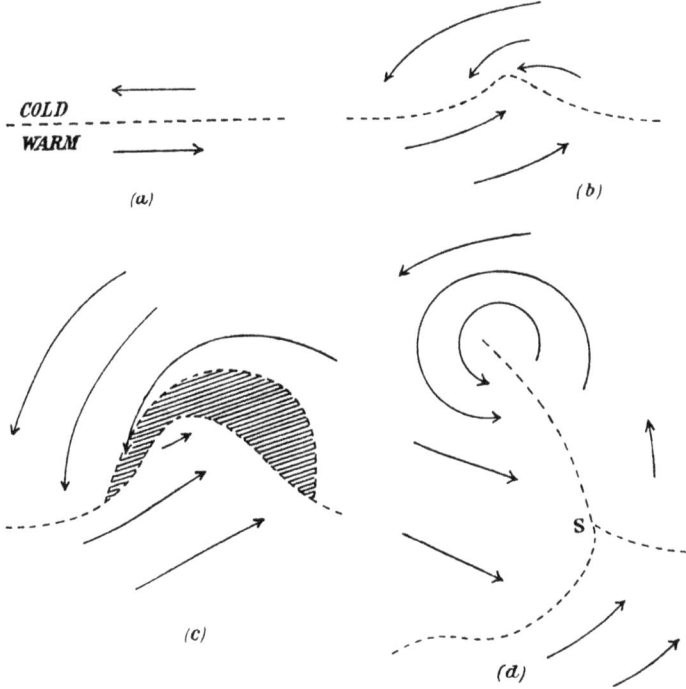

Fig. 78 *a–d*. The development of a depression at a polar front
between opposing currents.

polar front as encircling the earth, having easterly winds to north, and westerly winds to south, of the front. The development of the depression at such a front will first be considered, before we proceed to the more frequent happening in middle latitudes, in which depressions form at fronts separating a cool westerly current from a warm westerly current to south of it. The earlier picture of the development of Bjerknes and Solberg (*loc. cit.*), is most readily explained by reference to fig. 78 *a–d*. In this figure, diagram *a* represents a portion of the undisturbed polar front, diagram *b* represents the distortion of the front by the intrusion northward of the warm air, and diagram

* Bjerknes, V., Bjerknes, J., Solberg, H., and Bergeron, T. *Physikalische Hydrodynamik, Berlin*, 1933, pp. 565–621. A slightly simplified account will be found in a paper by J. Bjerknes and C. L. Godske, "On the theory of extratropical cyclones", *Astrophysica Norvegica*, **1**, No. 6.

c represents a later stage in the growth of the distortion. At stage *c* there is a well-marked depression centred at the most northerly point of the tongue of warm air. The bulge in the surface of separation, together with the newly formed cyclone, moves eastwards with the warm current. The part of the polar front along the eastern edge of the warm tongue is called the *warm front*,

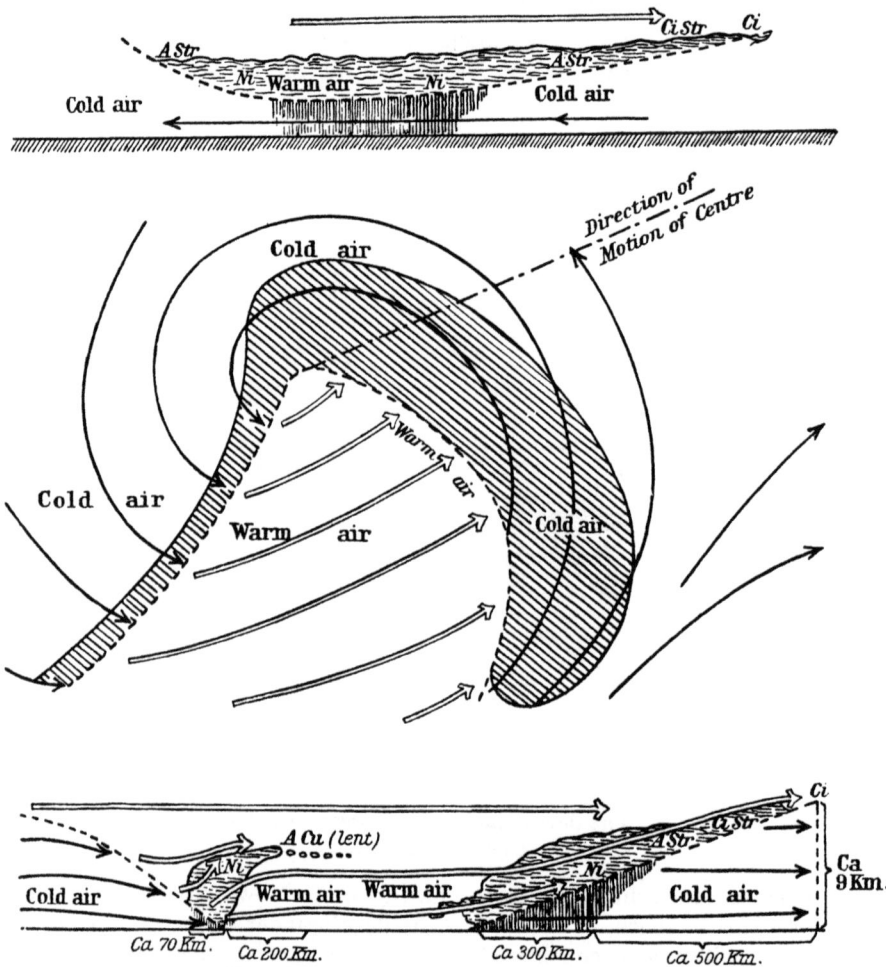

Fig. 78e. The typical polar front depression.

and the part along the western edge the *cold front*. (This refers to a depression moving in an easterly direction. In a depression moving in any other direction the positions of the fronts relative to the path of the centre remain unaltered.) At the warm front the warm air climbs up over the cold air, giving precipitation over a wide area, the rain falling through the cold air.

At the cold front the cold air pushes under the warm air, lifting it, giving precipitation over a less extensive area than at the warm front. The ascent of air at the warm front is usually steady, and the rainfall has the character of steady rain; but at the cold front the ascent is much more violent and intermittent, accompanied by squally winds, and the clouds are of the cumulus or cumulo-nimbus type, with frequently a long roll of cloud over the front. This distribution of weather phenomena and of the types of cloud in different parts of the depression is represented in fig. 78e, in which the central diagram represents the typical depression in active development, while the upper and lower diagrams represent vertical West-East sections taken north and south of the path of the centre respectively. It is to be noted that when the warm and cold fronts are strongly marked the isobars are refracted at the fronts, as shown in fig. 79.

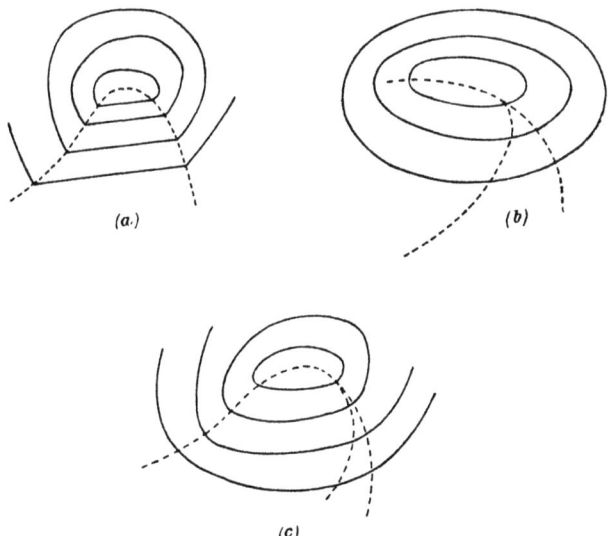

Fig. 79. The development of a back-bent occlusion.

A vertical section through the right-hand side of the lower diagram in fig. 78e from the cold air up through the front into the warm air will show the following characteristics. Near the ground the lapse-rate in the polar air will be high, possibly approaching the dry adiabatic, except over land in winter, when, in continental air, or in air cooled at the surface by radiation, a considerable inversion may occur. At the frontal surface the temperature in the warm air is higher than that in the cold air at the same level. There will therefore be a check in the fall of temperature, possibly amounting to a pronounced inversion, the change taking place in a layer whose relative humidity is high, approaching saturation.

On account of the polar current being heated by the earth's surface and the tropical current being cooled by the earth's surface, it follows that in the lower layers the lapse-rate is greater in polar than in tropical air. The difference in temperature is accordingly more accentuated at moderate heights than it is at the actual surface. This is very clearly borne out by observations. Douglas* has produced direct evidence showing that in disturbed westerly conditions

* Q.J. Roy. Met. Soc. 47, 1921, p. 23.

round the British Isles the difference of temperature between cold and warm air masses in juxtaposition is more than twice as great at 4 km as at the surface. There is however no direct evidence that at heights above 1 or 2 km the lapse-rate in genuine polar air is appreciably higher than that in genuine tropical air.

In the course of time the cold air in the rear pushes forward, raising the warm air and narrowing the tongue of warm air at the ground. Eventually the cold front overtakes the warm front, lifting the warm air up from the ground, and the depression is then said to be *occluded*. The occlusion begins at the centre, where the cold front has a shorter path to cover before overtaking the warm front, and works progressively outwards (see fig. 78*d*). So far as the surface layers are concerned the depression then consists entirely of cold air, though there will still be a warm sector in the upper air for some time after occlusion has taken place at the surface. The occlusion progresses upwards as the warm air is displaced still higher, until eventually the depression consists entirely of cold air, and has approached to a nearly symmetrical vortex. Continuous rain may fall for 12 to 24 hours after the surface occlusion, on

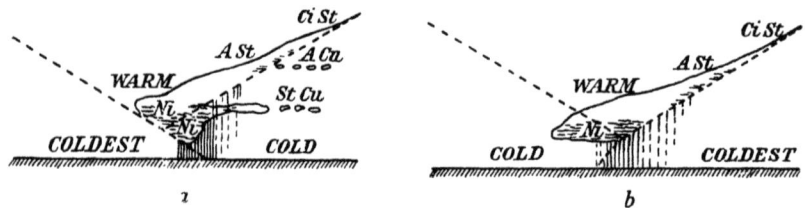

Fig. 80. Types of occlusion.

account of the progressive rise of the air in the warm sector at high levels. With the cessation of the continuous rain the depression begins to decay, since it no longer has a store of potential energy to draw upon, and friction and turbulence produce a steady diminution of kinetic energy.

In general there will be a difference of temperature between the cold air in advance of the warm front and the cold air in the rear of the cold front, so that when occlusion takes place there will still be some difference of temperature at the line of occlusion. Two kinds of occlusion may therefore arise, according as the cold air in the rear is colder or warmer than the cold air in front of the line of occlusion. These are shown in fig. 80*a* and *b*. The first is essentially a cold front with a narrow rain belt, while the second is a warm front with a broader belt of rain.

The somewhat simple picture of fig. 78*e* may be considerably complicated by the appearance of secondary cold fronts within the cold air in the rear of the cyclone. If the temperature contrasts are slight at these secondary fronts, the only result is the production of narrow rain belts, but if one of the secondary fronts shows a very marked contrast of temperature it will have the effect of enlarging the effective warm sector, so that all the air between this front and the warm front acts as a warm sector, giving correspondingly greater supplies of energy to the cyclone.

Some of the earlier papers on the polar front picture the development of the occlusion as shown in fig. 81, in which an island of warm air is isolated at the centre. This development is not typical of the depression as we know it, and is limited to regions where orographic causes hold up the travel of the warm front in such places as off the coast of Norway. This type of development is now known as "seclusion".

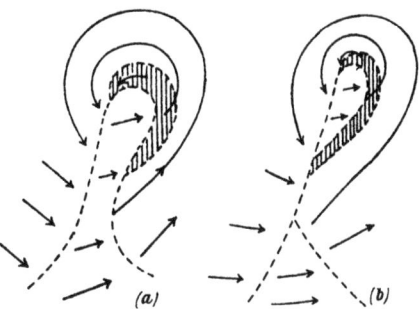

Fig. 81. The seclusion of depressions.

Fig. 78 above visualises the depression as originating at a straight front separating easterly and westerly currents. There is, however, no *a priori* reason why a front should be straight unless the isobars in the two currents are straight; but strictly straight isobars are exceptional rather than the rule. Most developments over the Central and Eastern Atlantic and round the British Isles occur at discontinuities in a westerly current, the course of the development being as shown in fig. 82 *a* to *d*. Initially there is no east wind,

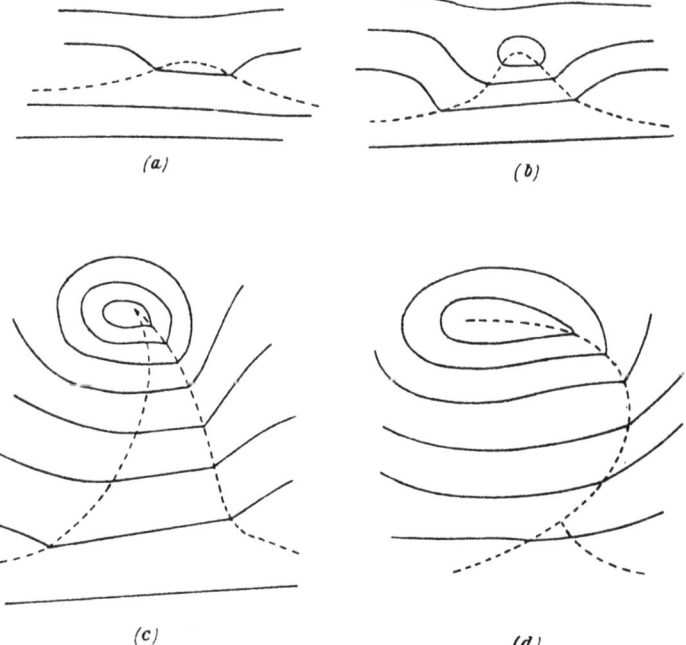

Fig. 82. The development of a depression at a boundary between two westerly currents.

and east winds only develop as the depression develops, as part of the circulation round the centre of low pressure. A good example of this is shown in the development of the depression discussed in detail by Bergeron and Swoboda.* This depression originated at a front separating two westerly currents, the more northerly of which was maritime polar air, while the more southerly was genuine tropical air which had come round the Azores anticyclone.

Shaw has suggested† that the initial condition for the development of a depression is not the occurrence of two currents in opposite directions, but rather of two currents at right angles to one another, associated with the occurrence of a right-angled kink in the isobars. In a large number of cases it is undoubtedly possible to trace such a scheme as Shaw suggests, which accords with Exner's view that the depression of middle latitudes is due to the invasion of the cold air of polar regions into the zone of westerly winds (see § 205 below), but in the two cases illustrated by Shaw the bulge in the isobars is not the initial stage, the depression being then definitely in existence.

The cyclone moves with the warm current or in the direction of the isobars in the warm sector, with a speed which never exceeds, and is usually a little less than, that of the warm air. There is no evidence that the speed of travel ever exceeds the wind speed at 10,000 feet, though in some depressions, more particularly in summer, it may exceed the wind speed at 2000 feet. After occlusion the depression usually becomes nearly stationary, and if it is isolated from other depressions it remains stationary in its dying stage. Frequently, however, the dying depression gets caught up in the circulation around other and more vigorous systems.

The fact that the depression moves in the direction of the isobars in the warm sector is confirmed by an examination of the data of observation, even in cases where the motion has differed widely from the mean tracks as usually laid down in the older textbooks. The growth of the typical Bjerknes depression is in the direction from marked dissymmetry toward symmetry. In its initial stage the depression shows a marked warm sector, but as the depression deepens the warm sector shrinks, in most cases with great rapidity. In its final occluded state the depression is seen to be of the nature of a whirl, surrounded by a close approach to a solid current, which protects the centre from the encroachment of any new currents. Only in very special circumstances can a new current approach the centre of an occluded depression, and any new development in an occluded depression is likely to be of the nature of the development of a new centre.

The brief account which has been given above of the frontal structure of a depression must be regarded as only an outline. The phenomena can be further complicated when the air masses involved are not homogeneous, or when their motion is affected by mountain ranges, which may hold up the portion of a warm front in the lower layers for a considerable time, giving rise to heavy rainfall.

The description given in the preceding paragraphs follows in the main the lines of the description given by Bjerknes and Solberg (*loc. cit.*) in 1922. Since that time no very considerable advance has been made along these lines, though J. Bjerknes and others have shown in a series of papers that the pheno-

---

* *Veröff. Geoph. Inst. Leipzig*, Ser. III, **3**, p. 63.
† *Manual of Meteorology*, **4**, p. 289, fig. 75.

mena are in many cases more complicated than was assumed in the original description. When unexpected developments occur they can as a rule be explained by the nature of the development of the fronts, though it is rarely possible in such cases to forecast the developments in detail.

Depressions sometimes form entirely within a polar current, but the air in the southern portion of these depressions will have moved over a longer trajectory in middle latitudes than the air in the northerly portion. Thus even within these depressions there may occur fairly marked contrasts of temperature. An example was provided by a depression which on November 3, 1910, was centred over England. Upper air soundings at Pyrton Hill, Berlin and Vienna showed a marked rise of temperature with the approach of the centre of the depression. It is of some interest to note that on November 5 when this depression had passed away eastward, a very deep depression developed rapidly off North-west Ireland, entirely in polar air. This rather suggests that when conditions in the upper air have become favourable for the formation of depressions in polar air, those conditions tend to persist. Douglas cites several other cases, notably the depressions of March 3 and 5, 1909, which also formed entirely in polar air. In the first of these, the temperatures in the central part of the depression were below the normal, but in the outer regions temperatures were still lower.

In a later paper J. Bjerknes* has shown that the phenomena of occlusion may lead to a subsidiary cold front. As the occlusion first develops near the centre of the depression, there is, in some depressions at least, a tendency for the centre of the cyclone to move along the occlusion to the angle of the warm sector. The occluded front then tends to become a subsidiary cold front as shown in fig. 79c above. Such a front is usually referred to as a "back-bent occlusion". In addition there is in such cases a tendency for the depression to be strengthened by reason of its re-acquiring a warm sector, as well as by reason of the part of the cold air between the main and the subsidiary cold fronts acting as a warm sector and helping to supply energy to the depression.

It is not practicable in the space here available to illustrate all the points brought out in the above description by means of synoptic charts selected for the purpose, and only certain typical situations can be illustrated. Fig. 83 shows a typical polar front depression, having well-marked warm and cold fronts. With the observations available on the chart it is not possible to place with complete certainty the accurate position of the warm front. Fig. 84 reproduces the autographic records for Holyhead. The thermograph shows that up to about 3 h Holyhead was in the cold air in advance of the warm front. In passing through the warm front the temperature rose about 2° F in about 1½ hours, then rose 3° F in the course of a few minutes, after which there was a further rise of 1° F in 3 hours, the highest temperature occurring just before the advent of the cold front. The fall of temperature at the passage of the cold front was in two very distinct stages, separated by an interval of about 50 minutes, indicating that the front was double at Holyhead, as in fact it was at several stations. The lowest pressure was recorded from about 6 h to 7 h,

* *Geophys. Mem.* No. 50.

corresponding to the time of highest temperature. On the anemogram is shown the arrival of the warm front at about 3 h 20 m, with an increase of wind and a change of direction from ESE to S. This occurred at the same time as the sudden rise of temperature. The first cold front arrived at about 6 h 30 m, with a drop of wind, and a veer to W by S, and the second cold front arrived at 7 h 20 m, with a further veer to NW and an increase of wind. The times of the arrival of the two cold fronts as fixed by the anemogram agrees with the two separate falls of temperature shown by the thermogram. The hyetograph

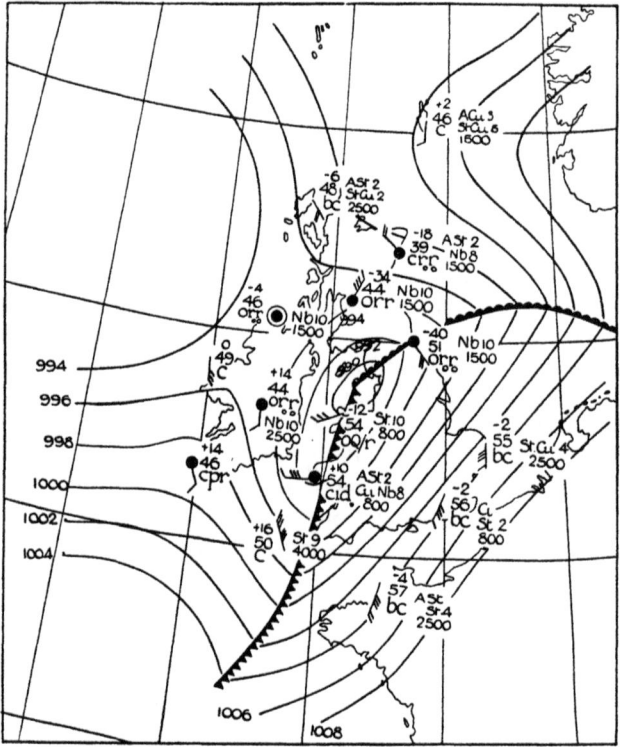

Fig. 83. The depression of October 22, 1932.

indicates that rain fell continuously from 19 h on the 21st to 7 h on the 22nd, being heaviest from 2 h to 4 h on the 22nd, when the warm front was practically over the station. The total rainfall during the whole period from 19 h on the 21st to 7 h on the 22nd was 15 mm, of which 5 mm fell between 2 h and 4 h.

The further motion of the depression was parallel to the isobars in the warm sector; at 13 h it was centred slightly east of the Firth of Forth, and by 7 h on the 23rd was centred over the coast of Norway in latitude 64°, the pressure at the centre being 984 mb.

Fig. 85 shows the upper air observations made at Duxford at 7 h 30 m on the 22nd. It will be seen that the lapse-rate was stable up to the level of about 700 mb, even for saturated air. At this time Duxford was not more than 140 miles from the warm front.

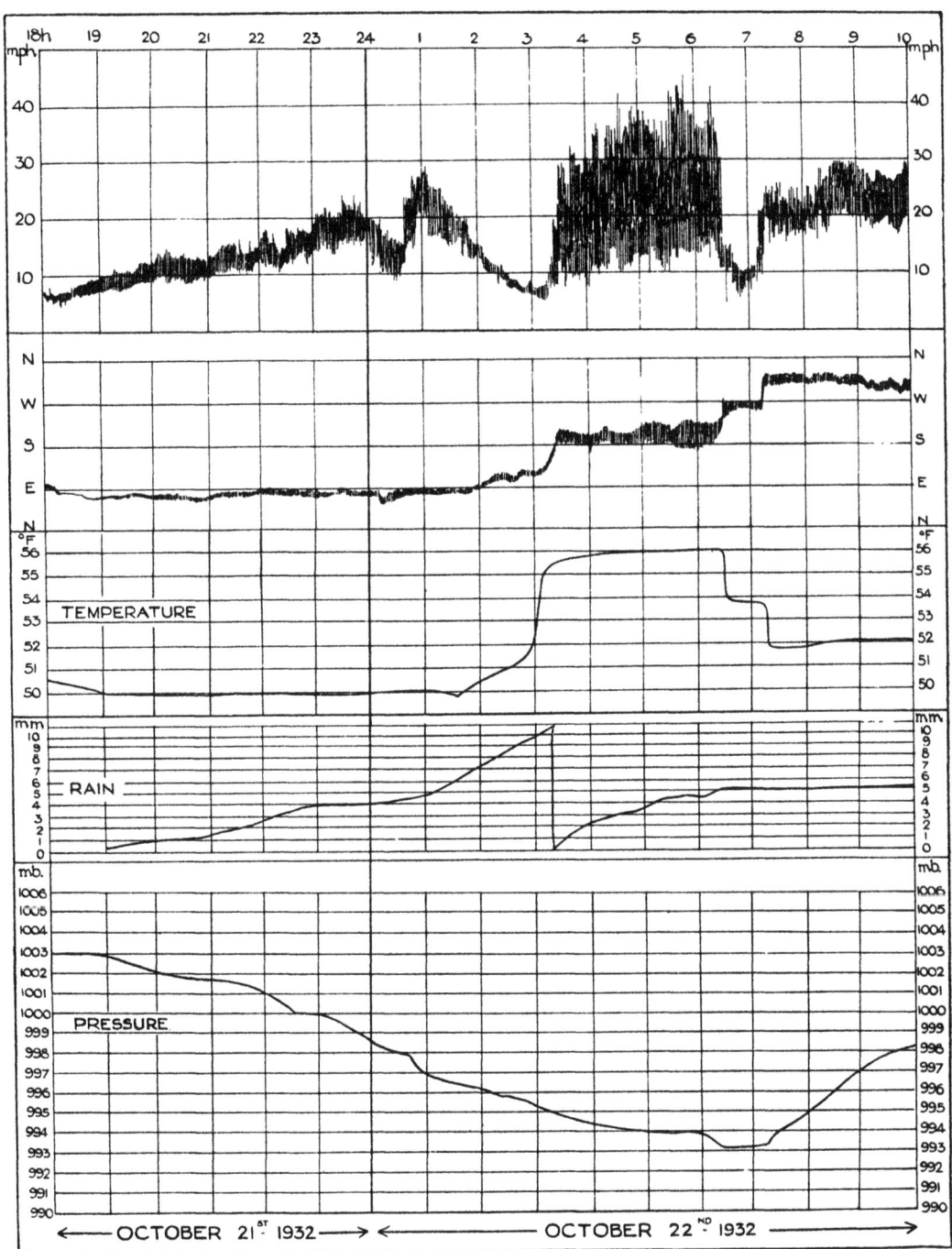

Fig. 84. Autographic records at Holyhead, October 21–22, 1932.

The reader is recommended to examine the charts in *Geophysical Memoir*, No. 50, in which J. Bjerknes has illustrated a number of interesting phenomena. One depression selected from the three discussed in this memoir is represented in figs. 86 and 87, which reproduce the synoptic charts for 7 h and 13 h on January 23, 1926.

The 13 h chart shows that the depression has moved roughly parallel to the isobars in the warm sector. These two charts present some typical features of

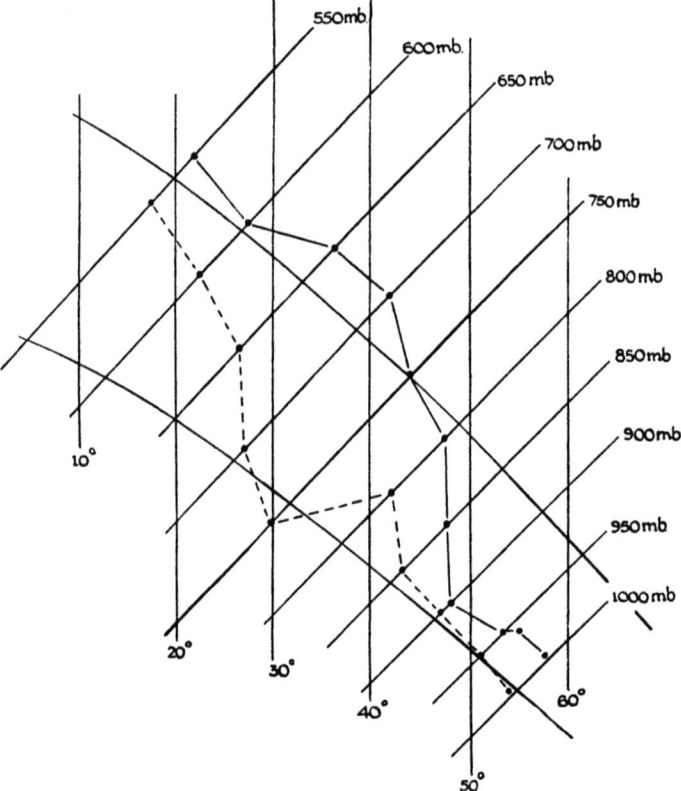

Fig. 85. Upper air observations at Duxford on October 22, 1932.
Wet bulb temperatures are shown by the broken line.

polar front depressions. The association of the heavy rain with the fronts is very clearly marked. On the 7 h chart the cloud in the part of the warm sector over Wales is St. and St.-Cu., but a subsidiary attenuated warm front running from about Plymouth to Spurn Head is marked by Nimbus, as is the whole of the principal front. The 13 h chart shows, in addition to the principal warm and cold fronts, a back-bent occlusion running from the centre in a SSW direction across the East of Ireland, a subsidiary cold front from Anglesey to Plymouth, and a subsidiary warm front crossing the warm sector, shown on the 7 h chart.

The association of different features of the weather with the fronts is best studied by means of autographic records. Those for Valentia, Holyhead, and Eskdalemuir are shown in figs. 88, 89 and 90. Consider first the records at

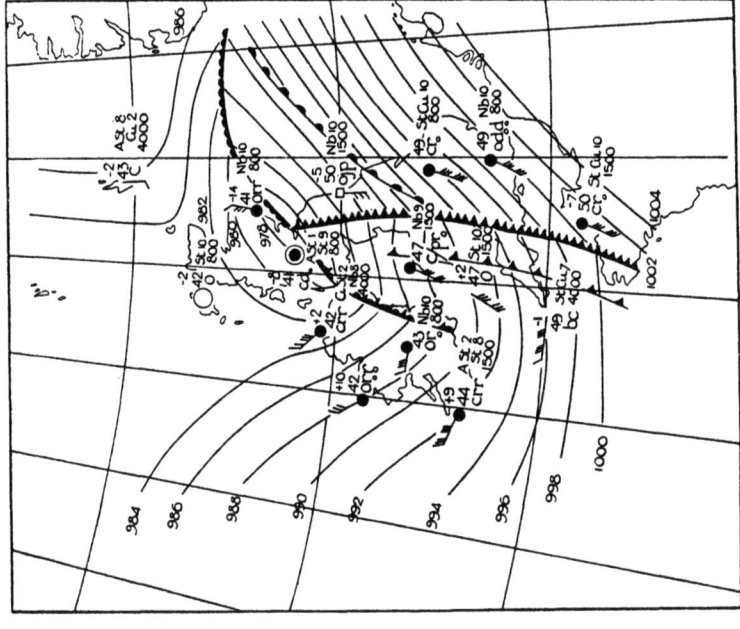

Fig. 87.

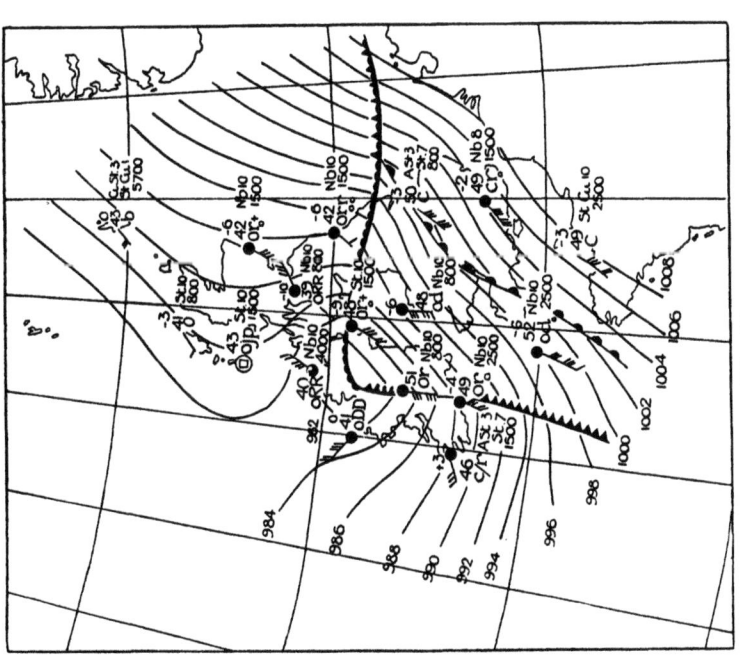

Fig. 86.

Figs. 86, 87. The depression of January 23, 1926. Charts for 7 h and 13 h. In charts showing fronts, the warm front is indicated by rounded teeth, the cold front by pointed teeth, an occlusion by a combination of rounded and pointed teeth, and secondary or attenuated fronts by the appropriate teeth spaced out.

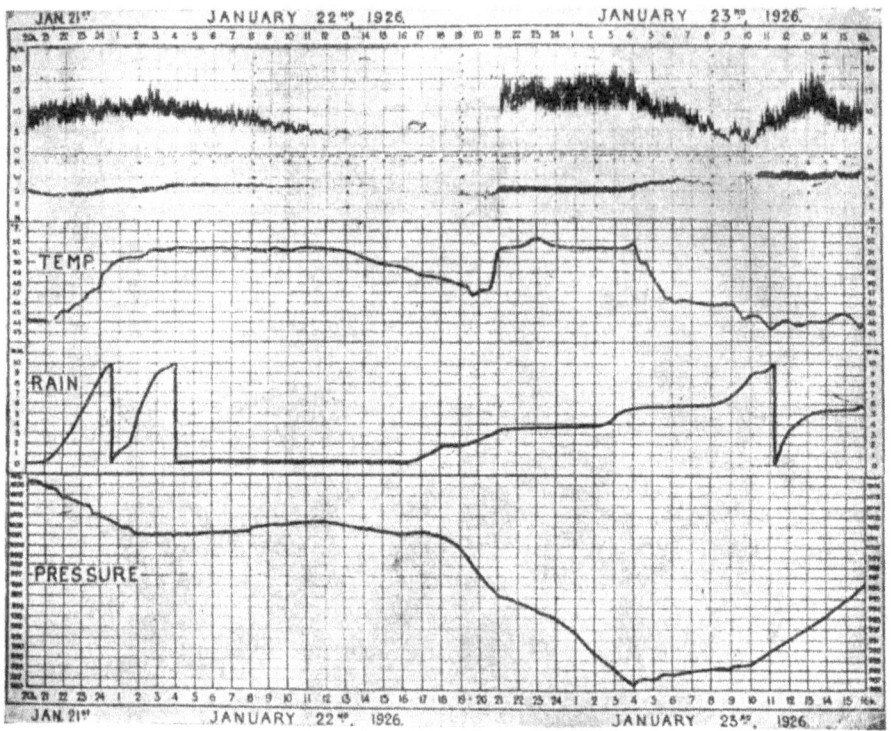

Fig. 88. Autographic records for January 22–23, 1926, at Valentia.

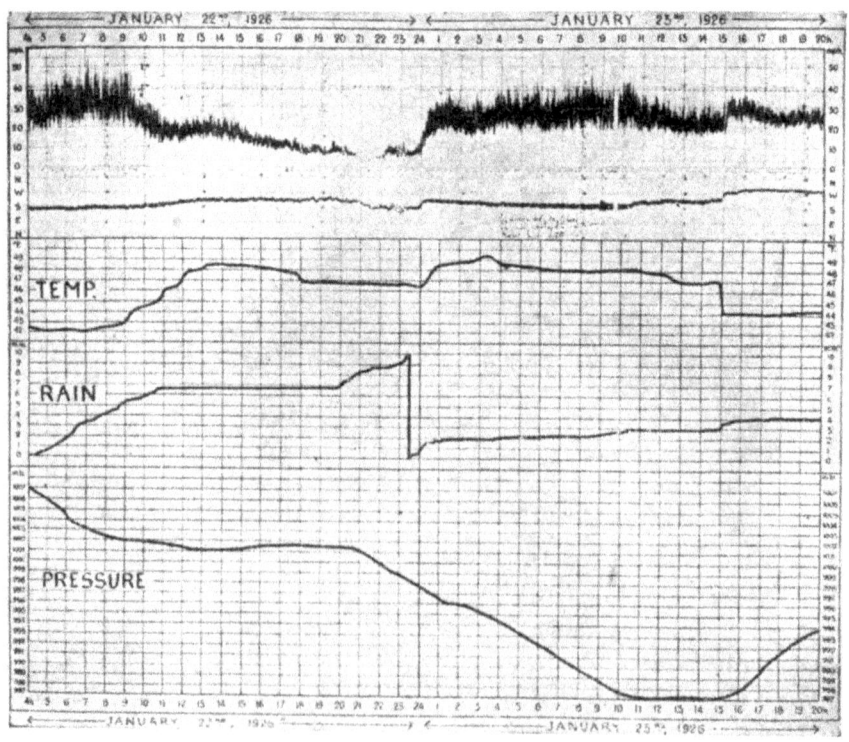

Fig. 89. Autographic records for January 22–23, 1926, at Holyhead.

Valentia (fig. 88). The cold front passed Valentia at 4 h on the 23rd, and the back-bent occlusion passed about 10 h. These times are best determined from the barograph. At 4 h the wind veered and decreased in strength, and a steady fall of temperature started and continued afterwards for several hours. The rain preceded the arrival of the front, commencing about 3 h. The passage of the back-bent occlusion was marked by no pronounced change of surface temperature, but there was a well-defined increase in the strength of the wind, and heavy rain fell for several hours.

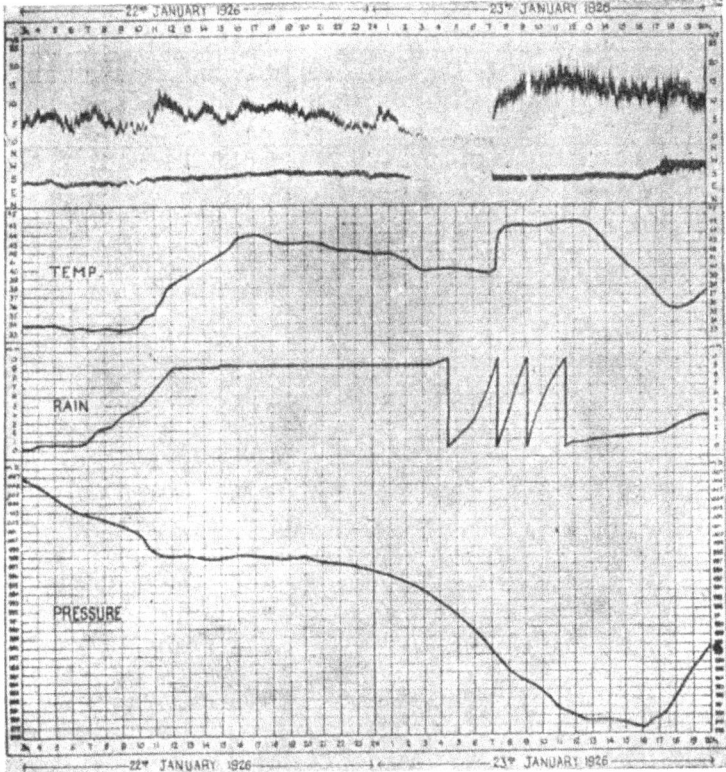

Fig. 90. Autographic records for January 22–23, 1926, at Eskdalemuir.

The "principal" cold front passed Holyhead at about 11 h on the 23rd; the secondary cold front, which in fig. 87 runs down the West coast of Wales, passed Holyhead soon after 13 h; and the back-bent occlusion passed about 15 h. An examination of fig. 89 shows that the fall of surface temperature at 11 h, when the principal cold front passed, was very slight, as was the fall at the passage of the subsidiary cold front. The greatest fall of temperature occurred at the passage of the back-bent occlusion at 15 h, when a drop of 3° F occurred. The rain which fell in the warm sector in advance of the cold front fell at Holyhead between 8 h and 11 h, after which no further rain fell until 15 h, when for a short time there was heavy rain, beginning at the same time as the sharp fall of temperature, while the wind veered from SW to W, and increased in strength. The fall of pressure was checked at 11 h, the rise beginning at 15 h.

Eskdalemuir (fig. 90) was in the cold air north of the warm front at 7 h, and by 13 h was in the rear of the cold front, having in the interval passed through the warm sector. Fig. 90 shows that the warm front passed at about 7 h 20 m, the temperature rising rapidly 5° F, while the wind increased rapidly in strength from the West. The cold front passed about 12 h 30 m, but the fall of temperature was distributed over more than 5 hours. Very heavy rain started

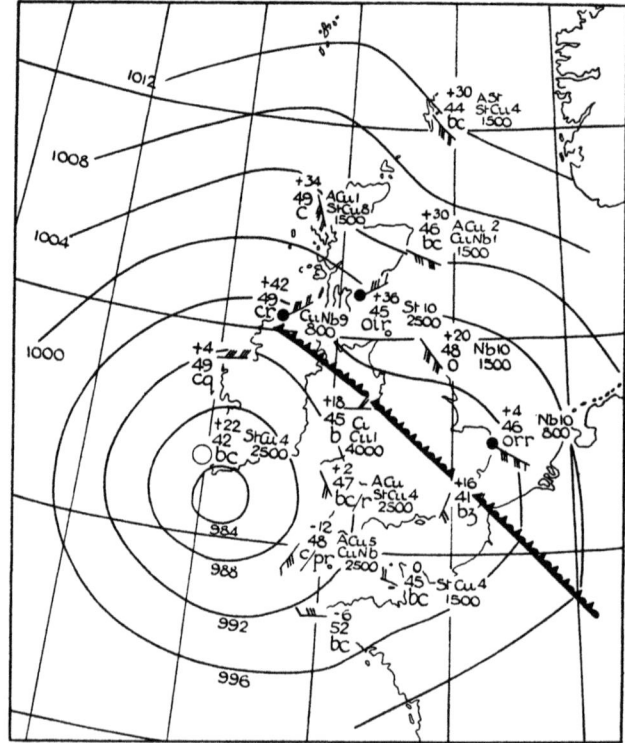

Fig. 91. An occluded depression, November 15, 1933.

some three hours before the passage of the warm front, and continued during the passage of the warm sector. Much lighter rain fell after the passage of the cold front.

The importance of the back-bent occlusion in the scheme of fronts is borne out by the changes which occurred at Valentia at 10 h on the 23rd, at Holyhead at 15 h, and at Eskdalemuir at about 16 h. At Valentia, heavy and continuous rain occurred at the passage of the back-bent occlusion; the rain was heavy for a short time at Holyhead, and less heavy but lasting for a longer period at Eskdalemuir. At Eskdalemuir heavy rain lasted from 4 h 30 m till 11 h 30 m on the 23rd, when the station was near the centre of the depression.

Such rain must have required the bodily ascent of enormous quantities of damp air.

In fig. 91 is shown an occluded depression of November 15, 1933, the chart reproducing the conditions at 18 h. The line of occlusion is shown running from the north of Ireland across central England down to about Dijon. There is a clearly marked line of rain running parallel with the surface occlusion, while south of the line of occlusion conditions were generally fair. There remains some discontinuity of temperature at the front, the cold air which has curved round in the rear of the depression being warmer than the cold air in advance. The discontinuity of surface wind at the line of occlusion is also clearly marked.

The line of occlusion had been clearly marked on the synoptic chart since 13 h on the 14th, when it ran roughly North-South some 100 miles west of Ireland. The depression had moved slowly southward, the pressure at its centre increasing very slowly. Subsequently to the time represented in fig. 91 the depression continued to move southward, and to fill up very slowly. The motion was in the direction of the strongest winds, which were NNW, force 9, in lat. 50° N, long. 20° W, and are not shown in fig. 91. By 18 h on the 17th it was centred over Portugal, the pressure at the centre being about 996 mb. Its later history was complicated by the formation of fresh centres over southern Spain.

The motion of the line of occlusion is not without interest. As the depression moved southward the line of occlusion swung round the depression counterclockwise, and by the morning of the 16th had ceased to be a clearly marked feature of the system.

In fig. 92 are shown the upper air ascents at South Farnborough at 11 h 50 m, at Lindenberg at 6 h, and at Munich at 9 h on the 15th. The South Farnborough observations were made entirely in the southerly current. The Lindenberg ascent apparently penetrated into this current at about $2\frac{1}{2}$ km, and since Lindenberg was some 500 km from the surface line of occlusion, the front was sloped forward about 1 in 200, the form of the occlusion being that shown in fig. 80 (b). Munich was in the cold surface current, but from about 1 km to $2\frac{1}{2}$ km was in the current which had come around in the rear of the occlusion, while above that height there appeared to be warm tropical air. It is doubtful whether this air is to be interpreted as the remnant of the warm sector. It was more probably tropical air which had recently encroached into the depression.

Such differences between the two air masses on the two sides of an occlusion are not always found. J. Bjerknes and Palmén* have given upper air data for the depression of March 28–31, 1928, which show that except in the lowest kilometre the temperature conditions were practically identical in the two masses on the two sides of the warm sector, the tropopause being at 7 km above each. Palmén† suggests that above occluded depressions the tropopause is normally very low for the latitude, say at 5–8 km over occluded depressions over NW Europe.

* *Beitr. Phys. fr. Atmos.* **21**, 1933, p. 53.
† *Mitteilungen Met. Inst. Helsingfors*, No. 25, 1933.

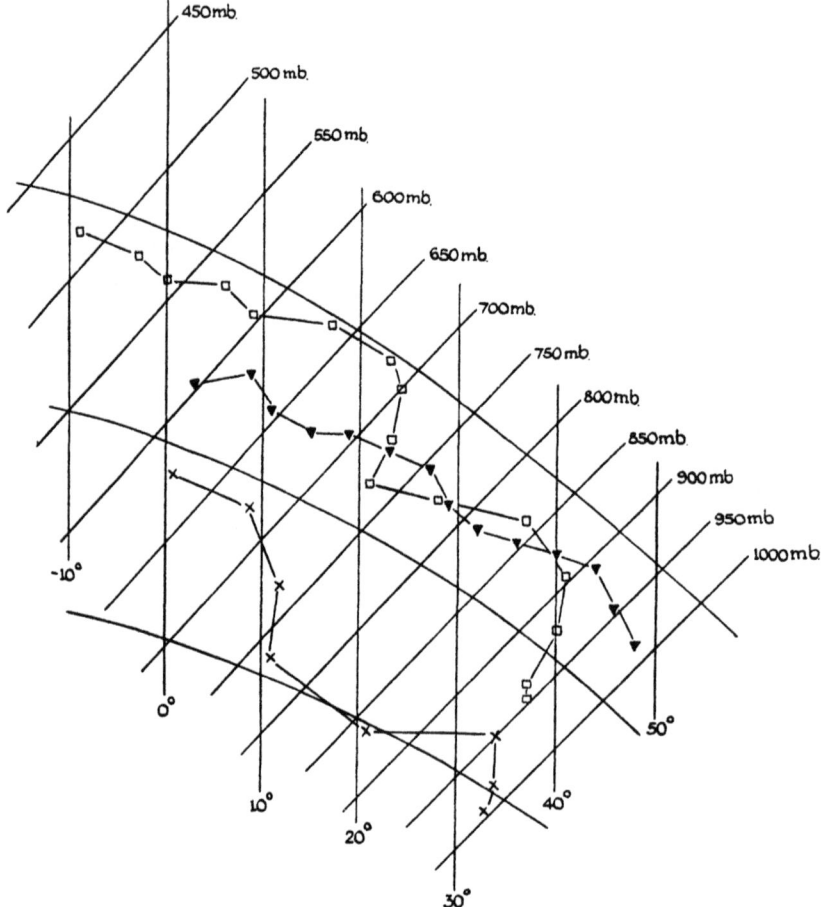

▼——▼ South Farnborough, November 15, 1933, 11 h 50 m.
×——× Lindenberg, November 15, 1933, 6 h.
□——□ Munich, November 15, 1933, 9 h.

Fig. 92. Upper air observations on November 15, 1933.

§ 191. *The formation of secondaries*

When a depression has become occluded, the cold front trails behind the
depression as in fig. 78 *d*. Secondaries tend mainly to form at a bend in the
trailing front. Every trailing front does not give rise to secondaries, nor do
secondaries invariably form only at trailing fronts. They can on occasion form
within the polar current itself, and then usually at a secondary front in the
cold air.

There are some regions in which secondaries tend to form at the angle of
the residual warm sector in an occluded depression, as at *S* in fig. 78 *d*. The
phenomenon is in part orographic, due to the warm front being held up by

mountain ranges. This is particularly liable to happen over the Skagerrak, when a depression centre passes to north of Scandinavia, and such secondaries are known as "Skagerrak cyclones".

Charts a and b of fig. 93 illustrate a situation in which secondary depressions develop readily. The secondary depression shown over Southern England in chart a has disappeared by the time of chart b. Later during the 9th a further secondary developed, as shown in chart c, which represents conditions at 7 h on the 9th. The two centres rotated counter-clockwise round each other, as seen from chart d, which represents conditions at 18 h on the 9th. This rotation was continued, and by 7 h on the 10th one centre was over the North Sea, and the other over Brest. Observations at Duxford on the 8th, in the southerly current, showed an isothermal layer from 950 to 900 mb, and above this, up to at least the height of 500 mb, a lapse-rate almost exactly equal to the dry adiabatic.

§ 192. *Families of depressions*

A series of depressions may form along the same surface of discontinuity. The easternmost depression is the oldest and most fully developed, and the other depressions form in turn, each on the trailing front behind its predecessor. Each successive depression passes a little farther south than the preceding one, until a stage is attained when a broad northerly current behind a depression sweeps on into the trade wind zone. This ends the series or family of depressions, and the next depression will form usually on a more northern track, and on a new front.

J. Bjerknes and Solberg state that when a cyclone family passes, on the average four depressions are observed at any particular place which is situated centrally in the depression belt. The number is, however, rather variable, and single depressions may occur.

The family of depressions described by Bjerknes and Solberg occurs at the south-eastern edge of a cold current which bursts out in a south-westerly direction from the polar regions. The outbursts are roughly periodic, with a mean period of about 24 days.

§ 193. *The regeneration of depressions*

In general the occlusion of a depression is followed, after an interval of 12 to 24 hours, by gradual decline in its intensity. But it is not an infrequent occurrence for a depression to be revived after having reached the stationary or dying stage. In fig. 79c it was suggested that the centre of the depression moved along the occluded front until it reached the angle of the warm sector. When this occurs, the depression has again acquired a warm sector in its central region, and it may then go on growing in intensity until it has again become occluded.

Another method by which a depression may be revived is by the approach of a new stream of cold air from a distance. It has already been suggested on p. 324 above that this can only be brought about with difficulty. The occluded depression being surrounded by a ring of solid current, it cannot in normal

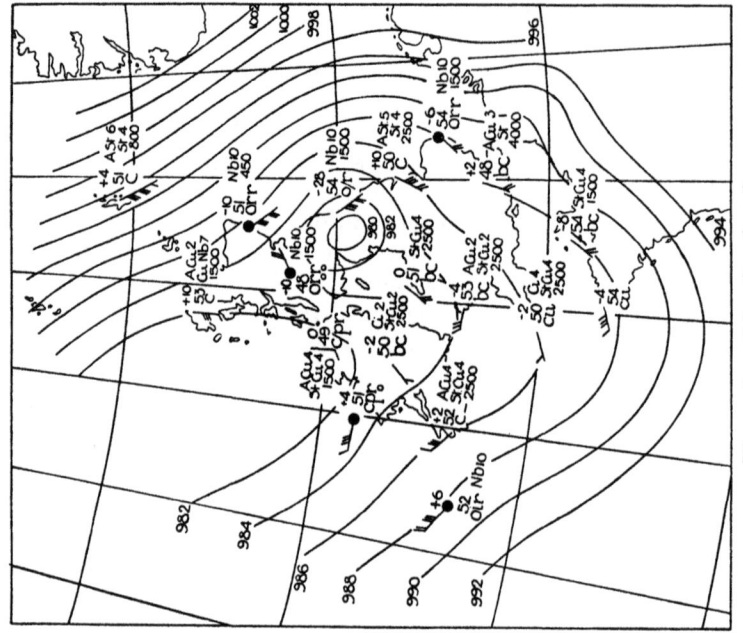

Fig. 93b. 18 h., October 8.

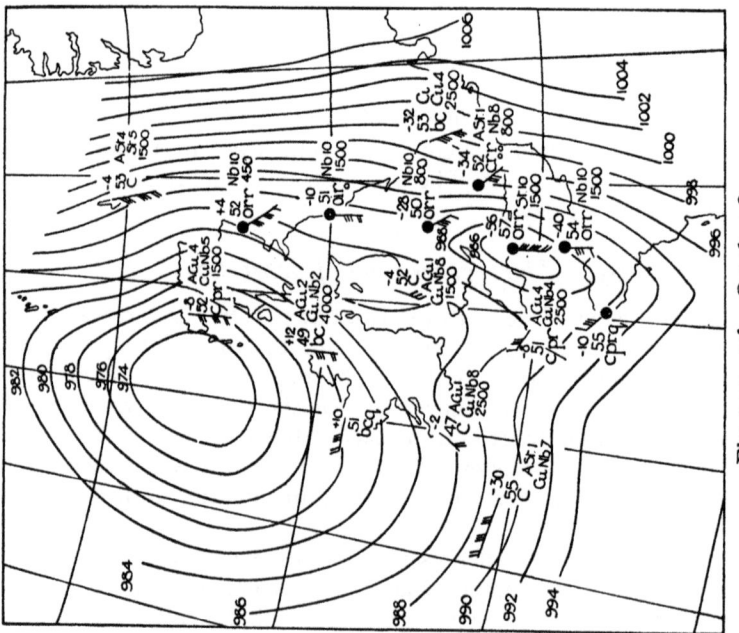

Fig. 93a. 7 h., October 8.

Fig. 93. Charts for October 8–9, 1932, showing the amalgamation of a depression and a secondary, and the rotation of two centres counter-clockwise.

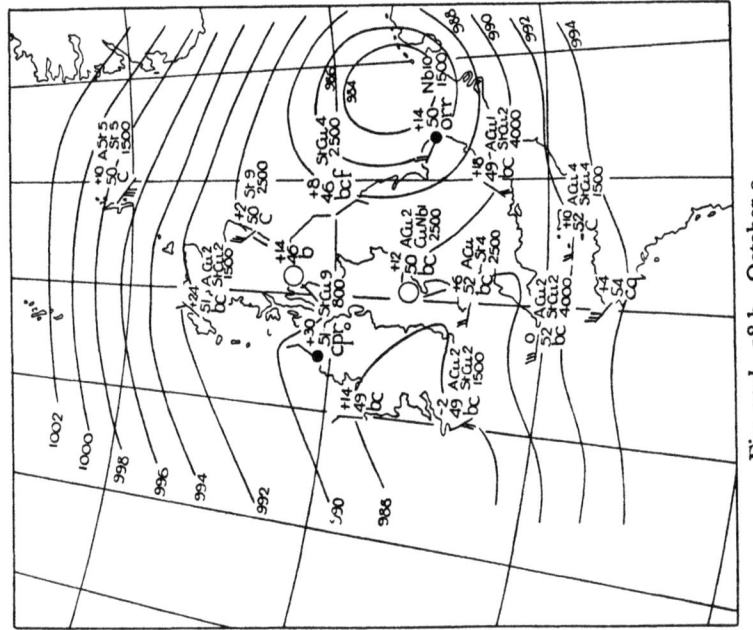

Fig. 93d. 18 h., October 9.

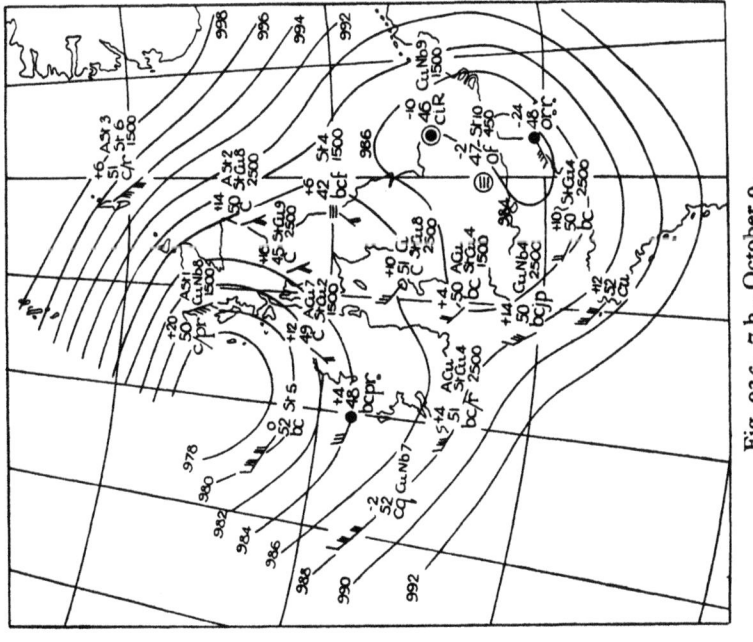

Fig. 93c. 7 h., October 9.

circumstances be penetrated by a fresh current. Any new development implies the breakdown of the solid rotation. This is likeliest to happen where the cyclic circulation is weakest, in the outer region of the depression. It has been suggested that when the front which marks the advance of cold air reaches the centre of the depression or near it, this front, together with the line of occlusion, can mark out a new warm sector which will furnish the necessary supply of potential energy to regenerate the depression. Examples have been worked out in detail by Schröder.* A particularly interesting example is furnished by case 6 in the "Life-history of Surface Air Currents", in which a depression moved up the west coasts of Ireland and Scotland, becoming occluded and stationary over the northern coast of Scotland. Later a cold current flowing across southern Scandinavia and the Baltic caused the depression to intensify and to move across the North Sea in a south-easterly direction.

There is some uncertainty as to what really happens in such cases. Maps at very short intervals might show the development of a new centre, which absorbs the old one, so giving the appearance of regeneration of the original centre. The original working charts drawn by Shaw and Lempfert for their case 6 mentioned above rather suggest the formation of a new centre over southern Scandinavia and the subsequent rotation of the two centres round one another, after the manner of the depression shown in fig. 93 c and d. It is not suggested that regeneration of depressions never occurs, but special care is required in any effort to make practical use of this idea.

§ 194. *Some general aspects of polar front depressions*

The picture of the life-history of the typical depression, as presented by J. Bjerknes and Solberg, is of great value in that it unifies into one single picture a number of facts which had previously remained isolated. It is not to be questioned that cyclones can form at boundaries between warm and cold currents and that their motion follows the direction of motion of the warm current. The phenomena of occlusion are readily found on the synoptic chart, and the value of the polar front method of analysis of charts is of the greatest practical value in forecasting, particularly in detailed forecasting for such periods as 6 hours.

Difficulties begin to arise, however, when we try to draw a mental picture of the processes involved in the birth of the depression. The surface of discontinuity as depicted by Helmholtz is stable so long as the slope of the surface is adjusted to the velocities. It is true that, as Helmholtz pointed out, the mathematical surface of separation does not correspond precisely to the physical reality, which usually presents not a mathematical surface but a layer of transition from warm to cold air. Helmholtz† added without formal proof the suggestion that the layer of transition might be unstable, and might therefore be forced to ascend. But the action which Helmholtz visualised is only a very feeble one. It is clear that the ascent of the layer of transition is not in itself sufficient to produce the observed fall of pressure,

* *Veroff. Leipzig*, Ser. II, 4, h. 2.
† *Ges. Abh.* 2, p. 302.

and moreover there is not in general a tendency for the layer of transition to be attenuated as the depression deepens.

The mathematical treatment of the problem becomes extremely complicated, once the hypothesis of steady motion is abandoned, and vertical components are introduced into the equations of motion. The most plausible way of stating the case is perhaps to suppose that the originally straight polar front surface is buckled or corrugated through the action of some irregularities in the meteorological conditions, and that the corrugation forms a channel along which the warm air is conducted upwards. The horizontal circulation is generated as a result of the convergence of air to take the place of that which ascends, as described in Chapter xv. Some evidence has been adduced by C. K. M. Douglas[*] to show that the centre of an incipient depression is a thermal centre. If this is once accepted, the further development of the depression will be facilitated by the distortion of the front by the circulation around the newly formed centre.

The Norwegian writers associate the development of a depression at the polar front with the evolution and breaking down of a wave in the front itself, though there is no necessary connection between the work of J. Bjerknes and others on the frontal methods of analysis and the theoretical work of V. Bjerknes. The evolution of a depression does not appear to be very closely analogous to the formation of a wave. In a wave there is an alternation of potential and kinetic energy. But the depression visualised by J. Bjerknes changes potential energy into kinetic energy, which is finally dissipated by turbulence, and reappears as thermal energy. The process is therefore irreversible.

The polar front depression shows an evolution from the non-symmetrical to the symmetrical, from lack of spin to something approaching pure rotation. It is likely that an approach to purely rotational motion occurs at a fairly early stage in the development shown in fig. 78, and this should be taken into account in drawing the fronts. For example, the back-bent occlusion of fig. 79 c will rotate round the depression and will suffer distortion on account of this motion. Recent discussions of depressions have tended to emphasise the frontal aspect and to neglect the rotational aspect. It should be remembered, however, that the ascent of air along a restricted portion of the front must produce some convergence towards the centre, leading to rotation about the centre, as was suggested in Chapter xv. Any complete view of the depression must take account of the rotational as well as of the frontal aspect, and must take into consideration the tendency of the fronts to rotate about the centre.

No satisfactory explanation of the fact that depressions tend to become occluded has yet been put forward. If the whole system of fronts moved with the geostrophic wind velocity, there should be little tendency for the depression to occlude. In practice it is found that while the cold front advances at about the speed of the gradient wind (and occasionally faster than the gradient wind), the warm front moves slower than the gradient wind by an amount which on the average is 10 m.p.h., and which may be as much as 20 m.p.h.

[*] *Q.J. Roy. Met. Soc.* **50**, 1924, pp. 339–63; esp. p. 357.

The advance of the cold front is not to any appreciable extent retarded by friction at the ground. It is true that the layers of cold air in immediate contact with the ground may be retarded more than the air at some distance above the ground, so that the cold front overhangs in the lowest layers. But the cold air at say 2000 feet is not affected by friction, and moves forward so as to overhang the warm air, producing an unstable arrangement which breaks down from time to time, giving squally winds. After such a breakdown the overhanging front begins to build up again, and the process may be repeated again and again, without holding up appreciably the advance of the effective cold front.

The warm front cannot be propagated in quite the same way, and the surface layers of air both in front of and behind the warm front may be retarded by surface friction. At higher levels surface friction can have no appreciable effect, and the front at say 2000 feet should advance freely so far as surface friction is concerned. The only way in which the retardation of the warm front can be explained kinematically is by the assumption that the cold air has an acceleration down the slope of the front, sufficient to produce the difference between the rate of advance observed and the gradient wind.

In an effort to discuss the occlusion of the warm sector along the lines suggested above, by considering separately the conditions at the warm and the cold front, we are at a disadvantage in not discussing the depression as a whole. The fact that a depression forms at a front, and continues to deepen, one of the features of the deepening being the occlusion, indicates some deep-seated instability in the whole system. No complete explanation of the occlusion of the depression is likely to be evolved until the nature of this instability has been found. Some further evidence bearing on this point is adduced later in connection with the discussion of rainfall in depressions (see § 198).

§ 195. *The drawing of fronts*

A front as drawn on a synoptic chart should separate masses of air which have some definite differences in physical characteristics of temperature, humidity, transparency, or motion. Some or all of these characteristics may differ for the two masses. In actual synoptic practice only fronts defined with reference to differences of temperature have been formally defined and are in general use, the other factors mentioned being treated as subsidiary aids in defining the position of the fronts. It was pointed out in § 127 above that a surface of discontinuity between two air masses in steady motion must intersect the surface either at a trough of low pressure, or in a line having stronger gradients of pressure on the high-pressure side than on the low-pressure side. When the surface of discontinuity is not parallel to the isobars but cuts across them at a finite angle, it is clear that there must be a sharp bend in the isobars where they cross a surface of real discontinuity. If the surface of discontinuity is replaced by a zone of transition, the sharp refraction of the isobars is smoothed out; and if the zone of transition is wide, the isobars at the zone may be rounded as were the isobars on the charts drawn in "pre-frontal" times. But granted the presence of a clear-cut surface of discontinuity, the element which should most clearly define its position is obviously pressure, provided observations are available from a close network of stations. The isobars will distinguish the sharp discontinuity from the wide frontal zone by the sharp bend in the former, as compared with the rounded form appropriate to the latter.

Surface temperatures are notoriously unreliable for tracing fronts. Temperatures at say 2000 feet would afford a much more satisfactory criterion, but such temperatures are not usually available. In actual practice the forecaster has to rely very largely on indirect evidence. Over the Atlantic observations of wind are of value in fixing fronts, and over the western coasts of the British Isles the barometer tendency is also frequently useful. The tendency usually increases with the approach of a cold front, and after the passage of the front may either be steady, or show a rise or a slower fall. The pressure usually rises after the passage of a cold front, and a positive tendency may indicate that a front has passed the station, though this is by no means universally true, and a rise of pressure may occur in tropical air, being due to the advance of an anticyclone or a general surge of pressure. Dew-points are also of use in fixing fronts, though they are apt to mislead if deduced from observations when rain is falling. Even when rain is not falling, water-vapour content is not a very conservative property. Willett* has stated that when northerly currents in winter pass over the Gulf of Mexico the water content at the surface may rise from 1 to 15 grammes per kilogramme of dry air in 36 to 48 hours, the corresponding dew-points being 0° F and 68° F. Other useful factors are the types and heights of clouds, and the presence of precipitation. The evidence in favour of the association of rain and fronts is so overwhelming that if it is known from other evidence that warm and cold air masses are in juxtaposition in a certain region, it is legitimate to connect with this fact any rain which falls in that region. Considerable caution is necessary however in making such use of precipitation alone to define a front, since the rain often runs ahead of the front as shown on a mean sea level chart, and also since a few casual showers may appear in line on the chart without being due to an occlusion.

It will be gathered then that the drawing of fronts on a synoptic chart cannot be regarded as a purely impersonal scientific operation, unless a close net of pressure observations is available. The use of indirect methods cannot be wholly impersonal, since each individual will have his own system of weighting the indirect factors he uses. As a result of this, the drawing of fronts is full of pitfalls for the unwary. It is all too easy to draw fronts over the Atlantic representing depressions as having warm sectors with sharply bent isobars in regions where no observations are available, and where rounded isobars are equally likely. Next to this in attractiveness is the drawing of occlusions in regions where no evidence of the previous existence of a warm sector is available. A genuine occlusion as defined by J. Bjerknes and H. Solberg originates in the passing away of a warm sector, and though an occlusion may have steady rain associated with it for some time after the occlusion appears it does not of necessity follow that any three rain-dots in line on the synoptic chart should be joined by an occlusion or other form of front. The criticisms which can be brought against the methods based on fronts and air

* "American Air-Mass Properties", *Mass. Inst. Tech., Met. Papers*, **1**, No. 4 (esp. p. 4).

masses are mainly due to the complexity of the atmosphere. J. Bjerknes and others have shown that cases occur in which large masses of air may be treated as homogeneous, but one finds, not infrequently, that the air over a limited region cannot be treated as a single mass, but can more appropriately be described as a patchwork of masses of air of different life-history. In such a case the practical meteorologist has to effect a compromise between the complexity of the actual atmosphere and the simplicity which is desirable in any representation which shall be of practical use in the analysis of the weather.

The difficulties in the way of drawing fronts are confirmed by an examination of the charts showing fronts issued by different European meteorological services. The differences are astonishing, not merely in the details of the fronts, but in their main outline, and the forms of isobars based on the same observations frequently show wide variations.

Frontal analysis is closely related to the movements of air as shown by trajectories. A skilful use of fronts and trajectories together cannot fail to assist the forecaster in understanding the processes which are taking place. The difficulty in the way of the general use of these two methods jointly is that there is no absolutely certain method of drawing a trajectory of air at 2000 feet, the gradient wind being often unreliable. Such a trajectory, when it can be drawn, represents the motion of a far greater mass of air than does the surface trajectory. If the trajectory of the warm air apparently crosses a well-defined front, it can be concluded that the warm air has certainly gone up. It is therefore advisable to check the trajectory at different stages with the fronts.

§ 196. *Sharp and diffuse fronts*

The records of temperature reproduced above in figs. 84, 88, 89 and 90 indicate that very wide differences in the sharpness of fronts are to be seen, and it is of importance to obtain some indication of the causes which underlie this variability. We shall consider some of these causes in turn.

(a) SUBSIDENCE

When a mass of air subsides, i.e. descends and spreads out laterally, it is dynamically heated and its lapse-rate is also modified. In § 23, p. 44 above, it was shown that if a mass of air at a mean pressure p covers a horizontal area S, then the difference between its lapse-rate and the dry adiabatic is proportional to Sp. Thus when air subsides the difference between its lapse-rate and the dry adiabatic increases, its initial stability or instability being increased. Further, since in the absence of rain its absolute humidity cannot change, the relative humidity decreases with increasing temperature. The occurrence of very low relative humidity may usually be taken as an indication that the air has subsided. The reverse of this is not true, and all air which has recently subsided will not of necessity have low relative humidity. Rain falling from a higher level and partially or wholly evaporating in the subsiding air will maintain high relative humidity and will lower the temperature of the subsiding air both by evaporation and by direct conduction. In such cases the wet-bulb potential temperature will be a more useful criterion of the origin of the air (see § 54), though this is not entirely satisfactory, being vitiated by the effect of cooling by direct conduction from the air to the raindrops. Subsidence normally occurs in cold air, and may occur either at a warm or a cold front. If the cold air adjacent to the front subsides, it has at the surface a higher temperature than the original cold air. J. Bjerknes* suggests that this results in the front originally separating the warm and cold masses being replaced by two fronts, one separating the undescended cold air from the cold air which has subsided and another separating the cold air which has subsided from the warm air. The contrast of temperature may in such a case be concentrated at the first of these two surfaces. The air in the transitional zone will show marked dryness if no rain has fallen through it. The double front may at a later stage develop into a single front through the ascent of the air warmed by subsidence, the ascent involving convergence and precipitation.

Bergeron has suggested the name *frontolysis* to denote the smoothing out of a front from discontinuity to continuity, and the name *frontogenesis* to denote the reverse process.

The subsidence of air behind a cold front on Oct. 29, 1923, was very clearly described by M. A. Giblett in a note on "Upper Air conditions after a line squall" in *Nature*, **112**, 1923, p. 863. In this case, two aeroplane ascents were made, one immediately before the arrival of the cold front, and one some hours after the cold front had passed, and the results are plotted in fig. 94. Between the two ascents, the temperature fell at all heights below 3 km. In the second ascent, the cold air showed an adiabatic lapse-rate to about 1·3 km, while between this level and 3 km conditions were approximately isothermal, with extreme dryness. The air between 1·3 km and 3 km can only be explained as air descended from the upper part of the cold wedge.

The phenomenon of descent of cold air at the cold front is of very frequent occurrence. In the autographic records it should, according to J. Bjerknes, show itself as a duplication of the cold front, the transitional air being usually drier than either the warm or cold currents which have not descended. But, as noted above, extreme dryness will not always occur as the evaporation of falling raindrops may lead to high humidity. A good example of a double cold front over England on September 1, 1925, has been described by R. S. Read in the *Q.J. Roy. Met. Soc.* for October 1925. In this case, as in that cited by Bjerknes, the rain fell mainly at the first cold front, and the humidities in the air behind this were still high, though rather lower than in the warm air.

It is by no means established that subsidence is the cause, or even the main cause, of double fronts. Some of the factors enumerated below, notably (*e*) and (*f*), may be of considerable importance. J. Bjerknes (*loc. cit.*) states that air which has subsided near the cold front "may have been heated by descending, but kept wet by the rain". This is not consistent with the known effects of the evaporation of water into air. If the rain falls continuously through the subsiding air, the latter will descend along a saturated adiabatic,

* M.O., *Geophys. Mem.* No. 50.

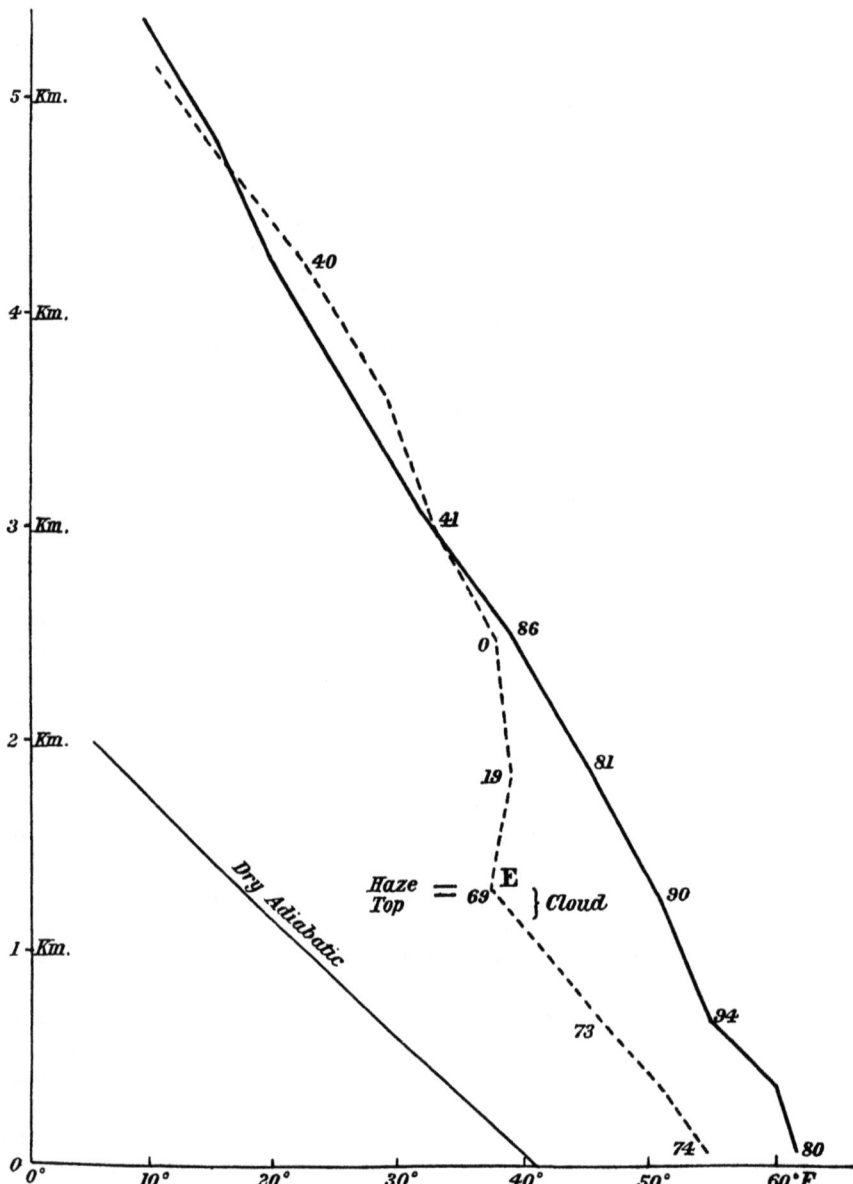

Fig. 94. The effects of subsidence; upper air conditions behind a line squall.
The numbers shown against the curves indicate relative humidity.

retaining its wet-bulb temperature unchanged. The lapse-rate within an appreciable mass will then vary much less than when the air is dry. If the air first subsides, and afterwards is cooled and wetted by rain falling through it, then, apart from effects due to turbulence, the result will be precisely the same as before, since the wet-bulb potential temperature of each element of air will remain unchanged, and the whole mass should approach saturation.

(b) INCLINATION OF THE FRONT TO THE ISOBARS

It was shown in Chapter XII, p. 257, equation (102), that friction at the ground causes a drift of air across the isobars, and that if a front is inclined at an angle β to the isobars, the frictional drift of air across the front, which is assumed to move with the geostrophic velocity, is proportional to $\cos(\alpha - \beta)$, α being the angle between the surface wind and the isobars. The frictional drift will help to keep the warm air up close to the front, and will tend to keep the front sharply defined, particularly if the angle between the front and the isobars is greater in the cold than in the warm air. The effect is greatest when $\beta = \alpha$, and diminishes as β increases. When the angle between the front and the isobars is large, the effect of surface friction in checking the motion of the warm air may cause the slope of the front to become extremely small, and turbulence will then tend to smooth out the front more effectively. Thus fast-moving warm fronts which lie across the isobars at a considerable angle will usually be more diffuse than those which lie nearly along the isobars. This is illustrated in figs. 89 and 90, which show that the warm front was sharp at Eskdalemuir, where the angle between it and the isobars was small; while at Holyhead, where the angle between the front and the isobars was large, the front was less sharp.

In the case of a cold front it is less important that the surface air should keep up with the front, since if the cold air comes in first at the 600 metres level it will normally descend owing to thermal instability. Thus a cold front should normally be sharper than a warm front, except when subsidence from considerable heights occurs, in which case the cold front may become smoothed out as discussed in (a) above.

(c) PRECIPITATION

Reference has already been made to the effect of precipitation on subsiding air. The cooling effect of rain and snow, both by evaporation and direct conductive cooling, may be of very considerable significance in the formation or accentuation of fronts. In an air-mass which is reasonably homogeneous initially it is possible for precipitation to produce large local differences of temperature. An example was shown over the British Isles on May 9, 1932. Croydon in the rain area had at 13 h a temperature of 40° F, while Cranwell, in the same air-mass but outside the rain area, had a temperature of 52° F. Since at the warm front the rain falls through the cold air, which is usually relatively dry, the effect of cooling by the rain is to enhance the difference of temperature. At cold fronts the rain falls to a greater extent through the warm air, and the cooling by rain extends to the warm air as well as the cold air. As a result, the cold front may become considerably more diffuse when judged by temperature records than when judged by wind records. Cases in which the warm air before a cold front is cooled by rain occur mainly in summer, when the lapse-rate is steep, and the air dry before the rain starts. It will be recalled that the limit to which the temperature can be lowered by evaporation is set by the initial wet-bulb temperature (see § 52).

The contrasts of temperature at the front in fast-moving line squalls is sometimes diminished in this manner. In such systems there is frequently some rain before the front, a brief heavy shower at the arrival of the cold air, and then a clearance. A wider belt of rain behind the cold front occurs more frequently with slow-moving fronts, though fronts show considerable individual variation. Fronts which show contrasts of temperature as great at the ground as in the free air can probably be explained in part by the accentuating effect of rainfall, and some fronts in polar air may possibly be explained as the effects of rainfall.

(d) Molecular and eddy diffusion

If initially a perfectly sharp mathematical surface of discontinuity existed between warm and cold masses, the effect of molecular diffusion would be to cause the front to become less clear cut, though molecular diffusion is so ineffective that the result would be inappreciable. In the atmosphere the effects of wave motion in the front, combined with the diffusing power of eddies formed in the warm and cold currents, will lead to the formation of a zone of transition. The computations made in Chapter XII, on the height to which turbulence can extend its influence in a given time, show (see p. 227) that within a time of 6 hours the zone of transition may become as much as $\frac{1}{2}$ km in depth, and therefore some 50 km in horizontal extent. Reference was made on p. 242 above to the demonstration by Douglas that there is an upper limit to the angle of slope of the front, beyond which turbulence increases rapidly, and tends to produce a wide zone of turbulent mixing instead of a sharp discontinuity. The growth of turbulent mixing is likely to be rapid if the air-masses are not homogeneous. There is usually a component of velocity parallel to the front, which results in variations of acceleration normal to the front if the masses are not homogeneous.

(e) Variation of wind with height

One effect of the variation of wind with height is to cause the warm front at say 2000 feet to be carried ahead of the front defined by surface temperatures, which is fixed by the air which lags behind on account of surface friction. The slope of the front near the surface is then extremely small, and the rain is associated with the steeper part of the front. In such a system it will be the front at some such height as 2000 feet which will be associated with the normal frontal changes in tendencies, with the cessation of rain, and with the trough indicated by the barograph. The passage of the front indicated by the surface observations may also produce a feeble trough on the barograph. The conditions near the surface will favour turbulent mixing, and the sharp front will tend to be replaced by a zone of transition. Variations of wind with height above the level of say 2000 feet may introduce additional complexity, and will promote turbulent mixing. It is thus clear that at a warm front the conditions may be extremely complex. There is no accepted method of fixing the front, though most meteorologists appear to accept the rain as the best criterion for its position. As already pointed out in § 195, the use of the rain for this purpose may involve arguing in a circle, unless there is independent evidence for the existence of a front.

The normal increase of wind with height from the surface up to say 500 metres causes the cold air at a cold front to overrun the front at the surface, giving the cold wedge an overhanging nose. In one or two cases the backward slope between the warm surface air and the overhanging cold air has been directly observed. The system so postulated is unstable, and the cold air must tend to fall to the surface, leading to some turbulent mixing, and giving two fronts separated by an intermediate zone. The form of cold wedge with a nose at say 500–600 metres above the ground might account for the rain frequently observed in advance of a cold front, but it cannot be said that any adequate explanation of the latter has yet been found.

(f) Variations of temperature within the warm or cold air

When there are conditions favourable to the warming or cooling of the warm or cold air or both, frontal phenomena may be complicated. The case which suggests itself is that of the warming of a shallow mass of cold air by passage over warm land or sea. The difference of temperature between adjacent warm and cold masses may then be practically annihilated, so that the existing front becomes so changed as to be undetectable from surface temperature records alone. The night cooling of air by ground cooled by radiation, depending as it does on the cloud amount, leads to distributions of temperature at the surface which may simulate fronts, though the phenomena may be purely superficial.

The absence of large variations of absolute humidity is probably the best check in such cases. But it is often difficult, on the basis of observations at 7 h, to analyse with certainty the air-masses shown on the chart.

The factors enumerated above possibly do not exhaust the list of factors which may affect the form and sharpness of fronts, but they suffice to show that each front has its own special features, and that general principles are not easily reached. So far no dynamical discussion of the factors which affect the sharpness of fronts is available, and the arguments given above are of necessity in very general terms.

§ 197. *Upper air conditions above depressions*

Sharply defined surfaces of subsidence are often observed in the polar air within depressions, but the inversions at these surfaces are usually small by comparison with the differences between air-masses.

Observations show that in the upper air over depressions sharp surfaces of discontinuity between air masses of different origin are rare, and that the nearest approach to sharpness is a discontinuity smoothed through a range of 0·5 to 2 km of height, which can frequently be traced to heights of 6 km. Probably the commonest occurrence is a wide frontal zone of transition, which is not of necessity to be explained as a degenerate front, but is in all probability the normal form of transition from warm to cold air at higher levels. When therefore we refer to fronts in the upper air, it is to be understood that these fronts are not of necessity visualised as sharp discontinuities. Such fronts may have a very important bearing on rainfall, and on the development of depressions. An occlusion with similar polar currents on the two sides comes into the category of upper air fronts.

In discussing the very important question of the three-dimensional structure of depressions, we must be on our guard against certain vagaries which in practice complicate the subject. A well-marked front will cut the surface at a trough of low pressure, or, less frequently, parallel to the isobars, with higher gradient of pressure on the side of high pressure than on the side of low pressure. Some meteorologists appear to regard the converse proposition as true, that every trough is to be interpreted as associated with a front, even when there is no observational evidence of any difference between the currents on the two sides. Such cases could equally well be interpreted as rotary systems carried round the main centre in the circulation of the main depression. It is by no means uncommon for a dying primary depression to swing round a newer and more active centre (cf. fig. 93 c, d). The phenomena then observed at all heights indicate approximate symmetry about the old centre, though on the isobaric chart the dying depression may appear as a trough of low pressure. Cases also occur where a vortex in the polar current causes distortion of a front, with subsequent accentuation of the depression at the centre. There are also occasions when a development on the front itself superposes some vorticity on the polar current near it, the result appearing as a rounded trough. There is always some tendency for rain to occur at a trough, on account of frictional convergence, and some fronts drawn across the isobars are troughs of this type, in which the effects of rainfall may simulate closely the phenomena at a true front.

Douglas[*] showed that in well-marked currents, leaving anticyclones out of account, the temperature conditions in the upper air depended very largely on the eventual origin of the currents, going backwards not just a few hours but for 3 days or even a week. The variables which showed a close correlation with the southward displacement in the previous three days included the pressure at 9 km, the height of the base of the stratosphere, the temperature at the base of the stratosphere, as well as the temperature within the tropo-

[*] *Q.J. Roy. Met. Soc.* **51**, 1925, p. 229.

sphere. Douglas's results suggest that polar currents extend to well within the stratosphere, and that they therefore bring low temperatures at all heights within the troposphere, and carry with them the low and warm stratosphere of high latitudes, while genuine tropical currents bring high temperatures in the troposphere, and carry with them the high and cold stratosphere of low latitudes. This accords well with the results derived by Gold,[*] who found that the stratosphere was high and cold over the southerly current in front of a depression, and low and warm over the northerly current in the rear of a depression. These investigations show that when currents of air of widely different origin are brought into juxtaposition, steep temperature gradients will occur in the upper air at all heights within the troposphere.

Above the surface of discontinuity a large increase of wind is a regular feature. Cirrus velocities range up to 150 m.p.h., and frequently exceed 100 m.p.h. in winter, and occasionally do so in summer. This accords with the effects to be anticipated from the temperature gradients in the neighbourhood of the front (see Chapter IX).

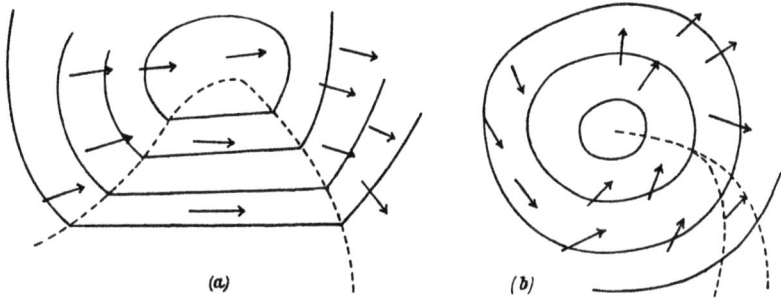

Fig. 95. Motion at cirrus levels above depressions.

The trough of lowest pressure is farther west in the upper air than at the surface. When a depression moves eastward the cirrus may continue to move rapidly from a southerly direction for some time after the wind at 15,000 feet has veered and fallen off in velocity. Also, after the passage of a cold front the tops of the showers attain heights of 20,000 feet or more. These observational facts, together with the correlation coefficients derived by Douglas (*loc. cit.*), indicate that the effects of cold air extend beyond 20,000 feet.

The motion at the cirrus level above a depression has been given in diagrammatic form by several writers. Fig. 95 reproduces the diagrams of Douglas.[†] Fig. 95 *a* represents the conditions in a depression with a large warm sector, and fig. 95 *b* represents a depression well advanced towards the dying stage. The first diagram indicates that in a young depression it is only in the warm sector and the central region that the cirrus motion is in the direction of motion of the depression. At cirrus levels the thermal wind defined in Chapter IX is so large that the cirrus motion over the polar air near the front tends to be parallel to the front. This tendency is perhaps the most marked feature of fig. 95. Some depressions have in their initial stages a warm sector extending through 180°, and having the front running as an approximately straight line through the centre. In such cases there is across the central region of the depression a current of cirrus parallel to the front, and in the direction of motion of the centre.

Douglas's diagrams show that the structure of the depression changes even at the cirrus level as the depression progresses towards occlusion, and that no single diagram can be drawn to represent the whole circulation in a depression. The detailed structure probably varies from one depression to another, even at comparable stages of development. Douglas indeed hints in one of his papers that depressions only have one completely common feature, a centre of low pressure with a counter-clockwise circulation of winds.

The diagrams of fig. 95 a and b above, together with the diagrams of the surface circulation already given in figs. 78 and 82, yield a picture of the three-dimensional structure of the typical depression. The motion at the level of the alto-clouds is intermediate between the surface and the cirrus motions, but nearer to the latter as a rule. In the early stages of the development of the depression there may be no closed cyclonic isobars at the cirrus level, but closed isobars show up in the later stages, the lowest pressure occurring to North-west of the surface centre. As a result, it is only far behind the surface centre of low pressure in a depression having a warm sector that the wind at cirrus levels veers to NW. In fig. 95 (a) no NW current is shown in the rear of the depression at cirrus levels, since its inclusion would give a misleading impression of a net flow toward South. The displacement to West and North-west of upper air features, relative to the surface features, leads to one of the most troublesome features of forecasting. It has the result that when a depression moves eastward, an observer at a fixed point observes changes in the lower troposphere before those in the upper troposphere or the stratosphere. Thus upper air observations do not in general give an earlier indication than surface observations of changes which are taking place; in fact the surface indications come first. The lag between changes at the ground and at the cirrus level is very clearly shown when a wedge of high pressure moving eastward is followed by a new depression. The upper wind will then in general continue from North-west for some time after the surface wind has backed to South-west. Thus apart from the information which they yield concerning the stability or instability of air masses, and the consequent chance of rain or thunderstorms, observations of upper winds and temperatures have been far less useful to the forecaster than was anticipated some 20 years ago. Such observations are more valuable for the subsequent investigation of depressions and other phenomena than for use in day to day forecasting of changes over periods of a day or more.

It is usually difficult to obtain observations of wind or temperature in the central region of a depression, and on account of the relatively short duration of the phase in which a depression has a well-marked warm sector it is not surprising that so far little information relating to warm sectors has been obtained by sounding balloons. The best method available for studying warm sectors is the examination of continuous records from high mountain observatories. The observatory on the Pic du Midi in the Pyrenees, at a height of 2859 metres, lies near the Atlantic, but south of the main cyclone tracks, and only relatively few depressions affect its pressure records, though its temperature records show fairly frequently large fluctuations associated with the passage of depressions across the British Isles. Douglas* collected data for 18 shallow winter depressions during the years 1895–1914 which had well-defined warm sectors and whose centres passed within 300 miles of the Pic du Midi. When the centre was nearest to the Pic du Midi the temperature was on the average 1° C above normal (after allowing for diurnal and seasonal variations), and apparently showed remarkable uniformity in the different cases. At the extreme boundary of the system, the temperature at the Pic

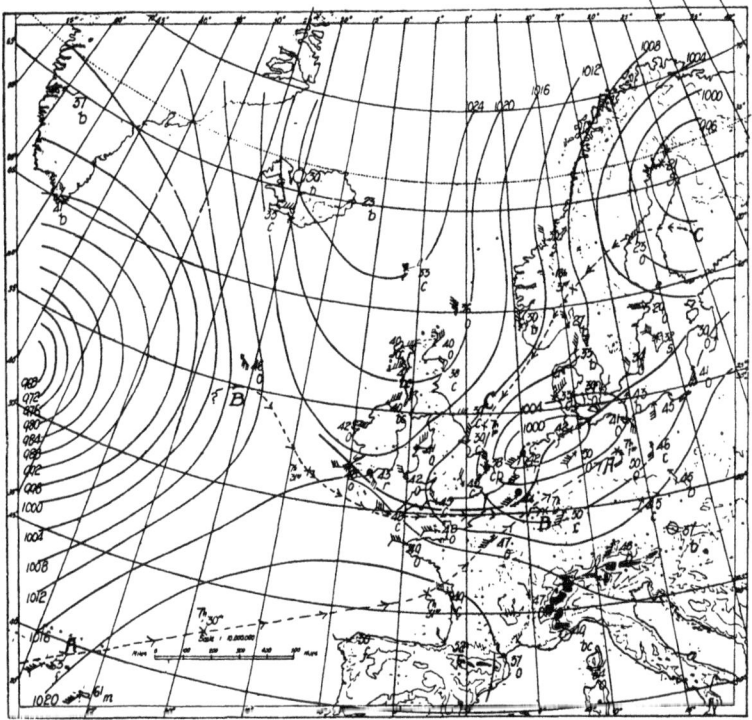

Fig. 96. A depression with a marked equatorial sector, April 1, 1909.

du Midi averaged 5° C below normal, though occasionally reaching 10° C below normal; the highest temperature was a little in advance of the centre; and after the passage of the centre a pronounced fall of temperature, averaging 7° C, was shown. These figures give a picture of the conditions at 3 km in a shallow depression having a warm sector. Douglas quotes in the same paper another interesting case. On April 1, 1909, a small depression centred over the northern coast of Germany (fig. 96) showed a marked equatorial sector. Gradient wind trajectories showed that the air in the warm sector had come from south of latitude 40°, passing across the southern part of the Bay of Biscay. The air in the northerly current in the

* *Q.J. Roy. Met. Soc.* **50**, 1924, p. 339.

rear of the depression had come from the Northern Baltic, while the air in the region just in advance of the cold sector had come from the Atlantic, from about latitude 55° N, longitude 20° W, with doubtful earlier origin, and had swept over the English Channel. Upper air soundings were made in these three currents at Berlin, Pyrton Hill and Paris, respectively. These are shown in fig. 97. At Berlin, in the equatorial current, the temperature was above the April normal, and was much higher than at Pyrton Hill at all levels within the troposphere, the difference amounting to 18° C at 6–8 km, but at Berlin the tropopause was higher and the stratosphere colder than at Pyrton Hill. On April 2 a kite ascent at Berlin showed that the temperature at 3 km, now behind the centre, had fallen 11° C, but no observations could be made to appreciable heights in the intervening period. The records reproduced in fig. 97 show how great the difference can be between air masses of different origin brought into juxtaposition, though the surface temperatures may differ only slightly. They also demonstrate in a striking manner the fact that the warm sector is warmer than the surrounding polar air at all levels within the troposphere. Confirmation of this will be found scattered through very many papers by Douglas and others.

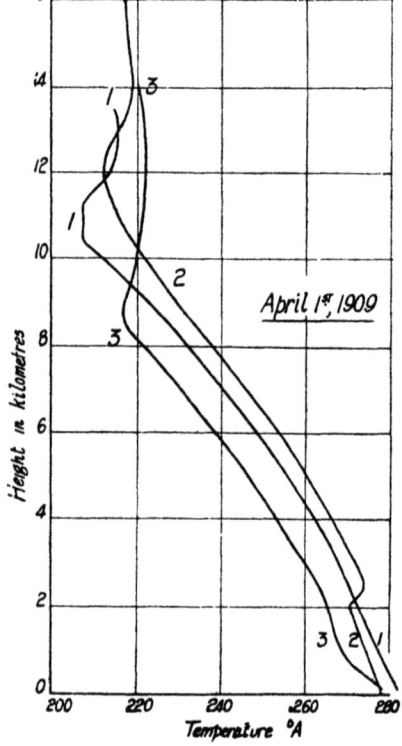

Fig. 97. Upper air observations on April 1, 1909. 1. Berlin, 8 h. 2. Paris, 4 h. 3. Pyrton Hill, 17 h.

§ 198. *Rainfall in depressions*

The fact that the presence of a front is not by itself sufficient to produce rainfall was emphasised by Brunt and Douglas,[*] who also pointed out that in a depression in which the winds were everywhere geostrophic there could be no appreciable rainfall, since no convergence of winds is possible so long as the geostrophic condition is satisfied.

At a sharp front where the warm air rises up over the cold air, the rainfall can be computed for any assumed values of the slope, of the differences in horizontal velocities on the two sides of the surface, and of the temperature and humidity of the warm air. Let the components of horizontal velocity at right angles to the front be u_1 and u_2 in the warm and cold air respectively, and let w_1 and w_2 be the corresponding vertical velocities. We assume $w_2 = 0$. If

[*] *Mem. R. Met. Soc.* **3**, No. 22.

ψ is the angle of the slope, then from equation (18), Chapter x, p. 210, taking $u_1 - u_2 = 4$ m/sec, tan $\psi = 0\cdot01$, and accordingly $w_1 = 0\cdot04$ m/sec, the rate of rainfall at different distances from the front is shown in the lower part of fig. 98 for conditions shown in the upper part.

It is assumed that all the condensed water falls as rain. The warm air is assumed to be saturated and in convective equilibrium at all heights, so that the maximum rainfall must occur at the front itself. If the warm air fell short of saturation at all heights, there would be a narrow rainless band in advance of the front. Both cases are to be found in practice, as well as intermediate cases in which the air is saturated at some levels and unsaturated at others. A layer of about 6 km is assumed to take part in the upward motion.

It was pointed out in Chapter x that the winds in the warm and cold masses would have slightly differing components at right angles to the front. Let

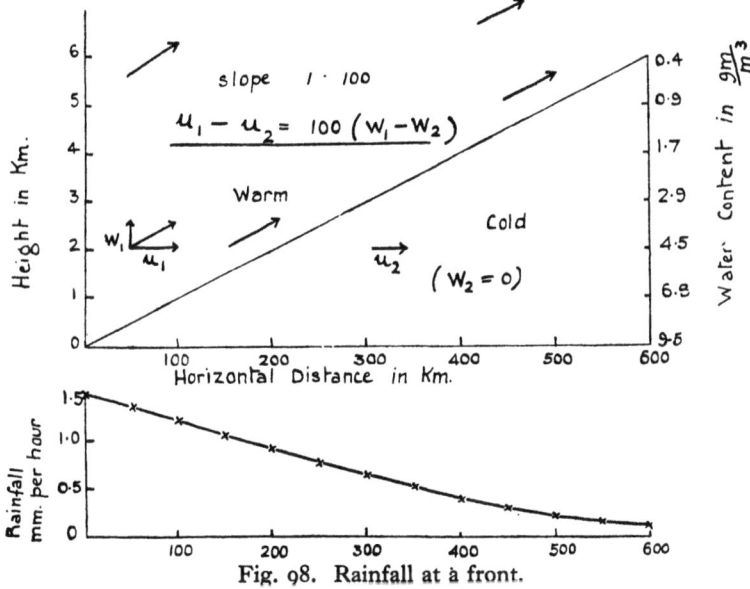

Fig. 98. Rainfall at a front.

A and B be two points on a front, and let p_A, p_B be the corresponding pressures. The component of pressure gradient along the front is $(p_A - p_B)/AB$. For a discontinuity of $3°$ C the density is 1 per cent less in the warm air than in the cold air, and if the winds are geostrophic $u_1 - u_2 = 0\cdot01u_2$. Let u_2 be 20 m/sec, and take the other variables as in the preceding computation. The resulting rainfall is now less than $0\cdot1$ mm per hour, a value considerably less than is observed at actual fronts. The mere presence of a front, acting through the density effect, is thus not capable of producing heavy rain, though it may produce cloud and slight rain in advance of a warm front. At a cold front it can only produce slight subsidence.

When a system of V-shaped isobars moves across without change of form, there is an inflow of warm air across the front as a result of surface friction. It has been shown in § 153 above that when the geostrophic wind velocity is

10 m/sec, $\alpha = 22\frac{1}{2}°$, and $K = 10^5$, the total frictional inflow across a front inclined at an angle β to the isobars is about 15 cos $(\alpha - \beta)$ m³ per second per cm of the front. If the slope of the front is 1 in 100, and the initial state of the air is as shown in fig. 98 above, then it is seen from that figure that the air in rising through $\frac{1}{2}$ km loses 1·5 grammes of water per cubic metre, so that 15 m³ lose 22 grammes of water in so ascending. The air moves 50 km in a horizontal direction, and the 22 grammes of water per cm of front is distributed over a strip $\frac{1}{2}$ cm wide and 50 km long. The amount of rain per cm² per sec is thus $4 \cdot 10^{-6}$ cc, which is equivalent to 0·14 mm per hour. Such rain would be very light drizzle, and even in the most favourable circumstances could not yield rainfall in amounts comparable with that due to isallobaric convergence, discussed below. The rainfall which has been computed above is moreover the maximum which corresponds to $\alpha - \beta = 0$, or corresponding to a front parallel to the surface wind. When the angle between the front and the isobars is large, the rainfall is diminished in proportion to cos $(\alpha - \beta)$.

The above computation assumes that the warm air ascends a uniform slope. There is definite observational evidence that the form of cold fronts is as shown in fig. 99, the trough line being below the nose of the cold air. If the warm air ascends rapidly from the ground to the height of the nose, which may possibly be as high as 500 metres above the ground, the amount of rainfall

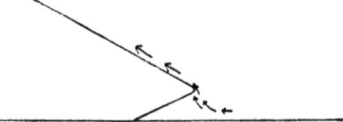

Fig. 99. Form of the cold front and its effect on rainfall.

which we have assumed to be distributed over a belt 50 km long will be concentrated, giving a narrow belt of heavy rain along the trough, with a wider belt of light drizzle behind it.

At an earlier stage it was pointed out that in a system of geostrophic winds no rain can occur. We can now supplement this by saying that the deviations from geostrophic winds in the surface layers, due to surface friction, cannot produce more than light rain at a front. Heavy rain in depressions must therefore be produced by deviations from geostrophic winds other than those due to surface friction.

The work of Brunt and Douglas quoted in § 117 indicates that in an isallobaric low converging winds having a component of 5 m/sec across the isallobars may occur. The same writers gave an estimate of the amount of rainfall produced by convergence of this magnitude over a circular area of radius 200 km, the convergence being assumed to extend up to 3 km. The total inflow of air across the cylinder, assuming an inward velocity of 5 m/sec, or 0·005 km/sec, is

$$2\pi \times 3 \times 200 \times 0\cdot005 \text{ km}^3 \quad \text{or} \quad 6\pi \text{ km}^3.$$

Assuming a mean water content of 6 grammes per m³ (see fig. 98), we find that the air brings in $6\pi \times 6 \times 10^9$ grammes of water per second; and if it is assumed that this is all deposited in the form of rain over an area of $\pi (200)^2$ km² which forms the base of the cylinder, the rate of rainfall is 10^{-4} cc per cm² per second, or 10^{-3} mm per second, or 4 mm of rain per hour. This is the right order of magnitude for fairly heavy rain, and since the amount of inflow into an area is proportional to its circumference, while the resulting rain is distributed over the whole area, it follows that with a given rate of convergence the rainfall is inversely proportional to the radius of the area over which the convergence takes place.

The computations which have been briefly summarised above show that heavy rainfall can only be produced by deviations from geostrophic winds leading to convergence. Convergence of the isallobaric components of winds (see § 117 above) at a trough is produced when the gradient is steepening or when a trough is becoming more elongated, and it is well known that in such cases the rainfall is heavy. The argument used above must be applied with caution to such cases, since the angle of slope of the surface of discontinuity is likely to vary.

When a depression is growing deeper, with the gradient of pressure everywhere increasing, the isallobaric component is everywhere at right angles to the isobars, and in fig. 100 (a) and (b) are shown the isallobaric components in different parts of a deepening depression at two stages of its existence. In stage (a) the convergence is great at both warm and cold fronts, causing the warm sector to shrink, with heavy rainfall at both fronts. In stage (b) there

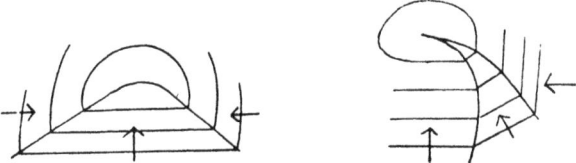

Fig. 100. Isallobars in a deepening depression.

is little convergence at the cold front, and there is little rain at the cold front at this stage, the frictional rain discussed in an earlier paragraph being then also small, on account of the large angle between the front and the isobars. At the warm front the convergence may still be considerable. The isallobaric component frequently exceeds 5 m/sec in the warm sector and in the cold air in advance of the warm front. The isallobaric effect at the latter (see fig. 100) manifests itself in the fact that the rate of advance of the warm front is usually less than the geostrophic wind; the advance of the cold front is accelerated by the isallobaric component in the early stages of development, as shown in fig. 100(a), but this effect diminishes with time, and by the time the stage of development shown in fig. 100 (b) is attained this effect has practically disappeared.

The use of isallobars requires a word of caution. In so far as the isallobars are due to development and not to the drift in a general current, the argument used above is sound; but it may require emphasising that a depression always formed of the same mass of revolving fluid and carried in a general current would produce a system of isallobars, though there would be no convergence or divergence within it. Unfortunately it has not been possible to separate out the parts of the isallobaric distribution due to motion from that due to development.

It is assumed in most of the arguments used above that the warm air is damp at all heights. If the warm air is dry, convergence may fail to produce heavy rainfall. After a spell of fine weather it is frequently noted that rain does not readily occur in the subsequent depression, and this is to be explained by the depression being fed with warm dry air at heights usually about 5000 feet.

Another feature of depressions in relation to rain must be mentioned at this stage. The lapse-rate in the warm sector, even in depressions which produce very heavy rainfall, may be less than the saturated adiabatic up to considerable heights. The occurrence of rain in such cases cannot therefore be attributed to thermal instability of damp air at low levels. Examples of this stability are shown in figs. 85 and 97. Douglas* has collected details of a number of cases in which heavy rain has occurred while the lapse-rate was stable for saturated air. This is of course not universally the case, and other cases could be cited in which heavy rain was associated with a lapse-rate unstable for saturated air. But as rain which is continuous for any considerable period must be due to the bodily ascent of large air masses, and not to local movements within the mass, there is in reality no special reason for expecting that the lapse-rate should be unstable for damp air at times when rain is falling, though instability extending to great heights must have important effects.

J. Bjerknes and Palmén† have also shown that in the depression of March 29–31, 1928, the mean lapse-rate up to 6 km was stable for saturated air, while in the polar air the lapse-rate was almost exactly equal to the saturated adiabatic up to 6 km.

The conclusion appears to be inevitable that rainfall in a depression is associated with convergence on a large scale, the convergence being brought about by large scale convection, which in turn is due essentially, not to thermal instability but to dynamical instability. The assumption of some essential dynamical instability would account for the forced ascent of damp air even when it is not thermally unstable, and would at the same time afford a reasonable method of explaining the fact that depressions become occluded. It is perhaps all too easy to think of a depression as consisting of a warm front and a cold front, which can be treated separately. This is obviously unsound, and the depression should be looked upon as one entity.

§ 199. *The vertical structure and extent of depressions*

While the large depressions which occur in middle latitudes appear to extend well into the stratosphere, small depressions frequently occur whose circulation is restricted to a height of 3 km or less. Figs. 101 and 102 reproduced from a paper by Douglas‡ give an interesting case of a depression on November 17, 1910, centred over the Bay of Biscay and moving rapidly eastward to the Alps. The centre of this depression passed near Bordeaux, where the barograph indicated a fall of pressure of 23 mb (fig. 102), with a pronounced minimum of pressure, while the thermograph indicated a narrow warm sector. The records at Pic du Midi (2859 m), 140 miles farther south, showed practically

* *Q.J. Roy. Met. Soc.* **55**, 1929, p. 127; also *ibid.* **60**, 1934, p. 143.
† *Beitr. Phys. fr. Atmos.* **21**, 1933, p. 53.
‡ *Q.J. Roy. Met. Soc.* **50**, 1924, p. 339.

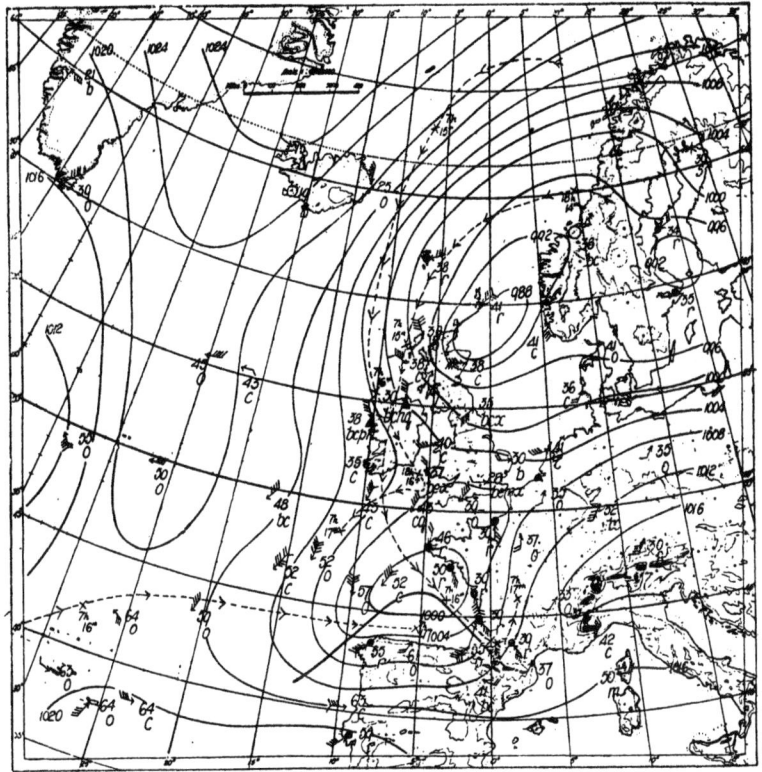

Fig. 101. The depression of November 17, 1910.

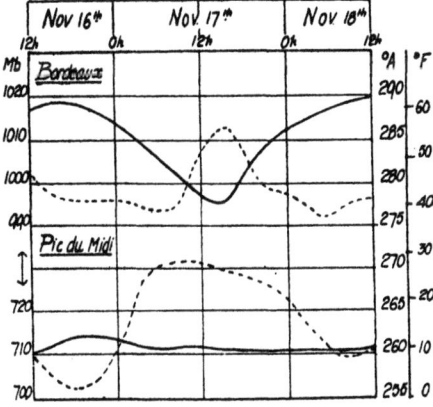

Fig. 102. Pressure (full line) and temperature (broken line)
at Bordeaux and Pic du Midi, November 16–18, 1910.

no change of pressure, but a broad warm sector with no pronounced dis continuity was indicated by the temperature records. This depression could scarcely be said to exist at all at a height of 3 km.

It is probable that all depressions with large warm sectors are superficial structures of this type, having no cyclonic circulation over them in the upper troposphere. Douglas* showed that the depression of November 2–4, 1925, which had a well-marked warm sector and produced a brisk fall of surface pressure of 12 mb, left the pressure at 4 km unaltered, while above 5 km pressure over the centre of the depression was higher than the pressure in the polar air at the same level the next day.

The large depressions of middle latitudes normally develop from smaller ones, which in the initial stages do not extend beyond a few kilometres of height. It is difficult to conceive of the development which then occurs as a mere progressive extension upwards through the troposphere. This question is discussed in a later section in connection with the advection of low pressure at high levels.

§ 200. *A brief review of the wave theory of the formation of extra-tropical cyclones*

The theory of the formation of cyclones as unstable waves at a surface of discontinuity, given in *Physikalische Hydrodynamik*, pp. 565–621, is primarily due to H. Solberg, who showed that there exists a class of free waves at such surfaces which is comparable with the cyclone wave shown on synoptic charts. A wave whose amplitude remains constant is a stable wave, and if initiated with a small amplitude will always retain that small amplitude and so escape observation by usual methods of synoptic analysis. But a wave whose amplitude increases with time, i.e. an unstable wave, will, even if initiated with a small amplitude, eventually grow and become observable on the synoptic chart. It is therefore among the unstable free waves at a front that we must look for the explanation of the origin of the extra-tropical cyclone.

There are several factors whose influence must be considered.† (*a*) The effect of compressibility alone is to generate sound waves, which travel with a velocity of about 300 m/sec. The velocity of travel of the cyclone is of the order of 1/10th of the velocity of sound, and it appears likely that the effects of compressibility may be neglected. (*b*) Gravitation acting on a heterogeneous fluid medium produces gravitational waves, essentially of the character of the waves which form on the surface of the sea. It has been shown by V. Bjerknes that in an atmosphere which is statically stable only stable waves with a constant amplitude can form, while in a statically unstable atmosphere, of the instability waves which can then occur, those with the shortest wave-length will increase in amplitude most rapidly. The disturbances produced will then be small in linear dimensions, and will be in no sense analogous to the cyclone. (*c*) Inertia due to the earth's rotation was shown by V. Bjerknes to have an effect similar to, though greater than, gravitation. The

---

* *Q.J. Roy. Met. Soc.* **55**, 1929, p. 123 (esp. para. 3. 1).

† Bjerknes, J. and Godske, C. L., *Astrophysica Norvegica*, **1**, No. 6.

waves which arise as an effect of the displacement of masses relative to the axis of rotation are of the nature of stability waves, and the combined effects of gravity and inertia also yield stable waves with constant amplitudes. (*d*) When there is a discontinuity of wind in an otherwise homogeneous fluid, waves can form in this surface. Such waves are always unstable, even if the atmosphere is everywhere statically stable, but it is found that the waves of shortest length have the greatest rate of growth.

We are thus obviously led to the conclusion that the cyclone wave is to be explained, if at all, by some combination of two or more of the four factors considered above. The mathematical treatment is too lengthy and involved to be reproduced here, but the essential results can be readily described. It has been shown by Solberg that shear at a surface of discontinuity on a rotating earth leads to an inclination for the surface of discontinuity as given by Margules (see equation (4), p. 204 above). The character of the waves formed at such a shearing surface will depend on the length of the waves. Short waves, of length of the order of a few kilometres, are instability waves, the shearing instability overpowering the effect of the static stability. Longer waves, but with wave-length less than a value of the order of 1000 km, will be stability waves, while still longer waves will be instability waves. Beyond this is a range of long stability waves, with wave-lengths greater than several thousand kilometres. Thus, dynamically, we are led to the conclusion that there should occur at a surface of discontinuity a class of instability waves of the right wave-length to account for the cyclone, as well as a short instability wave of the order of magnitude of the so-called "Helmholtz waves" deduced by Helmholtz from the classical theory of hydrodynamics, to explain clouds which take the form of long parallel billows. It is perhaps not without interest that small wave-like disturbances which often show on records of microbarographs can be shown to be disturbances of a few kilometres in diameter, by comparison with simultaneous wind records.

The theory is far from complete, and the mathematical difficulties are very great. The waves discussed are of the nature of infinitely long waves, whereas in the depression of middle latitudes the pressure gradients are of the same order of magnitude all around the isobars. But the theory has achieved one important result, in showing that long instability waves cannot occur in an atmosphere free from discontinuities of wind. This is of fundamental importance. The "cyclone wave" is a free oscillation, so that the barrier proposed by F. Exner (see § 205 below) is not necessary for the formation of a depression. The theory cannot be regarded as applicable beyond the initial stage of the development of the cyclone, and the later stages, including the explanation of the occlusion of the cyclone, cannot be said to have yielded to mathematical treatment. Further development will require the detailed investigation of the structure of individual depressions, before it will be possible to formulate the mathematical problems to be discussed in a sufficiently clear form.

The reader who desires to study the details of the mathematical treatment, the results of which are very briefly summarised above, is referred to the discussion in *Physikalische Hydrodynamik* and in the paper by J. Bjerknes and C. L. Godske already referred to, and to a paper by H. Solberg in the *Procès-*

Verbaux de l'Association de Météorologie de l'Union Géodésique et Géophysique Internationale, Edinburgh, 1936.

§ 201. *The causes of formation of depressions*

In § 109 above it was suggested that the only effective cause of production of horizontal circulations on a large scale in the atmosphere is convergence. Further, in Chapter xv it was suggested that a cyclonic circulation could be set up in a disc of air by the removal upwards of air from its centre. No attempt was then made to suggest any cause of removal of air, but it is probable that when depressions form in polar air, the procedure is similar to that visualised in Chapter xv, and that the air which is removed upward ascends partly on account of thermal instability. In a polar current the temperature is initially low at all heights in the troposphere. As the current attains lower latitudes it is continually warmed from below by contact with the surface of the earth, until instability is set up through a considerable range of height. The ascent of the lower unstable air is presumably brought about by some kind of trigger action, and the wind circulation is in the main produced by convergence towards the region of ascent.

If the depressions in polar air are to be explained as suggested above, the tendency for more than one such depression to occur is readily explained. For the conditions in the main polar current are likely to persist so long as the current itself persists, and the instability should therefore also persist. But it cannot be claimed that the details of the process of formation of such depressions are understood. The amount of upper air information relating to the early stages of their development is extremely meagre, and we can only guess at the nature of the trigger action involved. Moreover, an analysis of relevant synoptic charts shows that in the polar air there is usually a secondary cold front between different portions of the polar current which have travelled varying distances over a sea surface. In a current of maritime polar air sweeping first southward and then eastward the southern portion has had a longer trajectory over a sea surface usually warmer. The precise effect of the horizontal temperature gradients thus set up is not fully understood.

As was shown in § 200 above, the formation of depressions at polar fronts is not completely understood. The study of synoptic charts shows that while depressions frequently form at polar fronts depressions can also form in polar air (possibly through action at secondary fronts), and that fronts can exist without formation of depressions.

Thus depressions and fronts appear to be capable of separate existence, as well as of existence associated together. The initial stage of formation of the depression at a front appears to be capable of explanation as an instability wave. Once a depression has formed, its interaction with the front will lead to distortion of the surface of discontinuity in a manner which forms a corrugation in the front, and the corrugation will form a channel up which the warm air can readily ascend. The variation with height above the ground of the magnitude of the discontinuities of wind and temperature at the front may lead to the surface of discontinuity, even when sharply defined, not being a plane surface. This adds a further complication to the theoretical discussion.

One of the most difficult aspects of the problem concerns the lateral removal of air in sufficient quantity to produce the observed fall of pressure. Shaw suggested that the removal could be brought about by a strong current in the upper air, where velocity was much in excess of that of the currents in the lower troposphere. Such a current must exist when deep layers of warm and cold air are in juxtaposition. In a note in *Q.J. Roy. Met. Soc.* Jan. 1931, Douglas has given some consideration to this particular question. He cites December 14, 1929, as an example of the occurrence of strong cirrus motion above a front at which no depression developed. The gradient wind over South Scotland was 50 m.p.h., while cirrus velocities were estimated at 190 m.p.h. at Renfrew, and 150 m.p.h. at Leuchars (assuming a height of 5 miles). No depression developed, and the rainfall at the front was small. Somewhat similar conditions prevailed on October 24, 1929, above a front which gave heavy rain, exceeding 200 mm in 12 hours in places. This appears to indicate that the combination of convection and a strong upper current does not suffice to produce a depression.

There remains a possibility that the nascent cyclone has a purely thermal centre, a vortex formed by convergence towards a centre of convection, and that the further development is due to the interaction of the vortex and the front. A centre of convection is equivalent to the sink of hydrodynamics, and tends to set up a circulation represented by a simple vortex (vr = constant) over the region surrounding it. The interaction of the vortex and its surroundings can be readily seen from fig. 103, which comprises charts taken from a paper by Kaye and Durst.* In the first of these charts there is shown a depression formed in polar air, centred over Lake Superior, while a front sweeps from the Gulf of Mexico across the Atlantic to a depression south of Greenland. In the second chart, which shows conditions 24 hours later, the depression has moved to Newfoundland, but 12 hours later it has moved rather more slowly to the north-east of Newfoundland. The interest of these charts is in the motion of the front, which is drawn in towards the centre of the depression so that the depression has in the third chart a well-marked warm sector, while pressure has fallen very considerably. In the next 12 hours the centre moved to latitude 53° N, longitude 35° W, so that the motion had become much more rapid, and was now in the direction of the isobars in the warm sector. The depression had deepened very considerably.

In this case we have a weak cyclonic circulation strengthened by the interaction with a front, and it is possible that this represents the initial stages of all depressions which form at polar fronts. There is a further possibility that the spin of the depression is not produced in the lower atmosphere as in the case cited, but is present in the upper atmosphere, being brought to the neighbourhood of the front by advection. This view has found favour with many writers. Exner first put forward this view, and Douglas, as the result of the examination of many individual cyclones, appears to have reached a somewhat similar conclusion.

The argument put forward by Douglas is as follows. Since in the warm sector the horizontal temperature gradient is not generally steep, the possi-

* *Q.J. Roy. Met. Soc.* April 1932.

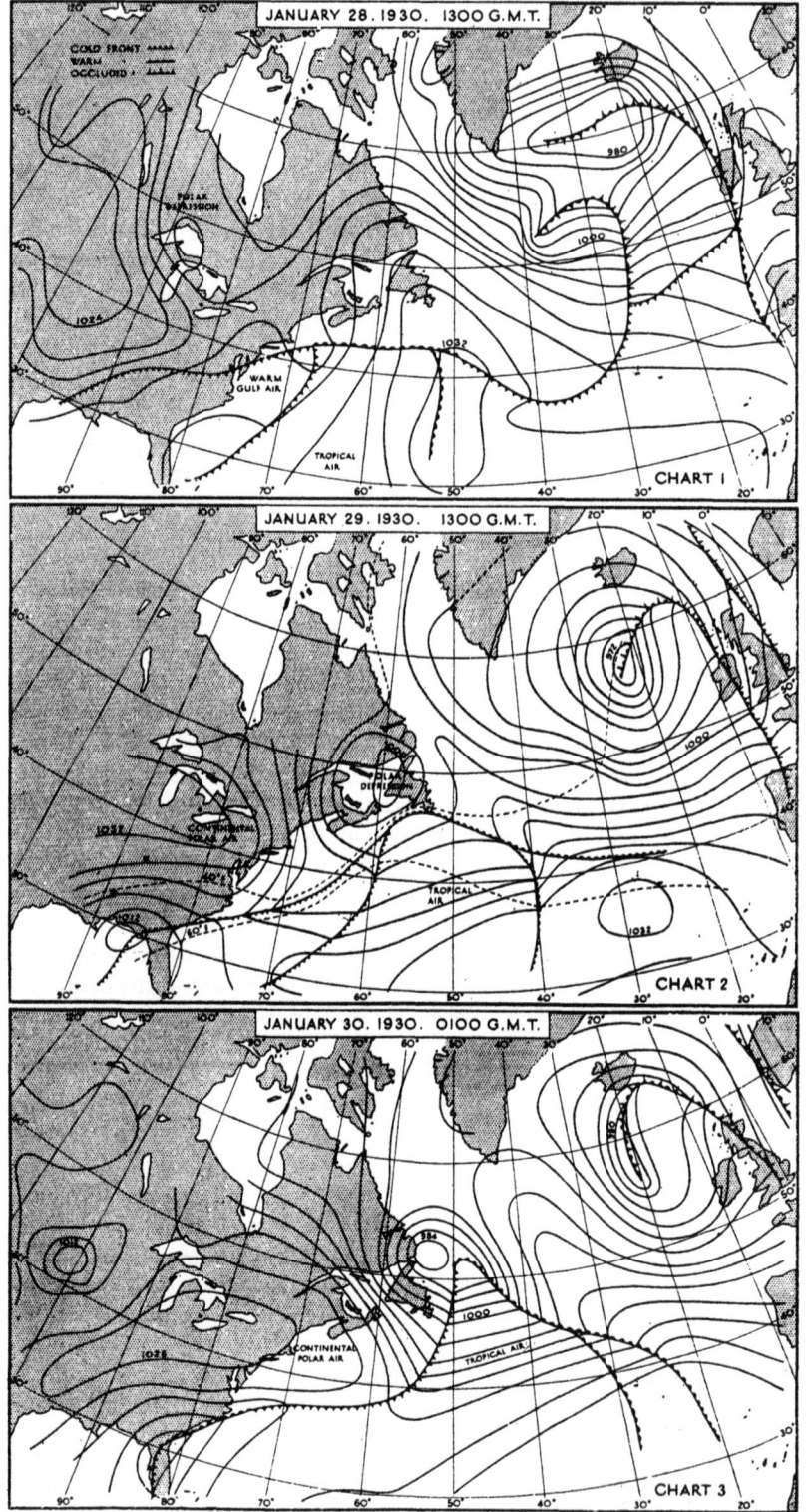

Fig. 103. Formation of depressions over the Western Atlantic.

bility of producing a fall of pressure by drawing in still warmer air is distinctly limited. Further, as shown in Chapter xv increasing cyclonic circulation implies convergence, so that when a depression deepens the required withdrawal of air takes place right above it. Over the developing cyclone observation shows the existence of strong cirrus motion on the polar side of the front, so that there appears to be no possibility of a supply of cyclonic angular momentum being available in the upper currents. Douglas in fact regards the circulation at cirrus level above a Bjerknes depression as anticyclonic. He therefore falls back on the only obvious alternative, an advectional effect with air spreading over at high levels, bringing with it lower temperature and pressure in the troposphere, and a low and warm stratosphere. The rapid deepening of a large depression is then to be regarded as associated with the spreading of polar air right over the system, simultaneously with the diminution and final disappearance of the warm sector. It is well known by observation that a pronounced polar current behind a depression favours its deepening. In the earlier stages of development of the depression the "high depression" is not of necessity closed at the centre, though in the later stages closed isobars may develop. This accords with the ideas put forward in § 197 above in connection with the cirrus motion at different stages of the development of the depression. The "low depression" formed at a polar front is not in itself capable of spreading up through the whole troposphere, and the large depressions which extend up to the base of the stratosphere are on this theory regarded as due to the amalgamation of a high level depression with a low depression, the former shearing across the latter. The rapid fall of pressure in big depressions is largely due to the advection of low pressure in the upper air, while the circulation at low levels is built up mainly through convergence towards the centre produced by the displacement of the warm sector by polar air. It has not yet been possible to treat this complicated problem mathematically. It appears however that when the advection of low pressure in the upper air superposes lower pressures on the lower air, some convergence into the area of diminished pressure must take place, with a growth of cyclonic circulation.

Douglas points out that in no single case has observation indicated the existence of a symmetrical vortex above a depression with a genuine warm sector. He suggests that the upper cold depression may possibly be a vortex travelling in the general current, but in the early stages of the development the high level depression lies altogether outside the system shown on the surface chart. It is possible that the lower system influences the upper one, by some mechanism not yet understood, but the whole problem is at the present time unresolved, and cannot be resolved until more upper air data over nascent and young depressions become available.

J. Bjerknes in a recent memoir* has suggested that the changes in height and temperature of the tropopause above a given place can be accounted for as due to the lateral movements of the tropopause, to North and South, produced by the effect of the surface circulation. The warm air flowing upwards over a calotte of cold air is forced to diverge on account of the restriction of the range of height in which it can flow, and an anticyclonic circulation is produced in

* *Geof. Publ.* **9**, No. 9. See also *Physikalische Hydrodynamik*, §§ 182-4.

the upper troposphere, while at the other side of the calotte of cold air descent of the warm air is associated with convergence and the growth of cyclonic circulation. Here again the phenomena visualised are too complex for mathematical treatment, and no complete theory has yet been evolved.

It is certain that the type of motion visualised by J. Bjerknes sometimes occurs. That warm air rising at a cold front can show vertical contraction, and therefore horizontal divergence, was demonstrated by Hewson* by the use of the wet-bulb potential temperature for the identification of air. If, however, the same mass of air goes up and then comes down again, the motion will involve the growth of spin during ascent, and the loss of this spin during the subsequent descent. Hence we should expect the anticyclonic curvature of paths of air over the warm front to be more pronounced than the cyclonic curvature over the cold front. A better view of the phenomenon is obtained when we postulate that the ascent and descent are in different air masses, the ascent being of warm air masses, and the descent being of cold air masses. The tendency for cold air masses extending originally through the whole troposphere to subside and diverge is well established by observation; and the growth of circulation in this manner is in accordance with the Margules scheme of growth of kinetic energy. If circulations can be thus generated at high levels, they should be capable of continued existence for a relatively long time, since the turbulent dissipation should be small in such systems. It is only in the lowest kilometre of the atmosphere that frictional effects are considerable, and at high levels these effects are slight (see § 171 above). In J. Bjerknes' scheme the upper air circulations would not be long-lived, but it is certain that upper air cyclic circulations which are set up in the atmosphere are capable of continued existence for fairly long periods. Dines found that the standard deviation of pressure was 10·4 mb at 9 km, and 11·1 mb at the surface. This evidence tends to confirm the existence of upper-air circulations. Systems already in existence high up would complicate the phenomena, and it is possible that the mechanism suggested above is unduly simplified.

It is not suggested that the depressions and anticyclones shown on the surface charts normally *originate* at high levels. All the researches of the last 20 years conflict with this view, and suggest a low-level origin for these systems; the high-level system amalgamates at a later stage of development with the low-level system (see also § 202).

In some of his earlier papers on revolving fluid Shaw suggested that an ascending current of warm air would undergo turbulent mixing with its environment, so that an increased amount of air could be removed from the core of the nascent depression by turbulence, the process being known as "eviction". The cyclonic circulation is in this theory a direct product of convergence. The evicted air was visualised by Shaw as carried away in a current whose motion was more rapid than that of the depression. The convection current cannot *generate* angular momentum; it can only re-distribute it; and if over the area of the depression the angular momentum is to increase in the cyclonic sense and the pressure is to diminish, the air which rises in the convection current must be carried away in the upper air, as the effect either

* *Q.J. Roy. Met. Soc.* **63**, 1937, p. 7.

of a shearing current, or of diverging currents. In either case there must be a growth of anticyclonic motions at high levels, but either cause would suffice to distribute this anticyclonic motion, and the excess of pressure associated with it, over such a wide area as to show no appreciable effect at the surface. R. A. Watson* has quoted evidence of anticyclonic motion at cirrus levels above tropical cyclones, and it is possible that such motion represents the high-level counterpart of the tropical cyclone.

§ 202. *The structure of depressions at high level*

Some extremely interesting results have been obtained by J. Bjerknes† and by J. Bjerknes and E. Palmén‡ by the detailed examination of individual depressions. It cannot yet be said that sufficient information has been accumulated to make it possible to draw theoretical conclusions regarding the origin and structure of depressions, and still further researches of the same character are desirable. In the first paper referred to, J. Bjerknes inferred that when warm air ascends over the western slope of a calotte of cold air it undergoes horizontal divergence, and vertical shrinking, while the tropopause is elevated beyond its normal level; and that when the air descends over the eastern slope of a calotte of cold air there is horizontal convergence, and vertical extension of air columns, while the tropopause is depressed. It is interesting to note that the horizontal divergence and vertical shrinking referred to were demonstrated for the ascent of warm air at a warm front, by E. W. Hewson,§ using the wet-bulb potential temperature as a means of identifying the air.

J. Bjerknes infers that the sinusoidal deformations of the tropopause shown by the observations which he analysed are to be explained by horizontal motions at high levels rather than by vertical motions, a crest on the wave-like structure of the tropopause being a region to which air has moved from lower latitudes. It is not yet clear, however, whether the phenomena are not in part due to vertical motion. The idea that deformations of the tropopause are to be explained by horizontal motions was first put forwards by Schedler.||

In a paper read before the Meteorological Association of the I.U.G.G. at Edinburgh in 1936, J. Bjerknes considered the nature of these upper perturbations in greater detail. He there discussed the nature of the flow associated with a train of wedges of high pressure and troughs of low pressure travelling from west to east. Fig. 104 represents the system of isobars, the isobars to extreme north and south of the diagram being drawn straight to indicate the limitation of the extent of the perturbations along the meridian. The air at cirrus level moves with about twice the speed of the perturbations, and so enters the perturbations from behind and passes out again in front, being curved cyclonically while overtaking the trough of low pressure, and anticyclonically while overtaking the wedge of high pressure. The curvature of the path must, however, be less than that of the isobars. The wind at the

* Q.J. Roy. Met. Soc. **53**, 1927, p. 446.
† Geof. Publ. **9**, No. 9; **11**, No. 4.
‡ Beitr. Phys. fr. Atmos. **21**, 1933, p. 53; Geof. Publ. **12**. No. 2.
§ Q.J. Roy. Met. Soc. **63**, 1937, p. 7.
|| Beitr. Phys. fr. Atmos. **9**, Heft 4.

northern and southern limits, i.e. at the straight isobars shown in fig. 104, is from west to east and there is no inflow across these isobars into the regions between them. The gradient wind components at right angles to AB and CD respectively are U_{AB} and U_{CD}, where

$$U_{AB} = -\frac{\frac{1}{\rho}\frac{\partial p}{\partial y}}{2\omega \sin \phi} + \frac{\frac{u^2}{r}}{2\omega \sin \phi}$$

$$U_{CD} = -\frac{\frac{1}{\rho}\frac{\partial p}{\partial y}}{2\omega \sin \phi} - \frac{\frac{u^2}{r}}{2\omega \sin \phi}.$$

The second terms on the right-hand side are the cyclostrophic components, which have opposite signs at AB and CD, the radius of curvature r being counted positive in both cases. The inflow across AB between the isobars $p = p_1$, and $p = p_1 + dp$, and the outflow across CD between the same isobars

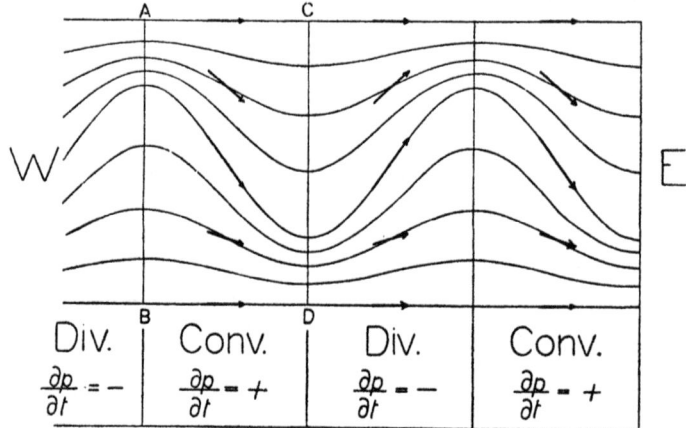

Fig. 104. Schematic pressure field of the upper perturbations after J. Bjerknes.

can be readily evaluated. The gradient wind inflow of mass per second through a vertical rectangle $dy\,dz$ is:

across AB
$$\rho U_{AB}\,dy\,dz = \frac{-dp\,dz}{2\omega \sin \phi} + \frac{\rho \frac{u^2}{r}\,dy\,dz}{2\omega \sin \phi}$$

across CD
$$\rho U_{CD}\,dy\,dz = \frac{-dp\,dz}{2\omega \sin \phi} - \frac{\rho \frac{u^2}{r}\,dy\,dz}{2\omega \sin \phi}.$$

The process can be repeated for any other pair of successive isobars, and the relative magnitude of the two terms for these flows can be evaluated, showing that the variation from

$$\frac{\rho \frac{u^2}{r}\,dy\,dz}{2\omega \sin \omega}$$

on the wedge to the same quantity with sign reversed at the trough is of greater importance than the variation of $\dfrac{dp\,dz}{2\omega \sin \phi}$ due to difference in latitude for the same pair of isobars on adjacent wedges and troughs. It follows that the air transport between any two neighbouring isobars, and consequently between the extreme isobars p_1 and p_2, will be greater across AB than across CD, so that $ABCD$ is an area of convergence. Similar arguments lead to the result that the area bounded by a trough in the west and a wedge in the east is an area of divergence. The wind arrows in fig. 104 have been drawn to fit in with these ideas.

The result described above can be stated thus. There is convergence east of, and divergence west of, a wedge of high pressure, so that the wedge must move eastwards. For similar reasons, the troughs must also move eastwards. The eastward speed will be less than that of the air, since only the cyclostrophic component is effective in producing their motion.

Bjerknes suggests that the upper perturbation is started by an impulse from a polar front wave and travels with it, the upper perturbation possibly being able, in the later stages, even when the depression is occluded, to grow independently as a free oscillation of the tropopause. On this view, the polar front wave is the primary wave, the tropopause disturbance being a secondary effect. This view stands in direct conflict with the views of the Frankfurt meteorologists, who regard the stratospheric disturbance as the primary phenomenon, and the tropospheric disturbance as secondary. The truth is probably intermediate between these two views. Some changes of pressure distribution appear unquestionably to be mainly due to stratospheric advection. Schmiedel,* in an excellent summary of the views of the Frankfurt and Leipzig schools of thought, quotes the demonstration by Thomas,† a pupil of von Ficker, that the change of pressure of about 47 mb which was observed over Germany from January 13th to 17th, 1930, was due to an outburst of cold air in the stratosphere. The Leipzig school regards much of the weather as due to the combinations of a number of periodic terms. It is found that, at least in the lower troposphere, the amplitudes of some of these harmonic terms increase with height, from which it is implied that the primary cause is to be found at high levels.

Further progress demands the detailed study of still more individual depressions, particularly as regards the distribution of temperature, humidity and wind. But it is probable that in the end it will be found that all depressions cannot be represented by one single picture. It is more probable that large differences of structure and development will be found between individual depressions, and that while some atmospheric disturbances are primarily of tropospheric origin, others are primarily of stratospheric origin, or may be the effect of the mutual reactions of disturbances in the troposphere and stratosphere. It is in fact certain that phenomena in the troposphere and stratosphere must be coupled together in some manner. So much can be inferred even from the statical equation and Buys Ballot's law.

* *Veroff. Geoph. Inst. Leipzig*, Bd. **9**, Heft. 1.
† *Sitzber. Preuss. Akad. Wiss., Phys. Math. Klasse*, 1934, XVII.

§ 203. *The energy of depressions*

A theory of the origin of depressions must take account of the necessity of explaining the development of colossal amounts of kinetic energy, and the removal of colossal amounts of air to provide for the fall of pressure. All theories which have been put forward to explain the genesis of depressions are eventually based on the existence of horizontal differences of temperature. It is because the horizontal gradients of temperature from equator to pole are greater in winter than in summer that depressions are more frequent and more intense in winter than in summer. It should be added that the air contains more water-vapour in summer than in winter, so that any theory which would explain depressions purely on the basis of water-vapour should demand greater frequency and greater intensity in summer than in winter.

Margules ascribed the development of kinetic energy in the depression to the readjustment of masses in unstable equilibrium, which liberates large amounts of energy. Some of Margules' computations are summarised in Chapter xiv above, where it is shown that differences of temperature of 10° C, which are relatively common in the atmosphere in winter, involve the existence of sufficient potential energy to supply the required kinetic energy by redistribution of the masses. The computations of Margules afford the only available criterion for the amount of energy which can be developed in the atmosphere, and the agreement found between the maximum possible development of kinetic energy and the observed kinetic energy in depressions, with differences of temperature of say 10° C, is, to say the least, very striking. The actual processes which Margules considers are more nearly analogous to squall phenomena than the development of depressions, but if there are sloping surfaces of separation between warm and cold masses of air it is not possible even to form a mental picture of the horizontal and vertical motions associated with the readjustment of potential energy, much less to represent these motions analytically.

There can however be little doubt that the formation of the depressions is to be explained by the displacement of warm air by cold air. In the growth of the circulation of the depression convergence probably is a factor of prime importance. The growth of an intense depression is accompanied by heavy rainfall over a wide area, in itself evidence of the displacement of large masses of warm damp air by heavier air, and of the liberation of large quantities of potential and thermal energies.

It is probable that in the genesis of the depression, the existence of horizontal instability, i.e. of currents of air of widely different densities flowing side by side, is the most important factor, and that vertical instability only becomes fully effective after the cyclone has come into being.

There is a difficulty in the way of applying the ideas of Margules to a Bjerknes depression to which the writer's attention has been drawn by C. K. M. Douglas. It was seen in equation (9) of Chapter x that the slope of a surface of discontinuity for a given difference of temperature is proportional to the differences of velocity. The liberation of potential energy would require a decrease in the slope of the surface, whereas a development of kinetic energy requires an increase in the slope.

It should be noted that there is no essential contradiction between the work of Margules and the idea that a depression may originate by pure convection. The latter involves a diminution of potential energy, and in the end the kinetic energy developed by convergence must amount to the potential energy released by the convection.

§ 204. *Kobayasi's theory of the formation of fronts in a vortex as a result of horizontal temperature gradients*

Kobayasi[*] considered the effect of horizontal variation of temperature in the general field upon the distribution in an originally symmetrical cyclone. In order to discuss the question mathematically he assumed a distribution of pressure in the cyclone giving everywhere a constant gradient of pressure in the central or principal part of the cyclone, whose outer boundary is a circle of radius R, while outside this region the transverse velocity is given by

$$vr = \text{constant}.$$

He further assumed that at a given level the air drifts across the isobars at an angle α, which decreases from the ground upwards, and that the cyclone moves through a field of uniform horizontal temperature gradient. The convergence of air to the centre involves the ascent of air at the centre, and the result is that air within a region which is of the form of a parabola (*EF* of fig. 105) with its focus at the centre of the cyclone is eventually removed upwards, while air currents from the two opposite sides outside this region are brought into juxta-position along a line which trails backward from the centre to the right of the path of the cyclone. This line is accordingly a line of discontinuity of temperature. Kobayasi further showed that the line of discontinuity is advanced in the direction of motion of the cyclone as α, the angle of drift across the isobars, is diminished. Hence the surface which is traced out by the discontinuity at different levels slopes forward and produces a system in which the colder air at say 500 metres is brought in above warmer air at the surface. This provides a mechanism at the cold front capable of giving the squally winds which are actually observed, and provides a simple explanation for the existence of a front in the rear of a cyclone.

Kobayasi further suggests that the temperature discontinuity left behind by one cyclone may act as steering line for the next cyclone. This portion of his argument is not as readily acceptable as the earlier portion. Subsequent researches have shown that cyclones do not move along the direction of the front in advance of the centre, but along the direction of the isobars in the warm current. Further, Kobayasi does not prove that what he calls the "steering line", the line of discontinuity produced by one cyclone and drawn into the ambit of the next, will have any steering properties.

The ideas developed by Kobayasi have not been applied to any Atlantic depressions. It may probably be assumed with safety that they are not applic-

[*] *Q.J. Roy. Met. Soc.* **49**, 1923, p. 177.

able to the ordinary depression with a well-marked warm sector, but the possibility of these phenomena occurring in the depressions which form in polar air deserves further study.

It is further possible that Kobayasi's ideas would find closer application in tropical cyclones. Horiguti* points out the absence of lines of discontinuity in the typhoon area. Father Gherzi† expresses the same view. Mal and Desai‡ have however traced fronts to very great distances from the centre of a tropical cyclone, but Father Gherzi suggests (*loc. cit.*) that these fronts are

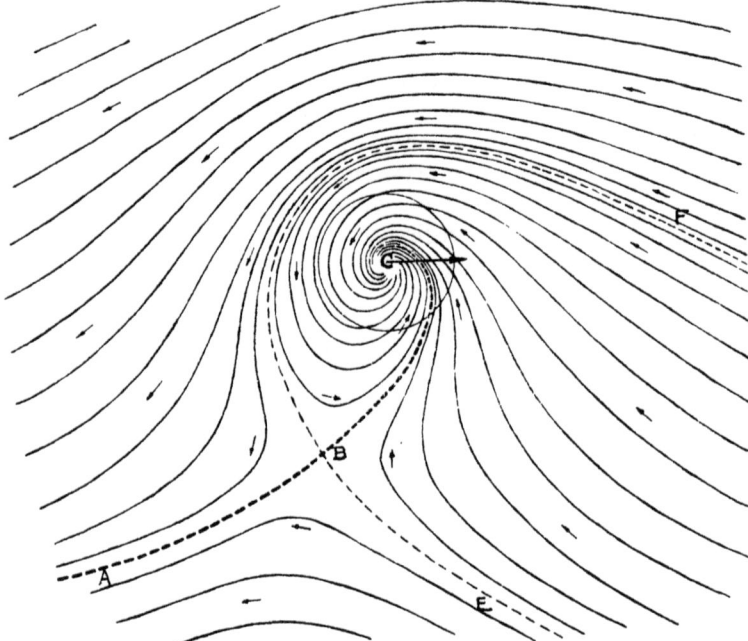

Fig. 105. Kobayasi's diagram to illustrate the formation of fronts.

phenomena of the surrounding region rather than of the true typhoon central region. A recent paper by Desai and Basur§ points to the inner structure of tropical cyclones not being symmetrical.

§ 205. *Exner's barrier theory of depressions*

Exner has outlined in his *Dynamische Meteorologie* (2nd edn.), Chapter XII, a view of the origin of depressions which is in some respects similar to the Norwegian view, in other respects contradictory to it. The cyclone is regarded as analogous to the whirl which forms in the lee of a rock jutting out into a stream, or in the lee of the pillars of a bridge. Exner suggests that the place of the rock or bridge pillar may be taken by high land, particularly near the ocean, and cites in particular the massif of Greenland, which, in his view, inter-

* Y. Horiguti, *Mems. Imp. Marine Obs. Kobe, Japan,* 2, No. 3, 3, Nos. 2, 3.
† *Q.J. Roy. Met. Soc.* 58, 1932, p. 303.
‡ *Ind. Met Dept., Sci. Notes,* 4, No. 39, 1931.   § *Beitr. Geoph.* 40, h. 1, 1933.

feres with the westerly current of those latitudes, producing a drop in pressure east of the southern projection of Greenland. This gives the westerly current a component of motion to north, and produces the Iceland minimum.

Exner points out, however, that most depressions do not form at such projections, but at the limits of currents of air. The initial stage of development is the outbreak of a cold mass of air from polar regions into the westerly current of middle latitudes, the cold air sweeping southward in the form of a tongue, as shown in fig. 106 (a) below.

The tongue of cold air C cuts off the supply of warm air from the corner A, causing a fall of pressure at A. The pressure gradients set up produce the circulation shown in fig. 106 (b), the cold tongue C being drawn round the centre of low pressure as shown. The bursts of cold air over North America, which are frequently accompanied by depressions on their eastern edges, are regarded by Exner as essentially similar to the scheme outlined above, and a map showing such an outburst in January 1895 is reproduced by Exner (loc. cit. p. 340).

Exner devised an experiment to illustrate his theory. A circular vessel which can be rotated about its centre has a cylinder of ice at its centre and is heated at its periphery. Colouring matter is placed at the centre, so that the motion of the cold water can be traced. A photograph reproduced by Exner (loc. cit.

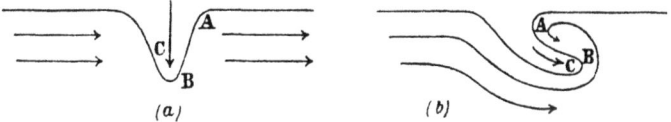

(a) (b)

Fig. 106. Exner's barrier scheme of formation of depressions.

p. 341) shows intermittent outbursts of cold water into the warm outer regions, showing considerable resemblance in their subsequent motion to the cyclonic and anticyclonic circulations of the atmosphere.

Exner's scheme for the formation of a cyclone resembles the Norwegian scheme in that it demands the interaction of warm and cold currents, but in other ways it is in direct contrast with the Norwegian views. It replaces a continuous polar front by discontinuous outbursts of cold polar air, and, if the present writer has completely understood it, demands that the actual centre of the cyclone should be filled with warm air, so that the "warm sector" should enclose the centre of the cyclone. This is not in agreement with the facts of observation. The location of the outburst of cold air should show a marked tendency to occur east of Greenland, over Nova Zemblya, east of the western mountain ranges of North America, and over the eastern portion of the northern coasts of Asia.

The production of the cyclone is regarded as due to dynamical interaction of the warm and cold air, but the original outburst of cold air is regarded as produced by a purely thermal cause—the difference of density due to differences of temperature in the two air masses; and the energy of the cyclone is eventually derived from the thermal energy of the meridional circulation of the atmosphere. Exner made no claim that his theory was completely developed. He could give no estimate of the extent of the fall of pressure in the lee of the cold tongue, and it has been objected that the fall of pressure so produced must be insignificant. Until some arithmetical estimate has been given of this fall of pressure, the situation as between the adherents and opponents of the theory remains unresolved.

There is however a marked contrast between this theory and that developed by J. Bjerknes and Solberg. The latter also regard the first stage in the development of a cyclone family as the outburst of a mass of cold air from the polar basin; but a *family* of cyclones forms along the front which separates the cold mass from the warm air to the east of it. Exner however would have *one* cyclone form at the northern corner of the warm air east of the tongue of cold air. On either scheme the cyclone becomes a definite feature of the exchange of air between different latitudes.

Exner's theory is a logical outcome of the study of the outbursts of cold air over Asia and Western Europe which formed the subject of a series of papers by Ficker. In a paper in the *Meteorologische Zeitschrift* for March 1923 Ficker gives a very complete bibliography of his own papers and of those of other authors on the subject. Ficker explicitly states that outbursts of cold or warm air could only explain the shallow cyclones and anticyclones which do not extend beyond the lower troposphere. He regards the depression which extends to the stratosphere as the combination of a depression in the lower atmosphere with a depression at high levels, the latter being formed by high level outbursts of warm air from the North.

It would be an error to regard the theory of Exner as an alternative to the Norwegian polar front theory. The depressions which form over the Atlantic and over the region of the British Isles generally form at the boundary between two portions of the westerly current which have originated in different latitudes, and there is definitely no trace of an initial outburst of cold air into the warm westerly current. Most of the depressions which affect the British Isles are therefore Norwegian rather than Austrian depressions. But there is not so marked a contrast of temperatures in the cold waves which move south over Western Europe as in the cold waves over America or Asia, and it is not clear whether Exner's views may not be closely applicable to other regions of the earth. In any case we have to guard against thinking of the depression as something which always has a typical form, and arises in a typical way. There is only one fundamentally typical feature common to all depressions, a centre of low pressure. Other features can show almost every conceivable variation, and it is possible that different depressions arise in different ways, some as the result of large-scale convection in unstable air, some at polar fronts as pictured by Bjerknes, and some in the lee of cold outbursts as pictured by Exner. The meteorologist has to bear these different possibilities in mind, as alternatives which shall apply in individual cases.

Reference should be made to a recent book on *Weather Analysis and Forecasting*, by Sverre Petterssen (McGraw Hill, 1940), for a detailed treatment of the applications of the ideas outlined above to the practical problems of forecasting.

CHAPTER XVIII

ANTICYCLONES

§ 206. *Types of anticyclones*

WE have already seen that an anticyclone is a centre of high pressure in which the wind circulates clockwise about the centre. Consideration of the appropriate solution of the gradient wind equation showed that the winds in the anticyclone spin round the centre at a slower rate than the rotation of the earth beneath them, so that, regarded as a circulation in space, the anticyclone is a slow counter-clockwise circulation. The angular velocity about the centre in an anticyclone has thus an upper limit, equal to the angular velocity of the horizon, $\omega \sin \phi$. There is a corresponding limit to the pressure gradient. We have also seen (p. 189) that the geostrophic wind is an underestimate of the gradient wind in an anticyclone, and an overestimate in a cyclone. Thus for a given gradient of pressure, the gradient wind is greater in an anticyclone than in a cyclone, or, conversely, a given gradient wind velocity requires a lower gradient of pressure in an anticyclone than in a cyclone.

The argument used in Chapter xv that convergence is necessary in order to produce a cyclonic circulation can be applied to show that divergence is required in order to produce anticyclonic circulation in the lower atmosphere. This divergence is most readily visualised as due to the accumulation of air at high levels causing a general settling down at low levels, with a gradual spread outwards from the centre. This would give the required clockwise circulation relative to the earth in the levels in which divergence occurs. But the transport of air to a restricted region at high levels is not readily visualised. If it took place by a general convergence towards a point or a small area, we should expect to find a cyclonic circulation at the level of convergence. But this is not found by observation over the anticyclones of middle latitudes, though it is present above the anticyclone which forms over Asia in winter.

The earlier writers on the subject regarded an anticyclone as a region of cold air, the excess of pressure being due to the excess of density in the lower troposphere. But Hann, and later other writers, showed that on the average the anticyclone is warmer than the cyclone at levels of 3–8 km. The stratosphere is higher above the anticyclone than above the cyclone, and the fall of temperature with height is continued farther over the anticyclone, yielding a colder stratosphere.

Hanzlik* made a detailed study of European anticyclones, and found two distinct types, the cold and the warm anticyclone. The cold anticyclone is a relatively shallow vertical structure, whose excess of pressure is due to excess of density in the lower troposphere. It usually moves fairly rapidly. If it slows down or becomes stationary it tends, according to Hanzlik, to be transformed into the second type, the warm anticyclone. The warm anticyclone

* *Denkschr. Wien. Akad.* **48**, 1909, pp. 163–256.

extends into the stratosphere, in which region it is colder than the normal for latitude and time of year. The region of lowest stratospheric temperature is not in general directly above the highest pressure at mean sea level, but is usually displaced to west or north-west of the latter. The "axis" of a warm anticyclone is therefore inclined to west or north-west in the same way as the "axis" of a cyclone, though occasionally it is inclined to south.[*]

Hesselberg[†] suggested that the high temperature in the anticyclone can only be explained by the descent of air. Compression alone can only give a relatively small increase of temperature. Radiation is also insufficient to produce the rapid changes of temperature which accompany changes of pressure and there remains only descent of air as a possible explanation of the high temperatures which occur. An anticyclone which is warmer than the normal in the lower troposphere must have a slower diminution of pressure with height than the normal in that region, so that the excess of pressure above normal increases with height, and the high pressure can only be explained by the addition of mass in the upper layers of the atmosphere.

The following extract from Table XII of the memoir by W. H. Dines[‡] on the characteristics of the free atmosphere will give an idea of the extent of the differences between cyclonic and anticyclonic temperatures at different levels, the mean surface pressure being respectively 989 mb and 1026 mb.

Height in km	1	2	3	4	5	6	7	8	10	12	14
Mean temp. °A	277	273	268	262	256	249	242	235	223	220	220
Cyclone °A	276	270	263	256	249	242	234	228	225	225	224
Anticyclone °A	279	276	271	265	259	253	246	238	225	217	215
Anticyclone − mean °C	2	3	3	3	3	4	4	3	2	−3	−5
Anticyclone − cyclone °C	3	6	8	9	10	11	12	10	0	−8	−9

The difference between cyclone and anticyclone, and the difference between anticyclone and the mean conditions, are both greatest round about 6–8 km.

Dines gives (*loc. cit.* pp. 62–3) tables which indicate uniformity of pressure at 20 km in all latitudes and at all times, which would appear to indicate that any addition of mass must be below the level of 20 km, though above the level of 8 km. But the result given by Dines does not agree with that of Dobson[§] on the variation of wind with height. Dobson found that while on the average the winds over the British Isles fall off to zero at 20 km the winds at Lindenberg are still of considerable strength at that height.

§ 207. *The cold anticyclone*

The cold anticyclones of Europe appear to be mainly generated by cold currents from polar regions brought southward in the rear of depressions. There was an anticyclone over Scandinavia on March 5–6, 1931, with very high surface pressure, at the centre exceeding 1040 mb (see fig. 107). Upper air ascents at Kjeller, in the NE current on the eastern side of the centre, at 18 h, and at Duxford, in the SE current on the southern side of the centre, at 10 h, are shown in fig. 108. Temperatures were much higher in the SE current

[*] Runge, Leipzig Dissertation, 1931.   [†] *Met. Zeit.* **32**, 1915, p. 311.
[‡] M.O., *Geophys. Mem.* No. 13.   [§] *Q.J. Roy. Met. Soc.* **46**, 1920, p. 54.

than in the NE current, being about 34° F higher at 15,000 feet, and over 20° F higher at the ground. The surface inversion shown at Kjeller was due to radiational cooling of the ground. Above the first 2000 feet, the Duxford ascent showed a layer of practically steady temperature up to 8000 feet. In at least the central part of this isothermal layer the relative humidity was about

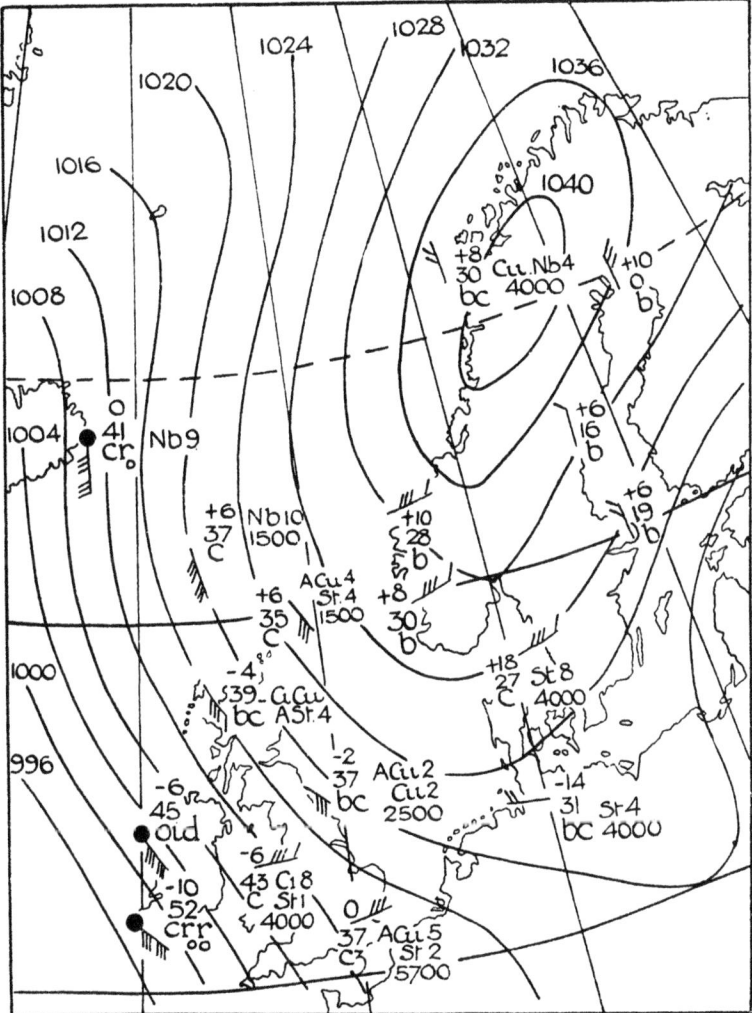

Fig. 107. The anticyclone of March 5, 1931.

70 per cent, while at 8000 feet, and also at 1000–2000 feet, it approached 100 per cent. At Kjeller the relative humidity was reported as 35 per cent through the whole of the ascent, a value which indicates that the air had probably subsided. It is readily seen from a Hertz diagram or tephigram that air which had relative humidity 35 per cent at the ground might have started saturated from a level of about 2 km and descended adiabatically.

By 18 h on the 6th pressure at the centre of the anticyclone was about 1044 mb, but after that it diminished slowly. The anticyclone was however maintained for some time, and drifted slowly westward to Iceland. At 7 h on the 9th it was centred over the north-eastern end of Iceland, and the pressure at the centre was a little above 1032 mb. Duxford was now in the NE current,

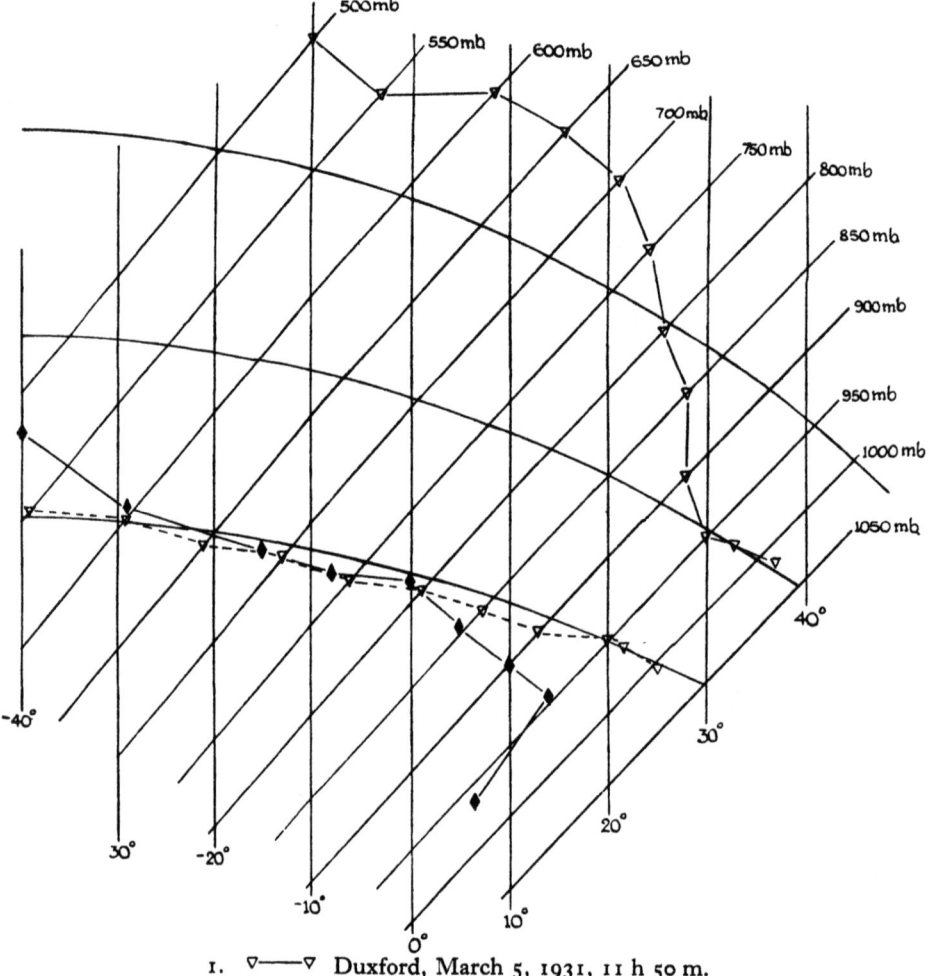

1. ▽———▽ Duxford, March 5, 1931, 11 h 50 m.
2. ▽-----▽ Duxford, March 9, 1931, 9 h.
3. ◆———◆ Kjeller, March 5, 1931, 8 h.

Fig. 108. Upper air observations, March 5, 1931.

and the conditions in the upper air resembled closely those at Kjeller on the 5th. These conditions are represented by curve 2 in fig. 108. The lapse-rate was very slightly in excess of the saturated adiabatic, and a little less than that observed at Kjeller on the 5th, except in the surface layers, which no longer showed an inversion, but a marked lapse-rate, on account of the heating of the surface layers in passing over the North Sea. The surface layers developed considerable instability, which on the 8th gave rise to heavy falls of snow over the British Isles.[*]

The changes in the lapse-rate in the NE current represented by the differences between curves (2) and (3) in fig. 108 are in the direction of decreasing stability, and are such as would be produced by surface heating. The resemblance of curves (2) and (3) indicate however a remarkable conservatism of the state of the upper air in the NE current. At Duxford on the 5th there was a pronounced inversion of 2° F between 900 mb and 860 mb (3200 to 4400 feet), with drier air (R.H. 70 per cent) above damp air (R.H. 95 per cent).

§ 208. *The warm anticyclone*

The warm anticyclones which extend into the stratosphere are different in their relation to their environment. They are warmer than their environment in the troposphere, and colder in the stratosphere, which is higher and colder than the normal for the latitude and time of the year. The excess of pressure above normal is thus to be explained by the excess of density at high levels. On account of the high temperature in the troposphere it follows that the excess of pressure above normal increases with height in the troposphere. The correlation coefficients between different variables in the upper air which were evaluated by W. H. Dines (see p. 21 *ante*) bear out the statements made above. These coefficients show that high pressure at 9 km is associated with a high and cold stratosphere, and a warm troposphere. The results derived by W. H. Dines are confirmed by Schedler,[†] who evaluated similar correlation coefficients for data accumulated in the course of a series of daily observations at Lindenberg. Schedler's coefficients were all slightly lower than those of Dines, but the confirmation of the general results is striking. Schedler investigated the contribution of different layers of the atmosphere to the total changes of pressure, and found that, on the whole, the first three kilometres above the ground contribute slightly to the net surface change of pressure, the next 5 or 6 kilometres act weakly against it, so that the changes of pressure at 8 or 9 km are at least as great as those at the ground. Thus effectively the whole pressure variation must be accounted for by the changes above 9 km. Schedler has estimated that 40 per cent of the variation was due to the layer from 9 to 14 km, and the remaining 60 per cent to the layers above 14 km.

The problem of explaining the warm anticyclone thus reduces in the main to that of explaining the occurrence of low temperatures at high levels. Two alternatives are possible. The air at high levels may be cooled *in situ* by radiation, or the cold air at high levels may be brought in from lower latitudes. The first alternative is difficult to discuss on account of our lack of knowledge of the time scale of radiation phenomena in the highest layers. The active agent which effects the changes we are considering may be the water-vapour in the lowest layers, or the occurrence of sheets of cloud. From the point of view of radiational phenomena the fundamental difference between low and

---

[*] *Vide* note by W. R. Morgans, *Met. Mag.* April 1928, p. 53.
[†] *Beiträge Phys. fr. Atmos.* **7**, p. 88.

high latitudes consists in their respective high and low water-vapour content of the lower troposphere. The effect of the water-vapour at these levels is to warm the lowest layers at the expense of the highest. But the water-vapour content of the atmosphere is not in general as great in anticyclones as in cyclones, and this line of approach to the problem is not promising. Indeed the main problem appears to be far removed from the possibility of explanation by radiation. This is borne out by the phenomena in cyclones, in which the occurrence of cloud sheets must lead to the reflexion back into space of a considerable part of the incoming solar beam, as a result of which the amount of long-wave radiation passing out into the stratosphere is considerably reduced. Yet the temperature of the stratosphere above a cyclone is higher than the normal, showing that the cyclone is not to be explained as a radiational phenomenon. The validity of this argument is borne out by the observations of temperature of the stratosphere over India. During the monsoon the stratosphere is colder than in winter, though the surface temperatures are higher, the reason being that solar radiation is sent back into space by the cloud sheet associated with the monsoon rainfall. The surface temperatures are high on account of the current of warm air from lower latitudes.

The reader cannot fail to notice that there is a close apparent analogy between the distributions above cyclones and anticyclones, and those above polar and equatorial regions. Over cyclones and in polar regions the troposphere is cold, and the stratosphere low and warm. Over anticyclones and in equatorial regions the troposphere is high and the stratosphere cold. It is thus an obvious step to ascribe the formation of anticyclones to the transfer of air from equatorial regions to high latitudes, by means of solid currents which extend high into the stratosphere. This suggestion is so obvious that it is perhaps advisable to add a word of caution that its obviousness may be in no sense physically justifiable.

The curves of fig. 109 are based on the figures given by Wagner* of the distribution of pressure over the globe in winter and summer. The differences between the mean surface pressures in different latitudes are so slight that it is obvious that no large differences of pressure in middle latitudes could be produced by the bodily transfer from low to high latitudes of sections of the atmosphere. Round about 20 to 22 km the pressure attains approximate uniformity over the whole globe, and the increase of pressure which could be produced in latitude 60° by the transfer from latitude 10° to latitude 60° of a section of the upper air reaching from the top of the atmosphere down to some chosen level in the troposphere can therefore be readily evaluated from the diagram by taking the differences of pressure in latitudes 10° and 60°, at that level. These differences are approximately as follows:

Height in km	0	2	4	6	8	10	12	14	16	18	20	22
Pressure Difference } Summer	2	5	11	17	23	23	21	16	11	6	3	2
Winter	4	11	19	24	27	27	22	15	8	3	1	0

Thus in latitude 60° if the portion of the atmosphere above about 9 km could be replaced by the corresponding portion of the atmosphere from latitude 10° a change of pressure of from 23 to 27 mb would be produced.

* *Handbuch der Klimatologie*, **1**, Teil F, "Klimatologie der freien Atmosphäre, p. F 67.

It was suggested by Gold* that some anticyclones may be due to the bulging northward of the cold equatorial current in the stratosphere, and subsequent writers have advocated somewhat similar ideas. Such outbursts to northward of cold air in the upper troposphere and stratosphere would provide an in-

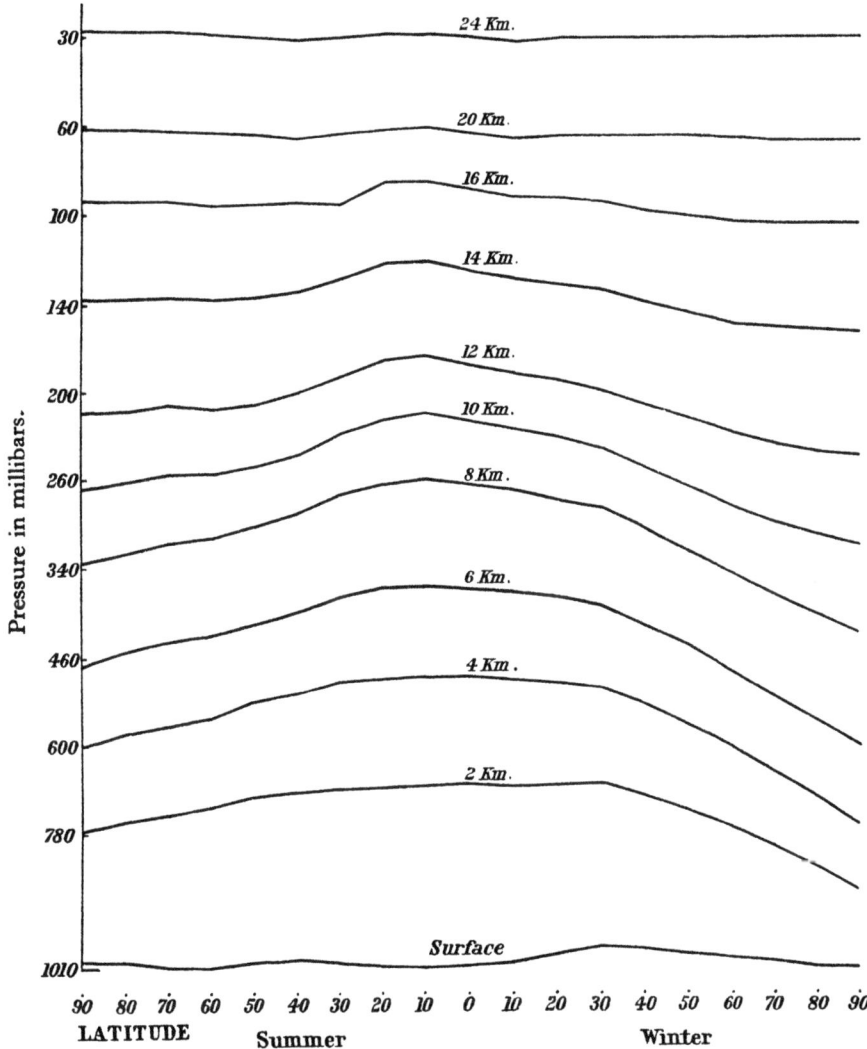

Fig. 109. The distribution of pressure at different heights. On the vertical scale of variation of the curves the interval between the successive numbers from 780 upwards shown on the vertical axis is 40 mb.

crease of pressure of the right order of magnitude, but it must be emphasised that if the outburst to northward includes the whole of the troposphere, the increment of pressure due to cold air at high levels is destroyed by the diminution due to the high temperature at lower levels.

* "International Kite and Balloon Ascents", M.O., *Geoph. Mem.* No. 5.

Exner* suggested that a warm southerly current of great breadth might extend into the stratosphere and bring with it the cold high stratosphere of more southerly latitudes. The arguments adduced above make it difficult to explain by this means the increment of pressure which produces the anticyclone as we observe it. The suggestion of an outburst of cold air at high levels is more plausible. The phenomena must become more complicated in the lower troposphere. For if an anticyclonic circulation is set up in lower levels, it will draw into its ambit warm southerly air on its western side and cold northerly air on its eastern side. The contribution of the lowest layers to the pressure distribution will thus be unsymmetrical relative to the centre of high pressure in the upper troposphere, the maximum pressure at the surface being displaced to the east. This point was brought out by Exner[†] in a discussion of some of Schedler's results.

The explanation of the cold anticyclone does not present any insuperable difficulty. The accumulation of a subsiding mass of cold air explains most of the phenomena, and the development is facilitated when the surface of the earth cools rapidly by radiation, as when the ground is covered with snow. The crucial difficulty is the explanation of the growth of the cold anticyclone into a warm one. Recent writers tend to the view that the apparent growth up into the stratosphere of an originally shallow anticyclone is to be explained as an amalgamation of a low-level anticyclone with a high-level anticyclone. The occurrence of closed isobaric systems at high levels is indicated by the far from negligible frequency of cirrus motion from the east. Such systems correspond to islands of warm or cold air which observations have shown to exist. Islands of cold air at high levels are to be explained as high-level outbursts of air from low latitudes. Douglas[‡] distinguishes three main phases in the life-history of the typical anticyclone or depression. "Firstly, the systems are confined to low levels; secondly, the (still unclosed) high-level systems are two or three hundred miles to west of the (often closed) low-level systems; finally, the upper systems are roughly coincident with the lower systems." These words present a picture which is in its main outlines similar to that of many other writers. The high-level system, whether depression or anticyclone, is in the main regarded as the effect of advection from high or low latitudes. Perhaps the strongest evidence in favour of advection as a fundamental cause of anticyclones is to be found in a letter to *Nature* from L. H. G. Dines[§] in which are given the correlation coefficients between the potential temperature at the tropopause and certain other fundamental variables. The variables correlated, and the suffixes used to denote them, are shown below:

	Suffix
Pressure at 9 km	3
Height of the tropopause	4
Temperature at the tropopause	5
Pressure at the tropopause	6
Potential temperature at the tropopause	7

* *Dynamische Meteorologie*, p. 358.
† *Met. Zeit.* 1921, p. 21.
‡ *Q.J. Roy. Met. Soc.* **59**, 1933, p. 62.
§ *Nature*, **127**, 1931, p. 815.

The correlation coefficients obtained are shown below, the column headed "smoothed" giving values derived by using departures from smoothed monthly means instead of the observed values:

		Smoothed
r_{34}	0·82	0·82
r_{36}	−0·70	—
r_{37}	0·89	0·82
r_{47}	0·82	0·81
r_{56}	0·79	—

The standard deviation of the potential temperature at the tropopause was nearly 10° C, and since departures may range up to twice the standard deviation or more, it is implied in these results that the potential temperature at the tropopause may alter by 20–25° in a week or so. So big a change in potential temperature (or entropy) is not readily explained by any known process, if it is assumed that the change referred to is a change in the condition of a fixed mass of air, and we are forced to the conclusion that the change is in reality the replacement of the original air by a fresh supply from some other latitude. If we consider the potential temperature at the tropopause in an anticyclone, to find air of the same potential temperature in a cyclone we should require to penetrate well into the stratosphere, and to find elsewhere air of the same potential temperature at approximately the same level, we should require to go to much lower latitude. The conclusion that the air at the tropopause in an anticyclone has recently come from lower latitudes appears inevitable, particularly as the extremes of potential temperature in depressions and anticyclones are well within the range provided by the poles and the equator.

One serious problem is the growth of anticyclonic rotation at high levels. One plausible explanation is that of J. Bjerknes (see § 201, p. 363 above) that it is produced where warm air diverges in the upper troposphere after flowing up a warm front surface.

In the present state of our knowledge of what occurs when anticyclones develop or die away, it is not possible to give any complete theoretical treatment, and future advance will only be possible by the discussion of actual observations. Much of the argument hitherto used in discussing anticyclones is rather of the nature of guesswork.

It was suggested by Hanzlik that the warm anticyclone is a stable system liable to persist for considerable periods, and this suggestion has frequently been used as a basis in forecasting, though it is probably much exaggerated. An exceptionally warm and apparently well-established anticyclone on September 15, 1932, collapsed completely within 48 hours, and other examples could be quoted to show that Hanzlik's suggestion cannot with safety be adopted as a basis of practical forecasting.

§ 209. *Some observational data*

A number of individual warm anticyclones have been investigated by different writers. Mügge,* using data accumulated on two international days—May 7, 1909, and May 19, 1910—found it possible to draw streamlines of vectorial

* *Veröff. Geoph. Inst. Leipzig*, **3**, h. 4.

mean winds at three levels, from 1 to 4 km, from 4 to 8 km, and from 8 km to the tropopause.

On May 7, 1909, an intense anticyclone covered the North Sea and Northern Europe, giving clear weather over a wide area, so making upper air observations possible over most of Western and Central Europe. The lines of flow at the three chosen levels showed a point of divergence of air-flow, which for the level 1 to 4 km was slightly north of Bergen on the West coast of Norway: for the level 8 km to the tropopause the point of divergence was over the North Sea in latitude 55°; and for the intermediate layer the point of divergence was rather indeterminate, but probably situated somewhere between the points found for the upper and lower levels. The distribution of temperature at the tropopause was represented by isotherms which marked out a clearly defined centre of low temperature over Holland, with temperature increasing outwards on all sides of this centre. The marked limitation of the cold centre, particularly to the south, appears to exclude the possibility of explaining the anticyclone as the effect of a continuous flow of cold air from the south, and the divergence of air-flow at all levels in the troposphere indicates that the inflow of cold air into the region took place well within the stratosphere. The general run of the isotherms suggest that the outburst of cold air from lower latitudes came from south-west of the observed centre of lowest temperature. The observed winds over the central region of the anticyclone were however generally from a northerly or north-easterly direction, and the theory suggested previously does not appear to fit the facts at all closely. But it should be borne in mind that the details of the lines of flow given by Mügge are by no means definitely fixed by the observations. A good deal of latitude is possible, using these observations, which are barely sufficient for the purpose. Cave gives (*Structure of the Atmosphere*, etc. p. 105) the details of two balloon ascents on May 7, 1909. At 6.30 p.m. the wind at the ground was 12 m.p.h. from ENE, and there was no marked change of velocity or direction up to 5 km. Beyond this height the velocity diminished, and from 12 to 15 km it was light and very variable in direction.

In Mügge's second case, May 19, 1910, there was an anticyclone centred over Northern Europe, and a second over Italy and the Adriatic. The temperature distribution at the tropopause indicated a region of low temperature stretching over Denmark and Germany to the Mediterranean. The lines of flow in the upper troposphere above 8 km indicated a southerly wind at the south-western edge of the cold tongue, sweeping eastwards over Central Europe. There was no indication (possibly on account of lack of observations) of the limitation to the south of the cold area, and the observations of May 19, 1910, appeared to fit reasonably closely the idea of an outburst of cold air in the stratosphere.

Khanewsky[*] studied in detail the anticyclone of September 30–October 1, 1908, which was centred over Germany. He found that on the eastern side of the anticyclone the wind was northerly or north-easterly up to heights of 19 km, while at Petersfield on the western side of the anticyclone the wind was south-westerly up to 16 km. The surface temperatures were lowest in the

[*] *Met. Zeit.* **46**, 1929, p. 81.

south-eastern section of the anticyclone, but in the lower stratosphere the lowest temperatures occurred in a closed region just east of Denmark, with, apparently, an increase of temperature to the south of this cold island. The stratosphere was very high and cold over Lindenberg, the temperature being $-73\cdot9°$ C at the tropopause (13·6 km).

The table of pressures at different heights evaluated by Khanewsky showed that the highest pressure at 15 km occurred at Petersfield in the southerly current, though the temperature at that level was lower at Lindenberg. Cave[*] has given diagrams showing the variation with height of the wind velocity and direction as observed by sounding balloons on September 30 and October 1. On the second of these days the wind direction was round about southerly up to nearly 18 km; the velocity increased from 4 m/sec at the surface to about 27 m/sec at 12 km, and then diminished to about 8 m/sec at 17 km. On September 30, the wind was southerly at the ground, and backed towards NE and fell to a calm at 2·5 km, beyond which the direction was about SSW, and the velocity 7 m.p.h. up to 5 km, after which there was a rapid increase up to 17 m/sec at 6 km. At Lindenberg, near the centre of the island of lowest temperatures in the stratosphere, the wind was between N and ENE up to 15 km on both days. The north-easterly current was colder than the south-westerly current practically throughout the whole range of height.

The points which appear to be of greatest importance in Khanewsky's analysis are (1) the "island" of low temperature sharply limited to the south, (2) the coldness and depth of the north-easterly current, and (3) the fact that the highest pressure at 15 km occurred in the south-westerly current. The last of these suggests that an outbreak of cold air from the south might have been the primary cause of the anticyclone, and that the subsequent drawing of the cold northerly current into the circulation distorted the form of the anticyclone at the surface, and displaced the highest pressure to the east. Khanewsky ascribes the development of this anticyclone to the thrust of the cold north-easterly current towards low latitudes against the general westerly drift of those latitudes, and suggests that this is probably the explanation of all anticyclones.

Runge[†] studied an anticyclone which was somewhat similar to Khanewsky's, that of December 5, 1912. The anticyclone covered Czecho-Slovakia, Austria and Hungary, while there were depressions to NE and SW of Iceland. In the period December 2 to 5, the anticyclone in the far North had brought in its rear a cold northerly current. The lowest surface temperatures were in the central region of highest pressure, but at 3 km and 5 km the centre of the anti-cyclone was warmer than its surroundings. At 3 km the highest pressure occurred to north-west of the surface centre, and at 5 km the highest pressure was about WSW of the centre at 3 km. Here as in the anticyclone investigated by Khanewsky there were two currents side by side, a relatively cold north-easterly current, and a warm south-westerly or westerly current. The difference of temperature between Hamburg and Lindenberg, in the warm and cold streams respectively, was about 4° C at the ground and over 15° C at 8 km. The isotherms of the stratosphere were roughly parallel, with lowest tem-

[*] *Loc. cit.* pp. 130–3.   [†] *Met. Zeit.* **49**, p. 131.

perature to the south, while the height of the tropopause exceeded 14 km in places.

The details quoted above for individual anticyclones do not directly confirm the theoretical views advanced earlier in this chapter, nor can they be said definitely to contradict those views. Rather do they emphasise a point which arose also in connection with the study of depressions, that the detailed structure in the upper atmosphere has a life-history, and that one instantaneous picture of the anticyclone can help but little in determining the course of that life-history.

One further piece of observational evidence remains to be presented. In a recent paper C. S. Durst* has collected values of the rotational velocity about the centre, in anticyclonic areas, showing that with increasing height there is a definite increase in this rotational velocity. In the typical anticyclone in which such a distribution of velocity occurs, subsidence brings to a given level air moving faster than is required by the gradient at that level, and accordingly this air tends to flow in towards the centre. This procedure will tend to maintain the anticyclone against attrition by friction. Durst points out that the vertical distribution of velocity described above will in general be that appropriate to a warm anticyclone, and he explains in this way the persistence of the warm anticyclone. This conclusion is open to doubt to the extent that the observations used do not of necessity represent the distribution of winds all round the anticyclone.

§ 210. *Subsidence and divergence in anticyclones*

When cold air subsides it spreads out laterally, and develops an anticyclonic circulation. It was shown by Brunt and Douglas† that there is divergence from a region of high positive isallobars. It is a fact of observation that when an anticyclone is developing the rise of pressure extends over a very considerable area, so that the gradient of the isallobars is very weak, usually corresponding to a diverging velocity of rather less than 1 m/sec, if averaged over periods of the order of 24 hours. The velocities involved in the divergence within a growing anticyclone are therefore not more than about 1 m/sec on the average. The pressure in the upper air rises with the surface pressure, and we may assume as a first approximation that neither the velocity of divergence nor the velocity around the isobars changes with height up to say 3 km. On some occasions both may increase with height.

Now take a cylinder of height h and radius r, and consider what happens to the air within it when the whole mass is given a mean downward velocity w, and a mean velocity of divergence outward v. Let ρ_1 be the mean density from the ground up to height h, and ρ_2 the density at height h. The variations of ρ_1 and ρ_2 with time will be neglected. The equation of continuity gives

$$2\pi r h \rho_1 v = \pi r^2 \rho_2 w,$$

$$w = \frac{2hv}{r} \frac{\rho_1}{\rho_2}.$$

* *Q.J. Roy. Met. Soc.* **59**, 1933, p. 231.
† *Memoirs R. Met. Soc.* **3**, No. 22.

Let $v = 1$ m/sec, $r = 400$ km, $h = 3$ km, $\rho_1/\rho_2 = 1 \cdot 2$; then

$$w = 0 \cdot 018 \text{ m/sec} = 65 \text{ metre/hour} = 1 \cdot 6 \text{ km/day}.$$

This is probably an overestimate of the average value over developing anti-cyclones, but a value of 1 km per day is probably of the right order of magnitude at the level of 3 km.

In a stationary unchanging anticyclone the only divergence and subsidence is that due to surface friction. Sir Napier Shaw* has estimated that in a large anticyclone with light winds the subsidence is only about 80 metres per day, so slight that the horizontal history of any element of air must far outweigh the vertical history in importance. Such subsidence can have little or no effect in producing increased stability. The relative humidity of the air above an anticyclonic inversion, say at $1 \cdot 5$ km, is usually below 30 per cent, and some-times below 10 per cent. If the air were saturated when it started to descend, a relative humidity of 25 per cent at $1 \cdot 5$ km would imply a descent of more than 2 km, and a relative humidity of 10 per cent at $1 \cdot 5$ km would imply a descent of about $3 \cdot 5$ km. The adiabatic descent of air through $3 \cdot 5$ km would imply, with normal conditions as to lapse-rate, an increase of temperature of about $25°$ F at the level finally reached.

The inversions found above anticyclones at relatively small heights are frequently referred to as "surfaces of subsidence". The name is misleading in that it implies a surface of separation between air which is descending and air which is not descending. The inversion frequently found occurs at a sur-face which is horizontal or nearly so, and which does not in general reach the ground, the inversion simply dying away before the surface is reached. It is much more probable that the discontinuity is developed *within* a mass of descending air. Many factors could be suggested to account for its formation. The effects of turbulence, variations in water content, and radiation, combined with subsidence, would in many circumstances suffice to produce inversions at approximately horizontal surfaces, of the magnitude of those observed.

A case of some interest was discussed in detail by Giblett,† using observa-tions at Cranwell on October 29, 1923. These observations showed an inver-sion within the polar air behind a cold front, the equatorial air being found at higher levels, with no inversion at the lower boundary, which was distinguished by an increase in humidity (see fig. 94). Giblett explained the form of the lower part of the continuous curve in polar air as the effect of the upward diffusion of heat from the surface of the ocean during the passage of the air over the North Atlantic. The stabilising effect of the subsidence would oppose the diffusion upward of turbulence, but in spite of this, turbulence would extend to an increasing height as the surface temperature rose. That the point E represented the limiting height reached by turbulence was confirmed by the haze-top discovered at that height. The air below E had a lapse-rate which approached the dry adiabatic, it was relatively humid, and it had a sheet of cloud near its top, all indicative of thorough mixing. While the inversion dis-cussed by Giblett was associated with a depression, the phenomena were in many respects similar to those observed in anticyclones.

* *The Air and its Ways*, 1923.   † *Vide*. p. 343 above.

When a mass of air containing a sheet of cloud subsides, the lapse-rate just above the cloud sheet becomes more stable, if it is initially stable, while the lapse-rate within the lower part of the cloud, if initially greater than the saturated adiabatic, becomes greater. The latter effect tends to maintain convection, and the cloud sheet will be maintained if enough moisture is carried up to counteract the effect of the slow subsidence. An examination of fig. 22 shows that if the descent were from 700 mb (say 3 km), at temperature 20° F, with no inversion initially, to 875 mb (say 1·5 km), an inversion of about 14° F would be formed. If later the clouds were dissipated, the inversion could persist, though probably in a weakened form, for some time. The mechanism thus suggested, associating the formation with the presence of clouds, is perhaps the easiest to understand physically, and is in all probability the commonest cause of inversions.

Other factors can also operate to form inversions, provided a rapid change of water-vapour content with height exists, as for example at the top of the layer affected by surface turbulence. Radiational effects have already been mentioned. If rain falls through a mass of air in which there is such a variation of water-vapour content, and is evaporated in falling, the cooling due to evaporation will vary according to the variations of wet-bulb temperature, and if all the air is brought to saturation thereby, each element of air will be cooled to its initial wet-bulb temperature. The course of events suggested would actually lead to instability.

Sufficient has been said to show that the formation of inversions does not require a sloping surface of subsidence, and that inversions can be formed in other ways than by simple subsidence, as for example by subsidence combined with some inequality of distribution of water-vapour or condensed water in the form of drops, or by turbulence. In view of the number of the factors which may be operative, it is necessary in any particular case to proceed with considerable caution in interpreting the existence of an inversion.

CHAPTER XIX

THE GENERAL CIRCULATION OF THE ATMOSPHERE

§ 211. *The surface conditions over the earth*

THE general circulation of the atmosphere has been mentioned briefly in an earlier chapter, and we shall now endeavour to sum up the main facts of observation, with a view to considering the physical causes underlying the phenomena. Some idea of the facts of observation can be derived by a study of figures which represent the mean conditions in the months of January and July. The marked feature which appears from a casual study of these charts is the tendency for high pressure to occur over the continents and low pressure over the oceans in the winter, while in summer this tendency is reversed. In both winter and summer the Southern hemisphere has a well-marked belt of high pressure centred about latitude 30° S, and more marked over the oceans than over the continents (see fig. 5–8). The January chart shows a similar distribution in the Northern hemisphere, but there are well-marked centres of low pressure in the Northern Atlantic (the Icelandic low) and the North Pacific (the Aleutian low), a very extensive anticyclone centred over Asia, and another but less extensive anticyclone over the north-west of North America. In July the sub-tropical anticyclones in the North Atlantic and North Pacific show marked increase in intensity and in extent, while the winter anticyclone over Asia is replaced by a depression centred over North-western India.

The greater wind systems of the atmosphere can be inferred from the pressure distribution by the use of Buys Ballot's law (see also figs. 10, 11). On each side of the equator are winds which have an easterly component, the North-east and South-east trades, extending from the high-pressure belts nearly to the equator. They are conventionally regarded as separated by a belt of calms known as the *doldrums*, but the doldrums are not clearly marked all round the equator, or at all times of the year. Brooks and Braby[*] showed that in the equatorial Pacific the phenomena appeared to be most easily explained by the presence of a sharp surface of discontinuity between the North-east and South-east trades, particularly in the region east of longitude 180°. In this region, where the South-east trade meets the North-east trade, the former rises over the latter, giving heavy rainfall. Beals[†] has shown that the doldrum zone of calms in the equatorial Pacific is restricted to the shore ends of the tropical belt over that ocean. Durst[‡] has shown that in the region between 25° W and 40° W in the Atlantic, the doldrums fluctuate in width and position, and that the heaviest rainfall is in the zone of light winds. At times the belt of doldrums over the Atlantic diminishes to vanishing point giving practically a surface of discontinuity separating the North-east and South-east trades.

* *Q.J. Roy. Met. Soc.* 47, 1921, p. 1.
† *Monthly Weather Review*, 55, 1927, p. 215.
‡ M.O., *Geoph. Mem.* No. 28.

The phenomena over the continents are somewhat different. The trade winds do not exist over the continents on account of the absence of the tropical high-pressure belts, which cannot form over heated land. The diurnal variation of temperature of the earth's surface gives rise to thermal instability over large areas, giving heavy rainfall over much more extensive areas of the continents than the rainfall zones over the oceans.

The central belts of the sub-tropical anticyclones are regions of calms, and on the poleward sides of these calms the winds are westerly, corresponding with the fall of pressure towards the poles. The westerly winds of middle latitudes extend to about latitude 65° to 70°. They are not steady winds, but are disturbed by the occurrence of travelling depressions and anticyclones. The long tract of the North Atlantic covered by the mean isobar of 1005 mb (figs. 7, 8) indicates the position of the mean track of depressions across the North Atlantic. Nearer to the North Pole pressure again increases, though in the inner polar regions conditions are not symmetrical on account of the irregular distribution of land and sea. Indeed over neither polar region are observations sufficient in number to enable us to say with certainty what the mean conditions are.

In winter the winds over almost the whole of Asia are explained by a clock-wise circulation around the Asiatic anticyclone, while in summer conditions over Southern Asia are entirely dominated by the depression which forms to North-west of India, and gives rise to the winds of the South-west monsoon. The South-west monsoon sets in during June, and lasts until about the end of September.

The above brief description gives the salient features of the mean surface conditions over the earth. The conditions from day to day vary considerably, particularly in those regions in which travelling depressions and anticyclones occur, and the mean condition to which we give the name "general circulation" may never occur *in toto*.

One marked feature is to be noted in the charts for both January and July. The flow of air is not everywhere symmetrical along the parallels of latitude. The flow is partly poleward, partly equatorward, as demanded by notions of continuity of air mass, and the substantial constancy of the pressure over the earth's surface. For though the pressure is everywhere of the order of 1000 mb, the limits of pressure found on the earth (at mean sea level) are everywhere within about 30 mb of the mean value. There is therefore no factor in the atmosphere capable of heaping up air in one region at the expense of the rest of the earth.

§ 212. *The circulation in the upper air*

We might begin the consideration of conditions in the upper air by the study of charts of pressure for different heights, such as are given by Shaw, *Manual of Meteorology*, **2**, but before doing so we shall first consider the mean pressure for different latitudes and for a range of heights from 0 to 24 km, as represented in fig. 109. This diagram is based on the estimated mean pressures given by Wagner in the *Handbuch der Klimatologie*, **1**, Teil F, "Klimatologie der freien Atmosphäre", p. 67. The pressures were computed on the basis of the temperature distribution given by Ramanathan.*

The curves in fig. 109 represent conditions in the Northern hemisphere in summer and winter. The winter conditions are represented on the right-hand side of the diagram, and continuous lines have been drawn across the whole diagram. It is not possible to say that the winter conditions represent the winter conditions of the Southern hemisphere, so that the diagram must not be interpreted as giving the pressure distribution over the whole earth. The observations for the Southern hemisphere are not sufficient to enable a mean curve to be drawn, and it is in any case certain that in the Antarctic the mean pressure distribution does not agree with that in the Arctic.

The curves in the diagram represent the variations with latitude of the pressures at 0, 2, 4, 6, 8, 10, 12, 14, 16, 20 and 24 km. The curves for the upper-most levels are not reliable in detail, as the number of observations of temperature available for these heights is very small.

At the surface the maximum pressure in summer is in latitude 40°, the tropical minimum occurring at about 15° and not at the equator. These correspond respectively to the sub-tropical high pressure, and the doldrum minimum. A second minimum in latitude 65° corresponds to the zone of occurrence of the depressions of middle latitudes. There is a maximum at 90° corresponding to an anticyclone over the North Pole. At 2 km the sub-tropical high-pressure belt has disappeared, as also has the maximum at the North Pole, and the form of the curve at 4 km is almost identical with that at 2 km, indicating a steady rise of mean pressure from the pole to the equator. Beyond 4 km the position of the maximum pressure is shifted to about latitude 13°, and this feature persists to about 20 km. The zone of the depressions in middle latitudes, which is indicated by a minimum in the curve of surface pressures, is represented at heights of 2 to 8 km by a check in the general rise of pressure from the pole to the equator, and a secondary minimum in latitude 65° at 10 km. This secondary minimum shifts nearer to the equator with increasing height, and appears in latitude 40° at 20 km.

Winter conditions differ considerably from the summer conditions described above. At the surface the sub-tropical high-pressure belt is centred about latitude 30°, and from this point to the pole there is a steady decrease of pressure. There is a flat minimum of pressure at the equator. The sub-tropical high-pressure zone is only feebly indicated at 2 km, and at greater heights, up to 12 km only, is represented by a slight check in the fall of pressure from the equator towards the pole. Above 12 km, there is a definite flattening out of the pressure curve between latitudes 25° and 10°.

The pressure distribution can be readily interpreted in terms of the corresponding winds. Let us consider first the summer conditions. At the surface the winds should be easterly from the doldrum region to latitude 40°, beyond which westerly winds should predominate up to latitude 65°; beyond this to the pole, winds should be predominantly easterly. At 2 km the winds should be predominantly westerly from the pole to the sub-tropics, the winds in the tropics being very light. At greater heights the westerly winds should extend

* *Nature*, **123**, 1929, p. 834. See also fig. 12, p. 18 above.

from the pole to nearly 10° N, beyond which there should be only light winds, with a tendency for easterly wind directions. The curves indicate that at 12 km the westerlies should be interrupted by a zone of easterly winds in latitudes 60° to 70° and that at still greater heights this zone should extend to the North Pole.

In winter the circulation around the sub-tropical anticyclones disappears before 4 km is reached. Westerly winds should predominate at this level in all latitudes, and this condition should persist up to between 12 and 16 km. At the 16 km level zones of calm appear to be possible in latitudes 10° to 25°, and near the poles.

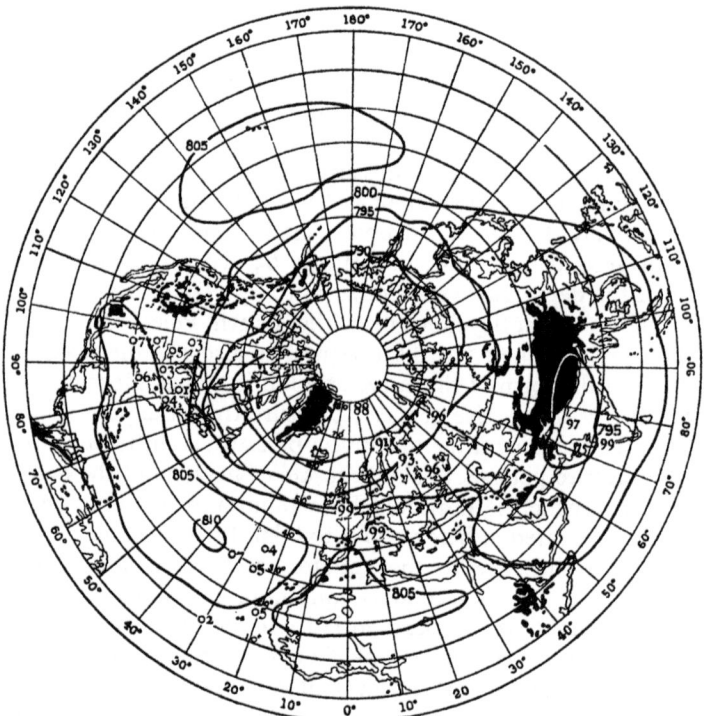

Fig. 110. Normal pressure at 2 km, Northern hemisphere, July.

It is necessary to reiterate the qualification that the curves for upper levels are based on few or no observations at many points, and that the deductions drawn from them above are perhaps not to be relied upon in detail beyond 8 or 10 km at the highest.

We shall next consider the forms of the pressure distribution at the ground, at 2 km, 4 km, and at 8 km, with a view to amplifying or correcting the impressions deduced from fig. 109. The distribution of pressure at 2 km in July for the Northern hemisphere, shown in fig. 110, indicates that the anticyclones over the North Atlantic and the North Pacific, and the depression over North-west India, still persist, though much flattened out, while over North-east Africa there is also a weak anticyclone. At 4 km (fig. 111) the

North-west African anticyclone is more marked, the North Atlantic belt of high pressure and the depression over North-west India are both much weakened, and the North Pacific belt of high pressure has disappeared. A new feature at this level is a small anticyclonic belt just north of the Himalayas.

At 8 km (fig. 112) the African anticyclone is still further strengthened, while the North Atlantic anticyclone is displaced farther west, and extends from longitude 50° to 110°, the highest pressure being in latitude 30° over the southern part of the North American continent. The anticyclone which appeared north of the Himalayas at 4 km now extends across from the Persian

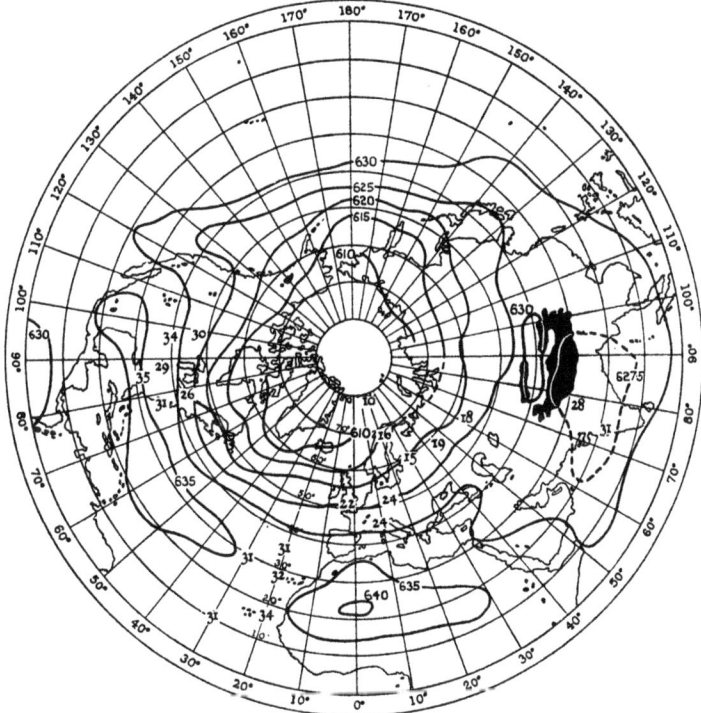

Fig. 111. Normal pressure at 4 km, Northern hemisphere, July.

Gulf to Western China. At all heights from 2 to 8 km the North Pole appears to be a centre of low pressure, and west winds should prevail from the pole down to about 30° N. At 8 km easterly winds should predominate over tropical Africa, North America, and India, while over the Pacific and Atlantic the winds in the tropics cannot be deduced from the charts on account of the fewness of observations, and possibly very light and uncertain gradient of pressure. The deductions from the charts of pressure distribution for July conflict with the deductions from fig. 109 to the extent that whereas fig. 109 indicates that the sub-tropical anticyclones do not appear at 2 km or any higher level, figs. 110, 111 and 112 indicate that in summer the sub-tropical anticyclones are clearly marked up to 8 km, with a shift of position from the oceans to the land at a level of 2–4 km. The summer depression over South-west Asia is very weak at 4 km, and is replaced by an anticyclone at 8 km. The contrast between the pressure distribution over the Northern hemisphere in July at the surface and at 8 km, represented in figs. 8 and 112 respectively,

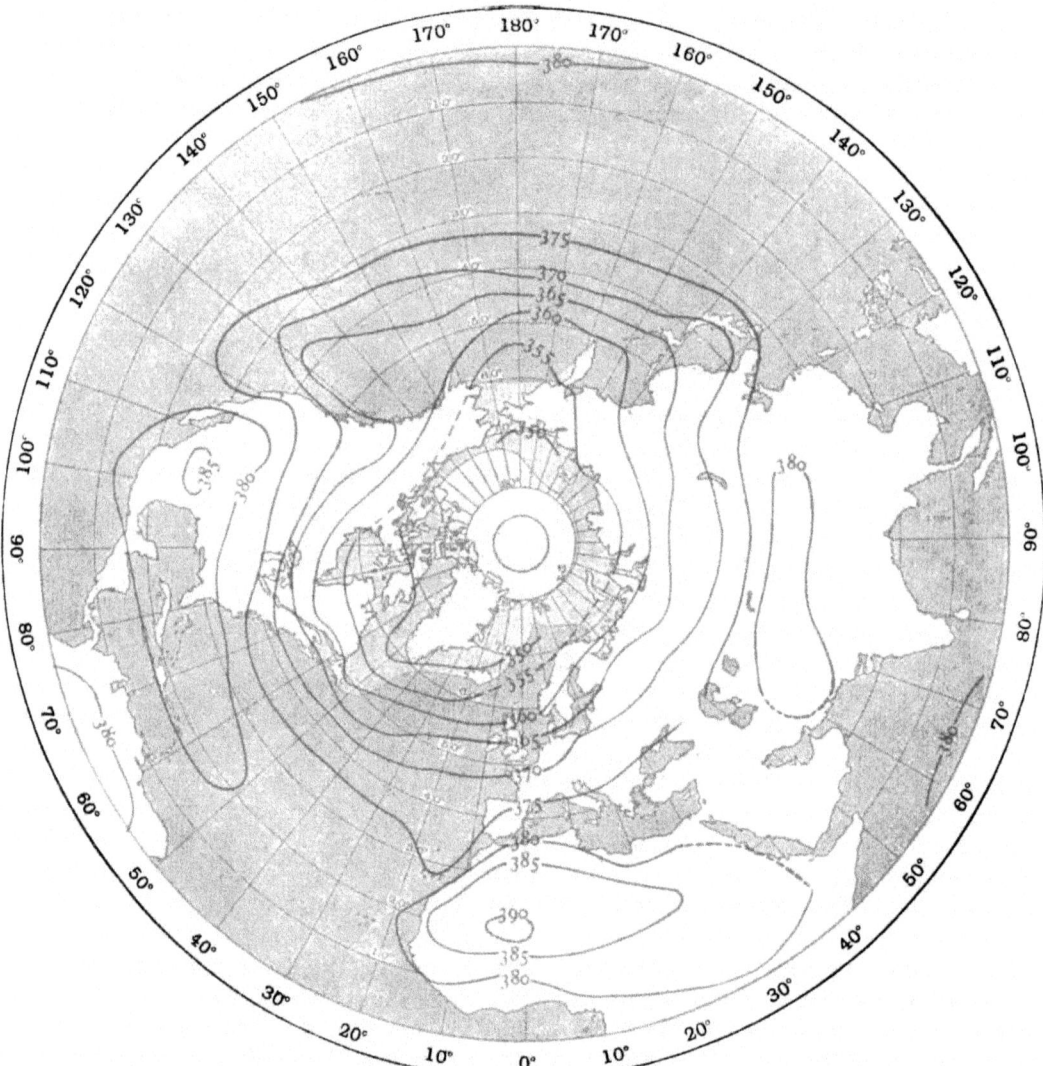

Fig. 112. Normal pressure at 8 km, Northern hemisphere, July.

is very remarkable. At the surface the pressure is highest over the oceans, while at 8 km it is highest over the continents.

It has been noted in fig. 7 that in January there is an intense anticyclone centred over Asia, this being one of the most marked features of the surface pressure distribution. Fig. 113 reproduces the pressure distribution over the

Northern hemisphere in January at 2 km, indicating that at this height the
Asiatic anticyclone has already disappeared, and that the only deviations from
a regime of North-South pressure gradients are to be found in a weak remnant
of the sub-tropical anticyclones, which shows secondary centres over the
eastern North Atlantic, the Gulf of Mexico, and the eastern North Pacific.
Everywhere from the pole down to latitude 30° N the mean gradients for
January are for West winds. At 4 km, fig. 114, the gradient is everywhere
favourable for West winds. Thus the deductions drawn from fig. 109 are
substantially correct for January (winter).

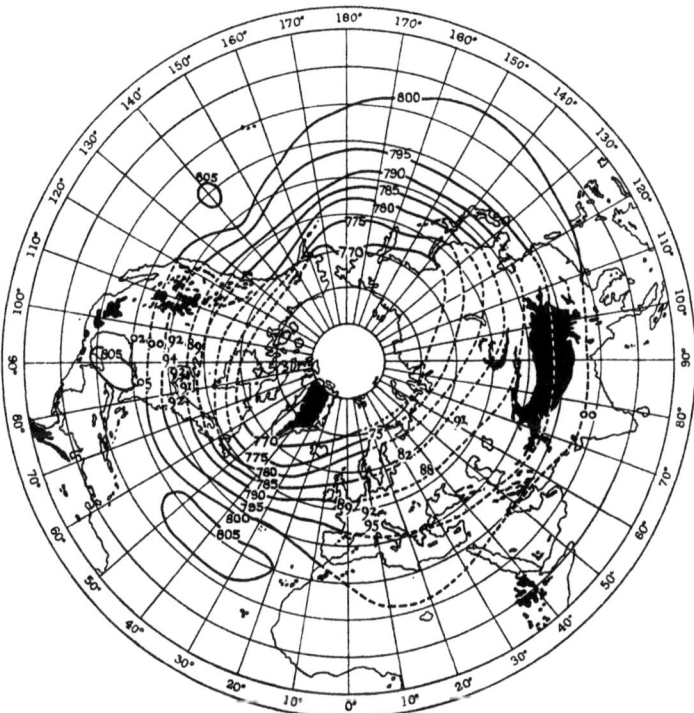

Fig. 113. Normal pressure at 2 km, Northern hemisphere, January.

Over the Southern hemisphere, the sub-tropical belt of high pressure shows
far more uniformity in all longitudes at the surface than is the case in the
Northern hemisphere (see figs. 5, 6). At 4 km the sub-tropical high-pressure
belt is barely appreciable in summer (fig. 116), and apparently has quite disappeared in winter (fig. 115), indicating the prevalence of westerly winds
everywhere above this height, and possibly at much lower levels, except in the
tropics, concerning which no reliable estimate can be made from the mean
distribution of pressure.

§ 213. *The observed distribution of winds in the upper air*

The observations of winds in the upper air are still insufficient to enable us to give a complete picture of the wind circulation in detail; but there are sufficient observations available to enable us to check the general conclusions which we

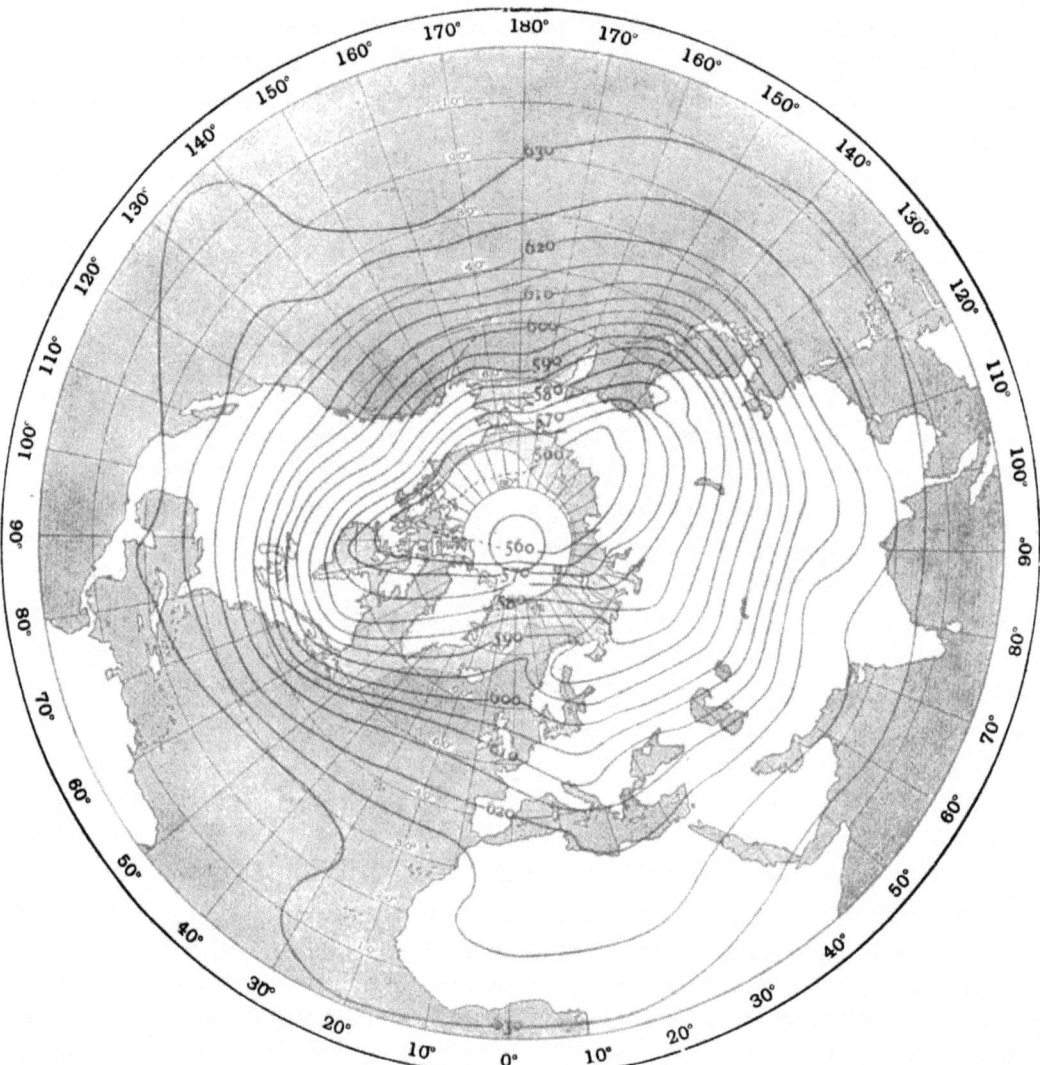

Fig. 114. Normal pressure at 4 km, Northern hemisphere, January.

have drawn from figs. 110 to 116. In the first place we shall use the monthly mean values of winds given by Wagner in his *Klimatologie der freien Atmosphäre*.

(a) THE TROPICS

At Batavia, in latitude 6° 11′ S, longitude 106° 50′ E, the wind in the lower levels is westerly in the months November to April. The westerly wind extends only to between 1 and 2 km in November, to 4 km in December, and to 5 or

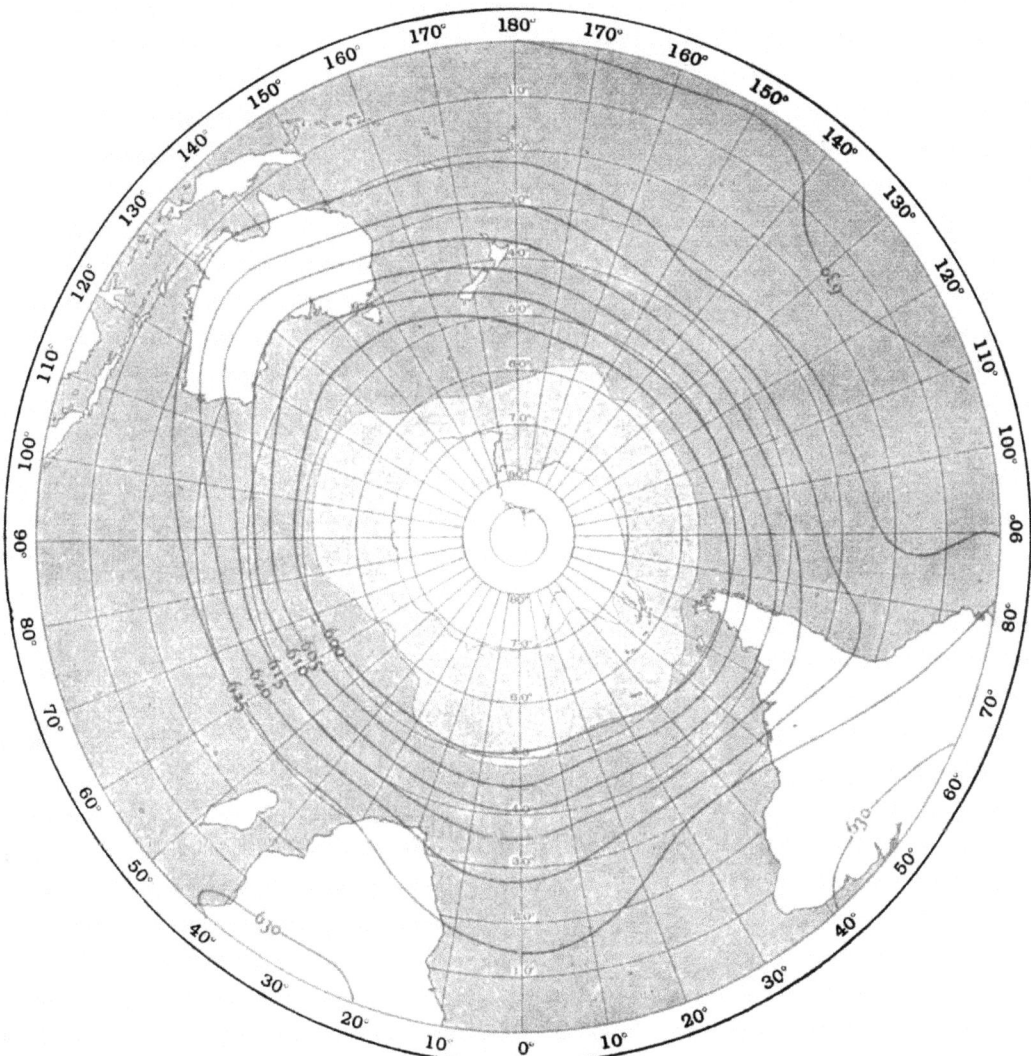

Fig. 115. Normal pressure at 4 km, Southern hemisphere, July.

6 km in March and April. In the months March to September there is a westerly or south-westerly wind at levels of about 18 to 22 km, with easterly winds at still greater heights. At the times and heights other than those mentioned, the winds are from an easterly direction. Thus in the southern summer the westerly monsoon extends at its maximum to a height of 6 km and has an

easterly wind above it. In the southern winter the winds are predominantly easterly at all heights up to about 18 km, usually with a slight northerly component of velocity. Van Bemmelen's* studies of cirrus motion in the tropics indicate in general easterly drift everywhere, and confirm the existence of a

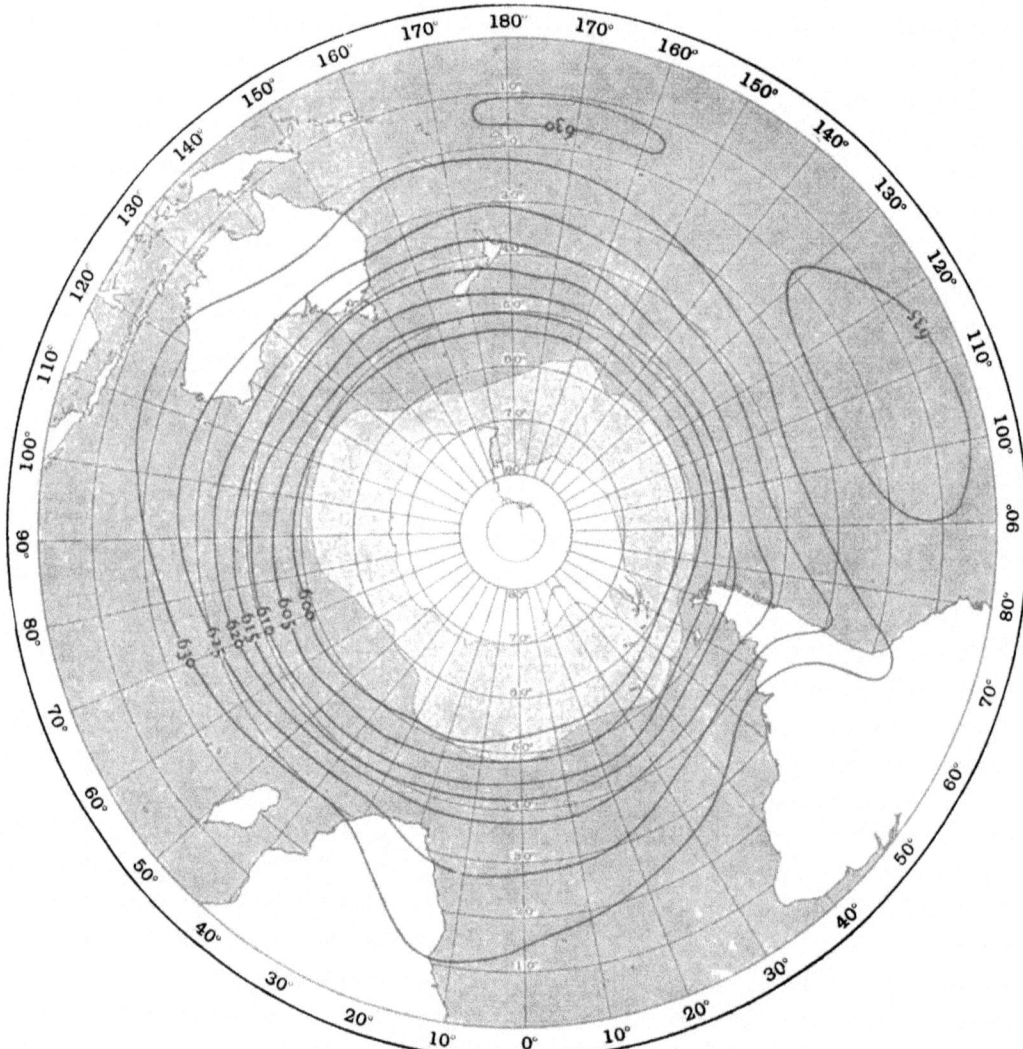

Fig. 116. Normal pressure at 4 km, Southern hemisphere, January.

northerly component in the motion at cirrus levels above Batavia (see figs. 117, 118). Berson and Elias at Lake Victoria Nyanza in tropical Africa (just south of the equator) found the South-east trade from the Southern Indian Ocean and above it a counter-trade wind with a northerly component, most

* *Proc. K. Akad. Wetenschap. Amsterdam*, **20**, 1918, pp. 1313–27.

frequently from NW but often from NE. With increasing height the wind took a more predominant southerly component, and at 16 km the few observations obtained indicated a definite tendency to SE winds.

At Honolulu (21° 22′ N, 157° 57′ W) the surface winds are round about NE, veering with height, but remain generally easterly up to heights of 3 to 5 km, and in May and June remain easterly up to 9 km. Beyond these heights the mean wind is westerly or north-westerly.

The winds at Samoa (13° 48′ S, 171° 46′ W) are predominantly easterly in the lower layers, with a westerly wind from 7 km upwards in the southern winter, and from 3 km upwards in the southern summer.

At Mauritius (20° 5′ S, 57·5° E) the mean winds are generally from ESE to E in the lower layers, changing to SW or W at 2½–3 km in the southern winter, and at 4–5 km in the southern summer.

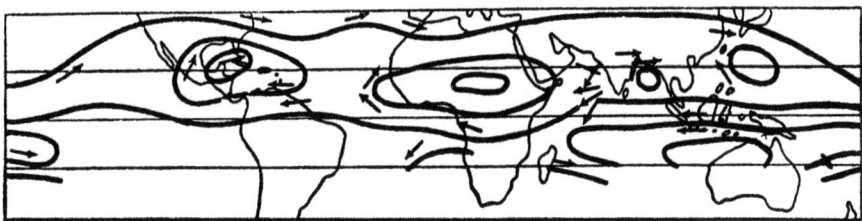

Fig. 117. Lines of flow of cirrus, December–February.

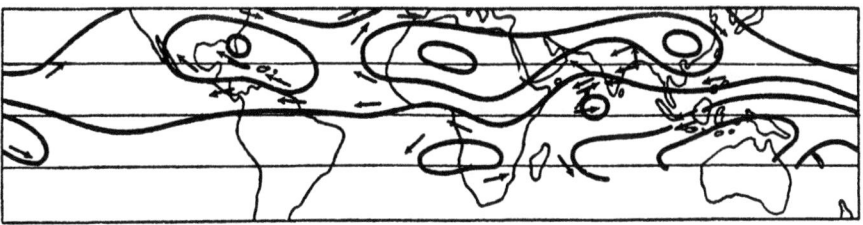

Fig. 118. Lines of flow of cirrus, June–August.

The trade wind region in the North Atlantic was investigated in great detail by Sverdrup* who found that the North-east trade extended only to a very limited height, which varied with both latitude and longitude. The upper boundary slopes downward from latitude 40° to latitude 20°, and from longitude 40° W to longitude 20° W. Above this limit is a zone of mixing or irregular winds, of thickness about 2 km, while above this is the counter-trade wind from a westerly direction.†

To summarise the above data very roughly, we may say that while the winds are usually easterly in the lower layers, there is a definite tendency to find westerly winds in the upper air, above heights which vary with locality from a few kilometres to 9 km.

* *Veröff. Geoph. Inst. Leipzig.* **2**, 1917.
† Recent observations over the whole of the Atlantic Ocean, obtained on the research ship "Meteor", have been discussed in *Wissenschaftliche Ergebnisse*, **14**, 2te Lief., Abschnitt B.

(b) The sub-tropics

Wagner gives mean wind directions for India for the three months of December, April and August, which may be taken as representative of the cool season, the hot season, and the monsoon season, respectively. In December the mean isobars run roughly across from West to East, with high pressure to the North, but the gradient is very slight on the chart of mean isobars for the month. In low latitudes, south of about 20° N, there is at low levels an easterly wind, which is replaced by a south-westerly wind at 8 km at Bangalore in latitude 12° 58′ N, but at heights of from 2 to 3 km farther north. At stations in latitudes 20° to 30°, the surface wind is northerly, backing to west within a range of 2 or 3 km. This agrees with the results derived by Harwood* for the cool season. Harwood's Plate 1 shows that over a strip along the north-eastern frontier of India, covering an area of nearly a half of India, the surface winds are north-westerly, becoming northerly over Burma, while over the rest of India the surface winds are about ENE. The direction of the motion of clouds at middle levels is westerly over the northern half of the peninsula and easterly over the southern portion. High clouds and pilot balloon observations at 9 km gave substantially the same drift as clouds at middle levels. In the hot season the easterly wind of the lowest layers is restricted to the extreme south of India, and the regime of northerly winds in the lowest layers, changing to westerly within a few kilometres, extends farther south. The gradients of pressure shown by the monthly mean pressure charts are very slight and irregular in April, though the Asiatic anticyclone still persists. Harwood's Plate 2 (*loc. cit.*) confirms these results.

During the monsoon season the surface winds are southerly over Burma, east-south-easterly over the strip of country north of the plain of the Ganges, and west or south-west over the remainder of India. At the 2 km level the motion is substantially the same as at the ground. At the level of high clouds, the winds are easterly everywhere east of a line joining Bombay and Agra, and westerly to the west of this line. The change from east to west wind occurs at varying levels, being at about 1 km at Agra, 3 km at Bombay, 5 km at Bangalore in the south. The final result is striking, in that at heights of 10–14 km there is a marked line of discontinuity across the peninsula from south-west to north-east.

Thus the winds of the Indian cold season (the north-east monsoon) are of the same nature as the trade winds of the North Atlantic, having a south-westerly current above them. The upper winds of the south-west monsoon season do not appear capable of such summary description. The east winds in the upper air over a large tract of the country are presumably part of the circulation around the anticyclone which Teisserenc de Bort found over the region of the Himalayas at 4 km (see fig. 111). Harwood suggests (*loc. cit.*) that most of the air of the south-west monsoon is carried off westward by this upper current, and that the greater part of this air passes northward over North-west India, and joins the circulation of middle latitudes, a portion

* *Indian Met. Memoirs*, **24**, pt 8.

however possibly descending over the Arabian Sea to join the westerly surface monsoon current.

(c) MIDDLE LATITUDES

In the zone of prevailing westerly winds of middle latitudes, conditions are difficult to summarise. The trade winds of low latitudes blow in the same direction with substantially the same velocity for considerable periods of time, and can therefore be described in relatively few words. In middle latitudes however the conditions are continually being disturbed by the passage of depressions and anticyclones, which affect the winds up to levels well in the stratosphere. Wagner* quotes numerous data for the mean wind-transport for places in middle latitudes. At Lindenberg the general direction of air transport is westerly in practically all months, at all levels up to 2 km, but with a component from North in summer and from South in winter. At Vienna the direction is between West and North in all seasons up to a height of 4 km. Mean values for Central Europe indicate westerly winds with a northerly component in summer, up to 18 km, and in winter westerly winds with a southerly component up to 10 km, and a northerly component at greater heights. Over Italy the winds are NE to E in the lowest kilometre, above which there is a rapid change to NW. At 3 km the mean air transfer is between W and NW over the whole of the Mediterranean, agreeing with the circulation around the anticyclone over North Africa at this level.

(d) UNITED STATES

Wagner quotes data for a number of stations in the United States. The results are similar in their main features to those for Europe. At 3 km a broad westerly current sweeps across the country from its northern limit down to latitude 30° N. A portion of this current is deflected southward over the Gulf of California, while a south-westerly current over the Gulf of Mexico joins the main stream moving across the Eastern Atlantic.

(e) NORTHERN HIGH LATITUDES

In the Arctic observations are far from sufficient to give a complete picture of the circulation of the winds. The mean of 252 pilot balloon ascents at Ebeltofthafen in Spitzbergen in 1912–13 gave winds between N and NE in the lowest kilometre, with northerly winds above, backing towards west at 7 km. Over Eastern Greenland the winds observed by A. Wenger were predominantly NW. Over the western coast of Greenland the winds were easterly at the surface veering towards south with increasing height, and reaching south at 8 km. At 10 km the wind direction is backed one point to S by E. The mean of 225 pilot balloon observations at Akureyri and Adalvik in Iceland at various times indicate SSE winds at the surface veering steadily to west at 8 km. Georgii† describes some pilot balloon observations in North-west Iceland (66° 22′ N, 23° 8′ W) which indicated for the periods 25th to 28th June 1927, and 17th to 21st July 1927, outbursts of cold air from about NNW, reaching to a height of at least 15 km, with velocities of 70 m/sec and 60 m/sec

* *Klimatologie der freien Atmosphäre*, pp. F 33, 34. † *Arktis*, **1**, h. 3/4.

respectively at 15 km, where the outflow had its maximum. The winds in the lowest 2 km were intermittently northerly or southerly, but at higher levels the north-westerly currents were clearly developed with strength increasing with height. Southerly winds at great heights were never so strong as the northerly outbursts referred to.

The aerological results of the Greenland Expedition of the University of Michigan* in 1926 and 1927–29 indicate that in a shallow layer near the surface the winds at Mount Evans on the west coast are east to south-east at all times. In the months of November, December, January and February the wind veers to SW or W at 7 km, and to about W at 10 km. In this season the winds show remarkable regularity in the variation of the monthly means with height. In all months the change to between south and south-west is reached in about 2 km. The change with height of the wind direction is generally similar to that in winter in the months of March to July, up to about 7 km, but at greater heights wide variation appears. In July and August 1927 the winds at 7 km were N to NW, but in the same months of 1928 the winds at 7 km were S to SW. In September and October 1927 the mean winds remained in the SE quadrant up to 7 km, and then backed towards north, but in September 1928 the mean wind was SSW at 6 km, NNW at 7 km, and remained in the NW quadrant to considerable heights above 7 km; while in October 1928 the mean wind remained southerly up to 8 km, and then backed through to N at 13 km.

Thus except in the winter months there are considerable variations in the mean winds above Greenland, though there is a predominance of winds in the south-west quadrant at 7 km. An examination of the individual pilot balloon ascents shows marked variations from day to day, or even from hour to hour. Thus on February 3 the wind at 7 km was 17 m/sec from south at 11.38 a.m., but was 35 m/sec from S by W three hours later.

(f) SOUTHERN HEMISPHERE

For the Southern hemisphere the available information is extremely scanty. Such observations as are available indicate that the winds in the inner tropics are in a generally easterly direction, and that in middle latitudes the mean winds are generally westerly. We have already referred above to the observations at Batavia and elsewhere in the southern Tropics, which indicate predominantly easterly winds. Wagner's table (*loc. cit.* p. F 63) shows that in Northern Australia, Willis Island (16° 18' S, 149° 59' E) the mean winds are ESE up to a height of 3 km, while at Melbourne, Adelaide, Sydney and Christchurch in New Zealand the mean wind direction is about NW up to a height of 3 km.

The observations obtained during the drift of the German Antarctic Expedition in 1911/12 gave mean velocities of 5–7 m/sec at the ground, increasing to 12–15 m/sec at 2 km. The mean directions, according to Wagner's brief summary, were not easterly at any time of the year, and the most frequent wind direction was south-south-westerly at the surface, backing about one point in the first 3 km, then veering through about 17° in the next 5 km;

* *Rep. Greenland Exped. Univ. Michigan* (1926–31), pt 1, Aerology.

beyond this height the veer became more rapid, so that at 10 km the wind was veered on an average 52° from the surface wind. On an average the wind was round about SSW in the first 6 km, increasing steadily with height, thereafter veering to west.

Observations made on the Scott Antarctic Expedition 1911 are not readily interpreted, on account of the marked topography of the continent, which forces the wind to follow certain directions independently of the free air conditions. The smoke of Erebus (4000 metres) was predominantly from a westerly direction when the surface wind was less than 20 m.p.h., but with surface winds of 20 m.p.h. or more there occurred a group of south-easterly winds at the summit of Erebus. Medium clouds showed precisely the same features as the smoke of Erebus, northerly and southerly winds being approximately equally frequent. When the surface winds were 20 m.p.h. or more, the motion of high cloud showed three directions of marked maximum frequency, NNE, ESE, and NNW, while with light surface winds the most frequent directions of motion of high cloud were N (21 per cent), NNW (11 per cent) and E (11 per cent). On the south polar plateau the observations of the Scott and Amundsen expeditions combined indicate a group of wind directions of maximum frequency centred at about S by E, and extending three points to each side of this. This result is obtained by taking South to be parallel to the meridian of 160° E. On the Western Plateau the most frequent direction for light winds (up to force 4) was WSW and for strong winds (force 5 or more) S or SSW. The diagrams reproduced by Simpson in the report of the expedition (*Meteorology*, 1) indicate that the blizzard winds usually blew from a southerly direction, and that the effect of the blizzard conditions was superposed upon the other conditions, which in themselves gave winds which might blow out of or into the Antarctic continent. The meteorological conditions over the plateau were therefore not steady, and did not correspond to a steady cyclonic or a steady anticyclonic condition at the South Pole.

We are now in a position to sum up our knowledge of the mean winds over the globe. Around the equator is a boundary between the circulations of the Northern and Southern hemispheres. The boundary is not coincident with the geographical equator itself, but is somewhat north of it at all seasons, but farther north in the northern summer than in the northern winter. The boundary is in places closely defined as a line of demarcation, as for example in the Pacific, as shown by Brooks and Braby; in places it is clearly defined as a zone of definite width and known as the doldrums; elsewhere it is ill-defined and possibly non-existent. Above this boundary is a zone in which the winds are easterly up to considerable heights. Above these east winds there may occur a belt of westerly winds above which there again occur east winds. Between this zone and the central lines of the sub-tropical high pressures are the North-east and South-east trade winds in the Northern and Southern hemispheres respectively. The trade winds blowing towards the equator are relatively shallow, with heights varying with time and with longitude. Above the trade winds are westerly winds, usually with a component from the equator towards the pole, known as the anti-trades, and above the anti-trades are at times, and in some places, upper-trade winds from an easterly direction, possibly an incursion of the equatorial belt of east winds   In middle latitudes the mean winds over most of the earth are westerly at all heights with a tendency for a component from the pole in summer, and towards the pole in winter. Embedded in these winds are regions of easterly winds which may form part of the circulation around such systems as the anticyclone which appears over North Africa at 4 km; also moving cyclones and anticyclones have at certain stages complete circulations around their centres extending up to great heights, and these may therefore temporarily introduce winds from any point of the compass at any heights within the troposphere.

In higher latitudes we can only guess at the nature of the phenomena, but such observations as are available indicate that easterly and westerly winds may alternate, particularly in the lowest kilometre or two; while at greater heights westerly winds predominate with a poleward or equator-ward component alternating irregularly, and from time to time there occur outbursts of cold air from within the Arctic circle, extending to great heights, and so bringing down to middle latitudes vast masses of very cold air.

§ 214.   *The problem to be solved*

Having in the last few paragraphs summarised, admittedly in a rough and ready manner, the observed winds of the globe, we are now in a position to state what is the nature of the problem which has to be faced if the general circulation of the atmosphere is to be explained physically. We require a thread which will connect the motions observed in the atmosphere with the amount of incoming solar radiation reaching different parts of the earth, and with the nature of the atmosphere itself and of the earth's surface. This is, in broad terms, the whole problem. But several subsidiary aspects of the problem deserve special mention. We have already seen (§ 170) that the kinetic energy of the earth's atmosphere is continually being degraded into heat as a result of turbulence and friction. Since on the average there is no observable change in its kinetic energy from year to year, the loss is continually being made good, through the agency of the only possible source of energy—solar radiation. We require to know the precise manner in which some of the energy of incoming solar radiation is used to maintain the kinetic energy of the atmosphere. Again we shall find that the flow of air in the trade winds towards the equator demands the ascent of large masses of air in the inner tropics, the ascended air flowing polewards at high levels. This air must eventually descend again to the earth's surface, and part of our problem is to determine first the agency which fits this air to come down to a lower level, and secondly the part of the earth's surface at which the descent takes place. Another aspect of the larger problem is the explanation of the origin and position of the sub-tropical high-pressure belts. Yet another is the relation of the moving cyclones and anticyclones to the general circulation. To what extent are these lesser circulations necessary for the maintenance of the general circulation; or, to what extent are they brakes on the general circulation which keep the latter from growing indefinitely?

We shall not be able to claim that the problem is completely solved until we have found the explanation of the origin; maintenance, travel and death of the cyclone and anticyclone, and have explained the mechanism which produces variations of pressure over large areas of the earth's surface. Nor at this stage can we say definitely that the cyclone, anticyclone and general circulation are not merely three aspects of one problem.

Unfortunately it must be admitted that no solution of the complete problem as stated above has yet been found. The most we can hope to do at the present stage of the development of meteorology is to enumerate more clearly the aspects of the problem mentioned above, and to connect up some of the observed phenomena with the results of some of the earlier chapters in this book. The present writer is not aware of any "theory" which fits the whole of the phenomena. Theories of the general circulation were rife in the latter half of the last century, but increasing recognition of the tremendous complexity of the problem has led to an increasing disinclination to attempt a general theory.

§ 215. *Some theoretical aspects of the general circulation*

The first point which we have to consider is the distribution of incoming radiation over the earth's surface. Apart from the geometrical considerations which affect this distribution, the most important controlling factor is the cloud amount, since (*vide* § 74) a sheet of cloud sends back into space nearly the whole of the incoming radiation. The figures and diagrams given by Brooks* show that over both land and sea the mean cloudiness is rather greater over the equator than it is within zones to either side of it, these zones showing an annual variation in position following the sun in its motion north or south of the equator. Simpson,† using Brooks' figures for mean cloudiness, estimated the total effective solar radiation in different months of the year, over the whole earth, and found that the maximum occurred in a zone which was approximately beneath the sun, and so was 20°–30° N in July, and 20°–30° S in January. These results are in a general way confirmed by Wüst's estimates of evaporation in different zones of the earth, according to which the evaporation is less over the equator than over zones some little distance to each side of the equator; but Wüst's figures for evaporation are *estimates* and not observations.

The curious way in which the highest temperatures avoid the equator, except at the equinoxes, makes it difficult to accept in all its simplicity the usual statement that on account of the maximum of insolation at the equator the air at the equator rises, its place being taken by air drifting in from both sides. Yet this is what appears to happen. In the Northern hemisphere the air which moves towards the equator to take the place of the ascended air should start as a northerly current, swinging round through NE and eventually to E, on account of the deviating force due to the earth's rotation. This current is the North-east trade-wind of the Northern hemisphere, and its counterpart in the Southern hemisphere is called the South-east trade-wind.

Since there is a continual transport of air towards the equator by the trade-

* *Memoirs R. Met. Soc.* **1**, No. 10.   † *Ibid.* **3**, No. 23.

winds, there must be an ascent of air in a region about the equator, and the air which ascends drains away poleward in the upper air; in the Northern hemisphere it will start as a southerly current swinging round through SW eventually to W, on account of the deviating force of the earth's rotation. Some of this air descends again to the earth's surface in quite low latitudes, as is shown by Sverdrup (*loc. cit.*) in his researches on the conditions in the North-east trade of the Eastern Atlantic. Sverdrup found definite evidence of the descent of air in a region 20°–25° N, in the poleward moving current of the anti-trade, showing that in the time required to drift to this distance from the equator some of the air is cooled sufficiently to enable it to descend to low levels. On the poleward side of the centre of the sub-tropical high pressure belt the poleward drift of air across the isobars gives generally south-westerly winds at the surface, with westerly winds above. The diminution of temperature poleward produces a gradient of pressure directed towards the poles, at all heights in middle latitudes, if we consider only the mean conditions. But we cannot take this result too literally, since it would involve a net flow of air towards the poles, which is naturally impossible if the mean pressure at the surface is to remain reasonably constant over the earth, and a vast accumulation of the air in high latitudes is to be avoided.

It was pointed out by Exner* that it is impossible to have the same distribution of winds all the way round a circle of latitude, since this demands that the gradient of pressure should be directed the same way round the whole circle, which is a physical absurdity. We therefore come to the conclusion that the zone of mean west winds must be broken by regions in which the flow is towards the equator. The synoptic charts drawn from day to day indicate that these regions are associated with travelling cyclones and anticyclones. In the rear of cyclones and in the front of anticyclones broad deep currents of air move equator-ward, carrying a compensating mass of air away from the polar regions. Thus the cyclones and anticyclones of middle latitudes are dynamically necessary for the maintenance of the general circulation.

It is frequently stated in meteorological treatises that if air moves from one latitude to another, retaining its original angular momentum (in space) about the earth's axis, then in its new latitude it will have enormous velocities along the circle of latitude. This statement is highly misleading. If there were no pressure gradients or other forces, a mass of air set in motion would be constantly deviated to the right in the Northern hemisphere and would move with constant velocity; if it were started in a northerly direction it would rapidly swing round until it had a motion to East, after which it would continue to swing round until eventually it had completed a circle. The last statement requires some slight qualification, since the path is only a closed circle if the variation of latitude is neglected, but as we shall see the radius of the path is so small that the argument is in no way affected by this approximation. Let V be the velocity of projection, and r the radius of the path. Then the equation of motion is

$$\frac{V^2}{r} = 2\omega \sin \phi \, . \, V.$$

* *Dynamische Meteorologie*, 2te Aufl., p. 216.

Assume $\qquad\qquad\qquad V = 20 \text{ m/sec} = 2000 \text{ cm/sec.}$

In latitude 60°, $\qquad\qquad r = \dfrac{V}{2 \times 7 \cdot 3 \times 10^{-5} \times 0 \cdot 866},$

or $\qquad\qquad\qquad r = \dfrac{2000 \times 10^5}{12 \cdot 6} = 1 \cdot 6 \times 10^7 \text{ cm} = 160 \text{ km.}$

Thus on a smooth earth with no forces, the maximum displacement could not exceed a few hundred kilometres in latitude 60°. In lower latitudes the displacement would increase in inverse proportion to $\sin \phi$, but even air projected in latitude 10°, with a velocity of 20 m/sec to North, would not travel to North more than a distance of the order of 1000 km, which is equivalent to about 9° of latitude.

The point of the argument given above is that air leaving low latitudes cannot travel to high latitudes unless it is guided by a suitably arranged pressure gradient.

If there is a pressure gradient in existence the motion of air projected in any direction over the earth's surface will oscillate about the velocity and direction corresponding to the pressure gradient, and the enormous velocities referred to above cannot come into existence. In practice the motion of a mass of air through a large range of latitude, while retaining its original angular momentum about the axis of the earth, can never arise.

It has been pointed out by many writers that it is impossible to derive a theory of the general circulation based only on the known value of the solar constant, the constitution of the atmosphere, and the distribution of land and sea. Not only are the laws which determine the transfer of energy by radiation too complicated to permit this, but the transport of heat by advection through the medium of the general circulation, and the inter-relationships of cloud amount, radiative transfer and the general circulation, whose precise nature are unknown, make it impossible to derive any simple theory. It is only possible to begin by assuming the known temperature distribution, then deriving the corresponding pressure distribution, and finally the corresponding wind circulation.

Jeffreys,[*] in a mathematical treatment of the problem, started off with the assumption that in a frictionless atmosphere the temperature variations can substantially be represented by a decrease of the annual mean temperature from the equator poleward. This involves a slower decrease of the pressure with height at the equator than elsewhere, so that there should be an outflow of air poleward in the upper air, and an inflow of air towards the equator everywhere along the surface. This evidently requires easterly winds everywhere at the surface, with westerly winds everywhere in the upper air. This is in contradiction with observations, which indicate that easterly winds occur only within a zone covering about half of the earth's surface, centred at the equator, the prevailing winds elsewhere, except possibly in the polar caps, being westerly.

[*] *Q.J. Roy. Met. Soc.* **52**, 1926, p. 85; a slight correction to this paper, by C. A. Coulson, is given in same Journal, **57**, 1931, p. 161.

Reference has been made above to Exner's demonstration of the im possibility of a zonal distribution of winds. Jeffreys discussed the same question from a different angle. The existence of a mean velocity which is the same around a parallel of latitude implies that the frictional effect at the surface is everywhere increasing or diminishing the angular momentum about the earth's axis of rotation, and if the mean conditions are to persist, the loss by friction must be compensated by some other agency. The only effective agency is the interchange of air with other latitudes. Consider air at a height z, above a point in latitude ϕ, whose distance from the earth's axis is ϖ. The angular momentum about the earth's axis of unit mass at height z is

$$(\varpi + z \cos \phi)^2 (\omega + \lambda), \qquad \ldots\ldots(1)$$

where λ is the rate of increase of longitude of the air, measured positive for eastward motion. Let the eastward velocity be u, and the northward velocity v. Then the total northward flux of angular momentum across the whole parallel is

$$2\pi \int_0^\infty \rho \, (\varpi + z \cos \phi)^3 (\omega + \lambda) \, v \, dz \qquad \ldots\ldots(2).$$

But since the total mass northward of lattitude ϕ is constant,

$$\int_0^\infty \rho \, (\varpi + z \cos \phi) \, v \, dz = 0. \qquad \ldots\ldots(3).$$

Now write
$$\int_0^z \rho \, (\varpi + z \cos \phi) \, v \, dz = q \qquad \ldots\ldots(4),$$

so that q is proportional to the total flow northward at heights less than z. The total northward flow of angular momentum is

$$M = 2\pi \int_0^\infty (\varpi + z \cos \phi)^2 (\omega + \lambda) \, dq \qquad \ldots\ldots(5),$$

taken through the atmosphere.

Integrating by parts we find, since $q = 0$ at $z = 0$ and at $z = \infty$,

$$M = -2\pi \int_0^\infty q \frac{d}{dz} \{ (\varpi + z \cos \phi)^2 (\omega + \lambda) \} \, dz$$

$$= -2\pi \int_0^\infty q \left\{ 2\omega \, (\varpi + z \cos \phi) \cos \phi + u \cos \phi + (\varpi + z \cos \phi) \frac{du}{dz} \right\} dz,$$

since
$$(\varpi + z \cos \phi) \, \lambda = u.$$

Now put $\varpi = a \cos \phi$ and neglect z by comparison with ϖ. Then

$$M = -2\pi \int_0^\infty q \cos \phi \left(2\omega a \cos \phi + u + a \frac{du}{dz} \right) dz \qquad \ldots\ldots(6).$$

The terms inside the brackets are of order 4×10^4, 2×10^2, and 4×10^5 cm/sec, so that the last is by far the greatest. Neglecting all but the last, we find

$$M = -2\pi a \int_0^\infty q \cos \phi \frac{du}{dz} \, dz = 2\pi a \cos \phi \int_0^\infty \rho u v a \cos \phi \, dz = 2\pi a^2 \cos^2 \phi \int_0^\infty \rho u v \, dz$$

$$\ldots\ldots(7).$$

The eastward frictional force per unit area is $-\kappa\rho_s u_s (u_s^2+v_s^2)^{\frac{1}{2}}$, where κ is the coefficient of skin friction, about 0·003. The eastward friction will produce forces on the air north of latitude ϕ whose moment about the axis will be

$$-\iint \kappa\rho_s u_s (u_s^2+v_s^2)^{\frac{1}{2}} a \cos\phi\, dS,$$

where dS = element of surface,

$$= -\int_0^{2\pi}\int_\phi^{\pi/2} \kappa\rho_s a^3 u_s (u_s^2+v_s^2)^{\frac{1}{2}} \cos^2\phi\, d\phi\, d\lambda$$

$$= -2\pi a^3 \int_\phi^{\pi/2} \kappa\rho_s u_s (u_s^2+v_s^2)^{\frac{1}{2}} \cos^2\phi\, d\phi.$$

If conditions are steady, the net loss by friction must equal the gain by advection. Hence

$$-\kappa a\int_\phi^{\pi/2} \rho_s u_s (u_s^2+v_s^2)^{\frac{1}{2}} \cos^2\phi\, d\phi = \cos^2\phi \int_0^\infty \rho uv\, dz \qquad \ldots\ldots(8).$$

If the motion is symmetrical about the axis and the winds geostrophic, this equation cannot be satisfied, even allowing for the usual modification near the surface to allow for friction. For in these conditions v is negligible except in the lowest kilometre, and there it is only about $\frac{1}{4}u$. Thus the ratio of the two sides is practically $\kappa a:\frac{1}{4}$ kilometre or 50 to 1. Even allowance for the variation of ϕ cannot reduce this discrepancy to less than 20:1. Thus such winds as are observed are incompatible with a steady symmetrical distribution of pressure, and the maintenance of the polar circulation against friction requires a greater supply of angular momentum from without than can be provided merely by the drift of air across the isobars produced by surface friction.

Exner showed that a pressure distribution in which the isobars cut across a circle of latitude at the same angle all round the earth is impossible. That does not exclude the possibility of the isobars coinciding with the circles of latitude, with the surface air drifting across the isobars as the result of surface friction. Jeffreys' discussion summarised above shows that this is also impossible, since friction would destroy angular momentum far more rapidly than it would be replaced by advection from other latitudes. It appears therefore that the general circulation must either involve no surface winds, and therefore no friction, or else must be unsymmetrical. The latter condition makes it possible to satisfy equation (8) above. If v is about equal to u instead of being $\frac{1}{4}u$, and if this condition applies to 6 or 7 kilometres of height, the equation can be satisfied. This is equivalent to assuming, in high latitudes, that the stream of air extends through most of the troposphere. The exchange of air between high and low latitudes is then to be explained as due to deep currents in opposite directions, over different parts of the same circle of latitude. And since the isobars must be closed systems, this demands the cyclones and anticyclones of middle latitudes as an integral part of the machinery which maintains the general circulation in being.

Jeffreys' theory was in the first place developed for a frictionless atmosphere, and, as its author pointed out, the introduction of friction demands that in any approximately steady state there must be a balance between East winds and West winds at the surface.

This argument does not bear out the suggestion that the depressions and anticyclones of middle latitudes represent either an instability of the general circulation, or an oscillation about a steady state. Jeffreys' theory, on the contrary, makes it clear that a steady general circulation is impossible *ab initio*, and regards depressions and anticyclones as the irregularities which are inevitable in any circulation when friction is taken into account.

It is not necessary to assume that the westerly circulation of middle latitudes should extend to the poles. Such an extension would be difficult to realise dynamically, as it would demand that the frictional drift of air poleward across the isobars should be compensated by the ascent of air at the poles. In view of the intense cooling of the earth's surface in the polar regions by radiation, the ascent of air there could not be expected. The general result stated by Jeffreys, that skin friction will destroy any circulation in a period of the order of one week unless fresh angular momentum of the same sense is brought in, does not preclude the possibility of anticyclonic conditions at the poles. It has been shown earlier (§ 115) that both cyclones and anticyclones have a rotation in space in the same sense as the rotation of the earth; and it is therefore possible to have an anticyclonic circulation maintained at the poles by the inflow at high levels of air whose subsequent subsidence provides angular momentum of the right sense to compensate for the losses by surface friction. There appears therefore to be every probability that the average conditions at the poles are anticyclonic.

In a note on Jeffreys' paper Whipple* suggested that if the trade winds represented the drift of air from the cooler parts of the earth to the warmer parts they would be developed in middle latitudes rather than in the tropics. There must therefore be some other cause than convection to force air into the anticyclonic belt. Whipple added that the most probable explanation is that in middle latitudes the strong West upper winds speed up the air at lower levels, dragging it onwards by turbulence on a large scale, and that this air is flung outwards and piled up in the anticyclonic belts. Whipple's suggestion is equivalent to the assumption of departures of the winds from geostrophic, giving a mean flow of air in the lower troposphere of temperate latitudes towards the sub-tropical anticyclones, more than sufficient to compensate for the surface frictional flow in the opposite direction.

The problem was taken up at this stage by C. K. M. Douglas,† who started from the mean pressure distribution over the North Atlantic in January, and estimated the mean frictional inflow to be of the order of 1 m/sec through a layer 600 metres thick. A flow of 0·15 m/sec up to 4 km would suffice to compensate for this loss of air, but something much greater would be needed in order to supply the necessary angular momentum.

In the ordinary theory of turbulence, if u is the component in the West-East direction, there is a term $K\dfrac{\partial^2 u}{\partial z^2}$ on the right-hand side of the equation for

---

* *Q.J. Roy. Met. Soc.* **52**, 1926, p. 332.   † *Ibid.* **57**, 1931, p. 423.

du/dt (see equation (85), p. 252), and if $\partial^2 u/\partial z^2$ is positive, the air is speeded up, and in consequence suffers a subsequent deviation to the right, or towards South. Douglas estimates that even with limiting values of 10^5 for K, and 5×10^{-9} cm/sec for $\partial^2 u/\partial z^2$, the southward velocity is only 0·04 m/sec, and is therefore much too slight to provide compensation for the frictional losses.

The deviations from geostrophic winds associated with changing pressure distribution, which are discussed in § 117 above, may range up to 5 m/sec, but are usually less than this. There is no *a priori* reason why such deviations should not be equally distributed in all directions, giving no tendency for a net flow in any one direction.

Douglas suggested that there might be a valve-like action at fronts, letting through the southward moving air, but forcing the northward moving air to ascend. Much of the displacement of tropical air is carried out at fronts running North and South, and these cannot contribute anything to deviations from geostrophic winds which shall have components to North or South. Using the Danish charts of the North Atlantic for December 1909 to February 1910, Douglas found that the maximum effect produced by fronts extending up to 4 km was to give a southward drift of at most 0·2 m/sec, which again is scarcely sufficient to compensate the frictional outflow. This agrees with what might have been anticipated from the theory of Brunt and Douglas. For, since the mean theoretical deviations from geostrophic winds are only of the order of 1–2 m/sec, with no apparent tendency to systematic direction, the mean flow in one direction, due to the influence of fronts, should be much less. We therefore come to the conclusion that in the lowest 4 km of the troposphere there is no net flow of polar air towards the sub-tropical anticyclone, and that the only possible conclusion is that of Jeffreys, that the exchange of air between different latitudes, which is required to make up the losses of momentum due to surface friction, is carried out by currents side by side, and not one above the other.

§ 216. *Jeffreys' theory of the monsoons and similar winds*

Jeffreys starts from an equilibrium condition in which pressure and density are constant in horizontal planes. Let the equilibrium values be p_0 and ρ_0, and let the deviations from these be p', ρ'. Then

$$p = p_0 + p', \quad \rho = \rho_0 + \rho', \quad \text{with } \frac{\partial p_0}{\partial z} = -g\rho_0 \qquad \ldots\ldots(9).$$

The deviations p', ρ', will be small by comparison with p_0 and ρ_0. The equations of motion may be written in the form

$$\rho \left(\frac{du}{dt} - 2\omega \sin \phi . v \right) = \frac{1}{R} \frac{\partial p}{\partial \phi} + F \qquad \ldots\ldots(10),$$

$$\rho \left(\frac{dv}{dt} + 2\omega \sin \phi . u \right) = -\frac{1}{\varpi} \frac{\partial p}{\partial \lambda} + G \qquad \ldots\ldots(11),$$

$$\frac{\partial p}{\partial z} = -g\rho \qquad \ldots\ldots(12),$$

where R is the radius of the earth, $\varpi = R \cos \phi$, and F, G are components of the viscous forces directed to South and East, as are the components of velocity u, v. The equation of continuity then reads

$$\frac{\partial \rho}{\partial t} - \frac{1}{R\varpi} \frac{\partial}{\partial \phi} (\rho \varpi u) + \frac{1}{\varpi} \frac{\delta}{\delta \lambda} (\rho v) + \frac{\partial}{\partial z} (\rho w) = 0 \qquad \ldots\ldots(13).$$

If only small deviations from the equilibrium condition are considered, u, v, w, and ρ' are all small, and their squares and products may be neglected. Thus $\frac{d}{dt}$ is everywhere replaced by $\frac{\partial}{\partial t}$, and ρu, ρv, ρw, by $\rho_0 u$, $\rho_0 v$, $\rho_0 w$.

When equation (13) is integrated from o to ∞, the first term gives $\frac{1}{g} \frac{\partial p_s}{\partial t}$ (see § 184, p. 308), while the last term vanishes. If we denote $\int_0^\infty \rho_0 u \, dz$ by U, and $\int_0^\infty \rho_0 v \, dz$ by V, the integral of equation (13) may be written

$$\frac{1}{g} \frac{\partial p_s'}{\partial t} - \frac{1}{R\varpi} \frac{\partial (\varpi U)}{\partial \phi} + \frac{1}{\varpi} \frac{\partial V}{\partial \lambda} = 0 \qquad \ldots\ldots(14).$$

Similarly, integrating equations (10) and (11), we find

$$\frac{\partial U}{\partial t} - 2\omega V \sin \phi = \frac{1}{R} \frac{\partial P}{\partial \phi} + \int_0^\infty F \, dz \qquad \ldots\ldots(15),$$

$$\frac{\partial V}{\partial t} + 2\omega U \sin \phi = -\frac{1}{\varpi} \frac{\partial P}{\partial \lambda} + \int_0^\infty G \, dz \qquad \ldots\ldots(16),$$

where $P = \int_0^\infty p \, dz$, and the last terms on the right-hand side of these equations represent the skin friction at the ground (see § 155, p. 259). The equation of state is written

$$p = R\rho T \qquad \ldots\ldots(17).$$

We may suppose that variations in humidity are taken into account by our defining T as the virtual temperature, on the absolute scale. Equation (12) may be written

$$\frac{\partial p}{\partial z} = -\frac{gp}{RT} \qquad \ldots\ldots(18),$$

whence

$$p = p_s \exp - \int_0^z \frac{g \, dz}{RT} \qquad \ldots\ldots(19).$$

In order to relate the time variations of P and p_s, we differentiate this equation:

$$\frac{\delta p}{p} = \frac{\delta p_s}{p_s} - \frac{g}{R} \int_0^z \delta \left(\frac{1}{T} \right) dz \qquad \ldots\ldots(20)$$

or, using the notation of relations (9) above,

$$\frac{p'}{p_0} = \frac{p_s'}{p_s} + \frac{g}{R} \int_0^z \frac{T'}{T_0^2} \, dz$$

whence

$$p' = p_s' \frac{p_0}{p_{0s}} + g\rho_0 T_0 \int_0^z \frac{T'}{T_0^2} \, dz \qquad \ldots\ldots(21).$$

Integrating this with respect to z, we find

$$P' = p_s' \frac{P_0}{p_{0s}} + \int_0^\infty g \rho_0 T_0 \int_0^z \frac{T'}{T_0^2} \, dz \, dz \qquad \ldots\ldots(22),$$

whence

$$\frac{\partial P'}{\partial t} = \frac{P_0}{p_{0s}} \frac{\partial p_s'}{\partial t} + \int_0^\infty g \rho_0 T_0 \int_0^z \frac{1}{T_0^2} \frac{\partial T'}{\partial t} \, dz \, dz \qquad \ldots\ldots(23).$$

$\frac{\partial p_s'}{\partial t}$ can now be eliminated from equation (14), yielding

$$\frac{1}{g} \frac{\partial P'}{\partial t} - \int_0^\infty \rho_0 T_0 \int_0^z \frac{1}{T_0^2} \frac{\partial T'}{\partial t} \, dz \, dz - \frac{P_0}{p_{0s}} \left\{ \frac{1}{R\varpi} \frac{\partial (\varpi U)}{\partial \phi} - \frac{1}{\varpi} \frac{\partial V}{\partial \lambda} \right\} = 0 \qquad \ldots\ldots(24).$$

Equations (15), (16), and (24) resemble those occurring in the theory of slow tidal motions in an incompressible ocean of uniform depth. The latter are

$$\frac{\partial u}{\partial t} - 2\omega v \sin \phi = \frac{g}{R} \frac{\partial}{\partial \phi} (\zeta - \bar{\zeta}) + F_1 \qquad \ldots\ldots(25),$$

$$\frac{\partial v}{\partial t} + 2\omega u \sin \phi = -\frac{g}{\varpi} \frac{\partial}{\partial \lambda} (\zeta - \bar{\zeta}) + G_1 \qquad \ldots\ldots(26),$$

$$\frac{\partial \zeta}{\partial t} - \frac{h}{R\varpi} \frac{\partial}{\partial \phi} (\varpi u) + \frac{h}{\varpi} \frac{\partial v}{\partial \lambda} = 0 \qquad \ldots\ldots(27),$$

where ζ is the height of the free surface, $\bar{\zeta}$ the height of the equilibrium tide, F_1, G_1 the components of the frictional force per unit volume, and h the depth of the ocean.

Equations (15), (16), and (24) will correspond to (25), (26), and (27) if U and V correspond to u and v, and P' corresponds to $g (\zeta - \bar{\zeta})$, F_1 and G_1 to $\int_0^\infty F \, dz$ and $\int_0^\infty G \, dz$, P_0/p_{0s} to h, and $P' - g \int_0^\infty \rho_0 T_0 \int_0^z \frac{T' dz}{T_0^2} \, dz$ with $g\zeta$. Thus for every atmospheric problem there is a corresponding tidal problem, whose solution, if known, will enable us to infer that of the meteorological problem.

In order to discuss the monsoons and similar winds, Jeffreys takes plane polar co-ordinates (r, θ) with origin near the centre of the region concerned, and neglects the variation of latitude within the region. He assumes all changes to be proportional to a factor $e^{i\gamma t}$. Then in polar co-ordinates, equations (15), (16) become

$$i\gamma U - 2\omega \sin \phi . V = -\frac{\partial P}{\partial r} \qquad \ldots\ldots(28),$$

$$2\omega \sin \phi . U + i\gamma V = -\frac{1}{r} \frac{\partial P}{\partial \theta} \qquad \ldots\ldots(29),$$

where U and V are now radial and transverse components at r, θ. From these equations

$$U = -\frac{1}{4\omega^2 \sin^2 \phi - \gamma^2} \left\{ i\gamma \frac{\partial P}{\partial r} + \frac{2\omega \sin \phi}{r} \frac{\partial P}{\partial r} \right\} \qquad \ldots\ldots(30),$$

$$V = -\frac{1}{4\omega^2 \sin^2 \phi - \gamma^2} \left\{ \frac{i\gamma}{r} \frac{\partial P}{\partial \theta} - 2\omega \sin \phi \frac{\partial P}{\partial r} \right\} \qquad \ldots\ldots(31).$$

Equation (24) now yields

$$i\gamma P' + gh\left\{\frac{1}{r}\frac{\partial}{\partial r}(rU) + \frac{\partial V}{r\partial\theta}\right\} = i\gamma g\int_0^\infty \rho_0 T_0\int_0^z \frac{T'}{T_0^2}\,dz\,dz \quad \ldots\ldots(32)$$

$$= i\gamma Q,\text{ say.}$$

Substituting in this equation from (30) and (31), we find

$$-\frac{gh}{4\omega^2\sin^2\phi-\gamma^2}\left\{\frac{1}{r}\frac{\partial}{\partial r}\left(r\frac{\partial P'}{\partial r}\right) + \frac{1}{r^2}\frac{\partial^2 P'}{\partial\theta^2}\right\} + P' = Q \quad \ldots\ldots(33).$$

Now Q can always be expressed as the sum of terms of the form

$$Q_n = A\mathcal{J}_n(\mu r)\exp(i\gamma t \pm in\theta) \quad\quad \ldots\ldots(34),$$

where A and μ are constants, and n is a positive integer. Then (33) is satisfied by making P' the sum of terms of the form

$$B\mathcal{J}_n(\mu r)\exp(i\gamma t \pm in\theta).$$

For if P' takes this form

$$r\frac{\partial}{\partial r}\left(r\frac{\partial P'}{\partial r}\right) + (\mu^2 r^2 - n^2)\,P' = 0 \quad\quad \ldots\ldots(35)$$

and (33) is then equivalent to

$$\left(\frac{gh\mu^2}{4\omega^2\sin^2\phi-\gamma^2}+1\right)B = A \quad\quad \ldots\ldots(36).$$

Thus the general solution is found, except in the special case $B=0$. If such a case were to arise, special treatment would be necessary. The coefficient is positive if γ is less than $2\omega\sin\phi$, i.e. if the period much exceeds a day. Equation (22) can now be written

$$hp_s' = P' - Q \quad\quad \ldots\ldots(37),$$

since $h = P_0/p_{0s} = $ depth of equivalent ocean,

or
$$hp_s' = -Q\Big/\left(1 + \frac{4\omega^2\sin^2\phi-\gamma^2}{gh\mu^2}\right) \quad\quad \ldots\ldots(38).$$

Two extreme cases arise, according as the second term in the denominator in (38) is large or small. If the horizontal extent of the disturbance is very great, so that μ is very small, hp_s'/Q is very small. Thus temperature disturbances over a large region give rise to small changes of surface pressure, and therefore to light winds. If the horizontal extent is small, and μ great, P'/Q is small, and equations (36) and (37) reduce to

$$p_s' = -Q/h \quad\quad \ldots\ldots(39).$$

Consider now the critical value of $(4\omega^2\sin^2\phi-\gamma^2)/gh\mu^2$ which is equal to unity. For latitude $45°$, we then have

$$\omega\sin\phi = 5.10^{-5}\text{ sec}^{-1}; \quad g = 981\text{ cm/sec}^{-2}; \quad h = 7\cdot3.10^6\text{ cm};$$

while for all motions of long period γ^2 is very small compared with $4\omega^2\sin^2\phi$. Thus

$$\frac{gh}{4\omega^2\sin^2\phi-\gamma^2} = 7\cdot1.10^{16}\text{ cm}^2 \quad\quad \ldots\ldots(40)$$

and the critical value of $1/\mu$ is $2\cdot7.10^8$ cm, or 2700 km.

A comparison of the theory with observation is only possible where the permanent general circulation of the atmosphere does not overwhelm the changes due to local heating. The best locality to try is probably Asia, and Jeffreys gave a comparison for this continent. The extent of Asia along a central meridian, excluding Southern India and Indo-China, amounts to about 55° of latitude. This is comparable with the size found above to be critical. It is so large that variations of latitude will probably produce appreciable effects; but as a first approximation, Jeffreys assumed a mean latitude of 45°, and supposed μ to have its critical value. Then we should have

$$p_s' = -Q/2h = -\tfrac{1}{2}Qp_0/P_0 \qquad \ldots\ldots(41).$$

Now
$$Q = g \int_0^\infty \rho_0 T_0 \int_0^z \frac{T'}{T_0^2}\, dz\, dz \qquad \ldots\ldots(42)$$

can be integrated if the annual variation of temperature at all heights is known. If it is assumed that the annual variation at all heights over Asia and over Canada are in proportion, the values given by W. H. Dines* for Canada can be used as a basis for a numerical integration of (42). This was done by C. A. Coulson,† who found for p_s', the range of pressure at the ground, a value of about 33 mb, as compared with the observed value of 30 mb.

Equation (21) can now be used to find the height at which p' vanishes and changes sign. Jeffreys, assuming a range of 41 mb for the surface change of pressure at sea-level, and 0·15 for T'/T_0 at all heights, found that p' vanishes at about 2·1 km. He used a slightly erroneous form for Q. When the correct form is used, as was done by Coulson, the height at which p' vanishes is slightly increased, but so far as observations are available to test it, the estimate of 2·1 km is not wide of the truth. The evidence of figs. 110, 111, that the depression over North-west India persists in a much weakened form at 2 and 4 km, must not be taken too literally, in view of the nature of the observations used to compute these charts.

It is thus seen that the theory developed by Jeffreys is capable of producing remarkable agreement with the observed facts, explaining both the magnitude of the annual variation of surface pressure and the height of reversal of the monsoon.

Jeffreys extended the theory to the general circulation, treating the latter as a slow motion regarded as the limit of periodic oscillation, when the period becomes very large. The result was that the surface winds should be everywhere easterly, while at no very great height there should be a reversal to westerly winds.

§ 217. *The present position regarding the theory of the general circulation*

It is not possible at present to put forward any satisfactory theory of the general circulation, in part because the details of the circulation have only been observed very incompletely. There are still very considerable gaps in our knowledge of these details, particularly over the oceans. The observations over

* M.O., *Geophys. Mem.* No. 13, 1919.   † *Q.J. Roy. Met. Soc.* 57, 1931, p. 161.

the Atlantic Ocean accumulated during the expeditions of the "Meteor" should place our knowledge of the circulation over the Atlantic on a surer footing.*

The older textbook explanation of why the air carried towards the equator in the trade-winds rises in the doldrums and spreads poleward in the upper air, is not readily acceptable to the modern meteorologist. The zone of highest temperature is centred in latitude 30° N in July, and in latitude 10° S in January. In July this zone is well within the sub-tropical high-pressure belt of the Northern hemisphere, as shown on the chart of mean pressure in fig. 8.

Georgii† has suggested that the source of energy of the tropical circulation, and possibly of the entire atmospheric circulation, is to be found in the sub-tropics. This is in accordance with the distribution of maximum temperature mentioned in the preceding section, which is to be explained by the greater cloudiness in the tropics than in the sub-tropics, that are estimated at 2–3 tenths and 7 tenths respectively. An appreciable fraction, probably one-tenth, of the incoming solar radiation in the trade-wind zones is used up in evaporation, and the ascent of the humid air brought equator-ward in the trade-winds is shown by the cloudiness of the inner tropics. The showery nature of the rain shows that the ascent is intermittent, not continuous.

Fig. 119 shows the distribution of potential temperature in the atmosphere, deduced from the distributions of temperature and pressure shown in figs. 12 and 109. The minor details of this diagram are not reliable, on account of the fact that the mean temperature and pressure in the upper air, particularly in the polar regions, are based on relatively few observations. But the general trend of the lines of equal potential temperature (isentropics) is probably fairly reliable, except at great heights, and in high latitudes. The most striking feature of fig. 119 is the very great increase of potential temperature with height through the whole atmosphere. If, as suggested by Shaw, air tends to move along the isentropic surfaces, it is not difficult to visualise the air which ascends in latitude 10° N as spreading slowly poleward, at first slowly ascending, while air in the polar regions could spread equator-ward down the appropriate isentropic surfaces.

Some of the air which ascends in the inner tropics appears to descend again to low levels even in latitude 23° N. Sverdrup‡ found in latitude 23° N that from 1–3 km the air is warmer than the air to north or south of it, and is very dry, a condition which can only be explained by its descent from high levels. An examination of fig. 109 shows that at all levels above 2 km there is a fall of pressure from equator to pole. At such levels any additional air interposed into this pressure field must tend to move down the gradient of pressure, and if it moves adiabatically it will slowly ascend at first, along the isentropic surfaces shown in fig. 119. But it is not legitimate to assume the motion to be adiabatic. Radiative processes will be active, and Sverdrup's results referred to above appear to indicate that radiation must produce rapid cooling, and enable some of the air so cooled to penetrate downward through the successive isentropic surfaces, and eventually to reach the surface in quite low latitudes.

---

* *Wissenschaftliche Ergebnisse*, 14A, 14B.   † *Arktis*, **1**, h. 3/4.
‡ *Veröff. Geoph. Inst. Leipzig*, **2**, 1917.

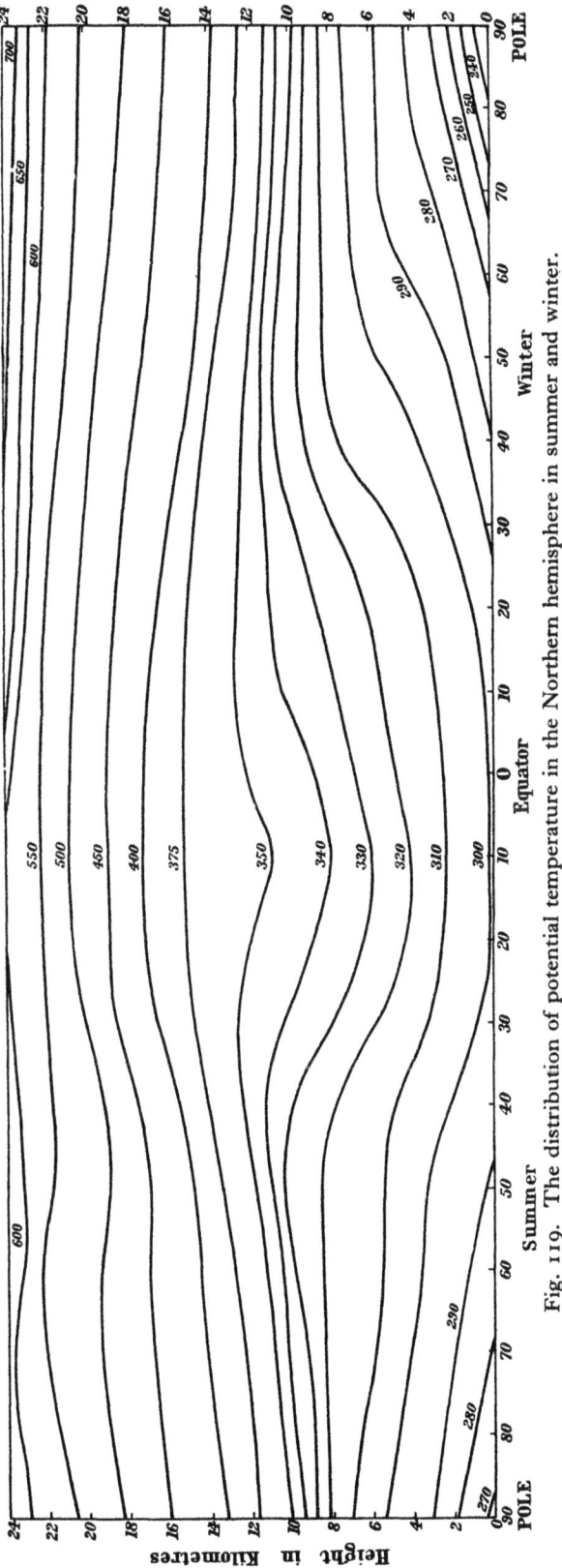

Fig. 119. The distribution of potential temperature in the Northern hemisphere in summer and winter.

This type of motion was referred to earlier, in § 179, but the assumption of adiabatic motion there made is obviously unjustifiable. If the motion were truly adiabatic, then air which had succeeded in penetrating upward to the level of even 2 km in latitude 10° N could never again come down to the surface. The conclusion appears to be inevitable that of the air which ascends in the tropics and flows poleward a part cools and descends in the sub-tropical anticyclones, sufficient at least to compensate the outward drift from these anticyclones towards the equator and towards the poles; the remainder of the upper current from the tropics continues beyond the sub-tropics into the zone of the westerlies. The current from the inner tropics starts as a southerly wind and veers through SW to W, when it forms part of the circulation around the upper air cyclone centred at the poles, and replaces the slow drift across the isobars into the polar centres of low pressure. There will be a corresponding sinking of air at the poles, produced by the intense radiative cooling at the earth's surface. The occurrence of periodical outbursts of cold air from the polar regions becomes a dynamical necessity if the pressure at the poles is to be kept from increasing indefinitely. We thus come to the view that the zone of westerly winds of middle latitudes will be filled partly by air which has come from the sub-tropical anticyclones, and partly by air which has come from the polar regions; and it is the clash of these two types of air which produces the travelling depressions and anticyclones of middle latitudes. The polar currents appear to penetrate considerably farther equator-ward in the Southern than in the Northern hemisphere, and the westerly winds of the Southern hemisphere are relatively stronger and more extensive. It is a well-known fact that in middle and high latitudes of the Northern hemisphere tropical air is a comparatively rare phenomenon, it being forced to ascend over the southward penetrating polar air. Much of this polar air curves around again, returning poleward as a south-westerly current, and penetrating back to the polar regions; and, as stated above on p. 409, in the lowest 4 km there appears to be no net flow of polar air towards the sub-tropical anticyclone of the Northern hemisphere.

The greater symmetry of conditions in the Southern hemisphere produces greater regularity in the eastward motion of anticyclones in sub-tropical and low temperature latitudes, and leads to a series of phenomena which can best be described as a series of waves rotating round the globe, each wave consisting of alternate bands of north-westerly and south-westerly winds. Where the eastern edge of the cold south-westerly wind meets the warmer north-westerly, the low pressure trough is marked by cold front phenomena.* Simpson† concluded that the weather of the Antarctic continent was controlled by a series of pressure waves issuing from the South pole or near it, but the possible connection between these waves and those of low temperate regions has not yet been elucidated.

The machinery of the atmospheric circulation visualised above demands a source of heat in the tropics or sub-tropics, to account for the upward motion observed in the tropics, and a source of cold at the poles to account for the

---

* See Kidson, *Q.J. Roy. Met. Soc.* **58**, 1932, p. 219; also **59**, 1933, p. 372.

† *British Antarctic Exped.* 1910–13, Meteorology, **1**.

descent of air which has penetrated from lower latitudes to the poles. It is not possible to specify with definiteness whether the source of cold is entirely at the surface, or is spread through a considerable range of height. These two sources of heat and cold cannot of themselves account for more than a small part of the observed complexity of the atmospheric circulation. It has been suggested that the cooling of air in the poleward currents occurs at high levels, and this could only be attributed to radiation. Further, the surface of the earth itself is a powerful source of heating for the polar currents which penetrate into middle latitudes. It is in fact impossible to describe the atmospheric engine by comparison with the ordinary heat engine, which has clearly definable sources of heat and cold. In the atmosphere the sources of heat and cold cannot as yet be completely specified, but considerable progress would be possible if a time-scale could be assigned to radiational effects in the upper atmosphere.

From time to time it has been suggested that the katabatic flow of air down cold slopes such as those of the Antarctic continent must contribute very considerably to the kinetic energy of the motion of the atmosphere. Such katabatic flows are usually extremely shallow, and are destroyed by friction and the heating effect of the earth's surface in a very short distance. Thus the British Arctic Air-Route expedition to Greenland under the late G. H. Watkins observed at the base camp on the east coast of Greenland a wind of 129 m.p.h., while at Angmagsalik, a few miles away, only light or moderate winds were observed. From this it may be inferred that these winds are very shallow, and are rapidly destroyed by friction and the effects of turbulent mixing. Very strong katabatic winds also occur along the edge of the Antarctic continent, but it is probable that they do not penetrate far from the coast. There are also other great masses of land which can produce katabatic flow on a large horizontal scale, such as the Himalayas, and the mountains along the western coast of the continent of America, but it is difficult to visualise them as producing any marked contribution to the general circulation of the atmosphere.

§ 218. *Epilogue*

In the last three chapters an attempt has been made to summarise the known facts concerning depressions, anticyclones, and the general circulation. The summary is of necessity incomplete, on account of the paucity of certain types of observations, particularly in the upper air. We are not yet in a position to specify the necessary facts regarding the circulation of the upper winds, or the distribution of temperature, and until we are in a position to do so it will not be possible to specify why the atmosphere behaves as it does. This is particularly the case as regards the general circulation, but it is also to a great extent true of the depression and anticyclone. In each of the last three chapters of this book the difficulties have been largely due to lack of observations of the right type.

But although the three atmospheric features mentioned are still incompletely understood, it would be erroneous to suppose that no advance has been made. From time to time the introduction of new ideas has led to definite progress. During the last 20 years the greatest progress has been made by the frontal methods introduced by the Norwegian school of meteorologists. These methods have thrown new light on the phenomena in depressions, and though they have not yet produced an *explanation* of the depression, they have clarified and vivified the *description* of the depression. The latest writings on the frontal analysis of depressions, on anticyclones, and on the general circulation, all indicate a recognition of the complexity of the phenomena, and less readiness to attempt general theories. This is a step in the right direction, since it encourages the collection of data, and the discussion of facts.

Further progress must depend largely on the introduction of new ideas, as well as on the application of results derived in such fields as radiation and turbulence to the discussion of the general circulation and the local circulations. That meteorology to-day offers a plentiful supply of problems for research is very clearly shown in a series of papers on "Problems of Modern Meteorology" published in the *Quarterly Journal of the Royal Meteorological Society* during the years 1930–34. The reader is referred to this series of articles for summaries of some of the outstanding problems of modern meteorology. Such a series naturally does not exhaust the problems, many of which are not yet sufficiently advanced to be capable of specific statement.

APPENDIX

Table I

Corrections to be subtracted from actual heights in metres to give heights in dynamic metres

Latitude Height	0	10	20	30	40	50	60	70	80
1000	21·9	21·8	21·3	20·7	19·8	18·9	18·1	17·4	17·0
2000	43·8	43·5	42·6	41·3	39·6	37·9	36·2	34·9	34·0
3000	65·7	65·3	63·9	62·0	59·4	56·8	54·3	52·3	51·0
4000	90·1	89·5	87·7	85·1	81·7	78·3	74·9	72·3	70·5
5000	114	113	111	107	103	99	95	91	89
6000	137	137	134	130	125	120	115	111	108
7000	161	161	157	153	147	140	135	130	127
8000	185	184	180	175	168	162	155	150	146
9000	209	208	204	198	190	182	175	169	165
10000	234	233	228	222	213	204	196	189	185

Table II

Values of $\left(\dfrac{1000}{p}\right)^{0\cdot288}$ for values of p from 10 to 1090 mb

p	00	10	20	30	40	50	60	70	80	90
00	∞	3·767	3·085	2·745	2·527	2·370	2·249	2·151	2·070	2·001
100	1·941	1·888	1·842	1·800	1·762	1·727	1·695	1·666	1·639	1·613
200	1·590	1·568	1·547	1·527	1·508	1·491	1·474	1·458	1·443	1·428
300	1·414	1·401	1·388	1·376	1·364	1·353	1·342	1·332	1·321	1·312
400	1·302	1·293	1·284	1·275	1·267	1·259	1·251	1·243	1·235	1·228
500	1·221	1·214	1·207	1·201	1·194	1·188	1·182	1·176	1·170	1·164
600	1·159	1·153	1·148	1·142	1·137	1·132	1·127	1·122	1·118	1·113
700	1·108	1·104	1·099	1·095	1·091	1·086	1·082	1·078	1·074	1·070
800	1·066	1·063	1·059	1·055	1·052	1·048	1·044	1·041	1·038	1·034
900	1·031	1·028	1·024	1·021	1·018	1·015	1·012	1·009	1·006	1·003
1000	1·000	0·997	0·994	0·992	0·989	0·986	0·983	0·981	0·978	0·976

Table III

Table for conversion of temperatures from the Fahrenheit scale to the Absolute scale, from °F to °A

°F	0	1	2	3	4	5	6	7	8	9
					Degrees absolute					
− 60	221·9	221·3	220·8	220·2	219·7	219·1	218·6	218·0	217·4	216·9
− 50	227·4	226·9	226·3	225·8	225·2	224·7	224·1	223·6	223·0	222·4
− 40	233·0	232·4	231·9	231·3	230·8	230·2	229·7	229·1	228·6	228·0
− 30	238·6	238·0	237·4	236·9	236·3	235·8	235·2	234·7	234·1	233·6
− 20	244·1	243·6	243·0	242·4	241·9	241·3	240·8	240·2	239·7	239·1
− 10	249·7	249·1	248·6	248·0	247·4	246·9	246·3	245·8	245·2	244·7
− 0	255·2	254·7	254·1	253·6	253·0	252·4	251·9	251·3	250·8	250·2
0	255·2	255·8	256·3	256·9	257·4	258·0	258·6	259·1	259·7	260·2
10	260·8	261·3	261·9	262·4	263·0	263·6	264·1	264·7	265·2	265·8
20	266·3	266·9	267·4	268·0	268·6	269·1	269·7	270·2	270·8	271·3
30	271·9	272·4	273·0	273·6	274·1	274·7	275·2	275·8	276·3	276·9
40	277·4	278·0	278·6	279·1	279·7	280·2	280·8	281·3	281·9	282·4
50	283·0	283·6	284·1	284·7	285·2	285·8	286·3	286·9	287·4	288·0
60	288·6	289·1	289·7	290·2	290·8	291·3	291·9	292·4	293·0	293·6
70	294·1	294·7	295·2	295·8	296·3	296·9	297·4	298·0	298·6	299·1
80	299·7	300·2	300·8	301·3	301·9	302·4	303·0	303·6	304·1	304·7
90	305·2	305·8	306·3	306·9	307·4	308·0	308·6	309·1	309·7	310·2
100	310·8	311·3	311·9	312·4	313·0	313·6	314·1	314·7	315·2	315·8
110	316·3	316·9	317·4	318·0	318·6	319·1	319·7	320·2	320·8	321·3
120	321·9	322·4	323·0	323·6	324·1	324·7	325·2	325·8	326·3	326·9
130	327·4	328·0	328·6	329·1	329·7	330·2	330·8	331·3	331·9	332·4
140	333·0	333·6	334·1	334·7	335·2	335·8	336·3	336·9	337·4	338·0

Table IV

Mass in grammes of one cubic metre of dry air

Temperature °A

Pressure (mb)	200	210	220	230	240	250	260	270	280	290	300	310
100	174	166	158	152	145	139	134	129	124	120	116	112
200	349	332	317	303	290	279	268	258	249	240	232	225
300	523	498	475	455	436	418	402	387	373	360	348	337
400	697	664	634	606	581	558	536	516	498	480	464	449
500	872	830	792	758	726	697	670	645	622	601	581	562
600	1046	996	951	909	871	836	804	774	747	721	697	674
700	1220	1162	1109	1061	1016	976	938	903	871	841	813	786
800	1395	1328	1267	1212	1162	1115	1072	1032	995	951	929	899
900	1569	1494	1426	1364	1307	1254	1206	1161	1120	1081	1045	1011
1000	1743	1660	1584	1515	1452	1394	1340	1290	1244	1201	1161	1124
1040	1813	1725	1648	1576	1510	1450	1394	1342	1294	1249	1208	1169

For saturated air subtract the figures below from the table:

	200	210	220	230	240	250	260	270	280	290	300	310
	0	0	0	0	0	1	1	2	5	9	16	27

Table V

Saturation vapour-pressure over water (in millibars)

(After Holborn, Scheel u. Henning, *Wärmetabellen, Phys. Tech. Reichsanstalt*)

°A	0	1	2	3	4	5	6	7	8	9
250	—	—	—	—	—	—	—	1·75	1·91	2·07
260	2·25	2·44	2·64	2·86	3·10	3·35	3·62	3·90	4·21	4·54
270	4·89	5·27	5·68	6·11	6·57	7·06	7·58	8·13	8·72	9·35
280	10·02	10·73	11·48	12·28	13·13	14·02	14·98	15·98	17·05	18·18
290	19·38	20·64	21·97	23·38	24·87	26·44	28·09	29·84	31·68	33·61
300	35·65	37·80	40·06	42·45	44·93	47·55	50·31	53·20	56·23	59·42
310	62·76	67·26	69·92	74·10	77·79	82·00	86·40	91·01	95·84	100·87
320	106·1	111·6	117·4	123·4	129·6	135·8	142·9	150·0	157·4	165·1
330	173·1	181·4	190·1	199·1	208·6	218·4	228·5	239·1	250·1	261·5
340	273·3	285·6	298·3	311·6	325·2	339·5	354·3	369·6	385·5	401·9

Saturation vapour-pressure over ice

°A	0	1	2	3	4	5	6	7	8	9
240	0·27	0·30	0·34	0·39	0·42	0·46	0·51	0·57	0·63	0·70
250	0·77	0·85	0·94	1·03	1·11	1·25	1·37	1·50	1·65	1·81
260	1·98	2·17	2·37	2·60	2·83	3·10	3·38	3·68	4·01	4·37
270	4·76	5·17	5·62	6·12	—	—	—	—	—	—

At 230° A, 0·09;   at 220° A, 0·025;   at 210° A, 0·005.

Table VI

Density of saturated water-vapour

	Over water									Over ice			
°A	340	330	320	310	300	290	280	270	260	270	260	250	240
gm/m³	174·2	113·7	72·0	43·9	25·8	14·5	7·8	3·9	1·87	3·82	1·65	0·66	0·25

Table VII

Latent heat of water

(After Holborn, Scheel u. Henning, *Wärmetabellen*)

°C	0	5	10	15	20	25	30	35	40
L	594·9	592·4	590·0	587·5	585·0	582·4	579·8	577·2	574·5

°C	45	50	55	60	70	80	90	100
L	571·8	569·0	566·2	563·4	557·6	551·6	545·5	539·1

Up to 40° C, L may be represented with great accuracy by
$$L = 594\cdot9 - 0\cdot51t,$$
where t is the temperature in degrees C.

Latent heat of fusion of ice = 79·7 g-cal per gramme.

Table VIII

Temperature radiation of a black body at absolute temperature T

$E = \sigma T^4$ (where σ is Stefan's constant) in g-cal per cm² per min

°A	0	1	2	3	4	5	6	7	8	9
210	0·161	0·164	0·167	0·170	0·173	0·176	0·180	0·183	0·187	0·190
220	0·193	0·197	0·201	0·204	0·208	0·212	0·215	0·219	0·223	0·227
230	0·231	0·235	0·239	0·244	0·248	0·252	0·256	0·261	0·265	0·270
240	0·274	0·279	0·283	0·288	0·293	0·298	0·303	0·308	0·313	0·318
250	0·323	0·328	0·333	0·339	0·344	0·349	0·355	0·360	0·366	0·372
260	0·378	0·384	0·389	0·395	0·401	0·407	0·414	0·420	0·426	0·432
270	0·439	0·446	0·452	0·459	0·466	0·472	0·479	0·486	0·493	0·500
280	0·508	0·515	0·522	0·530	0·537	0·545	0·553	0·560	0·568	0·576
290	0·584	0·592	0·601	0·609	0·617	0·626	0·634	0·643	0·651	0·660
300	0·669	0·678	0·687	0·696	0·706	0·715	0·724	0·734	0·743	0·753
310	0·762	0·772	0·782	0·793	0·803	0·813	0·823	0·834	0·845	0·855
320	0·866	0·877	0·888	0·899	0·910	0·921	0·933	0·944	0·956	0·968

Table IX

Coefficients of conductivity of heat, k, of viscosity, μ, and of diffusion of water-vapour in air, D

Temp. °C	40	30	20	10	0	−10	−20	−30	−40	−50
$10^7 k$	642	625	608	591	574	556	539	522	505	488
$10^6 \mu$	190	186	181	176	171	161	156	151	148	146
$10^5 \rho$, at 1000 mb	111	115	119	123	128	133	138	143	150	156
κ (approx.)	0·24	0·23	0·21	0·20	0·19	0·17	0·16	0·15	0·14	0·13
ν (approx.)	0·17	0·16	0·15	0·14	0·13	0·13	0·11	0·11	0·10	0·09

The coefficient of diffusion of water-vapour in air, represented by D, is given by
$$D = D_0 \, (T/T_0)^{1.75},$$
where $D_0 = 0\cdot22$ (*Inter. Crit. Tables*, **5**, p. 62), and $T_0 = 273°$ A, and $p_0 = 1$ atmosphere (1013 mb).

Individual values of D_0 vary rather widely. Winkelmann (1884) gave its value as 0·198; Guglielmo (1884) as 0·231; Houdaille (1896) as 0·203; Le Blanc and Wupperman (1916) as 0·224; and Summerhayes (*Proc. Phys. Soc. London*, **42**, 1930, p. 218) as 0·252. The last of these is probably the most reliable value.

The values of $10^7 k$ and of κ have been computed from the values given by Hercus and Sutherland (*Proc. Roy. Soc.* A, **145**, 1934, p. 599) and by Kannuluik and Martin (*ibid.* **144**, 1934, p. 496).

Table X

Terminal velocities of fall of water drops in still air at normal pressure

Radius cm	Velocity cm/sec	Radius cm	Velocity cm/sec
0·0005	0·3	0·05	420
0·001	1·3	0·10	594
0·005	32	0·15	691
0·01	136	0·175	739
0·02	180	0·225	805
0·03	270	0·273	798
0·04	340	0·318	780

The velocities of fall of spherical drops of radius up to 0·01 cm are computed from Stokes's law, while the remaining values in the table above are values observed by Schmidt or Lenard, or means of values due to these two. The decrease of terminal velocities for the largest drops is associated with a distortion from the spherical form.

INDEX OF NAMES

Aitken, J., 52
Akerblom, 255
Aldrich, 112, 113, 132, 286
Ali, Barkat, 251
Angot, A., 24, 112
Ångström, 113, 127, 136, 143
Apjohn, 86
Apte, 255
Aschkinass, 120
Asklöf, 144
August, 86

Ballot, Buys, 12, 387
Baur, F., 137
Beals, 387
Bemmelen, K. van, 20, 396
Bénard, H., 219
Bergeron, T., 17, 53
Berson, 20, 396
Best, A. C., 218, 230, 244, 249, 268
Bezold, von, 57, 63
Bigelow, 318
Bilham, E. G., 56
Bjerknes, J., 68, 319, 325, 328, 333, 335, 339, 341, 343, 356, 359, 363, 365, 372, 381
Bjerknes, V., 287, 318, 358
Boyden, C. J., 311
Boylan, 53
Braby, 387, 401
Brooks, C. E. P., 387, 401, 403
Brunt, D., 27, 48, 49, 65, 136, 195, 220, 285, 300, 352, 354
Buisson, 110, 152

Cave, C. J. P., 200, 383
Chapman, E. H., 248
Chapman, S., 124, 154
Coblentz, 107
Coulson, C. A., 405, 413

Davies, E. L., 138
Defant, 53
Dennison, 118
Desai, 370
Dines, L. H. G., 18, 145, 380
Dines, W. H., 19, 21, 121, 136, 145, 159, 198, 374, 413
Dobson, G. M. B., 110, 152, 199, 255
Douglas, 195, 201, 242, 306, 316, 321, 339, 348, 350, 352, 354, 356, 361, 363, 368, 408
Dryden, 248
Durst, C. S., 238, 305, 361, 384, 387
Durward, J., 24, 26, 187

Elias, 275
Elsasser, 118
Emden, 149

Ertel, 310
Exner, F., 359, 370, 380, 404, 407

Fabry, 110, 152
Ficker, von, 318, 372
Findeisen, 53, 54
Fjeldstad, 57, 61
Fowle, 111, 124

Gallé, 27
Georgii, W., 414
Gherzi, 370
Giblett, M. A., 218, 272, 343, 385
Glauert, H., 184
Godske, C. L., 288, 359
Gold, E., 20, 147, 151, 190, 349, 379
Goldstein, 242
Gowan, 153
Griffiths, E., 275

Hann, 24, 131, 373
Hanzlik, 318, 373, 381
Harwood, W. H., 20, 398
Haurwitz, B., 230
Hellmann, 25, 262
Helmholtz, 203, 318, 338
Hertz, 57, 61
Hesselberg, 374
Hettner, 114
Hewson, E. W., 92, 364
Heywood, G. P., 26, 262, 281
Hildebrandsson, 20
Humphreys, W. J., 147

Idrac, 221

Jeffreys, H., 271, 276, 287, 288, 309, 312, 405, 407, 408, 409
Johnson, N. K., 138, 140, 144

Kármán, von, 245, 251
Kaye, 361
Kelvin, 50
Kendrew, W. G., 16
Khanewsky, 382
Kidson, E., 461
King, L. V., 122
Kirchhoff, 104
Kobayasi, 369
Köhler, 53
Kopp, 222
Koschmieder, 295

Ladenberg, 120
Lagrange, 162
Lamb, H., 172, 184
Lempfert, R. G. K., 304, 313, 318, 338

Lettau, H., 274
Lindemann, 20
Littwin, 295
Low, A. R., 220
Lummer, 107

Mal, S., 221
Mallock, 241
Margules, 293, 368
Möller, 196, 218
Montgomery, 247, 249, 261
Morgans, W. R., 377
Mügge, 157, 381
Mulholland, 27
Munday, 201

Neuhoff, 57, 58, 63
Niederdorfer, 53
Nikuradse, 245, 248
Normand, 86, 87, 89, 91, 99, 295

Owens, J. S., 52

Palmén, 333, 356, 365
Partington, 37
Patten, 141
Patterson, J., 20
Pearson, K., 214
Pekeris, 137
Penndorf, 153
Peppler, 20
Pettit, 111
Phillips, 137
Pick, W. H., 146, 316
Powell, 275
Prandtl, 230, 235, 242, 243, 244, 247, 249
Pringsheim, 107

Ramanathan, 17, 137, 389
Ramdas, 137
Randall, 114
Rayleigh, 219, 223, 242, 243, 300
Read, R. S., 343
Reger, J., 21
Relf, E. F., 56, 266
Reynolds, O., 213, 214, 215, 247
Richardson, L. F., 201, 226, 237, 242, 249, 264
Roberts, O. F. T., 135, 149
Rosenhead, 241
Rossby, 97, 247, 248, 249, 261
Rubens, 120

Runge, 383
Rykatchew, 20

Sandström, 287
Schedler, 22, 365
Schlichting, 243
Schmidt, 127, 230
Schmiedel, 367
Schröder, 338
Schumann, 56
Schuster, 123, 124
Schwarzschild, 149
Scrase, F. J., 217, 266
Shaw, Sir N., 77, 104, 187, 285, 297, 304, 308, 313, 318, 324, 338, 364, 385, 388, 414
Sieber, 196
Silberstein, 179
Simpson, G. C., 24, 55, 116, 141, 154, 403, 416
Skinner, 269
Smith, Warren, 146
Solberg, H., 335, 341, 358, 372
Squire, 243
Stanton, 251, 259
Steiner, 131
Stokes, 172
Sutcliffe, R. C., 261, 305
Sutton, O. G., 249, 251, 266, 272
Sverdrup, 250, 397, 404, 414

Taylor, G. I., 216, 223, 225, 230, 233, 239, 242, 243, 255, 259, 263, 269, 275
Tetens, 26
Thomson, J. J., 50
Tietjens, 243
Tollmien, 243

Waals, van der, 30
Wagner, 20, 281, 378, 388, 399
Walker, Sir G. T., 221, 222
Wallén, 127
Washburn, 103
Watkins, G. H., 417
Watson, R. A., 365
Weber, 114
Wenger, 399
Whipple, F. J. W., 20, 87, 103, 167, 408
Wigand, 53
Willett, 99, 341
Wright, H. L., 125
Wüst, 127, 403

Young 103

INDEX OF SUBJECTS

Absorption, at surface, 128
  coefficient of, 113
  in the atmosphere, 114
  spectrum of liquid water, 120
  spectrum of water vapour, 114, 118
Adiabatic, saturated, 65
  unsaturated, 41
Adiabatic ascent of damp air, 89
Adiabatic equation, 38
Adiabatic lapse-rate, for dry air, 39
  in geodynamic units, 45
Air, annual variation of mass of, 12
Air masses, 313, 315
Albedo of earth, 112
Aleutian low, 7
Altimetry, barometric, 35
Ångström unit, 105
Antarctic, 401, 417
  Scott Expedition, 401
  temperature on Barrier, 141
Anticyclone, formation of, 299, 301
  cold, 374
  subtropical, 387
  types of, 371
  warm, 377
Antitriptic winds, 278, 280
Appleton layer, 154
Atmosphere, composition of, 30
  heat balance in, 126, 154
  I.C.A.N., 36
  kinetic energy of, 285
  vertical transport of heat in, 158
Austausch, 231, 245
Auto-convection, 46

Baroclinic, 171
Barometric altimetry, 35
Barotropic, 171
Barrier theory of depressions, 370
Beer's law of absorption, 113
Bénard cell, 219
Black-body radiation, 106
Boundary layer, 257
Boyle's law, 30
Breezes, land and sea, 181, 280
Buys Ballot's law, 12

Carnot cycle, 69
Changing pressure distribution, effect of, 194
Circle of inertia, 167
Circulation, and vorticity, 172
  development of, 176, 299
  upper air, 388
Clausius-Clapeyron equation, 66, 101
  integration of, 103
Classification, of energy, 282
  of winds, 276
Cloudiness, diurnal variation of, 27

Cloudy skies, nocturnal radiation with, 142
Coefficient of absorption, 113
Cold front, 320
Condensation, nuclei of, 51
  at temperatures below freezing-point, 54
Continuity, equation of, 168
Continuous movement, diffusion by, 263
Convection, nature of, 221
Convection cell, 219
Convergence, 182, 300, 304, 311
Cooling of air from below, 227
Correlation of free air variables, 21
Criterion of turbulence, 237
Currents, life-history of surface, 313
Cycle of changes in the atmosphere, 297
Cyclic circulation, growth of, 299
Cyclone, development of, 318
  families of, 335
  formation of, 299, 319, 360
  kinetic energy of, 368
  rainfall in, 352
  regeneration of, 335
  trajectories of air in, 313
  tropical, 303
  upper air conditions above, 348
  vertical structure of, 356, 365
  wave at a front, 358
Cyclostrophic winds, 190, 279

Dalton's law, 31
Damp air, ascent of, 57
  density of, 30
Density, of damp air, 30
  of water vapour, 31
  variation with height, 45
Depression, see Cyclone
Deviating force of earth's rotation, 166
Dew point, 48
Diffuse reflexion, 122
Diffusion, by continuous movement, 263
  by eddies, 224, 226
Diffusivity, eddy, 225, 236
  of soils, 140
  radiative, 133
Discontinuity, of density only, 205
  of velocity only, 204
  slope of surface of, 204, 208
  surfaces of, 203, 319
Dissipation of kinetic energy, by turbulence, 285
  by viscosity, 185
Diurnal variation, of cloud, 27
  of pressure, 24
  of temperature, 22, 133, 229
  of wind, 24, 261
Divergence, 182, 299, 302, 373, 384
Doldrums, 387, 401

Drawing fronts on synoptic charts, 340
Dry adiabatic lapse-rate, 39, 45
Dynamical similitude, 214

Earth, physical constants of, 28
    rotation of, 164
    rotation of, deviating force, 165, 166
Eddies, 212, 219
    diffusion by, 224, 226
    energy of, 216
    Mallock's representation, 241
    Rosenhead's representation, 242
Eddy diffusivity, 225
    for different properties, 236
Eddy transfer of momentum, 230
Eddying energy, partition of, 216
Efficiency of heat engine, 75
Energy, equations of, 288
    liberated by vertical interchange, 79, 291
    of depressions, 368
Entropy, 69
    effect of conduction or mixing, 74
    formulae for, 75
    of mixture of damp air, water and ice, 85
    of moist air, 81
    of saturated air, 81
Entropy-temperature diagram, 76
Equation, adiabatic, 38
    Clausius-Clapeyron, 101
    continuity, 168
    energy, 288
    hygrometer, 84, 269
Equivalent potential temperature, 94, 96
Equivalent temperature, 94, 96
Eulerian winds, 277
Evaporation, 270
    from water surfaces, 127, 270
    latent heat of, 82
Eviction of air, 303, 364

Families of depressions, 335
Filaments, vortex, 175
Flow, along and across isobars, 256
    in pipes, 245, 246, 258, 259, 268
    stream-line, 215
    turbulent, 212, 215, 222
Föhn winds, 60, 91, 278
Form of R_{ξ}, 265
Friction, internal, 255
    skin, 259
Fronts, accelerations at, 211
    cold, 320
    drawing of, 340
    formation in a vortex, 369
    sharp and diffuse, 342
    valve-like action at, 409
    warm, 320
Frost problem, 146

Gas constant, 30
Gas equation, 30

General circulation, 16, 387
    in upper air, 389
General equations of motion, 160
    in Cartesian form, 163
Generation of circulation, 176
Geopotential, 29
Geostrophic wind, 189, 277, 279
    variation with height, 196
Gradient, auto-convection, 46
Gradient wind, 189
    changes in the pressure field, 310
    comparison with observed winds, 191
    equation, 188, 191, 193
Grey radiation, 106
    Emden's solution for, 149
Ground, nocturnal cooling of, 145

Hail, formation of, 56
Heat, mechanical equivalent of, 69
    transfer by radiation, 128, 129, 134
    transfer by turbulence, 224
Heat balance of the atmosphere, 126, 154
High-level depression, 363
Horizontal motion, equations for, 166
Humidity mixing ratio, 33
Hygrometer equation, 86, 269

Icelandic low, 7
Indicator diagram, 69
Inertia, circle of, 167
Insolation, variation with season and latitude, 112
Instability, latent, 99
    potential, 98
Internal friction due to turbulence, 255
Inversion, 19
    formation of, 386
Irrotational motion, 174
Isallobaric component of wind, 195
Isobaric surfaces, form of, 208
Isopycnic surfaces, 171
Isosteric, 172

Katabatic winds, 281, 417
Kennelly-Heaviside layer, 154
Kirchhoff's law, 105

Land breeze, 181, 280
Lapse-rate, 17
    adiabatic, dry, 39, 40, 45
    adiabatic, saturated, 65, 67
    ascending damp air, 65
    constancy of, 20
    effect of vertical motion on, 44
    super-adiabatic, 243
Latent heat, 82
Latent instability, 99
Life-history, of cyclones, 319, 349
    of surface air currents, 313
Lines, vortex, 176
Local circulations, 16
Logarithmic law of velocity distribution, 246, 247

Maintenance of pressure difference by heat, 296
Maritime polar air, 317
Metre, dynamic, 29
Micron, 105
Mischungsweg, 231, 245
Mixing of two masses of damp air, 47
Momentum, eddy transfer of, 230
Monsoon, South-west, Indian, 13
Monsoons, Jeffreys' theory of, 409
Mountain winds, 281

Neuhoff diagram, 61, 64, 78, 79
Nocturnal radiation, 136, 138, 145
   with cloudy skies, 142

Occlusion, 322, 341
   back-bent, 325
Ozone, absorption by, 152
   effect on temperature distribution, 153
   height of centre of gravity, 153

Planck's law, 106
Polar air, 316
Polar coordinates, 161
Polar front depressions, general aspects of, 338
Polar front methods, 319
Potential instability, 98
Potential temperature, 38
   equivalent, 94, 96
   wet bulb, 90, 92
Precipitated ice or water, loss of, 63
Precipitation, effect on sharpness of fronts, 345
   in depressions, 352
Pressure, distribution over the globe, 6
   diurnal variation of, 24
   variation with height, 33
Pressure gradient, variation with height, 196
Pressure inequalities, genesis of, 308
Pseudo-adiabatic changes, 63

Radiation, 105, 282
   absorption of long wave, 114
   black body, 106, 108
   diffuse, 149
   distribution of incoming, 112
   grey, 106, 149
   in the troposphere, 124
   long wave, 109
   nocturnal, 136
   phenomena in the atmosphere, 124
   range of wavelengths, 109
   reflection of, 113
   short wave, 109
   spectral distribution, 109
   variation with latitude, 112
   W-, 130
Radiative diffusivity, 133
Radiative equilibrium, 147
Radiative transfer of heat, equations of, 128
Rain, formation of, 55

Rainfall, distribution over the globe, 13
   in depressions, 352
Rainy seasons, 16
Reflection, diffuse, 122
   of short waves, 112
Regeneration of depressions, 335
Relative humidity, 32
   diurnal variation of, 27
Reversible processes, 72, 75
Revolving fluid, 299
Reynolds number, 215
Reynolds stresses, 213
Rossby diagram, 98
Rotational motion, 174

Saturated adiabatic lapse-rate, 66
Saturation, definition of, 49
   vapour-pressure, functional relation to temperature, 103
Scattering of light, 122
Sea breeze, 181, 280
Seclusion, 323
Secondary depressions, 334
Sharp and diffuse fronts, 342
Sheets, vortex, 175
Siberian anticyclone, 7, 13
Single layer, unstable, 293
Skagerrak cyclone, 335
Skin friction, 259
Snow, formation of, 55
   physical constants of, 141
Soils, physical constants of, 140
Solar constant, 111
Source of heat or cold, 416
Specific heats of air, 37
Spectral distribution of solar radiation, 109
Spectrum, absorption, of liquid water, 120
   absorption, of water vapour, 114
   effect of pressure on, 118
   infra-red, 114
Sphere, resistance of, 215
Stability, general conditions of statical, 100
   in dry air, 40, 237
   in saturated air, 66
   in unsaturated air, 41
   of motion, 222
Steady motion in two dimensions, 187
Stefan's law, 108
Stratosphere, 17, 147
   temperature in, 18, 20, 148, 152
Stream-line flow, 212
Stresses in viscous fluid, 184
Subsidence, effect on sharpness of fronts, 343
   and divergence in anticyclones, 384
Subtropics, upper winds in, 398
Supercooling of water, 54
Supersaturation, definition of, 49
Surface layers, of ground, changes in, 138
   velocity distribution in, 247
Surface of discontinuity, slope of, 203
   approximate equation, 205
   general equations, 208
Surface turbulence, vertical extent of, 256

Sutton's extension of Taylor's theory, 266
  theory of evaporation, 272
  theory of variation of wind with height, 267

Temperature, 1
  distribution in vertical, 17
  distribution over globe, 18
  diurnal variation, 22, 133, 229
  equivalent, 94, 96
  equivalent potential, 94, 96
  horizontal gradient, 196, 198
  in stratosphere, 18, 148, 152
  monthly mean, in free air, 19
  night minimum, 140
  potential, 38
  variation with height, 20
  virtual, 31
  wet bulb, 86, 89
  wet bulb potential, 90
Tephigram, 77
Thermodynamics, of atmosphere, 69
  of hygrometry, 85
Tornado, 303
Total flow across isobars, 256
Trade-winds, 401, 408
Transformations of energy in the atmosphere, 283
Tropical air, 316
Tropical cyclone, 303
Tropics, upper winds in, 395
Tropopause, 17, 19
Troposphere, 17
Turbulence, 224
  criterion of, 237
  dissipation by, 285
Turbulent flow, 212
Turbulivity, 227

Upper air conditions in depressions, 348

Valley winds, 281

Vapour pressure, control by temperature, 31
  diurnal variation, 27
  saturation, 31, 49, 103
Variation of wind with height, 246, 247, 249, 251, 252
Velocity potential, 183
Vertical motion, effect on lapse-rate, 44, 98
Vertical transport of heat, by radiation, 158
  by turbulence, 224
Viscosity, 184
Viscous fluid, stresses in, 184
Vortex lines, filaments and sheets, 175
Vorticity, 172
Vorticity-transport, 232
  in three dimensions, 234

W-radiation, 130
Warm front, 320
Water, suspended, in fog, 122
Water vapour, absorption by, 114,
  radiation by, 114
  spectrum, 114
Wave theory of depressions, 318, 358
Wet-bulb, hygrometry, 84, 269
  potential temperature, 90, 92
  temperature, 86, 89
Wien's law, 107
Wind, antitryptic, 278, 280
  cyclostrophic, 190, 279
  distribution over globe, 12
  diurnal variation of, 24, 261
  geostrophic, 189, 190, 196, 201, 279, 309
  gradient, 189, 190, 191, 193, 310
  katabatic, 281, 417
  observed upper, 394
  variation with height, 246, 247, 249, 251, 252
  zonal distribution of, 406
Winds, classification of, 276, 278
  upper air distribution, 394

Zonal distribution of winds, 406

For EU product safety concerns, contact us at Calle de José Abascal, 56–1°,
28003 Madrid, Spain or eugpsr@cambridge.org.

www.ingramcontent.com/pod-product-compliance
Ingram Content Group UK Ltd.
Pitfield, Milton Keynes, MK11 3LW, UK
UKHW051010240426
470322UK00018B/592